The Ecology of Freshwater Molluscs

This book provides a comprehensive review of the ecology of freshwater bivalves and gastropods worldwide. It deals with the ecology of these species in its broadest sense, including diet, habitat and reproductive biology, emphasizing in particular the tremendous diversity of these freshwater invertebrates. Following on from these introductory themes, the author develops a new life history model that unifies them, and serves as a basis for reviews of their population and community ecology, including treatments of competiton, predation, parasitism, and biogeography. Extensively referenced and providing a synthesis of work from the nineteenth century through to the present day, this book includes original analyses that seek to unify previous work into a coherent whole. It will appeal primarily to professional ecologists and evolutionary biologists, as well as to parasitologists.

ROBERT T. DILLON JR is Associate Professor in the Department of Biology, College of Charleston, South Carolina. He has published widely on the ecology, evolution and genetics of molluscs for 20 years.

The Ecology of Freshwater Molluscs

ROBERT T. DILLON, JR

Department of Biology, College of Charleston

CAMBRIDGE
UNIVERSITY PRESS

PUBLISHED BY THE PRESS SYNDICATE OF THE UNIVERSITY OF CAMBRIDGE
The Pitt Building, Trumpington Street, Cambridge, United Kingdom

CAMBRIDGE UNIVERSITY PRESS
The Edinburgh Building, Cambridge CB2 2RU, UK   http://www.cup.cam.ac.uk
40 West 20th Street, New York, NY 10011-4211, USA   http://www.cup.org
10 Stamford Road, Oakleigh, Melbourne 3166, Australia
Ruiz de Alarcón 13, 28014 Madrid, Spain

© Cambridge University Press 2000

First published 2000

Printed in the United Kingdom at the University Press, Cambridge

Typeface Monotype Bembo 11/13pt.   *System* QuarkXPress® [SE]

A catalogue record for this book is available from the British Library

Library of Congress Cataloguing in Publication data

Dillon, Robert T., 1955–
The ecology of freshwater molluscs / Robert T. Dillon, Jr.
  p. cm.
ISBN 0 521 35210 X
1. Gastropoda – Ecology.   2. Bivalvia – Ecology.   3. Freshwater
invertebrates – Ecology.   I. Title.
QL430.4.D55 2000
594.176–dc21   99–15476   CIP

ISBN 0 521 35210 X hardback

To the memory of my mother,

Viola Turner Dillon

Let every creature rise and bring
   Peculiar honours to our King;
Angels descend with songs again,
And earth repeat the loud Amen!

Isaac Watts

Contents

Preface

In the twentieth century we have witnessed a general movement away from the study of organisms and toward the study of processes. The old faculties of zoology, botany, and microbiology have reorganized themselves into departments of genetics, physiology, molecular biology, and ecology. Without question, there have been some benefits. For example, early in my career I remember my attention being called to a work by Brown and Davidson (1977) showing competition between such disparate groups as ants and rodents for seeds in 'desert systems'. The implication seems to have been that organismal biologists (entomologists, mammalogists, and botanists in this case) might each have missed important aspects of a phenomenon central to the composition of the desert biota. A biologist more broadly trained in ecological processes would seem to be required.

But I feel certain that the authors of that particular work on desert ecology would join ecologists generally in acknowledging the foundation of entomology, mammalogy, and botany upon which their insight stood. Thus the declines we have witnessed in the fortunes of organismal biology in the last decades are ultimately quite troubling.

This work is an unabashed advertisement for the study of organisms. In particular, here I offer the great diversity of freshwater molluscs as an untapped resource for the study of ecological questions of considerable generality and importance. It is my hope that this work will serve to open freshwater malacology to the larger community of ecologists worldwide, reviewing the known, and suggesting some directions for the future. Perhaps I can recruit a few collaborators.

The following dear friends and insightful colleagues have, in some cases without their direct knowledge, exerted special influence upon me in this endeavour: Fred Benfield, Ken Brown, George Davis, Sally Dennis, Bob McMahon, Bob Ricklefs, and Amy Wethington. Others contributing time and intellect have included B. Dennis, P. Dustan, A. S. Harold, A. D. McCallum, B. Stiglitz, and G. T. Watters. David S. Brown read an early version of the manuscript to excellent effect.

Many friends have been most forthcoming with data and access to their works both published and otherwise. Among these are K. Baba, A. G. Eversole, M. E. Havlik, D. M. Lodge, P. Jarne, J-G. Je, E. H. Jokinen, J. Monge-Najera, G. L. Mackie, G. Majoros, B. McKillop, R. W. Neck, T. D. Richardson, S. Szabo, C. R. Villalobos and S-K. Wu. D. J. Eernisse was a gracious host while I visited The University of Michigan libraries. I am especially indebted to S. A. Ahlstedt and I. J. Holopainen for furnishing large and valuable data sets, and to T. Roop for his computer skills.

This work would not have been conceived without the love and support of my father, R. T. Dillon (Sr), for 20 years the best field assistant a boy could have. It has come to fruition only through even larger measures of the same two commodities from my wife, Shary, and my children, Ginny and Bryan. I'll be home soon.

1 · *Introduction*

The invasion launched by elements of the molluscan army upon fresh water has aptly been characterized as 'desultory' (Deaton and Greenberg 1991). Dozens of offensives seem to have taken place, sporadically timed over hundreds of millions of years, some ultimately carrying the works, others now witnessed only in the fossil record. The diversity of freshwater molluscs is vast. Yet it is my thesis here that in their interactions with the environment and with each other, freshwater molluscs share enough similarities that some intellectual profit may be gained by examining their ecology together.

We might begin with a brief overview of the forces (Tables 1.1 and 1.2). The larger freshwater bivalves belong to the order Unionoidea, an ancient group of six families whose fossil record extends to the Devonian period. They are distinguished by a parasitic larval stage that is unique among the Bivalvia. The best-known families are the Margaritiferidae and the highly diverse Unionidae, both worldwide. The hyriids of Neotropical and Australian regions are less studied, while the Neotropical mycetopodids, and the Ethiopian mutelids and etheriids, remain rather obscure.

The smaller infaunal freshwater bivalves belong to the superfamily Corbiculoidea. This is a somewhat younger group, generally hermaphroditic, with a fossil record beginning in the Jurassic and Cretaceous periods. The two corbiculoid families may represent separate invasions of fresh water (Park and O'Foighil 1998). Bivalves of the worldwide family Pisidiidae hold developing embryos for extended periods, ultimately releasing juveniles quite large in relation to their own bodies. The family Corbiculidae (restricted to old world tropics and subtropics until recently) release juveniles as smaller 'pediveligers'. The Dreissenoidea is much less diverse than the Unionoidea or Corbiculoidea, the freshwater *Dreissena* being restricted to the Ponto-Caspian basins until recently.

Table 1.1. *A classification of some of the genera of freshwater bivalves more commonly mentioned in this work*

Class Bivalvia
Subclass Paleoheterodonta
Order Unionoidea
  Superfamily Unionacea
    Family Margaritiferidae
      Margaritifera
      Cumberlandia
    Family Unionidae
      Subfamily Anodontinae
      Anodonta
      Anodontoides
      Cristaria
      Subfamily Ambleminae
      Amblema
      Actinonaias
      Elliptio
      Fusconaia
      Lampsilis
      Unio
      Villosa
    Family Hyriidae
      Diplodon
      Hyridella
  Superfamily Etheriacea
    Family Etheriidae
    Family Mutelidae
    Family Mycetopodidae

Subclass Heterodonta
Order Veneroidea
  Superfamily Corbiculoidea
    Family Corbiculidae
      Corbicula
    Family Pisidiidae
      Pisidium
      Eupera
      Sphaerium
      Musculium
  Superfamily Dreissenoidea
    Family Dreissenidae
      Dreissena
      Mytilopsis

Source: modified from Vaught (1989).

They doubtless represent yet another separate invasion from the sea (their oldest fossils are Eocene), retaining the overall aspect of edible marine mussels. By virtue of their ability to spin strong byssal attachment threads, they have occupied the epifaunal habitat not exploited by unionoids or corbiculoids. They have also retained a planktonic larval stage in their development.

Among the bivalves, adaptation to fresh waters does not seem to have been rare nor restricted to the unionoids, corbiculoids, and dreissenoids. Freshwater or brackish/freshwater species are to be found among the Arcidae, Mytilidae, Trapeziidae, Donacidae and Cardiidae, to name but a few families. Although quite interesting from many points of view, the

Table 1.2. *A classification of some of the gastropod genera more commonly mentioned in the present work*

Class Gastropoda
Subclass Pulmonata
Order Basommatophora
  Family Acroloxidae
    Acroloxus

  Family Lymnaeidae
    Lymnaea
    Pseudosuccinea
    Galba
    Myxas
    Radix
    Stagnicola

  Family Physidae
    Physa
    Aplexa
  Family Planorbidae
    Planorbis
    Anisus
    Gyraulus
    Armiger
    Segmentina
    Biomphalaria
    Helisoma
    Menetus
    Planorbula
    Promenetus
    Bulinus

  Family Ancylidae
    Ancylus
    Rhodacmaea
    Ferrissia
    Hebetancylus
    Laevapex

Subclass Prosobranchia
Order Archaeogastropoda
  Superfamily Neritoidea
    Family Neritidae
      Neritina
      Theodoxus
Order Mesogastropoda
  Superfamily Viviparoidea
    Family Viviparidae
      Bellamya
      Cipangopaludina
      Viviparus
      Campeloma
      Tulotoma
    Family Ampullariidae
      Marisa
      Pila
      Pomacea
  Superfamily Valvatoidea
    Family Valvatidae
      Valvata
  Superfamily Rissoidea
    Family Hydrobiidae
      Amnicola
      Hydrobia
      Potamopyrgus
    Family Bithyniidae
      Bithynia
      Hydrobioides
    Family Micromelanidae
      Pyrgula
    Family Pomatiopsidae
      Oncomelania
      Pomatiopsis
      Tricula
  Superfamily Cerithioidea
    Family Thiaridae
      Melanoides
      Thiara
      Pachymelania
    Family Melanopsidae
      Melanopsis

Table 1.2 (*cont.*)

Family Pleuroceridae
Amphimelania
Goniobasis
Pachychilus
Paludomus
Brotia
Juga
Pleurocera
Semisulcospira

Source: modified from Vaught (1989).

freshwater members of such primarily marine groups are at present too poorly known to have much impact upon our discussions here.

Most species of gastropods belong to what has for many years been called the subclass Prosobranchia, a universally distrusted collection of taxa not fitting into other groups (Ponder 1988b). They share a few (probably ancestral) characters: they breathe through gills, carry an operculum, and are usually gonochoristic and occasionally parthenogenetic, but only rarely hermaphroditic. Prosobranchs have invaded fresh waters on at least as many occasions as the number of their superfamilies listed in Table 1.2, plus twice again for minor groups (buccinids and marginellids) not treated in this volume. Most of the families are effectively worldwide in distribution: the neritids, viviparids, hydrobiids, pomatiopsids, and pleurocerids. The valvatids are restricted to the northern hemisphere, while the melanopsids and bithyniids were both restricted to the eastern hemisphere until recently. The thiarids and ampullariids are circumtropical, with distributions reaching to the subtropics.

As their name implies, snails of the subclass Pulmonata have lost their gills and now respire over the inner surface of their mantle, effectively a lung. The four major (and several minor) freshwater pulmonate families belong to the order Basommatophora, so named because their eyes are at the base of their tentacles. (The primarily terrestrial Stylommatophora have eyes at their tentacle tips.) They seem to have derived from a single ancient invasion of fresh water, dating at latest to the Jurassic period (Starobogatov 1970). Minor families (including the limpet-shaped Acroloxidae and two others not treated here) are held to be the most ancestral of the freshwater pulmonates, by virtue of anatomical detail (Hubendick 1978). The worldwide Lymnaeidae, with their rather ordinary looking, dextral shells of medium to high spire, are believed the next most ancestral. Members of the primarily holarctic Physidae are distinguished by their inflated shells coiled in a left-handed ('sinistral') fashion, quite unusual in the class Gastropoda. Members of the worldwide Planorbidae are also sinistral, but often planispiral. The members of the worldwide Ancylidae are limpet-shaped.

Most freshwater pulmonates carry an air bubble in their mantle cavity, which they replenish periodically at the surface, and which they use to regulate their buoyancy. This allows many species to exploit warm, eutrophic habitats where dissolved oxygen may be quite low. Some (especially smaller, cold-water) species do not seem to surface-breathe, however, and their mantle cavities are found to be filled with water. Pulmonates typically have much lighter shells than prosobranchs, and lack an operculum. Other major, although less immediately apparent, pulmonate distinctions include radulae with many small, simple teeth per row and reproductive hermaphroditism.

Authors have sometimes held that the freshwater environment is more harsh than the marine environment (Macan 1974). Temperature fluctuation is typically more extreme in fresh waters, and freezing more likely. Water levels and current speeds are more unpredictable in fresh waters than in the ocean, as is the chemical composition of the medium. Given that all classes of molluscs evolved in the sea, and all share the same broad body plan, it is not surprising that their freshwater representatives, although quite diverse, display some broad points of resemblance.

Osmoregulatory adaptation is one area of striking similarity. Marine molluscs generally conform to sea water, osmotically equivalent to a 0.56 M solution of NaCl. Freshwater molluscs have evolved much lower body fluid concentrations, the equivalent of about a 0.040–0.070 M NaCl solution for gastropods, and 0.020–0.040 M for bivalves. The osmolalities of freshwater bivalves are among the lowest recorded for any animal (Pynnönen 1991, Dietz et al. 1996). By way of comparison, freshwater crustaceans, insects, and fish all generally show osmolalities in excess of 0.100 M as NaCl.

Fresh waters are extremely variable in their ionic concentration, but typically range about the order of 0.005 M NaCl, very much lower than any of the figures cited above. So since molluscan tissues are highly permeable to water, freshwater molluscs nevertheless have substantial osmoregulatory chores to perform. Their overall strategy involves active transport of ions from the medium and production of copious urine hypo-osmotic to their hemolymph. Reviews of water balance and excretion in the freshwater molluscs have been offered by Machin (1975), Burton (1983), Little (1981, 1985), Dietz (1985), and Deaton and Greenberg (1991).

Active transport of Na^+, Ca^{+2}, and Cl^-, even against sharp concentration gradients, has been documented in all groups of freshwater molluscs. It is not certain whether the process may be localized (the gills are often mentioned) or whether all body surfaces are involved. Ions may also be exported as required to maintain electrical balance, as for example H^+ or NH_3^+, may be exchanged to balance Na^+ uptake, and HCO_3^- exchanged for Cl^- (Byrne and Dietz 1997). Voluminous urine is produced by ultrafiltration of haemolymph across the heart wall of freshwater molluscs into the pericardial cavity, where it is conducted to the kidney. The lining of the kidney resorbs Na^+ and Cl^-, and perhaps Ca^{+2} and K^+ as well.

Osmotic regulation in the reverse direction, balancing salt concentrations higher in the environment than in the organism, is unknown in the Mollusca. Thus the process of adaptation to fresh waters seems to be irreversible. Although some freshwater mollusc populations can be found tolerating low salinities, none has apparently recolonized the sea.

I am not aware of any direct estimates of the energetic cost of osmoregulation in freshwater molluscs. But a wealth of indirect evidence (reviewed in Chapter 8) suggests that in many waters of the earth, the price of osmoregulation may be so high as to limit the success of molluscan colonists. It also seems possible that adaptation to fresh water is at least partly responsible for the suppression of the larval dispersal stages so common in marine molluscs. With their high surface-to-volume ratios and reduced shells, planktonic larvae may simply be unable to gather enough fuel to fire the machinery necessary for osmotic balance. With few exceptions, all freshwater molluscs pass their larval stages in the egg, or enfolded within their parents.

This brings to the fore a second point of general similarity over all groups of freshwater molluscs. They are poor dispersers in an environment notably difficult to colonize. The directionality of freshwater flow is in some places powerful, and in other spots negligible. But without exception, all successful molluscan colonists of fresh waters have adapted to directional flow at some point in their evolutionary history. Osmoregulatory barriers are certainly not the only conceivable explanation for the general absence of a conventional larval stage from the life history of freshwater molluscs. Their turbulent medium interferes with the external union of egg and sperm. Planktonic larvae are swept away.

The general suppression of the larval stage in the freshwater Mollusca has yielded a large group of obligately aquatic organisms that do not swim. Some groups (e.g. the unionoids) have evolved unique solutions to this problem. But on a fine spatial scale, freshwater molluscs appear generally immobile, helpless to avoid predators, parasites, or the ecologist's sampling device. On a coarse geographic scale, one must figure high the likelihood that freshwater mollusc distributions derive from vagaries of chance colonization.

Fretter and Graham (1964) pointed out that hermaphroditism seems to be more common in freshwater molluscs than in those that inhabit the sea. This third point of general similarity may be a consequence of the second, that their dispersal capabilities are so poor. 'Reproductive assurance,' the certainty that all simultaneous hermaphrodites can find at least one partner (themselves), has for years been a leading hypothesis for the origin of hermaphroditism (Heath 1977). I would strengthen Fretter and Graham's generalization a bit by adding that it seems to me that asexual reproduction is more common in freshwater than in marine molluscs. Not only can hermaphroditic pulmonate snails and corbiculoid bivalves self-fertilize, they often do. Parthenogenesis has evolved three times in freshwater prosobranchs, but not in their marine ancestors.

The success that freshwater molluscs have enjoyed, together with their relative immobility, constitute for me the most persuasive arguments for a unified treatment of their ecology. Bivalves and gastropods, pulmonates and prosobranchs, are easier to sample than just about any other animal. A biologist need only walk to the creek bank, or row to mid lake, and drop his or her sampling gear, and the molluscs below are as helpless as ferns. But in contrast to the situation in plants, a biologist can fairly assume that the molluscs inhabiting any patch of sampled habitat are not entirely a function of a passive process, for molluscs are not rooted. They could leave if they wished. Study of population biology or community ecology is thus greatly facilitated.

In the two chapters that follow, emphasis will often come to rest upon the undeniable biological diversity of freshwater molluscs. Touching on the broad themes of habitat, diet, and reproduction in filter-feeding and grazing organisms, the reviews of Chapters 2 and 3 will traverse most of the territory of benthic ecology. But the common elements of the biology of freshwater molluscs will be featured in the six chapters that form the body of this book. A general life history model is developed in Chapter 4, the consequences of competition, predation, and parasitism explored in Chapters 5–7, and all this material is placed into a general community-ecological context in Chapters 8 and 9.

2 · Bivalve autecology

In this chapter we will review a few of the basic attributes of the biology of freshwater bivalves. Although filter feeding might at first seem a relatively simple process, closer examination shows a wide discrepancy between the particles in the medium (often largely inorganic) and the food actually assimilated (diatoms, green and blue-green algal cells, bacteria, and organics both dissolved and suspended). Our discussion of bivalve feeding will be divided into sections on particle retention, ingestion, and assimilation. There is some large-scale diet and habitat specialization in bivalves; *Pisidium* seems to have become adapted to filter waters from within the sediments, sometimes deep in the profundal zone, and *Dreissena* has colonized the hard bottoms. But in general, we will see that all bivalve populations seem to live in about the same habitat and eat about the same food at the same time. In light of the evidence that large populations of bivalves may substantially depress the concentration of suspended particles in even the largest lakes and rivers, the potential for food limitation and both intra- and interspecific competition must be acknowledged.

The freshwater bivalves are quite diverse in their modes of reproduction. We will see that unionoids are gonochoristic (although their mechanisms of sex determination are unclear) with widespread hermaphroditism. Their adaptation to hold developing larvae ('glochidia') and impose them parasitically upon fish hosts constitutes one of the more interesting natural history sagas of which I am aware. The corbiculoids are generally hermaphroditic and often self-fertilize, although dioecious populations are common in some groups, and brooding or delayed release of juveniles is again the rule. Thus in the freshwater bivalves we will find examples of both great biological unity and striking biological diversity.

Feeding and digestion

Digestive anatomy

Although morphological details are quite various in the diverse taxa of freshwater bivalves, in broad outline the digestive anatomy in typical unionoids, *Corbicula* and *Dreissena* can be discussed together (Morton 1983). The posterior margin on the mantle of these taxa is modified into an incurrent siphon fringed by sensory tentacles and a smaller, more dorsal excurrent siphon nearby. The term 'siphon' may be somewhat misleading, for these structures are very short and sometimes indistinct, more like holes than tubes. In fact, the posterior mantle of margaritiferid mussels does not form distinct siphons at all, although the path of water flow is the same as in other unionoids. Specialized batteries of cilia on the gills draw water slowly into the mantle cavity, along with any suspended particles not so large as to irritate the tentacles or other sensory devices, moving it anteriorly. Most bivalves have a second, longer set of cilia or cirri on the gills, the 'laterofrontals', acting either as mechanical particle filters or as modulators of fluid mechanical processes capturing particles (Jorgensen 1983). The movement of particles after their entrapment on the gills has been nicely illustrated by Huca *et al.* (1982), for the South American hyriid *Diplodon*, and by Avelar (1993) for the mycetopodid *Fossula*. Filtered water is expelled back posteriorly, often with increased force, presumably to minimize the likelihood of refiltration.

The pattern of water flow described above is the most common in bivalves generally. The corbiculoids have a pair of short, fairly elaborate, well-differentiated posterior siphons. The siphons of *Sphaerium* are fused into a proper tube and may extend 50–100% of the animal's shell length. But interestingly, *Sphaerium* and *Musculium* are apparently also able to take in water through the mantle margin some distance from the siphons.

The unusual manner by which *Pisidium* feeds has attracted a good deal of attention (Mitropolskij 1966a,b, Meier-Brook 1969, Holopainen 1985). Unlike most other bivalves, these little clams often do not maintain direct contact with the water column. They bury beneath the surface and then move through the substrate with umbo down, drawing a current through the ventral mantle margin (rather than the posterior) and ejecting excurrent water and pseudofaeces through the posterior siphon. So although water flow may loop about inside the mantle cavity in a path at least as elaborate as that seen in typical bivalves (Lopez and Holopainen

1987, Holopainen and Lopez 1989), filtered water is not expelled into the same region from which unfiltered water is drawn. The adaptive value of this water flow pattern to a bivalve living below the sediments in a lake is fairly apparent. Water circulation would be negligible in such an environment, so that an individual *Pisidium* would be in danger of refiltering its own medium if it had two siphons side-by- side, as in the usual arrangement. By this reasoning, one would predict that in general, bivalves of any taxon inhabiting more lentic habitats would need some sort of adaptation to minimize refiltration, for example well-developed siphons that may be directed in opposite directions. And one can understand why the margaritiferid mussels, with no siphons at all, might be restricted to areas where good water circulation occurs naturally.

In the last several decades, biologists have become aware that because of the mechanical properties of water, small aquatic organisms experience their environments quite differently from larger ones. Water is effectively more viscous to a small organism and much more difficult to move. In addition to the cilia on its gills, *Pisidium* apparently has rows of cilia on its foot to help it move the syrupy water. So even setting aside microhabitat choice, one would not be surprised to see unidirectional water flow in any very small bivalve. Churchill and Lewis (1924) reported that siphons have not developed in unionid juvenile mussels at 0.2–2.0 mm length, and that incurrent water passes through the anterior mantle of the mussel, about where the foot protrudes. Yeager and colleagues (1994) have described pedal-sweep feeding behaviours in *Villosa* recently shed from their hosts.

In all freshwater bivalves, mucus and particles trapped on the gill are moved forward toward the labial palps by yet a third set of specialized cilia. The palps convey material to the mouth, dropping any excess onto the mantle for expulsion as 'pseudofaeces'. A short oesophagus leads to an elaborate stomach, with typhlosoles, grooves, ciliated regions, and a rotating crystalline style regulating the flow of mucus and food. Among other functions, the style is a primary source of digestive enzymes. Material is either moved into digestive diverticulae, where digestion seems to be largely intracellular, or to intestine and anus.

Commonly, then, the particles assimilated are only a subset of those ingested, and the particles ingested are only a subset of those retained. If any competition for food occurs between bivalves, it is for particles retained, but a bivalve's health and success are a function of the particles assimilated. In a discussion of bivalve feeding, it is important to keep the distinction between retention, ingestion, and assimilation in mind. These topics will be discussed separately.

Particle retention

Recent advances in our understanding of particle retention in bivalves have come primarily from work with marine species. It is important to note at the outset that the majority of bivalves have gill filaments equipped with large laterofrontal cirri. These bivalves, including such diverse families as the Mytilidae, Veneridae, Carditidae, and Myidae, generally completely retain all food particles larger than 4 μm (Riisgård 1988, Møhlenberg and Riisgård 1978). Efficiency varies somewhat below this level, with some species retaining as much as 70% of 1 μm particles, and others as little as 20%. A second set of bivalves, including the oysters (Ostreidae) and scallops (Pectinidae), have reduced or absent laterofrontal cirri. Most of these species can retain particles of 5–6 μm with 100% efficiency, but show reduced efficiency for small particles, and may be unable to retain 1 μm particles.

The most complete work on particle retention in a freshwater bivalve is that of M. Sprung on *Dreissena polymorpha*. Given its similarity to the marine mussels, one would not be surprised at the finding of Sprung and Rose (1988) that *Dreissena* has large laterofrontal cirri, and is capable of some retention of particles down to 0.7 μm. The adult mussel retains 5–35 μm particles (at minimum) with 100% efficiency, and is about 60% efficient below 5 μm. The larvae retain particles only between 1 and 4 μm in diameter, however – an unusually narrow size range compared with larvae of marine bivalves (Sprung 1989). In addition to laterofrontal cirri, Way and his colleagues (1989) have described 'frontal cirri' associated with frontal cilia in *Corbicula* and *Musculium*.

Unionid laterofrontal cirri have been described by Tankersley and Dimock (1993), and evidence from several studies suggests that they retain small particles with admirable efficiency. Charcoal slurries have been experimentally fed to the European *Unio pictorum* and *Anodonta anatina* (= *piscinalis*) by Bronmark and Malmqvist (1982) and to the Asian unionid *Cristaria discoidea* by Dudgeon (1980). The former authors compared particle size distribution in the guts with that in unfiltered slurry, while Dudgeon compared filtered and unfiltered suspensions. Both studies found high apparent retention rates of 1–2 μm particles, Dudgeon suggesting that small particles might be preferentially retained over larger particles. He also reported that the size of the individual mussel affected the distribution of particle sizes retained. Although selective ingestion of small particles is fairly commonly observed (discussion to follow), the selective retention of smaller particles by a bivalve the size of *Cristaria* is unprecedented as far as I know. Since the data are relative frequencies, they may reflect selection against large particles.

Rather than concocting an artificial suspension of particles, Paterson (1984, 1986) used a Coulter counter to compare the size distribution of particles naturally occurring in vessels of lake water with and without the unionid *Elliptio complanata*. Paterson's results on *Elliptio* are consistent with results from the more efficient marine species. The mussel seems to retain all particles larger than 1.59 μm with high efficiency (maximum efficiency at about 4 μm), with a significant drop-off in the 1.26–1.59 μm size class.

A most interesting investigation of particle retention in freshwater bivalves is the work of Cohen *et al.* (1984) on *Corbicula* in the freshwater tidal region of the Potomac River, Maryland. The river along this 40–50 km stretch produced high concentrations of phytoplankton during every summer low-flow period of the 1960s and 1970s. But a 40–60% reduction (or 'sag') in one 6 km region was noticed in the summers of 1980 and 1981 (Figure 2.1), corresponding to a recently introduced population of *Corbicula*. *Corbicula* was first recorded in the freshwater tidal Potomac in November of 1977. Average densities of the clam increased from $1.2/m^2$ to $425/m^2$ in 1979, expressed as 159 g wet weight/m^2 in Figure 2.1. The wet weights shown in Figure 2.1 translate to average densities of 1400 and 1467 clams/m^2 in 1980 and 1981, respectively, after which a die-off occurred. Cohen and his colleagues effectively ruled out zooplankton, unionid clams, toxic substances, nutrient limitation, and peak discharges from dams as explanations for the phytoplankton sags of 1980 and 1981, convincingly demonstrating that filter feeding by *Corbicula* can noticeably reduce phytoplankton levels in large rivers. From their known population density, filtration rate, and water flow, Lauritsen (1986b) suggested that *Corbicula* might have a similarly large impact on the Chowan River in North Carolina.

Using calculations similar to those of Lauritsen, Stanczykowska and colleagues (1976) estimated that unionids, sphaeriids, and *Dreissena* combined to remove about 8% of the primary production of a 460 ha Polish lake during a typical growing season. Madenjian (1995) estimated that the recently introduced *Dreissena* population of western Lake Erie may have removed 26% of the primary production of 1990. The zebra mussel

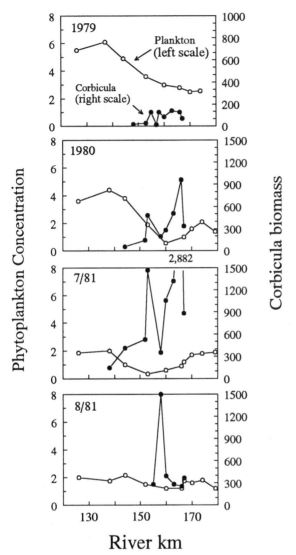

Figure 2.1. Samples taken at four summer dates over a 60 km length of the Potomac River, Maryland. Depth-integrated phytoplankton concentration ($\times 10^7$ cells/l) is plotted with open symbols and *Corbicula* biomass (mean wet weight per transect, in g/m^2) is plotted with closed symbols. (Data of Cohen *et al.* 1984.)

invasion of the early 1990s seems to have had dramatic effects on the water column of the Great Lakes. The phytoplankton productivity of inner Saginaw Bay was lowered approximately 38% by zebra mussel colonization, while increased underwater irradiance increased benthic primary productivity (Fahnenstiel *et al.* 1995a,b). The mussels seem to remove diatoms (*Cyclotella*), autotrophic flagellates (*Cryptomonas*) and a variety of protozoans very efficiently, while having no significant effect on the blue-green *Microcystis* (Lavrentyev *et al.* 1995). They also seem to be able to remove bacterioplankton, although in some cases they may positively affect bacterial abundance by nutrient or organic carbon excretion (Cotner *et al.* 1995). Caraco and colleagues (1997) have offered convincing evidence that the spread of zebra mussels to the Hudson River in 1992–94 resulted in a 'massive' decline in phytoplankton biomass similar to that demonstrated by Cohen for *Corbicula*. The complex and far-reaching ecological effects of zebra mussel invasion will be re-examined in Chapter 5.

Particle ingestion

Because of the bivalve capacity to form pseudofaeces, one cannot be certain that particles retained will be ingested. And it has often been found that bivalves remove particles from suspension indiscriminantly, without regard to their value as food (Jorgensen 1966, Haven and Morales-Alamo 1970). Especially in silty environments such as one might find in turbid rivers, it would seem possible that organic particles might comprise such a small fraction of the total retained that a bivalve might starve. One mechanism by which bivalves can adjust their intake of particles is by adjustment of their filtration rate (reviews by Jorgensen 1975, Kryger and Riisgård 1988). Morton (1971) reported that the rate at which *Dreissena* filtered water increased when ambient concentrations of algae or bacteria were increased, reaching a maximum at a turbidity of 0.075 g collodial graphite/l. Filtration rate was markedly reduced as turbidity reached 0.15–0.2 g/l. Way and colleagues (1990) reported an inverse relationship between filtration rate and the ambient concentration of suspended particles in three *Corbicula* populations from rivers of the southeast United States.

Early workers assumed that in addition to reducing their filtration rates when presented with undesirable particles, bivalves could exert some selection between retention and ingestion, so that undesirable particles tended to be preferentially rejected as pseudofaeces. Selection by specific gravity was usually suggested to remove inorganic particles. This view has been criticized on quite strong grounds in the past. For example, some of the earliest workers pipetted distasteful particles such as carmine directly into the incurrent siphons of unsuspecting mussels. As Allen (1924) wrote, 'Such . . . material, however neutral, of whatsoever dilution, or however administered through the respiratory water, was never found subsequently in the alimentary canal.' Allen then showed that mussels fed a mixture of carmine and the known foodstuffs *Gleocapsa* and *Spirogyra* did not sort the mixture, but rejected all particles as pseudo-faeces.

Nevertheless, it now seems clear that some selective removal of undesirable particles can in fact take place, possibly mediated by the labial palps (Kiørbe and Møhlenberg 1981, Newell and Jordan 1983). Payne and his colleagues (1995) have reported increased palp-to-gill area ratios in populations of *Dreissena* and *Corbicula* from habitats where the concentrations of suspended particles are especially high. Strong, albeit indirect, evidence of differential sorting has been obtained in laboratory feeding experiments on *Corbicula fluminea*, famous for inhabiting silty rivers, canals, and basins. Foe and Knight (1985) monitored tissue growth in 8 *Corbicula* cultures, 4 inorganic silt concentrations (up to 150 mg/l) by 2 phytoplankton concentrations (*Ankistrodesmus*). Clam tissue growth was independent of silt concentration but significantly affected by phytoplankton concentration.

One gets the impression that the contents of a typical freshwater bivalve gut generally duplicate the naturally occurring population of particles available (Boltovskoy *et al.* 1995). The gut contents of the Korean unionid *Lamprotula*, for example, included detritus, filamentous algae, and diatoms mixed in with sand and calcareous debris (Chun 1969). About 60–70% (by volume) of the gut contents of *Corbicula sandai* in the Japanese Lake Biwa was detritus, mud, and sand (Hayashi and Otani 1967). Included among the remainder were 89 species of diatoms, 12 species of green algae, 14 species of blue-green algae, rotifers, protozoans, sponge spicules, etc.

The most informative of such studies compare gut contents to the particulate composition of the surrounding water. Wallace *et al.* (1977) measured tracings from 1626 particles from the stomachs of 3 individual *Corbicula fluminea* collected from the Altamaha River, Georgia. Setting aside diatoms, which they excluded from analysis, Wallace and colleagues found that 85% of the stomach content volume (1624 particles) was detritus, with a mean size of 100 μm^2 and range of 5 μm^2 to 20 000 μm^2. I would not have guessed that the particles would be that large, based on particle retention experiments. But the mean size of detritus particles in Altamaha water samples ranged from 157 μm^2 to 270 μm^2, a fair match to the *Corbicula* gut. Amusingly, the remaining 15% of the gut content volume comprised just two animal particles (unidentified), one 2000 μm^2 and the other 30000 μm^2. One has no way of knowing whether the removal of such a particle (perhaps approaching 200 μm on a side) could be an accident, or if the little clams have to filter 1624 worthless particles to get one good one.

Gale and Lowe (1971) compared the monthly gut contents of *Musculium transversum* in a Mississippi River pool with water samples taken at the surface. They found no evidence of selective ingestion among the various genera of the three major algal groups: greens, blue-greens, and diatoms. There was some evidence that *Musculium* preferred green algae over diatoms. But admittedly, the water sample at the surface may not reflect the composition of the plankton at the bottom. Ten Winkel and Davids (1982), on the other hand, compared stomach and midgut contents of *Dreissina* collected from the underside of a platform with the surrounding waters of the Dutch Lake Maarsseveen I. Samples were taken on five different dates from March to May. The data clearly suggest that the very large, colony-forming diatom *Asterionella formosa*, with clumps of long, slender cells 100 μm each, is not readily retained. It is fairly easy to see how such colonies might be excluded at the incurrent siphon. Ten Winkle and Davids also suggest a positive selection for smaller diatoms on the order of 15–45 μm, but again, since the data are relative frequencies this may be the inevitable consequence of selection against *Asterionella*. They also noted an apparent deficit of the smaller (about 15 μm) green algae *Phacotus* and *Trachelmonas* in the stomachs of *Dreissena*. This is quite at variance with other studies of bivalve feeding, including Morton (1969a) on *Dreissena*. Organic particles of this size are generally retained and ingested readily.

Assimilation

Evidence regarding the actual assimilation of ingested particles comes from several quarters. For example, assimilation has been inferred by careful comparison of carbon taken in and lost from individual bivalves. Walz (1978a) fed *Dreissena* a pure diet of the diatom *Nitzschia* and obtained a value of 40% assimilation. Stanczykowska and co-workers (1975) monitored how efficiently that naturally occurring foodstuffs were assimilated by *Dreissena* suspended in chambers in a Polish lake. Although the individual elements of the seston were not characterized, the authors noticed dramatic declines in efficiency on days when the large (215 μm) dinoflagellate *Ceratium* bloomed.

Radiolabelling, the most direct approach, has been used to investigate the digestive capabilities of *Corbicula* by Lauritsen (1986a). Lauritsen fed individual clams pure cultures of the spherical, unicellular green alga *Chlorella*, the sickle-shaped unicellular green *Ankistrodesmus*, and the filamentous blue-green *Anabaena*, all labelled with ^{14}C. Because of its shape and reputation for unpalatibility, I would have predicted very little assimilation for *Anabaena*. But remarkably, the assimilation rates for *Anabaena* (58%) and *Chlorella* (56%) were almost identical, with slightly reduced rates for *Ankistrodesmus* (47%). Intact cells of *Chlorella* and *Ankistrodesmus* were commonly found in clam faeces, apparently protected by their tough cell walls. *Anabaena* was never observed in faeces, but was the only foodstuff to appear in pseudofaeces. So although large, tangled mats of *Anabaena* must surely be safe from *Corbicula*, individual filaments may be swallowed quite handily. At the other end of the spectrum, Silverman and colleagues (1995) offered ^{35}S-labelled bacteria of several species (1.3 μm to 4.1 μm size range) to *Dreissena*, reporting assimilation in all cases. *Dreissena* seemed more efficient in clearing and assimilating *E. coli* (1.7–2.9 μm) than did *Corbicula* or the unionid *Carunculina*, an ability perhaps attributable to their laterofrontal cirri, which seem to be larger and present in greater number (Silverman *et al.* 1996).

We have already mentioned the unusual feeding method adopted by tiny clams of the genus *Pisidium*. Lopez and Holopainen (1987) found that although diatoms and chloroplasts were commonly observed in the guts of Finnish *Sphaerium corneum* and the larger (up to 13 mm) *Pisidium amnicum*, the stomachs of the smaller (2 mm) *Pisidium casertanum* and *P. conventus* were always empty. These latter two species live in loose, flocculent sediment (water content 60–90%) of high bacterial content in Lake Paajarvi, Finland: *P. casertanum* in shallow water and *P. conventus* in the profundal zone. Using the ^{51}Cr/^{14}C balance method (Calow and Fletcher 1972), Lopez and Holopainen showed that these two *Pisidium* can assimilate a natural suspension of interstitial bacteria at 40–60% efficiency. Absorption of bacteria on particles was much less efficient. The authors suggested that although larger sphaeriids quite probably do ingest and assimilate algae and/or detritus, the smaller *Pisidium* may feed exclusively on bacteria.

Yet a third method by which assimilation has been verified is by monitoring growth or reproduction in laboratory populations fed controlled diets. Foe and Knight (1986) measured 30-day tissue growth in cultures of *Corbicula fluminea* fed six genera of green algae in various combinations. Evidence suggested that *Selenastrum* emits some toxic substance inhibiting *Corbicula* filtration. But strong positive growth was noted from all trialgal combinations of *Ankistrodesmus, Chlamydomonas, Chlorella,* and *Scenedesmus,* as well as the dialgal combination of *Chlorella* and *Chlamydomonas.* Growth was slower, and sometimes negative, in other dialgal combinations. Interestingly, the maximum laboratory growth obtained on a defined algal diet was still only about 25% of the growth the authors saw when clams were cultured in water from their own environment, enriched to stimulate nanoplankton blooms.

The dietary requirements of larval *Dreissena* contrast strikingly with those of other freshwater bivalves, including adults of their own species. Sprung (1989) offered laboratory cultures of *Dreissena* larvae 47 different pure cultures of algae, green and blue-green, singly and in combination, plus two bacterial strains and yeast. Although some of the algal strains were too large to be ingested (recall *Dreissena*'s narrow particle size requirements), most foodstuffs were in fact retained. But larval shell growth was in all cases poor, and rearing success 'extremely sporadic'. Sprung manipulated density of larvae and food, light, temperature, aeration, and vessel, but could not achieve consistent rearing success. So although the evidence is not conclusive, the suggestion is that the food requirements of larval *Dreissena* may be exacting.

As long ago as 1948, Rodina found that *Sphaerium, Musculium,* and *Pisidium* could grow on a diet of pure bacteria at natural concentrations. To my knowledge, the most thorough laboratory study of freshwater bivalve diet was completed by Mackie and Qadri (1978) on *Musculium securis* from Ontario. These authors showed that *M. securis* would not grow in unsupplemented lake sediment or pond-bank soil, nor would they grow in plain dried leaves, but they would grow and reproduce in a combination of the two. Importantly, *M. securis* will not grow under any condition if the soil is autoclaved. Thus it would appear that nutrients from the leaves nourish micro-organisms from the soil, and micro-organisms from the soil nourish *Musculium.* But the little clam lives in regions where tree foliage is absent, and gut contents typically contain a wide variety of algae, especially the golden-brown filament *Vaucheria* and the blue-green filament *Lyngbya.* The correlation between the density of the little clams and the biomass of *Vaucheria* and *Lyngbya* on the bottom is striking. Thus it is clear that *M. securis* can probably retain, ingest, and assimilate a great variety of foodstuffs.

Similarly inferential evidence on assimilation in natural populations of freshwater mussels comes from the work of James (1987) with the New

Zealand hyriid mussel, *Hyridella menziesi*. He estimated that production by algae and aquatic macrophytes in a clear, oligotrophic lake might account for only about 5% of the calculated energy needs of its dense mussel population, and suggested that the ultimate source of energy for this population would have to be leaves blown or fallen into the lake, presumably with associated bacteria.

In spite of all the work done on particle retention and ingestion in unionoid mussels, and in spite of all the attention devoted to their biology early this century, surprisingly few controlled studies have been made of the food they actually assimilate. Early workers were able to rear mussels only in crates or troughs continually supplied by unfiltered, natural water (Coker *et al.* 1921). Imlay and Paige (1972) noted growth on a diet of commercial trout fry food, although this affords little insight into natural diet. Hudson and Isom (1984) reared *Anodonta imbecillis* from mature glochidia to about 5 mm juveniles in 74 days using filtered Tennessee River water with induced phytoplankton bloom, supplemented with silt (filtered pond substrate). Hudson and Isom listed 16 planktonic taxa common in their cultured river water, including green algae, blue-green algae, diatoms, and ciliates. Juvenile mussels would not grow on an artificial diet composed of three taxa of green algae, nor would they grow deprived of silt. Gatenby and colleagues (1996) also reported optimum growth and survivorship for juvenile *Villosa* and *Pyganodon* in a trialgal diet supplemented with sediment.

Non-particulate dietary components

As early as 1916, Churchill offered unionid mussels stained solutions of fat, starch, and protein (albumin) and noted direct uptake both through the alimentary system and the outer epithelium. In a medium nearly free of bacteria, Efford and Tsumura (1973) showed that *Pisidium casertanum* could actively transport and assimilate ^{14}C-labelled glucose and ^{14}C-labelled glycine from a solution of low concentration. Similar results were obtained from two other species of *Pisidium* and *Musculium lacustre* as well. Loss rates were also high, however, and considering all labelled and unlabelled organic molecules, it is not clear if inflow was net positive. Further, given known respiration rates for *Pisidium*, Efford and Tsumura calculated that typical levels of dissolved organics in interstitial mud bottoms could supply at most 4% of energy requirements. They suggest, however, that direct uptake of dissolved nutrients may be the method by which marsupial young are nourished in the mantle cavity of their mother.

Pardy (1980) observed that the shells of *Anodonta grandis* in Windy Pond, Massachusetts, are 'pocked, abraded, and on occasion completely perforated, thus exposing the underlying mantle . . . .' 'On occasion there were intense green areas . . . underlying these windows.' I have noticed this phenomenon myself in thin-shelled anodontines, but it is quite rare. Pardy found viable zoochlorellae in these greenish regions, with photosynthetic capability. He incubated living *A. grandis* with $NaH^{14}CO_3$ in dark tanks and in tanks illuminated by fluorescent tubes. Light-treated mussels showed a seven-fold increase in ^{14}C uptake, especially in gills and blood. It is not known what proportion of the total energy budget of these clams comes from this highly unusual symbiosis.

There is some inferential evidence that pisidiids may feed directly on deposited organic matter, rather than suspended particles. From estimates of suspended organic carbon and individual filtration rates, Burky and co-workers (1985a) estimated that a typical individual *Musculium partumeium* in an Ohio marsh pond would ingest approximately the same amount of carbon it is estimated to assimilate. Since assimilation efficiencies are never 100%, Burky speculated that additonal food may be obtained by deposit feeding, possibly using the cilia along the foot. Hornbach and co-workers (1984) also estimated suspended organic carbon levels, filtration rates, and annual energy budget for *Sphaerium striatinum* in an Ohio creek; they calculated that filter-feeding could account for only 24% of typical energy needs, and suggested deposit-feeding.

These results are not as striking as those of James (1987) on the New Zealand mussel population. James, Burky *et al.*, and Hornbach *et al.* all estimated energy budgets for natural populations of bivalves, but rather than comparing energy requirements with total annual primary production, the last two studies compared requirements with annual estimates of food available to the clams. There is much opportunity for error in such estimates; both Burky *et al.* and Hornbach *et al.* measured suspended organic carbon from samples taken at the middle of the water column, not at the bottom where the little clams live. The work of Lopez and Holopainen (1987) suggests that for the smaller *Pisidium*, at least, the cilia of the foot are an additional mechanism to move water, rather than deposits. I suspect that for particles in the 1 μm range, the distinction between being suspended over the substrate and deposited on the substrate is immaterial.

Feeding period

There has been some work on rhythm, or lack thereof, in unionid activity, but as Jorgensen (1966:103) has stated, 'the various authors are not in agreement'. Jorgensen mentioned three studies showing entirely different periods of activity for *Anodonta*. But the evidence is now fairly strong that any periodicity *Anodonta* may show in valve activity occurs at two different frequencies. Barnes (1962) described a rapid rhythm resulting in up to 20 partial closures per hour and a slower rhythm, showing perhaps three complete closures of several hours each per day in *A. cygnea*. De Bruin and David (1970) also noted regular interruptions of the water current in *A. cygnea* at a frequency of 6–40 per hour, but their experiment was not designed to detect longer-period cycles. And rapid, regular valve opening was mentioned yet again by Salbenblatt and Edgar (1964) in the North American *A. grandis*, although these authors were not equipped to quantify such short-period cycles.

Salbenblatt and Edgar reported that although *A. grandis* spent most of the time closed under their experimental conditions, different animals displayed cycles of 7–58 hours, depending upon light and temperature. Periods of cyclic activity were short-lived, however, and easily disrupted under laboratory conditions.

The situation with *Dreissena* seems to be similar. Morton (1969b) reported regular nocturnal periods of filtration in laboratory populations of *Dreissena*, followed by diurnal quiescence. Sprung and Rose (1988) noted a short-period phenomenon: pauses of some seconds alternating with several minutes of activity, depending on food concentration. But no regular activity was noted in the laboratory by Walz (1978a) or in the the wild by Stanczykowska and colleagues (1975). Many sorts of behaviour are extremely sensitive to experimental conditions – feeding in bivalves may be no exception.

It is the longer-period cycles that are more interesting from an ecological standpoint. Thus we are fortunate that the longest-term study of periodic feeding in unionids to my knowledge is also the study that seems most likely to be free of observer-induced error. Salànki and Vero (1969) attached a pair of small electrical devices to each valve of four individual *Anodonta cygnea*, allowing them to monitor valve opening after their release at a depth of 2 m in Hungary's Lake Balaton. The four *Anodonta* displayed irregular periods of filtering and quiescence (about 1–3 cycles per day). Daily fluctuations of the environment tended to synchronize the animals only slightly.

Salbenblatt and Edgar found the long-term pattern of valve activity to be similar among three genera of North American mussels, *Anodonta*, *Anodontoides*, and *Lampsilis* – usually one or two cycles per day. Intriguingly, the authors noted in passing that unlike *Anodonta*, *Lampsilis* filters during almost its entire cycle, and that there may be some diel differences in filtering activity among the species. The laboratory experiments of Chen and colleagues (1999) confirmed that the lentic *Pyganodon grandis* tends to gape nocturnally, while the lotic *Quadrula pustulosa* is a diurnal gaper. An *in situ* comparison of feeding schedules in co-occurring freshwater bivalves would be an interesting, if technically demanding, study.

Habitat

Unionoids

Any discussion of the habitats of freshwater molluscs must begin with the great review work of Boycott (1936). Briefly, Boycott considered that the four large British unionids (two *Anodonta*, two *Unio*) need firm, muddy bottoms. 'They will not live either on a hard stony or gravelly bottom or on one which is covered with a thick layer of soft humus in which they would have to disappear before they found firm ground' (pp. 132–133). Boycott would not hazard any absolute statements about the habitat of *Margaritifera margaritifera* on the other hand, but noted that 'A typical locus is a quick running river up to 3 or 4 feet deep with a mixture of boulders, stones, and sand; it can burrow in fine gravel and particularly likes the sand which accumulates behind large stones' (p. 133). Boycott referred to a study showing that *M. margaritifera* occurs only to the north and west of a line from Scarborough to Beer Head, and footnoted 'If a line be drawn from Hull to Gloucester and then on to Plymouth, it can be said roughly that all the country to the east and south of it was for Parliament, and everything to the west and north for Charles.' This I transmit for the benefit of the next generation of freshwater malacologists.

Figure 2.2 shows a more contemporary treatment of mussel habitat choice in Lake Matikko, Finland (Haukioja and Hakala 1974). Four species co-occur, all showing a pattern that is characteristic of lake-dwelling unionoids everywhere. From about $1/m^2$ in the shallows, mussel density peaks rapidly then falls gradually as depth increases. Haukioja and Hakala note that the depth distribution of *Pseudanodonta complanata* differs somewhat from those of the other three species, peaking at 2–3 m rather than at 1 m.

Similar patterns have been reported for unionoid populations in Poland (Lewandowski and Stanczykowska 1975), the United States (Fisher and Tevesz 1976), England (Stone *et al.* 1982), and Brazil (Henry

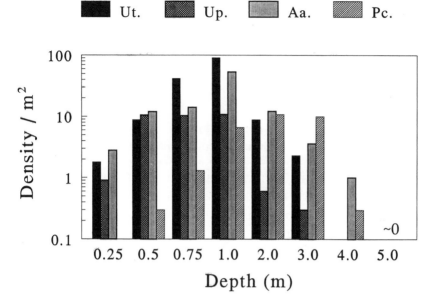

Figure 2.2. Average densities (N/m^2) of *Unio tumidus* (Ut.), *Unio pictorum* (Up.), *Anodonta anatina* (Aa.), and *Pseudanodonta complanata* (Pc.) at 8 depths in Lake Matikko, Finland. (Data of Haukioja and Hakala 1974.)

and Simão 1985). Because light, temperature, food, and flow rates would seem to deteriorate with depth, it seems in general that all mussels prefer to live as close to a lake surface as possible. (This does not seem as universally true for pisidiids, however, as we shall see.) But because mussels cannot predict low water, ice, and predation, population densities are reduced at depths less than 1 m by mortality. Thus as a rule, mussel populations of lentic habitats reach maximum abundances about 0.5 m to 2.0 m, varying only in the rates at which they disappear with depth thereafter.

This effect may not be as pronounced in clear, oligotrophic lakes. For example, Burla (1972) tabulated the abundance of *Anodonta* in SCUBA transects collected over a period of 1 year from Lake Zurich, Switzerland. The maximum density in any sample, about 50 individuals per square metre, occurred at 1 m in a June/July sample. But I would not judge that any significant decrease in abundance occurred in most samples until depths of 9–12 m. In some transects a (probably non-significant) maximum abundance occurred as deep as 8 m. Another unusual case is presented by the dense population of *Hyridella menziesi* inhabiting clear, oligotrophic Lake Rotokawau, New Zealand (James 1987). This lake was formed when an explosion crater filled with water 4000 years ago, and has very steep sides. Maximum mussel densities were not reached until depths of 5–10 m in some of James' transects, apparently because the bottom levelled off to a shelf at those depths. But once again, the maximum population density recorded in the lake (814/m^2!) was at 1 m in a transect taken where the sediment on the steep sides was exceptionally stable.

There would seem to be potential for space competition among lake-dwelling mussels at their optimum depths, in a fashion analogous to that well-documented in rocky intertidal environments. In addition to the high densities recorded by James, Fisher and Tavesz (1976) reported 15 *Elliptio complanata*/0.1 m^2 in a Connecticut Lake, and Bronmark and Malmqvist (1982) recorded about 100 mussels/m^2 in a Swedish lake outlet. A density of 161 mussels/m^2 was the average at 1 m in Lake Matikko (Figure 2.2). Haukioja and Hakala did not mention the possibility of competition between their four species of mussels, but I suggest that it may be intense, and that *Pseudanodonta* may be excluded from shallower water because of it. We shall return to this subject in Chapter 9.

Often it is the composition of the substrate that determines the depth to which mussel populations extend. Although modal mussel abundance again occurs at 1 m, the SCUBA transects of Cvancara (1972) found *Anodonta grandis* and *Lampsilis radiata* to depths of 8 m and 9 m on a side of Long Lake, North Dakota, where sand and gravel were prevalent. Cvancara speculated that the depth to which Long Lake mussels extend on coarse substrate may be a function of the metalimnion, the top of which varies from 4 m to 12 m in Long Lake in the summer. Mussels ranged only to a depth of 1 m on another side of the lake where mud and aquatic plants predominated.

Figure 2.3 shows the depth distribution of *Elliptio complanata* and *Anodonta grandis* in Lake Bernard, Ontario, as recorded by Ghent *et al.* (1978) over 11 circular plots using SCUBA. The authors noted that the substrate in Lake Bernard is typically fine, hard sand down to about 2.5 m, where it is generally replaced by soft silty mud between 3 m and 6m, so that the substrate is overlain by a 10 cm silt 'blanket' below 6 m. The distribution of *E. complanata* seems to correspond closely to firm bottom, with the usual density reduction nearshore. Fisher and Tavesz (1976) have noted that in life, *E. complanata* is found half-buried, anterior downward, and suggested that it cannot maintain this life position in the mud found on the bottoms of deeper lakes. *Elliptio complanata* is very compressed laterally, and I have rarely seen it lying on its side as the fatter *Anodonta* often

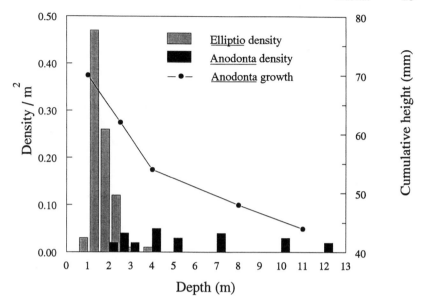

Figure 2.3. The bar graphs show population densities (N/m^2) of *Elliptio complanata* and *Anodonta grandis* at 11 circular plots down a depth gradient in Lake Bernard, Ontario. The line shows the cumulative growth to age 9 years (shell height) of *A. grandis* for combined samples from five depth ranges. (Data of Ghent *et al.* 1978.)

do, for example. Lying on its side, the siphons of *E. complanata* might not be 10 mm above the bottom. So taken with the results of Kesler and Bailey (1993) and the mark-recapture study of Kat (1982a), it appears that substrate may be more critical than depth *per se*, at least to this mussel.

Returning to Figure 2.3, one's first impression of *Anodonta grandis* is that it selects deeper, muddier sites in Lake Bernard. But on the right scale of the figure one finds recorded a striking inverse relationship between *A. grandis* growth rate and depth. Ghent *et al.* viewed the distribution of *A. grandis* in Lake Bernard as evidence that it is adapted to a soft, muddy bottom. This is true, but maladaption to a hard, sandy bottom may describe the situation just as well. Perhaps the 12 *A. grandis* from 0.5–2.0 m depth upon which the figure is based were collected from little pockets of softer substrate behind obstructions in an otherwise hard, sandy region.

The effects of depth and substrate on lake-dwelling unionids have been examined experimentally on several occasions. Narrow Lake, Alberta, is another lake in which *Anodonta grandis* occurs to some depth. Hanson and his co-workers (1988) could not detect any difference in the annual growth rates of mussels collected from 1 m to 7 m in depth, even though the deeper waters are significantly cooler. Samples of clams measured and stocked in small enclosures at depths of 1 m, 3 m, 5 m, and 7 m (using uniformly sandy substrate) did, however, show significant variance in growth rate, suggesting that migration between depths may be sufficient to keep population growth rates uniform. Inexplicably, no such migration seems to occur in Lake Bernard.

Hinch *et al.* (1986) found that a population of *Lampsilis radiata* from a sandy site in Long Point Bay, Lake Erie, grew faster than a population from a muddy area in the same bay about 10 km distant. They marked and transplanted samples of *Lampsilis* from sand to mud and mud to sand for a growing season. Somewhat surprisingly, no improvement was noted in the growth rate of the clams transferred from mud to sand over those remaining in mud. (Recapture rates may have been low for some categories, however.) Clams originally from the sand habitat seemed to be faster growers regardless of where they spent their four months, suggesting perhaps some genetic divergence between populations occupying different regions of Lake Erie. Further complicating the picture, a substrate-choice experiment performed by Bailey (1989) showed that all Inner Long Point Bay *L. radiata* prefer mud over sand. Individuals from the sand-dwelling subpopulation actually showed greater preference for mud than did mud-collected individuals, at least in artificial ponds. One wonders about the generality of these unusual findings.

Cvancara (1972) found a third unionid population in Long Lake, *Anodontoides ferrusacianus*, a species 'generally characteristic of the shallow water in creeks and rivers'. Although *Lampsilis* and *Anodonta* extended to 8–9 m, *Anodontoides* was not collected below 2 m anywhere in the lake. The species was not collected at all in a 1969 survey, suggesting that its existence in Long Lake may be tenuous. It seems clear that freshwater mussels show varying degrees of adaptation to the lentic environment. Lewis and Riebel (1984) did not notice any substantial differences between *A. grandis*, *L. radiata*, and *E. complanata* in burrowing speed or efficiency in sand, clay, mud, or gravel. But we have noted that the heavier, compressed shell of *Elliptio complanata* does not seem as stable in soft substrate as the rounder shells of *Anodonta grandis*.

Turning now to more lotic environments, Norelius (1967) and Bronmark and Malmqvist (1982) have studied populations of *Anodonta anatina* and *Unio pictorum* inhabiting Swedish lake outlets. The latter authors took 64 samples of 0.05 m² each, recording distance from the lake, distance from the margin, depth, current velocity, vegetation, percentage organic content of the sediment, and particle size, and performed several discriminant analyses. *Unio* tended to be found more toward the centre of the river, away from vegetation, and on coarser sediment than *Anodonta*. *Unio* was strikingly more common nearer the lake than *Anodonta*, and showed increased growth rates. Norelius found the trend just as strong but in the opposite direction: *Anodonta* more common near the lake and *Unio* increasing downstream. Bronmark and Malmqvist suggested that food quality (e.g. chlorophyll *a* content) may be improved immediately below their study lake.

Our discussion of mussel habitat choice thus far may give the impression that unionoid mussels prefer lakes and ponds. They do not. Worldwide, the great majority of mussel species are never found in lentic habitats. But ecological studies of truly riverine unionoids are fewer than those directed toward lake and pond populations. Huehner (1987) modified a substrate-choice technique developed by Meier-Brook, placing mussels on edge at an interface between sand and gravel. Most of Huehner's experiments involved two populations of *A. grandis* and three populations of *L. radiata*, lake-dwellers with which we are already familiar. These mussels always showed highly significant preference for sand. But Huehner included one truly riverine species, *Elliptio dilatata*, in his trials. *Elliptio dilatata* preferred gravel over sand by a margin of 42 to 32, which though not significantly different from 1:1 is certainly significantly different from Huehner's five populations of mussels more characteristic of slackwater.

Stern (1983) collected 28 species of unionids in his 14 SCUBA transects of the Lower Wisconsin and St Croix Rivers, Wisconsin. He divided his samples into four different depth categories from 1 m to 3.3 m and four different substrate categories (sand, mud–sand, mud–sand–gravel, and sand–gravel–boulder). Stern found mussels at all depths and all substrates except pure sand. The most striking feature of his data, however, was that only six of the mussel species were ever found on mud–sand: 60% of all *Anodonta grandis*, 40% of *Lampsilis radiata*, and four riverine species at trace frequencies. Here we see our two familiar lentic unionoids inhabiting the muddy parts of the Wisconsin and St Croix Rivers, while 26 other unionoid populations co-occur on coarser sediments.

One of the saddest 'natural experiments' we have witnessed in North America has been the impoundment of most of our free-flowing rivers during the last 50–60 years. Lake Ashtabula, North Dakota, was formed in 1950 by the impoundment of the Sheyenne River. Eight mussel species are known from the river above the lake, and 11 species below.

Cvancara and Freeman (1978) collected belt transects by SCUBA at 6 stations on Lake Ashtabula from 1 m to 6 m in depth at 1 m increments. They found alive 81 *A. grandis* and 12 *L. radiata*, both at a modal depth of 3 m. This seems odd in light of the fact that substrate at this depth is generally soft, organic mud. But two drawdowns of over 2 m had occurred at Lake Ashtabula in the five years previous to Cvancara and Freeman's study. The density of *L. radiata* was significantly higher in the Sheyenne River above the lake than in the lake and below. But population density of *A. grandis* in the lake is comparable to that in the river above and significantly greater than that below the dam. Of the river mussels, Cvancara and Freeman found two individual *Amblema plicata*, one *Lasmigona compressa*, and none of the other seven species.

One might speculate that the habitat choice of juvenile unionacean mussels might be different from that of adult mussels, and possibly more similar to that of the smaller pisidiids (a discussion of which follows shortly). But this does not seem to be the case, at least for riverine mussels of the North American interior, such as *Fusconaia*, *Pleurobema*, *Medionidus*, and *Villosa*. Neves and Widlak (1987) took 573 cm^2 circular bucket samples in Big Moccasin Creek, Virginia, a small stream averaging only 7 m wide and 0.2 m deep at low flow. They took 15 samples from each of three 'major habitats': pool (slow flow, silty bottom), run (moderate, laminar flow, coarse substrate), and riffle (swift, turbulent flow, coarse substrate), and two 'microhabitats': behind boulders (which occurred in riffle and run only) and stream bank (which occurred in all three major habitats). The age-0 mussels collected by Neves and Widlak averaged only 2.7 mm shell length, and the age-1 individuals but 6.4 mm, certainly within the size range of pisidiid clams. But Table 2.1 shows that while the pisidiids (*Sphaerium striatinum*, *Pisidium compressum*, and *P. casertanum*) preferred silty pools, unionids preferred riffle and run regardless of age. Juvenile mussels seem to differ from adults only when very young in their significant preference for the sandy areas behind boulders.

Corbiculoids

Boycott (1936) counted 4 species of *Sphaerium* and 15 species of *Pisidium* in Britain. He stated that 'As the species are as much (or more) affected by the nature of the bottom in which they live as by the character of the whole habitat, their preferences cannot always be described in ordinary terms . . . The species may therefore be extremely localised in a habitat, and collecting them satisfactorily is something of a special art.' He specifically noted substrate or depth preferences for only five species. Ordinary

Table 2.1. *Number, age group, and habitat of unionoid mussels and pisidiid clams collected in 75 quantitative samples from Big Moccasin Creek, Virginia, on nine sampling dates*

Habitat	Unionids, by age					Pisidiids, all ages
	0	1	2	3	4+	
Pool	0	4	3	1	0	1218
Run	4	3	3	1	7	717
Riffle	5	6	3	3	13	189
Boulder	15	6	7	7	8	664
Bank	1	1	0	1	8	557

Source: Neves and Widlak (1987).

terms apparently failed for the majority. But as a broad generalization, *Pisidium* are more characteristic of calmer (perhaps deeper) regions with fine sediment, *Sphaerium* more characteristic of coarse sediments and higher flows, and *Musculium* intermediate.

In some situations, the habitat preferences of corbiculoids are similar to those of unionoids. Hamill and colleagues (1979) took monthly Ekman grab samples over a period of one year from 19 sites on the Ottawa River, noting depth, sediments, current velocity, and vegetation, and counting the *Pisidium casertanum* collected. Their results (Figure 2.4) illustrate quite clearly a phenomenon apparent in our examination of unionoid habitats. Current speed and substrate are strikingly correlated (log $r = 0.71$). The inverse correlation between depth and current would be almost as strong. The 17 sites graphed in Figure 2.4 show that *Pisidium* densities are not greatest in the coarse sediments with rapid flow, nor in the finest sediments with very little flow. Rather, *P. casertanum* seems to prefer intermediate habitats, with fine sand in gently flowing waters. This is similar to the habitat choice of riverine mussels (Salmon and Green 1983).

I know of no field studies where an attempt has been made to separate analytically the effects of current, substrate, depth and/or macrophyte abundance on habitat choice in pisidiids. Toth and Baba (1981) found about twice the density of *P. amnicum* on clay–silt bottom in the Tisza River, Hungary, as on fine sand, with no individuals at all inhabiting coarse sand, pure clay, or pure silt. Harrison and Rankin (1978) reported an impressive correlation ($r = 0.95$) between population density of *P. punctiferum* and percentage organic matter in mud in a Caribbean marsh.

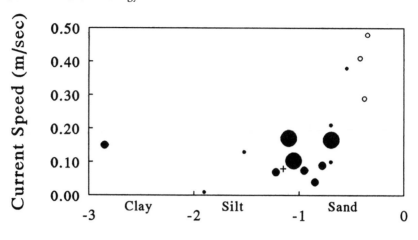

Log Mean Particle Size (mm)

Figure 2.4. Abundance of *Pisidium casertanum* in the Ottawa River, plotted by current speed (depth averaged velocity, summer) and log mean sediment grain size. Large circles designate sites where mean annual clam density of individuals was 208–234 per m^2, medium closed circles 120–187 per m^2, small closed circles 15–90 per m^2, and no clams were found in sites with open circle or '+' (polluted). (Data of Hamill *et al.* 1979.)

Eckblad and colleagues (1977) found a significant tendency for '*Sphaerium* sp.' to live in open water, as opposed to emergent macrophyte beds, in a pool of the Mississippi River. Bailey (1988) took quantitative samples from 41 sites in the same Lake Erie bay where mussel transplants were performed by Hinch *et al.*, finding 14 species of pisidiids. The principal component (PC) that explained 87% of the variance in Bailey's environmental measurements was correlated highly and positively with percentage sand and negatively with the organic content of the substrate. As high scores occurred near the mouth of the bay and low scores near the shore, Bailey interpreted this PC as an exposure gradient. But it could be considered a substrate or macrophyte gradient as well, since macrophyte species richness was negatively correlated with this PC at the $P<0.001$ level. However the PC may be interpreted, pisidiid species richness was positively correlated with it, also at the 0.001 level.

Unlike the unionoids, however, some pisidiids characteristically inhabit the profundal zones of lakes (Bagge and Jumppanen 1968, Thut 1969, review by Mouthon 1986). Using a dredge, Harman (1972) commonly found *Pisidium compressum* on a bottom of silt and decomposing organic matter in Otsego Lake, New York, to a depth of about 20 m. Variable population densities through this range of depths were due to fields of bedrock and boulders, which interfered both with the animals and the gear used to sample them. The population density of *P. compressum* diminished gradually from 5/m^2 to 0 between 20 m and 30 m, as the silt bottom was replaced by clay. And between 20 m and 30 m the density of *P. subtruncatum* gradually increased from 0/m^2 to 30/m^2, at which density it remained to the 50 m bottom of the lake.

Holopainen and Hanski (1979) have reviewed the biology of profundal *Pisidium*, especially emphasizing *P. casertanum* and *P. conventus*. The former species they aptly characterize as a 'generalist' – able to live in a great variety of environments, from streams to lakes, deep water and shallow (as may be recalled from the dietary study on *P. casertanum* and *P. conventus* by Lopez and Holopainen 1987). Although *P. casertanum* can range down to the profundal zone, in Lake Paajarvi, Finland, it extends only to a depth of 5 m. Ranging from 5 m to 70 m in Lake Paajarvi one finds *P. conventus*, characterized by Holopainen and Hanski as 'arctic cold stenothermic'. A key factor seems to be that temperatures in the littoral zone of this lake vary from 1 °C to 23 °C, while in the profundal the variation is only from about 2 °C to 10 °C. The authors suggest that *P. casertanum*, which has a large clutch size but reproduces only in July, cannot compete in the profundal zone with *P. conventus*, which reproduces at a lower rate but all year around. This is a small victory – judging from its higher population densities, respiration and production rates, *P. casertanum* still monopolizes the superior habitat. Holopainen and Hanski do not speculate on whether *P. conventus* is excluded from the upper 5 m of the lake physiologically or competitively. This would be an interesting investigation.

Dale Hollow Reservoir, Tennessee, is unusual among the impoundments of the southern United States in that the water has remained fairly clear. It supports a large population of *Corbicula fluminea*, extending to a remarkable depth. Abbott (1979) took quantitative samples of *Corbicula* from 8 m (in the July epilimnion, approximately 22 °C) and 12 m (about the top of the hypolimnion, 15 °C in July). Juveniles comprised a negligible percentage of the samples – Abbott believed that juvenile recruitment had been unsuccessful for several years at these depths in Dale Hollow, and that these samples were of nearly uniform age. Just as was the case with *Anodonta* in Lake Bernard, individuals from deeper sites were significantly smaller. Abbott suggested that reduced growth rates in the colder hypolimnion might be responsible, although substrate (much finer at 12 m) or population density (much greater at 12 m) cannot be ruled out. Abbott did not examine population densities or age distribution in shallower (and more typical) depths.

Dudgeon (1983a) took seven samples rather irregularly spaced down a depth transect in Plover Cove Reservoir, Hong Kong: 5 cm, 50 cm, 60 cm, 70 cm, 5 m, 8 m, and 10–12 m. *Corbicula fluminea* was quite common in all except the deepest sample. Although the author considered his sampling gear inadequate to estimate density, the age structure of the four shallower samples was much younger than that of the two deeper samples. (Abbott may well have obtained a similar result had he sampled shallower regions of Dale Hollow Reservoir.) The phenomenon may once again be traceable to either substrate differences (deeper samples had finer sediments, as usual) or to the periodical dewatering of shallower regions of Plover Cove.

Pisidiids lend themselves more readily to laboratory experiment than do the larger unionoid mussels. Perhaps the most influential investigations of habitat choice in these little clams were completed by Meier-Brook (1969). He placed about 60 individuals of three *Pisidium* species from Lake Titisee, Germany, in a Petri dish on the interface between sediments of two different particle sizes. *Pisidium lilljeborgii* significantly preferred the finer sediments, while *P. hibernicum* and *P. nitidum* significantly preferred the coarser. These results correspond to observations on the horizontal distribution of the three species in Lake Titisee – *P. lilljeborgii* substantially deeper. Meier-Brook also noticed that only *P. hibernicum* commonly had dense growths of blue-green algae on its shell. He placed individuals of this species along with three other Lake Titisee *Pisidium* species under approximately 1 cm of mud in two Petri dishes, exposing one dish to diffuse daylight and keeping the other dark. Regardless of light regime, twice as many *P. hibernicum* were seen on the surface of the substrate as any other species – a significant difference. Perhaps unexpectedly, however, this amounted to only 27.3% of the *P. hibernicum* in the dark and 8.5% in the light. So the presence of algae on the shells of most *P. hibernicum* in Lake Titisee suggests either that the little clams are repeatedly emerging and re-burying at the surface, or that their behaviour may be somewhat modified in the Petri dishes.

Most other laboratory experiments on pisidiids have involved single species only. Larger and characteristically occurring in lotic environments, *Sphaerium striatinum* offers a striking contrast to the *Pisidium* of Meier-Brook. Gale (1973) placed *S. striatinum* from a creek into a circular pen within which mud, sandy-mud, and sand were arranged in a checkerboard fashion. Although juveniles selected mud over the other substrates, adults showed no preference. The similar experiments of Gale (1971) involving *Musculium transversum* suggested that clams of all ages significantly preferred mud over sandy mud, and sandy mud over sand. But in Pool 19 of the Mississippi River, *M. transversum* appears equally common on all substrates.

Incongruence between laboratory results and field observations has been a common experience in studies of pisidiid ecology (e.g. Hinz and Scheil 1976). Mackie and Qadri (1978) placed five newborn *Musculium securis* in Petri dishes under a variety of experimental treatments and, over a period of weeks, recorded both growth and mean natality. As touched upon previously in this chapter, *Musculium* will not grow or reproduce in pure sediments taken from a depth of 2 m in its native Britannia Bay of the Ottawa River, Canada, nor will it grow on pure crushed white elm leaves, but it will grow on a combination of the two. Mackie and Qadri sieved Britannia Bay sediments into four categories: <0.05 mm (silt–clay), 0.05–0.125 mm (fine sand), 0.125–0.5 mm (medium-fine sand), and 0.5–2.0 mm (coarse sand). Added to white elm leaves, they found average natalities of about 20 offspring per dish in silt–clay and fine sand, but less than the replacement rate of five per dish in medium-fine or coarse sand. Other experiments showed an effect due to type of tree litter, with elm, cedar, willow, oak and birch better than spruce or aspen, which were better than red maple.

Frustratingly, however, Mackie and Qadri do not generally observe that *M. securis* shows any preference for sediment particle size in Britannia Bay, and it often occurs where tree foliage is absent. Kilgour and Mackie (1988) reported that *M. securis* was most common at 4 m on a bottom of about 0.16 mm mean particle size, much more coarse than the clam seems to require in the laboratory. Mackie and Qadri suggested that finer (more preferable) sand is generally restricted to the shallows where *Musculium* may be killed by the summer heat. The type of *in situ* experiments required to address this question would generally be quite difficult, but the approach of Mackie and Qadri is promising.

The densities of some North American populations of *Corbicula* (thousands per m^2) are such that the clam on occasion seems to inhabit all substrate up to and including smooth bedrock (Graney *et al.* 1980). But Belanger and co-workers (1985) used substrate-choice experiments similar to those of Meier-Brook to establish that *Corbicula* prefers sterilized fine sand (0.25–0.7 mm) over a 1:1 mixture of fine sand and finely ground leaves. Either of these was preferable to coarse sand

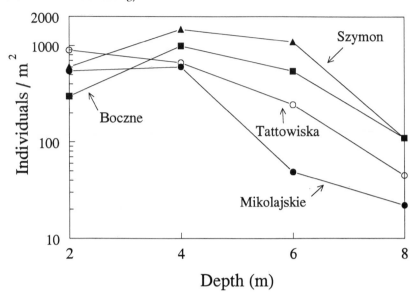

Figure 2.5. The density of *Dreissena polymorpha* in four Polish lakes as a function of depth. (Data of Stanczykowska *et al.* 1975.)

(2.5–4.5 mm), and any substrate was preferable to none at all. Field distributional data from a South Carolina stream corroborate this: maximum abundance on sand, lesser in 'sand/muck', and absent from gravel habitats (Leff *et al.* 1990). Thus the substrate preference of *Corbicula* seems similar to that of *P. casertanum* in the Ottawa River, or *M. securis* in the experiments of Mackie and Qadri. Its substrate choice seems quite different from that of *M. transversum* as determined by Gale (1971), however.

Dreissena

This is clearly a subject quite separate from habitat choice by other freshwater bivalves. After perhaps 1–3 weeks in the plankton, *Dreissena* must find a suitable surface upon which to settle and attach a byssus. Stanczykowska (1978) noted that the following regions have generally been found unsuitable: the hypolimnion (where oxygen is insufficient), the shore and shallow littoral (where strong water currents may not allow settlement), and areas of inappropriate substrate (e.g. flocculent silt). Figure 2.5 shows some typical data on the depth distribution of *Dreissena*, extracted from a larger study of 13 Polish lakes by Stanczyskowska and colleagues (1975). Maximum abundances are usually about 4 m, somewhat deeper than our experience with unionacean and sphaeracean bivalves has led us to expect. Ranges are much deeper in the exceptionally clear waters of lakes Geneva and Zurich (Stanczykowska 1977).

Dreissena depth distribution is probably more a function of current speed and substrate than depth *per se*. Once settled, *Dreissena* shows optimum growth in shallower regions with at least some natural water movement, as we have seen in the other bivalves (Smit *et al.* 1992). Although its larvae tend to settle most frequently in regions of calm, Stanczykowska noted in her 1977 review that larvae cannot settle where fine silt may bury them or clog their gills, preferring macrophytes, stones, shells, and other hard items uncommon in deep water. Thus low density populations are found in patches on solid substrates, not in the shallowest water but not especially deep, and high density populations often tend to form continuous belts. Stanczykowska notes that populations migrate from shallower areas in the winter, not significantly extending their lower depth ranges but rather condensing.

Settlement in regions of otherwise soft, inhospitable bottom is no doubt facilitated by a significant tendency for *Dreissena* individuals to aggregate (Stanczykowska 1964). Lewandowski (1976) placed a large number of small (4–10 mm) individuals in an aquarium with large, living mussels, dead mussel shells, and pebbles. After 20 days, the small *Dreissena* were significantly more common on the living individuals of their own species.

Reproduction

A large portion of our knowledge regarding unionoid reproduction derives from the U.S. Bureau of Fisheries which, under the influence of an expanding pearl button fishery, maintained an active program of mussel propagation for many years (Lefevre and Curtis 1912, Coker *et al.* 1921). More recent reviews have been offered by Fuller (1974) and Kat (1984). The reproduction of spharaeceans has received considerably less attention, although Heard (1965, 1977) has compiled a fair summary for *Pisidium*, *Sphaerium*, and *Musculium*. Burky (1983) and Mackie (1984a) have reviewed bivalve reproduction generally.

Unionoid gonochorism

Sexes are separate in most unionoids. In some taxa the posterior shell margin of the female becomes enlarged to accommodate bulging marsupia, resulting in one of the most striking examples of sexual dimorphism in the Mollusca. But as is true for bivalves generally, in most unionoids gender can be distinguished only by histological examination of the gonads, which lie in close proximity to the digestive gland in the central visceral mass. One might imagine, given the commercial importance of such marine bivalves as oysters, clams, and scallops, that bivalve mechanisms of sex determination might be understood. Alas, although some progress has been made recently (Guo *et al.* 1998), we are as ignorant of these matters for most commercial bivalve species as we are for the relatively obscure species of fresh waters. The discovery that mitochondria show 'double uniparental inheritance' in unionoids, as previously reported for marine mytilids, has led to some speculation that cytoplasmic factors may be involved in bivalve sex determination (Skibinski *et al.* 1994, Hoeh *et al.* 1996, Liu *et al.* 1996).

Large random samples from unionoid populations usually contain ratios of male:female not significantly different from 50:50. Table 2.2 shows a sample of nine studies (20 populations) involving collections of at least 30 adult unionoids made without conscious regard to size or sex. The sex ratio in 18 of these populations is not significantly different from equality, and rejection of the 1:1 hypothesis in a nineteenth (*Lampsilis radiata*, Kat 1983) may be a type I statistical error. The data of Cruz and Villalobos (1984) on sex ratios in the Limon population of the Neotropical mycetopodid *Glabaris luteolus* are anomalous. They reflect a highly significant male bias, even after Bonferroni correction. The authors were hard put to explain this phenomenon. But it may reflect phenotypic plasticity of sexual expression, as we shall see.

Samples of unionoids often show some variation in sex ratio when sorted by size or apparent age class. Smaller individuals seem more likely to be male. This phenomenon has been interpreted as evidence for protandry; some individual mussels may pass through a male stage in their development to female reproductive function, while other individuals remain male (Tudorancea 1969, Kat 1983, Downing *et al.* 1989). Other workers have interpreted such data not as evidence for protandry, but simply a result of sexual dimorphism with respect to size (Smith 1979, Cruz and Villalobos 1985). There is some anecdotal evidence of protogyny in a captive population of the Malayan unionid *Contradens ascia* (Berry 1974). Clearly our understanding of sex determination in the unionoids remains in its infancy.

Unionoid populations vary widely in the timing of their gametogenic cycles. Gametogenesis of both sorts proceeds year round in Chilean *Diplodon* (Peredo and Parada 1986) and in four Virginia unionid popula-

Table 2.2. *Sex ratios (proportion male of gonochoristic individuals) for populations of unionoid mussels*

Reference, Population	Sample N	Hermaph.	Sex ratio
van der Schalie and van der Schalie (1963)			
Actinonaias ellipsiformis, Michigan	239	1	0.55
Tudorancea (1972)			
Unio tumidus, Poland	629	0	0.50
Unio pictorum, Poland	677	0	0.50
Yokley (1972)			
Pleurobema cordatum, Alabama	306	0	0.55
Heard (1979)			
Elliptio arctata, Florida	126	4	0.51
E. buckleyi, Florida	59	3	0.46
E. complanata, pop. A, Florida	140	2	0.49
E. complanata, pop. B, Florida	126	2	0.50
E. crassidens downiei, Georgia	30	0	0.47
E. hopetonensis, Georgia	32	0	0.53
E. icterina, pop. 1, Florida	54	3	0.49
E. icterina, pop. 2, Florida	211	3	0.51
E. icterina, pop. 3, Florida	310	5	0.50
E. mcmichaeli, Florida	36	0	0.56
Kat (1983)			
Lampsilis radiata (11 pops)	234	10	0.60[a]
Alasmidonta undulata (5 pops)	59	6	0.53
Cruz and Villalobos (1984, 1985)			
Glabaris luteolus, Costa Rica (Limon)	323	0	0.65[a]
G. luteolus, Costa Rica (Canas)	164	0	0.53
Yeager and Neves (1986)			
Quadrula cylindrica strigillata, Virginia	74	0	0.47
Avelar et al. (1991)			
Castalia undosa, Brazil	100	0	0.52

Note:
[a] indicates a significant difference from 1:1, by goodness-of-fit χ^2.

tions surveyed by Zale and Neves (1982b). Although active spermatogenesis occurs year round in Michigan *Actinonaias* and Hong Kong *Anodonta*, oogenesis is reduced during the colder months (van der Schalie and van der Schalie 1963, Dudgeon and Morton 1983). Both oogenesis and spermatogenesis decrease during the colder months in Florida populations of *Anodonta* (Heard 1975) and Alabama populations of *Pleurobema* (Yokley 1972). In any of these cases, gametes are ultimately released into the suprabranchial chamber, through which external water is being pumped for the purpose of feeding.

As is true for bivalves generally, the fairly lengthy gametogenic period displayed by individual unionoids is followed by episodic spawning. In a turn quite different from typical marine bivalves, however, it is only the male that 'spawns' in the ordinary sense, releasing sperm into the environment. I am not aware of any direct experimentation regarding the 'trigger' for spawning in unionoids, but observational evidence suggests that temperature is the primary factor, just as in the better studied marine bivalves. (This subject will be pursued in Chapter 4.) Sperm morulae (multinucleated masses) have often been reported, although their significance is not clear (Edgar 1965, Heard 1975, Jones *et al.* 1986). Sperm are taken through the incurrent siphon of the females, along with food particles and miscellanea. It is not clear how the female transports her eggs from her gonad to her gills. Fertilization may occur in the suprabranchial chamber itself, or perhaps in the water tubes of the gills (Mackie 1984a).

Unionoids are distinguished from all other bivalve species by the adaptation of their larvae to parasitic development, usually on fish, and the modifications of the female reproductive system that this entails. Embryos develop in special chambers in the female's gills, called marsupia. There they are held for varying lengths of time, developing into a dispersal stage called a glochidium in the unionids and hyriids, a haustorium in the mutelids, and a lasidium in the mycetopodids. The period of larval development and holding (during which the female is said to be 'gravid') as well as the phenomenon of ectoparasitism on fish, will be examined in greater detail both later in this chapter, and again in Chapter 4.

It has been speculated that the likelihood of fertilization in unionoid populations is exceptionally low. Downing and his colleagues (1993) used SCUBA and a fine system of co-ordinates to take detailed samples of the *Elliptio complanata* population inhabiting a 6 m × 7 m portion of Lac de l'Achigan, Quebec. Their effort was concentrated on two July days after

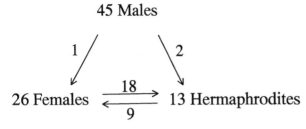

Figure 2.6. The result of Bauer's (1987) experiment involving 84 *Margaritifera margaritifera*. The arrows indicate net sex change after 7 weeks of displacement upstream.

the population had spawned, but prior to the release of mature glochidia. The marsupium of each female was examined to estimate percentage fertilization. Mussels were collected by hand, and as such 'endobenthic' (i.e. buried) individuals were not tallied. Nevertheless, their data seem to suggest almost complete fertilization failure where local mussel densities fell below about 15 individuals per square metre. This is much lower than typical mussel densities observed in the developed world today, and bodes ill for the future of our unionoid faunas.

Bauer (1987) reported the results of a remarkable experiment involving the sex characteristics of a population of *Margaritifera margaritifera* inhabiting a rapid, mountainous stream in Germany. He hypothesized that the displacement of a small sample of mussels upstream from a large bed prior to spawning ought to adversely impact their fecundity, by reason of fertilization failure. Such was not the case; glochidia developed abundantly even in his most upstream individual. Bauer repeated his experiment the following year, marking each of 84 mussels and assessing the sex of each via a needle-punch biopsy prior to displacement upstream. He was surprised to discover that a substantial fraction (15%) appeared to be hermaphrodites. He was even more surprised when he reassessed the sexuality of his 84 animals seven weeks later and found that 28 had changed sex (Figure 2.6). Only 3 of 45 males seemed to have changed, a result that Bauer thought might be attributable to initial errors in sexing. But Bauer considered his observation that 69% of the displaced females became hermaphroditic an adaptive response to a low concentration of exogenous sperm. It is not clear how general this phenomenon may be over the Unionoidea; many aspects of the biology of *Margaritifera* are unusual. But broader subject of hermaphroditism in unionoids will be pursued in the section that follows.

Unionoid hermaphroditism

As is true for bivalves generally, simultaneous hermaphrodites seem present at low frequencies in most unionoid populations. In addition, a few species appear predominantly hermaphroditic. The phenomenon was first reported in the unionoids by Sterki (1898). Van der Schalie (1970) surveyed 97 North American unionid species (1871 individuals), finding four species 'dominantly hermaphroditic' and detecting 'sporadic' hermaphroditism in 22 additional species, generally those with large sample sizes. Hermaphroditism is also well documented in Eurasian unionoids (Bloomer 1939, Nagabhushanam and Lohgaonker 1978) and in the hyriids and mycetopodids of South America (Avelar *et al.* 1991, Avelar 1993, Avelar and de Mendonca 1998). The gonads of unionoid hermaphrodites typically show either monoecious acini (producing both eggs and sperm) or closely intermingled male and female acini (Heard 1975). Sometimes testicular and ovarian tissues seem regionally distinct, but gonads are never entirely separate within a single hermaphrodite. Sporadically occurring hermaphrodites usually show differential development of the two gonadal types, such that the functionality of their hermaphroditism is dubious. This is not always the case in populations where hermaphrodites predominate.

During the 1930s, Bloomer (review 1939) performed an influential survey of hermaphroditism in English populations of *Anodonta cygnea*. A sample of 95 individuals from Bracebridge Pool included 20% males, 59% females, and 21% hermaphrodites. But a sample of 66 individuals from Highams Park included 38% hermaphrodites, with all the remainder apparently normal females. And populations from Beam Brook, Holmwood, Eardisland, and the River Frome appeared to be composed almost entirely of hermaphrodites. It is important to note that unionoids manifest their gender in two capacities: the gametes being produced in the gonad determine the 'visceral sex,' and the differentiation of the gill into a repository for glochidia the 'marsupial sex.' Bloomer found all combinations of visceral and marsupial sex among English *A. cygnea*. Most commonly, both eggs and sperm were made in the viscera, and the marsupium was female. But the occasional individuals showing male visceral sex and female marsupial sex prompted Bloomer to hypothesize that sex-reversal may occur in *A. cygnea*.

The situation is similar in several species of North American *Anodonta*. But Heard (1975) recognized only two sorts of hermaphrodites: 'female hermaphrodites' with ovarian tissue exceeding testicular tissue and gills modified into marsupia, and 'male hermaphrodites' lacking marsupia, with a prevalence of testicular tissue. Two populations of *A. imbecillis* comprised substantial fractions of both pure females and female hermaphrodites, and Heard was able to confirm the simultaneous presence of both egg and sperm in the gonads of the latter. A population of *A. couperiana* contained 31% male hermaphrodites, 6% female hermaphrodites, and 63% females. Heard also reported about 10% male hermaphrodites in an otherwise bisexual population of *A. peggyae*. The complete correlation between visceral and marsupial sex displayed by North American *Anodonta* led Heard to question Bloomer's hypothesis of sex-reversal.

Heard (1979) characterized the phenomenon of sporadic hermaphroditism in nine species of North American *Elliptio*. Among 1178 individuals examined, Heard discovered 22 hermaphrodites. Most of these data are shown in Table 2.2. There were 17 male hermaphrodites, 1 female hermaphrodite, and 4 individuals in which visceral sex and marsupial sex were not in agreement. As had been reported by Bloomer, Heard found two *Elliptio* with predominantly testicular gonads but gills differentiated into marsupia, and two *Elliptio* with predominantly ovarian gonads but undifferentiated gills. He did not notice any parasitism, although Kat (1983) reported an association between digenean trematodes and sporadic hermaphroditism in *Lampsilis radiata* and *Alasmidonta undulata*. None of Heard's 22 animals contained both mature eggs and sperm simultaneously. That observation, together with the rarity of hermaphrodites in most unionoid populations and their occasional tendency to display parasitic castration, suggested to Heard that the phenomenon of sporadic hermaphroditism might be of minor importance. But in an evolutionary sense, the capability for hermaphroditic reporduction is unquestionably available for selection in a large fraction of unionoid populations.

Mackie (1984a) suggested that protandric hermaphroditism might be the ancestral condition in the molluscan Class Bivalvia. The unionoids appear to be primitively dioecious, however, with predominant simultaneous hermaphroditism arising independently on several occasions (Hoeh *et al.* 1995). Van der Schalie (1970) credits the early German worker H. Weisensee with the first observation that hermaphroditism seems more common in the mussel populations of lentic waters than in lotic environments. Kat (1983) found that several especially high–density *Anodonta imbecillis* populations from lentic environments tended to have strikingly low numbers of males and/or lower proportions of testicular tissues. Bauer's (1987) results suggest to me that sexuality in the Unionoidea may be more plastic than previously suspected. Might the striking equality of sex ratios shown in Table 2.2 hide a non-genetic component? As we shall see in Chapter 4, many attributes of the life history of freshwater molluscs have a large environmental component; perhaps the production of eggs, sperm, or both is no exception.

The unionoid larval stage

Full understanding of the unionoid life cycle arrived piecemeal, over a period spanning several centuries (Coker *et al.* 1921). Tiny ectoparasites found attached to the gills of European fish were identified as bivalve larvae as early as 1695. But ironically, in 1797 the hypothesis was advanced that the tiny organisms found in the gills of European freshwater mussels were parasites of mussels (described as '*Glochidium parasiticum*') rather than mussel larvae. This 'glochidium theory' was overthrown in 1832, but the term 'glochidium' has persisted.

When packed with developing glochidia, the expanded water chambers within a mother's gills are termed 'marsupia'. In some taxa, all four gill demibranchs are modified and swollen into marsupia, and in other taxa only the two outer demibranchs so serve. Of this latter group, the entire outer gill may be modified into a marsupium, or only the posterior regions of the gill may be involved. In such circumstances the posterior of the shell may become enlarged to accommodate bulging marsupia, resulting in the striking sexual dimorphism alluded to at the outset of this section. Gills modified for brooding do continue to retain and process particles, but at reduced efficiency (Tankersley 1996).

Development in the marsupium generally seems to require about 4–8 weeks. Mature glochidia are quite simple structurally. Their bodies, almost featureless except for sensory hairs and (sometimes) a 'larval thread', are enclosed in a pair of bowl-like shells held by a single adductor muscle. Unionid and hyriid glochidia are of three general types. Those of the 'hooked' type are generally larger (0.2–0.4 mm, Bauer 1994), having hinged hooks at the ventral margins of their rather triangular shells, and generally seeming adapted to attach to skin, fins, and scales. The 'hookless' type have more rounded shells, generally ranging from 0.1 mm to 0.3 mm (Bauer 1994), and seem best adapted to attach to gills. A few species have 'axe-head' glochidia (0.15–0.25 mm), with oddly elongated shells, usually carrying unhinged hooks. Lasidium and haustorium larvae (of mycetopodids and mutelids, respectively) have uncalcified, univalve shells and peculiar morphologies (Parodiz and Bonetto 1963).

Glochidia are ready for release immediately after maturation in 'tachytictic' populations, while in 'bradytictic' populations, mothers hold their glochidia in a mature state for some months. The life history consequences of this dichotomy will be examined in Chapter 4. In either case, mature glochidia (or lasidia or haustoria) are passed from the marsupium to the suprabranchial canal, and discharged through the excurrent siphon. Expulsion may be triggered by mechanical stimulation or simple warming. Some species (anodontines generally) manufacture a web-like structure that seems to facilitate glochidial suspension, and may be attractive to fish (Haag and Warren 1997). Although glochidia are unable to swim, they may remain suspended in the water column for long periods by natural turbulence, aided in some cases by the larval thread. Plankton nets cast through rivers or lakes inhabited by unionoids may receive large pulses of glochidia at certain times of the year (Clark and Stein 1921, Neves and Widlak 1988). Glochidia are not specific in their attachment. Valve closure may be chemically induced by diluted fish blood or by various organic compounds (Heard and Hendrix 1964), or physically induced by tactile stimulation from paper tissue or film (Wood 1974).

In most cases, glochidia released into the environment will die in 10–14 days if they do not attach to a suitable fish host. *In vitro* glochidial culture typically seems to require some non-specific component of fish blood plasma, in addition to the usual amino acids, salts, and vitamins (Isom and Hudson 1982). Successful glochidia become encapsulated a few hours after attachment, and generally remain so for about 2–4 weeks, subject to considerable extension by cool temperatures. Development of internal organs is completed during encapsulation. While lasidium and haustorium larvae grow during their encapsulation (Fryer 1970), the great majority of glochidia do not. This observation, together with the fact that fish hosts generally seem unaffected by glochidial attachment, prompted Kat (1984) to characterize the fish/mussel association as that 'grey area between phoresy and parasitism.' Young mussels ultimately rupture the capsule in which they have been enclosed with their newly developed foot, and fall to the substrate, where they will remain.

There are a few documented instances of unionoid parasitism on amphibians (Seshaiya 1941, Howard 1951). A few unionid species seem to produce glochidia which can facultatively complete metamorphosis and survive without a host (Allen 1924). Completely aparasitic development has been reported in one group of South American hyriids (Parodiz and Bonetto 1963), and in the Lake Tanganyika unionid *Grandidieria* (Kondo 1990).

The adaptations shown by females of the genus *Lampsilis* to attract potential fish hosts by baiting have been appreciated for years (Coker *et al.* 1921). As the time approaches to release their glochidia, female *Lampsilis* develop elongated, brightly coloured flaps on their mantle margins, near the siphons. Spots may appear on each side, looking very much like eyes, and the mantle flaps wave spasmodically, mimicking in great detail the appearence and actions of small minnows (Kraemer 1970). The mussel pushes the marsupial edge of its gills out toward the siphons, extending even a bit outside the shell. Any fish attracted by what it believes to be a small prey item receives a mouthful of glochidia instead of a meal.

Developing glochidia are typically bound together in the water tubes of the marsupia by a mucilaginous matrix. This matrix usually dissolves before or during discharge to yield individual glochidia capable of suspension in the water. However, it has often been observed that gravid females of some species, when disturbed, may eject their larvae still bound together in 'conglutinates' roughly the size and shape of the expanded water tubes that held them. This was for years believed to be abnormal; essentially an abortion of the reproductive products. But Chamberlain (1934) pointed out that the conglutinates of *Cyprogenia alberti* contain glochidia that are quite viable. These conglutinates are 2–5 cm in length, bright red, and seem to possess a coiled body and wormlike 'head'. Partially extruded from the excurrent siphon, they constitute an effective bait for host fish. The conglutinates of *Pleurobema* and *Ptychobranchus* species in the southern United States are strongly reminiscent of dipteran larvae (Hartfield and Hartfield 1996) or larval fish (Barnhart and Roberts 1997, Haag and Warren 1997). Apparent adaptations of conglutinates for bait have now also been described in the hyriids (Jones *et al.* 1986).

Superconglutinates, first described by Haag *et al.* (1995), must rank among the greatest marvels of the natural world. Several species of *Lampsilis* inhabiting rapidly flowing headwater streams of the U.S. Gulf Coast package the entire contents of their marsupia (both gills, 16–29 conglutinates per gill) into a form 37–50 mm long, resembling a minnow even to the inclusion of eyespots. This bait is then tethered to a tough, transparent mucus strand 10–15 mm in diameter and up to 2.5 m in length. 'The strand is pliant, allowing the terminal end containing the glochidial mass to dart with the current in a manner resembling a small fish or fishing lure.' Haag and his colleagues reported that 60% of a small sample of basses taken from the region were infected with lampsiline glochidia.

It is unfortunate that little is known about the effect of unionid glochidial infection on fish hosts. Laboratory infection experiments suggest that intense exposures to the glochidia of the Thai unionid *Chamberlainia* may kill newly hatched fish fry (Panha 1993). But my impression is that glochidial effects are generally negligible. The scores of published artificial infection experiments have rarely included mention of fish morbidity or mortality. The intensity of unionid glochidial infection is typically low in the wild, perhaps averaging 5 to 10 glochidia per fish (Trdan 1981, Neves and Widlak 1988, Jansen 1991).

Host fish may acquire immunity to glochidial parasites. Although black bass are normally suitable hosts for *Lampsilis luteola*, Arey (1923, 1932) demonstrated that reinfected fish are more likely to slough off glochidia than 'naive' fish not previously exposed to the parasite. Immunity may not be detectable until the third, fourth, or even fifth infection, and may be complete or partial, depending on temperature. Watters and O'Dee (1996) exposed 15 species of fish wild-caught in small Ohio creeks to the glochidia of *Lampsilis fasciola* co-occurring with them, and noticed that untransformed glochidia were shed in two pulses. All the fish successfully infected shed glochidia during the first few days post-infection. In addition, the rosefin shiner and rainbow darter shed large numbers of glochidia over days 19–43. Watters and O'Dee suggested that this might constitute evidence of immunity acquired during the course of a single infection.

The development of margaritiferid glochidia differs in a number of respects from that of the more diverse and widespread unionids outlined above. Maturation time in the female margaritiferid is more brief (12 days in *M. margaritifera*, Murphy 1942), and glochidia are ready for release at a significantly smaller size (0.05–0.08 mm, Bauer 1994) than unionid glochidia. The intensities of margaritiferid glochidial infection are often hundreds per fish. Margaritiferid glochidia require longer development times on the gills of their hosts, do grow in size during attachment, and seem more likely to affect their host adversely than the glochidia of other mussels (Cunjak and McGladdery 1991). Young *et al* (1987) found no evidence that brown trout acquire immunity to *Margaritifera margaritifera*, even after multiple separate infections.

The effect of *M. margaritifera* glochidial parasitism on salmonids has been nicely explored in a series of papers by Meyers and Millemann (1977), Fustish and Millemann (1978), Karna and Millemann (1978), and Meyers *et al.* (1980). The authors focused their attention on hatchery-reared juvenile fish, in the 20–80 mm size range. They began by comparing six salmonid species in their susceptibility to glochidial infection of varying intensity, exposing 10 fish each to six concentrations of glochidia for three hours, and keeping 10 fish as unexposed controls. The salmonids varied greatly in their mortality rates during the first 48 hours post-infection. The lower half of Figure 2.7 compares the mortality of the most susceptible fish, the chinook salmon, with that of the least susceptible, the coho salmon, at the completion of glochidial development. The authors noted that at low exposure levels, coho salmon were able to slough off all glochidia, and that glochidial sloughing continued through the study period at higher exposure levels in coho. Chinook salmon tended to retain much higher proportions of the glochidia to which they were exposed.

The upper half of Figure 2.7 shows the result of a second experiment where 250 fish of each species were exposed individually to 15 000 glochidia in 250 ml of water for 30 minutes. The authors noted increased concentrations of leukocytes in the coho salmon soon after infection. They characterized the response of the coho gill epithelium as 'hyperplasia', an excessive growth resulting in the extension and fusion of gill lamellae, and their breakage (sloughing) with respiratory movement. Very little leukocytosis or hyperplasia was noted in chinook gills. There also seems to be a humoral component to the greater resistence to infection displayed by the coho. *In vitro* survival times of glochidia in mucus and plasma taken from coho salmon were less than those observed in the same chinook salmon fluids. But interestingly, the humoral differences in chinook and coho responses do not seem to be immunological. Agar gel plates disclosed no evidence of precipitating antibodies when glochidial antigen was exposed to mucus or plasma taken from either fish previously uninfected. Plasma (only) from both fish showed equal immunological response to glochidial antigen 8–12 weeks post infection.

Unquestionably, the relationship between a mussel species and its host may be quite specific. Virginia's Big Moccasin Creek is home to seven unionid species and 24 species of fish. But the extensive electrofishing surveys (Zale and Neves 1982a,b,c, Weaver *et al.* 1991) over five months and 4000 individuals, found only 13 fish populations normally serving as glochidial hosts. Artificial infection experiments, involving all combinations of four mussels and 14 prospective hosts (with an impressive number

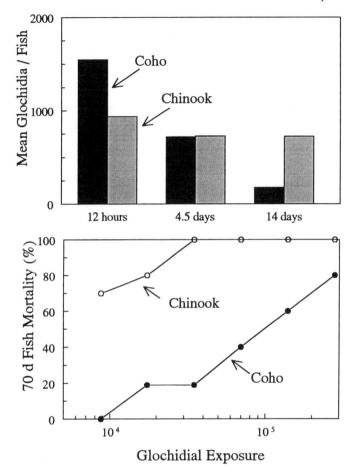

Figure 2.7. The top figure shows the persistence of *Margaritifera margaritifera* glochidia (initial exposure 15000 glochidia) on the gills of two salmon species (Fustish and Millemann, 1978). The bottom figure shows the impact of varying levels of glochidial infection on the survivorship of the fish. (Data of Meyers and Millemann, 1977.)

of replicates) confirmed that Big Moccasin Creek populations of *Lampsilis fasciola* require the smallmouth bass to complete glochidial development; *Medionidus conradicus* requires either the redline or fantail darter; *Villosa nebulosa* requires either the smallmouth or rock basses; and *V. vanuxemi* requires the banded sculpin. The host range of *Pleurobema oviforme* is a bit broader, encompassing at least five fish species.

Other species of basses not found in Big Moccasin Creek appear suitable as hosts for Big Moccasin Creek *V. nebulosa*, however, as do other non-indigenous species of sculpins for *V. vanuxemi* (Neves *et al.* 1985). Indeed, Neves and his colleagues observed complete development of Big Moccasin Creek *V. nebulosa* in mosquitofish from Florida. The successful metamorphosis of North American unionid glochidia on exotic fishes was first reported by Tompa (1979). The glochidia of Hong Kong *Anodonta woodiana* attach and metamorphose more successfully on the American mosquitofish recently introduced there than on three native fishes (Dudgeon and Morton 1984.)

The host range displayed by two Michigan populations of *Anodonta grandis* offers a striking contrast to that of most unionids of Big Moccasin Creek. Trdan and Hoeh (1982) identified 26 fish species inhabiting the two sites combined, representing eight families. Careful inspections of 21 of these fishes revealed *A. grandis* glochidial infections on 17. Artificial infection experiments involving 24 of the 26 species resulted in successful metamorphosis of *A. grandis* glochidia in 21. The host ranges of these particular mussel populations seem to be quite large indeed. Bauer (1994) has found evidence of a positive correlation between glochidial size and host range. *Anodonta* have large, hooked glochidia, apparently adapted for attachment on fins and the external surfaces of host fish. One might expect that this would allow *Anodonta* to exploit a wider variety of hosts than *Lampsilis*, *Villosa*, or *Medionidus*, with their smaller glochidia dependent upon gill attachment. Either group of unionids might be expected to have a broader host range than the margaritiferids, whose tiny glochidia require closer communication with their host fish and longer development times.

Presence or absence of glochidial hooks correlates with the two-subfamily classification of the Unionidae proposed by Lydeard *et al.* (1996): Anodontinae with hooks and Ambleminae without. Appendix A of Watters (1994) lists all known fish hosts for 67 species of amblemines, 24 species of anodontines, and 5 species of margaritiferids. As may be seen in the upper half of Figure 2.8, the modal number of known fish hosts for both amblemines and anodontines is one. The medians of all three groups of mussels seem essentially identical – between three and four known hosts for all. Some bias may have been introduced into these data by increased investigator interest in amblemines and margaritiferids, initially because their shells were more commercially valuable, and more recently because of conservation concerns. One might expect more hosts to have been documented for these groups than for the anodontines, whose propagation has attracted less interest. But in any case, the data illustrated in the top half of Figure 2.8 show no evidence of a relation-

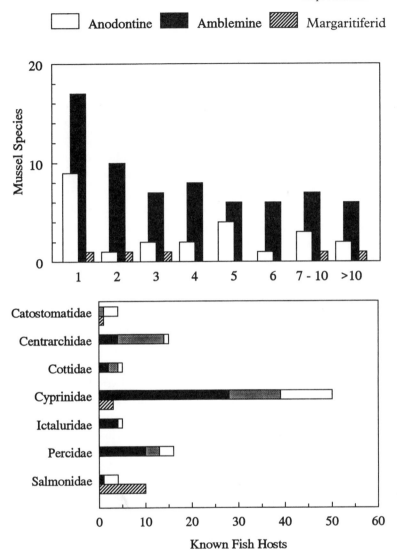

Figure 2.8. The top figure shows number of fish hosts reported in Appendix A of Watters (1994), for 96 unionoid species. The bottom figure shows number of fish hosts listed in Appendix B of Watters (1994) combining species within the 7 most commonly reported fish families. Shaded regions in the lower graph indicate fish species hosting both amblemine and anodontine mussels. Margaritiferid hosts are counted separately.

ship between glochidial morphology, or any other aspect of the biology of these disparate groups of mussels, and known host range.

Unionoid host range may not be a function of the number of fish species parasitized, however, but the number of higher taxa. Appendix B of Watters (1994) lists all known unionoid parasites for 196 species of fish, 162 of which host amblemines, anodontines, and/or margaritiferids. These 162 fish species represent 23 families. So within each family of fish, I counted the number of fish species known to host amblemines but not anodontines, known to host anodontines but not amblemines, known to host both unionid subfamilies, and known to host margaritiferids (separately, regardless of other unionoid parasites). Only seven of the 23 fish families had five species or more known to host unionid and/or margaritiferid glochidia (Figure 2.8 bottom). A relationship between margaritiferids and salmonids is obvious. Ten salmonid species are known to host margaritiferids. Three of these fishes are known also to host anodontines, and one also hosts an amblemine species. No salmonid species is known to host unionids but not margaritiferids. In addition to the margaritiferid/salmonid relationship, an obvious relationship was also evident between mutelids and cichlid fish in Watters' Appendix B, although these data were not tabulated.

However, in spite of the differences in glochidial morphology between amblemines and anodontines, and many other biological differences as well, the lower half of Figure 2.8 reveals no difference in their known fish hosts or host ranges. Both unionid subfamilies parasitize at least a few species of all seven commonly reported fish families. Combining data from the four less-represented fish families (Catostomatidae, Cottidae, Ictaluridae, and Salmonidae) into a single category, the value of chi-square from a test for independence was 2.35, not significant with three degrees of freedom.

Predictions regarding the hosts or host ranges of unionid mussels seem to be frustratingly difficult to make. For example, Weiss and Layzer (1995) surveyed both the fishes and mussels inhabiting a five kilometre length of the Barren River, Kentucky, at monthly intervals over the course of a year. Inhabiting the river were at least 43 species of fish (15 families) and 27 species of unionids. The authors found glochidia on 25 species of fish, 14 of which had no previously documented unionid parasites, while 7 species of fish predicted to host Barren River mussels nevertheless appeared uninfected. A key factor may be Weiss and Layzer's observation that only 4.1% of their 2510 fish were infected with glochidia, and that most of these 4.1% carried only 1–5 glochidia each.

The selection pressure exerted by unionid glochidia on fish populations may generally be negligible. The reason the other 2407 Barren River fish were uninfected with glochidia is almost certainly that their behaviours, habitat choices, and various aspects of their biology did not bring them into contact with mussels, and/or that their non-specific defences, such as tough integument or general hyperplastic, leukocytic, and humoral responses, rejected all glochidia with equal efficiency. Most host responses resulting in rejection of one glochidial species with apparently increased efficiency are likely individual, and acquired.

From the parasite's perspective, I suggest that whatever host specificity may be displayed by unionoids may primarily be a function of physical, ecological, and/or behavioural factors. Some mussel populations are clearly adapted for lotic habitats and others for lentic, some release glochidia in the spring and others in the autumn, and so tend to come into contact with subsets of potential host fish. Some mussels seem to be adapted to offer baits that may tempt some fish but not others. Nevertheless, such laboratory experiments as those of Zale and Neves (1982c) have demonstrated clear evidence of cellular or molecular adaptation by certain mussel parasites to specific fish hosts. Zale and Neves reported that all 50 banded sculpins they exposed to the glochidia of *Villosa vanuxemi* carried infections to metamorphosis, but that 30 sculpins shed all *V. nebulosa*, *M. conradicus*, and *L. fasciola* glochidia with 100% efficiency. One might predict that some sort of ecological or behavioural association might also exist between sculpins and *V. vanuxemi*, such as a temporal relationship involving glochidial release. We will return to Zale and Neves' mussel fauna in Chapter 4.

Watters (1997) has suggested that, to the extent that mussels become behaviourally and/or ecologically associated with particular subsets of hosts, cellular and molecular specializations may follow to counter special subsets of defences. If a mussel population has evolved a special lure that is only attractive to large basses, we might also predict special cellular and molecular adaptations to counter the basses' defences after glochidial attachment. An illustration of the Watters hypothesis may be provided by *Leptodea leptodon* and *Potamilus capax*, whose glochidia seem to successfully transform only on drum (of 24 and 29 potential fish hosts tested, Barnhart 1999). Both of these mussels have unusually large thin shells, and it has been speculated that glochidial infection occurs when drum crush gravid parents upon ingestion.

Dreissena

Dreissena polymorpha is gonochoristic, with sexes generally found in approximately equal proportions (Stanczykowska 1977). Antheunisse

(1963) reported 56% female, 40% male, and 4% hermaphrodite in a river near Amsterdam. Walz's (1978b) sample of 306 adult *Dreissena* from Lake Constance also included 56% female, with the remainder male, and no hermaphroditism in evidence. A ratio of 3 female to 2 male was reported by Mackie (1993) in North American Lake St. Clair. Genetic polymorphism is high, and although some heterozygote deficiency is occasionally reported, Hardy–Weinberg equilibrium generally prevails (Marsden *et al.* 1996).

Gametogenesis in Lake Constance begins in August and proceeds through the winter, terminating with May spawning. Although the *Dreissena* population introduced into Lake St. Clair between Michigan and Ohio also spawns in May, gametogenesis is not initiated until late winter or early spring. Some German populations spawn in mid–May and August, while others show a single summer spawn, dependent on temperature (Borcherding 1991). Males and females spawn into the water column, the ova containing species-specific chemoattractants (Ram *et al.* 1996). Fertilized eggs develop into planktonic veliger larvae, which may settle in as few as five days or as many as five weeks, with about two weeks of planktonic life commonly observed (Sprung 1993). A chemical cue may be released by adults to trigger larval settlement (Chase and Bailey 1996).

Corbiculoids

Most corbiculoids are simultaneously hermaphroditic, with gonads divided into ovarian and testicular regions. In some groups (*Pisidium, Corbicula*) the gonad shows medio-lateral specialization, while in other groups (*Sphaerium*) the specialization is from anterior to posterior. Testes mature before ovaries in pisidiids and in some Asian populations of *Corbicula*, but the reverse is true in the best-studied North American populations of *C. fluminea*. In any case, after maturity at least some oogenesis seems to occur year round, with seasonal peaks. Spermatogenesis seems to be more periodic, especially in North American *C. fluminea*. Sperm morulae dissociate into biflagellate sperm of poor mobility (Kraemer and Galloway 1986, Kraemer *et al.* 1986). Gametes are exported from the gonad through common gonoducts, showing no apparent specialization to separate egg from sperm. Araujo and Ramos (1997) observed self-fertilization within the ovary itself in a Spanish population of *Pisidium amnicum*.

Self-fertilization is generally believed to be widespread in the Corbiculacea. The Swedish biologist N. Odhner seems to have been the first to demonstrate asexual reproduction in the group. His report that *Pisidium conventus* could reproduce in isolation was expanded by Thomas (1959), who reared six generations of *Musculium partumeium* isolating all individuals at birth. Protein electrophoretic surveys have uncovered low levels of polymorphism and highly significant heterozygote deficiencies at almost all variable loci examined in wild populations of *Sphaerium* (Hornbach *et al.* 1980), *Musculium partumeium* (McLeod *et al.* 1981) and *Corbicula* (Hillis and Patton 1982, McLeod 1986, Kijviriya *et al.* 1991). The levels of heterozygosity at polymorphic loci are never zero, however, suggesting that some outcrossing may at least occasionally take place.

I am unaware of any description of spawning in either *Pisidium* or *Sphaerium*. King and colleagues (1986) were able to induce sperm release in North American *Corbicula* by a combination of thermal, mechanical, and salinity shock. Kraemer and her colleagues have also reported mid-summer observations of sperm being extruded between individual *Corbicula* on mucous strands. On the other hand, developing embryos are periodically observed within *Corbicula* gonads, suggesting intrafollicular fertilization. Such embryos appear to be successfully transported from the gonad via the 'gonoducts'. It has been suggested that North American *Corbicula* may produce both an outcrossed spring generation and a self-fertilized autumn generation (Kraemer and Galloway 1986, Kraemer *et al.* 1986).

In North American *C. fluminea*, the water tubes of the inner pair of gills are swollen and modified into marsupia for the brooding of developing embryos. All the ontogenetic stages one normally expects to observe in molluscs are demonstrated within the marsupium. Blastula, gastrula, trochophore, veliger and pediveliger stages are typically passed in 6–12 days, and early straight-hinged juveniles shed to the environment. Kraemer and Galloway noted that stressed animals may release trochophore and veliger larvae, which do not seem especially viable outside their mother's marsupium. Mothers may also retain their offspring for over two months, should the environment dictate.

A wide range of sexuality is displayed by the *Corbicula* populations of Asia. The Chinese population identified by Morton (1982) as *C.* cf. *fluminalis* was dioecious, with 4.5% hermaphrodites. Sex ratios in the dioecious fraction varied, however, from about 60% female among the younger individuals to 60% male in samples of older individuals. This particular population did not seem to brood its young, but rather spawned gametes openly into the water column. Asian populations of

C. fluminea may also contain dioecious individuals, although the situation here has been confused by inadequate sampling.

Smith and his colleagues (1979) reported high levels of genetic variability at allozyme loci in one Japanese *Corbicula* population of unknown species. I might speculate that this population may be related to the dioecious *C. sandai*, apparently reproducing sexually in Lake Biwa (Komaru *et al.* 1997). In contrast, Komaru and his colleagues reported that Japanese '*Corbicula* aff. *fluminea*', Japanese *C. leana*, and Taiwanese *C. fluminea* were all hermaphroditic, producing 'non-reductional' gametes identical in DNA content to somatic cells. They hypothesized that sperm may be necessary to initiate development in these species, but that reproduction might be effectively parthenogenetic.

In pisidiids the semi-enclosed spaces between the lamellae of the inner gill demibranchs serve as the marsupia. Fertilized eggs pass into these spaces, become enclosed by gill tissues, and form sacs in the marsupial wall. The typical ontogenetic stages demonstrated by *Corbicula* seem to be suppressed in pisidiids. Workers find it convenient to recognize primary, secondary, and tertiary brood sacs, containing embryos, foetal larvae, and prodissoconch larvae, respectively. Embryonic mortality (or 'intramarsupial suppression') is often substantial, and it has been suggested that some developing larvae may be sacrificed to yield space or perhaps even direct nourishment for others (Meier-Brook 1977). Young released from tertiary sacs remain in the marsupium for some period, either free or attached by a byssal stalk, as 'extramarsupial larvae'. Extramarsupial larvae may grow quite large, relatively speaking: typically 20–25% of the size of their parents and sometimes 50%. Heard (1977) confirmed early reports of J. Groenewegen and M. Thiel that extramarsupial larvae of many *Sphaerium* species have in fact begun active ('precocious') gametogenesis. It seems odd to apply the term 'larvum' to an individual much beyond its sexual maturity. The mother ultimately releases her young via the excurrent siphon.

Pisidium develop no more than one brood sac per demibranch at any time, all embryos enclosed are substantially the same size and age, and all are released simultaneously. The elapsed time from fertilization to birth may be no more than one month. Adult *Sphaerium* and *Musculium* commonly carry multiple brood sacs through much of the year, in various stages of development, and the young within each sac may be of different sizes. Newborn *Sphaerium* and *Musculium* are typically released a few at a time, even in populations considered semelparous. The total time elapsed from fertilization to release in *Sphaerium* and *Musculium* is typically several months, although exceptions have been noted (Morton 1985).

Total lifetime fecundities of most pisidiids are spectacularly low by comparison with other molluscs; e.g. 12 offspring over two years for *Sphaerium simile* (Avolizi 1976), and 10 over a single year for *S. striatinum* (Hornbach *et al.* 1982). The number, size, and developmental stage of offspring brooded by the various populations and species of corbicula-ceans has been the object of considerable interest, as has been the overall timing of their reproductive cycles. These topics will be pursued in Chapter 4.

Summary

The unionoids and *Dreissena* are gonochoristic, with sexes generally found in equal proportions. Yet most populations show some frequency of hermaphroditism, and several species are predominantly hermaphroditic. The unionoid populations of ponds and other possibly temporary habitats seem more likely to display hermaphroditism. The likelihood that sperm finds its way into the siphon of a mature female unionoid may be extrordinarily low, and some evidence suggests that sexuality may respond plastically to fertilization failure.

Dreissena produces planktonic larvae unique among the freshwater bivalves. The life history of the Unionoidea includes a remarkable dispersal stage, the glochidium, adapted to parasitize fish. Some specific associations have evolved between particular unionoid populations and populations of co-occurring fishes, while in other cases, little host specificity has been noted. Margaritiferid mussels generally seem to parasitize salmonid fishes. But no similar relationship seems to have evolved between any taxon of unionids and any group of fish, nor is any relationship apparent between unionid glochidial morphology and host range. Close host–parasite associations may only arise in exceptional cases, where cellular and molecular adaptations evolve as a consequence of an initial physical or ecological relationship between fish and mussel.

The corbiculoids, which include many taxa famously adapted to temporary habitats, are almost all hermaphroditic, and apparently self-fertilizing. Williams (1975) pointed out that across all kingdoms of life, asexual reproduction is often associated with colonizing ability. It would seem that bivalves are no exception. Corbiculoids are brooders; their larval stage is passed internally, and offspring (sometimes quite large relative to their mothers) are released directly to the environment. The connection between asexual reproduction, ovoviviparity, and polyploidy will be explored in chapters to follow.

The recurring theme of this chapter has been overlap, not just of unionid fish host, but of bivalve diet and habitat generally. In lentic environments, the ideal habitat for almost all bivalves seems to be as close to the surface as predation and disturbance will allow, perhaps 0.5–2.0 m deep. One might speculate that the differences in the life forms of the broad, light *Anodonta*, the narrow, hatchet-shaped *Elliptio*, and the epibenthic *Dreissena* have resulted from adaptation to different substrates in these same, shallow waters. Species of *Pisidium* may select sediments of different grain sizes, and possibly different depths in the sediment. A few *Pisidium* species seem to have become physiologically adapted to deeper water, apparently exchanging reduced competition for reduced lifetime reproduction. But for most lentic bivalve populations, adapted as they are for the shallows, both intra- and interspecific competition for space seems a real possibility.

Lotic waters typically offer a relatively greater proportion of suitable habitat for bivalves and, perhaps as a consequence, bivalve diversity is increased. There is little evidence of divergence among the species in microhabitat preference, however. Most species find a gravel substrate in a region of moderate current to be superior. So to the extent that the size of individual freshwater bivalve populations generally escapes the vagaries of the physical environment, the potential for space competition would seem substantial. This theme will be developed in Chapters 5 and 9.

3 · Gastropod autecology

Here we broadly review the diet, habitat, and reproduction of the freshwater gastropods, starting with general observations on their feeding mechanism. Snails are able to exert more control over their immediate environment and the materials entering their mouths than bivalves, and hence there is greater opportunity for ecological specialization. The diets and habitats of 13 freshwater gastropod families are discussed separately, compared and contrasted. Members of the eight prosobranch families tend to grow more slowly than those of the five pulmonate families, and to be found more commonly in lotic environments. Substrate preferences are striking and widespread. Among the groups we catalogue special adaptations and dietary preferences for suspended particles, detritus, bacteria, algal filaments, diatoms, macrophytes, and even carrion. Differential gastropod grazing has been demonstrated to impact algal and macrophyte community structure, as well as the macroinvertebrate community as a whole. Yet against the numerous striking examples of ecological specialization in freshwater gastropods, there is a general uniformity of body plan and habit that suggests the potential for interspecific, as well as intraspecific competition.

We will review a large literature on the complexities of pulmonate reproduction. Adults are generally simultaneous hermaphrodites, although they cannot mate reciprocally at the same time, a situation that may lead to mating conflicts of at least two sorts. Pulmonates can also self-fertilize, although usually at a sacrifice of fitness to both parent and offspring. Long-term sperm storage, multiple insemination, 'sperm sharing', mixed mating and aphally all contribute to the diverse reproductive strategies available to the pulmonate snails. By contrast, most prosobranchs are gonochoristic, with reproductive behaviours inconspicuous and/or poorly documented. Sometimes sex ratios are balanced, but often they are not, and dimorphism in growth or survivorship often does not seem to account for the bias. The actual mechanism of sex determination in the diverse prosobranch taxa is called into question. Parthenogenesis has arisen three times in the freshwater prosobranchs. In all three cases females produce apomictic ova, which are brooded and released viviparously. Males may be present (although rare) and sexual reproduction occurs at least occasionally in many parthenogenic populations. The valvatids are hermaphrodites, completing the general picture of reproductive diversity we will survey in the pages that follow.

Feeding and digestion

As a generality, snails crawl over surfaces, stuffing material into their mouths. Often there seem to be few cues to aid snails in finding anything nutritious. Streit (1981) has suggested that they follow stagies of 'area restricted searching' and 'giving-up time', moving rapidly until encountering a patch of food, then lingering until the food level falls below the average of the habitat (Charnov et al. 1976). A snail detecting some chemical evidence of food in a water current will immediately crawl into that current. In still waters, snails seem to orient at least partly by 'tropotaxis', measuring the difference in the concentrations of the attractant primarily at the bases of their tentacles (Townsend 1973, 1974). Upon arrival at a food source, many of the same chemicals that served as the attractants also stimulate feeding (Thomas et al. 1986), although food arousal may be short-lived (Tuersley 1989).

The first part of a grazing snail to make contact with the substrate is its proboscis, a muscular organ containing the trophic apparatus. As the proboscis folds over the section of substrate being grazed or material being eaten, the jaws open and the radula is protruded. The radula, a ribbon carrying hundreds of fine teeth of various types, is mounted on a tongue-like mass called an odontophore. (The radula and odontophore, together with their elaborate musculature, are called the 'buccal mass', which is said to move in the 'buccal cavity'.) A patch of substrate is scoured or material ripped off, the odontophore withdrawn, and the jaws closed (Hubendick 1957). The jaws of many taxa (e.g. *Lymnaea*, *Pomacea*) are covered with thick cuticle to facilitate the cutting or ripping of leaves, filaments, and other large particles. The opening of the jaw and protrusion of the odontophore is typically repeated rhythmically as the snail crawls, perhaps 24–60 times a minute.

The diversity of radulae among freshwater snails is striking. Almost all the major families of freshwater prosobranchs have radulae of the 'taenioglossan' type, showing seven teeth per traverse row. The major exception is the Neritidae, with 'rhipidoglossan' radulae 60–80 teeth abreast. Although some prosobranch taxa have quite simple teeth (e.g. *Campeloma*), most have teeth which are themselves studded with smaller teeth or cusps in various patterns. Pulmonates have a much larger and more variable number of teeth per traverse row (perhaps 50–150), but each tooth of the pulmonate radula is much smaller and somewhat less elaborate than those typically mounted by the prosobranchs. Teeth are held on a 'ribbon' which may be well over 100 rows long, although most of the rows of teeth are either incompletely formed or being held in reserve to replace the forward teeth as they wear. Teeth on the undisturbed, retracted, radular ribbon are flattened facing inward, and only when the ribbon slides over the extended odontophore do they erect and spread. Studies of radular feeding tracks show that snails are typically capable of a wide variety of stroking, gouging, and biting motions (Thomas *et al.* 1985, Hickman and Morris 1985). An individual snail may rasp with only centre teeth, only lateral teeth, or the entire width of the radular ribbon (Stiglingh and van Eeden 1970).

Material scraped from the substrate passes posteriorly through a long oesophagus, generally bound in mucus and moved by cilia, just as in the bivalve digestive tract. But unlike bivalves, digestion commences almost immediately after food is consumed, with the introduction of saliva from glands emptying near the junction of the buccal cavity with the oesophagus. Although this secretion may serve primarily to lubricate the passage of food, both amylase and protease activity have been detected. Further back, the oesophagus of some pulmonates and prosobranchs is enlarged to form a 'crop' for food storage. Dazo (1965) has provided a nice figure of a typical gastropod digestive system.

Gastropod stomachs are every bit as elaborate as those of bivalves, marked by folds, grooves, and ciliated sorting areas. Most prosobranchs have a crystalline style rotating against a gastric shield, controlling the movement of the mucus string and providing digestive enzymes. Although apparently identical in form and function to that of the bivalves, the prosobranch crystalline style is believed to have evolved independently. A style is absent from the pulmonate stomach. In prosobranchs, the term 'gizzard' is applied to a muscular chamber of the stomach where ingested material is titurated, usually with sand grains. The 'gizzard' of pulmonates such as *Lymnaea* is a bilobed region at the entrance to the stomach, nevertheless serving the same grinding function. Only the smallest particles (0.4 μm or less) pass from the stomach into the digestive diverticulae, where both phagocytosis and extracellular digestion have been reported.

Sand grains are commonly found in the stomachs of most freshwater gastropods, constituting an especially large fraction of the gut contents in many lymnaeids (Reavell 1980). Some evidence has been gathered that such material is actively ingested and used in the tituration of food (Carriker 1946, Frenzel 1979, van den Boom-Ort 1991). Storey (1970) reared two sets of *Lymnaea pereger* on a variety of diets for several months from their date of hatching. To the control set of culture containers he added acid-washed sand, but to the experimental set he did not. In all but one case, the experimental snails died within 2 months, while control snails appeared normal. Storey repeated the investigation, with the one modification that experimental snails were allowed to feed in a stock tank with sand for 2–3 weeks before isolation. After 9 weeks there was no substantial difference in the death rate of experimental and control snails. It is possible that the sand provided a source of, or home for, beneficial bacteria such as might remove ammonia, for example. But at least as likely an explanation is that to some species of freshwater snails, stored mineral particles are critical for the digestive process.

At least some cellulase activity was detected in the digestive tracts of all of 14 European species of freshwater snails tested by Calow and Calow (1975) and in all of 11 North American species tested by Kesler and Tulou (1980) and Kesler (1983a,b). Activity may be found in the stomach, digestive gland, and intestine. Some evidence suggests that the cellulase is produced endogenously, as has been demonstrated in land snails (Marshall 1973, Owen 1966), although from an ecological standpoint the actual origin of the enzyme is not especially important. Calow and Calow reported that little European planorbids and *Ancylus* showed cellulase activities an order-of-magnitude lower than the European lymnaeids and four species of prosobranchs. The Calows demonstrated a strong correlation between cellulase activity and efficiency in assimilating *Scenedesmus*, a green alga with thick cellulose walls. This suggests that cellulase activity might be useful as an indirect method of estimating dietary preference. Kesler, on the other hand, did not detect a great deal of cellulase variation among his diverse collection of North American gastropods, reporting very similar values in species believed to have different diets in the field. But if snails with similar cellulase activities have different diets, perhaps snails with different cellulase levels may have very different diets. I am intrigued by this area of research, and would like to see it extended to other digestive enzymes, in the fashion of Hylleberg (1976).

Undigested material passes out of the stomach and is compacted into faecal pellets in the upper intestine. In the pulmonates, Runham (1975) considers the uppermost intestine homologous to the prosobranch style sac. Three different types of faeces have been recognized in the well-studied *Lymnaea stagnalis*, the 'gizzard string' (coarse material passed directly from gizzard through stomach to intestine), the 'caecal string' (material 1–10 μm rejected at the entrance to the digestive gland), and the digestive gland string (very fine but undigestible material). No caecal string is apparently manufactured in the gut of *Pomacea*, although the other two types of faeces are distinguishable (Andrews 1965). Faeces are moved peristaltically through the intestine to the anterior, where they are expelled from the right mantle edge. Brendelberger (1997) has documented voluntary and regular coprophagy in *Lymnaea peregra*, demonstrating that *Lymnaea* faeces contain nutritional resources digestible by the snail. The phenomenon was not obseved in *Bithynia tentaculata*.

Finally it should be noted that freshwater snails, like the bivalves, can actively transport amino acids, sugars, and even particulate matter across their general body surfaces (Lewert and Para 1966, Gilbertson and Jones, 1972, Zylstra 1972). Thomas and colleagues (1984) showed that *Biomphalaria* can actively transport the short chain carboxylic acids commonly loosed by decaying macrophyte tissue into the medium. They estimated that the snail might absorb up to 6% of its energy requirements from environmental acetate alone. Active uptake of glucose and maltose was demonstrated in *B. glabrata* by Thomas and his colleagues (1990).

Pulmonate diet and habitat

Planorbidae

More is known of the medically important planorbids than any other freshwater mollusc taxon. The large literature on factors affecting planorbid distribution has been reviewed on several occasions (Malek 1958, Appleton 1978, Brown 1994). The following quotations, patched together from Malek, serve as a fair summary: 'The type of substrata almost always associated with bilharziasis intermediate hosts is a firm mud bottom, rich in decaying matter. The current should not be strong. Water plants . . . affect the number of snails, but are not essential for their occurrence.'

Helisoma anceps is rather unusual among the planorbids in that it seems to occur more often in river pools than in lakes or ponds. Cummins and

Lauff (1969) have reported an unusually complete investigation of substrate particle size selection in 10 macroinvertebrates inhabiting a small, shallow Michigan stream, among which they included *H. anceps*. The authors combined a quantitative field survey with artificial stream experiments on substrate choice over eight particle sizes (fine sand to pebble). Although strong particle-size selectivity was demonstrated in a number of aquatic insects, evidence was much less striking in the case of *Helisoma*. It is a shame that no investigation as thorough as that of Cummins and Lauff seems to have been completed for the more typical, lentic-water planorbids.

Liang (1974) has reviewed, I would estimate, approximately 100 published accounts of pulmonate diet in culture, the great majority involving medically important planorbids. Although most workers have used lettuce or fish foods of various complex formulation, Liang's review includes ten papers reporting at least some evidence that an algal diet may contribute to the growth and fecundity of planorbids. For his experiments, Liang cultured both *Biomphalaria* and *Bulinus* on pure diets of blue-green algae (*Nostoc* or *Fischerella*), inoculated into a mud substrate. In a more recent study, Jantataeme and colleagues (1983) found that *Indoplanorbis exustus* fed lettuce plus *Nostoc*, or lettuce plus the diatom *Achnanthes*, showed improved fecundity over cultures fed any of the three dietary items separately.

The only demonstrations that planorbids can be cultured on aquatic macrophytes appearing in Liang's (1974) review were six applications of watercress (*Nasturtium*, e.g. Chernin and Michelson 1957a,b). But more recently Thomas and colleagues (1983) have studied *Biomphalaria* feeding on duckweed (*Lemna paucicostata*) and two pondweeds (*Groenlandia densa* and *Potamogeton crispus*). Not surprisingly, assimilation rates for these macrophytes, calculated as the difference between dry weight consumed and dry weight of faeces over 24 hours, were all much lower than the assimilation rate for lettuce. But Thomas and colleagues did demonstrate short-term growth in *Biomphalaria* fed on these macrophytes.

Thomas (1987) expanded this research programme to include 33 genera of macrophytes, representing 23 higher taxa. Snails were fed two 1.7 cm discs (or the equivalent) of each of these plants daily and weighed after four days (experiment A). Clearly, more macrophyte tissue is typically available for consumption in the wild than was supplied to these snails. Nevertheless, snail growth was significant on most of the diets, even surpassing lettuce in the case of *Groenlandia*. Example data are shown in Figure 3.1. Thomas offered other snails homogenized plant tissues in a calcium alginate gel, again measuring four-day growth

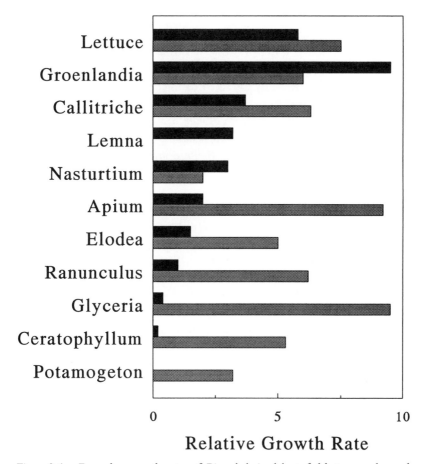

Figure 3.1. Four-day growth rates of *Biomphalaria glabrata* fed lettuce and samples from 10 aquatic macrophytes. Snails were offered leaf disks (Exp. A) or calcium alginate homogenates (Exp. B). Growth rate is expressed as (100×) the difference between ending weight and starting weight, as a proportion of (4×) starting weight. (Data of Thomas 1987.)

(experiment B). Although most macrophytes seem much more difficult to ingest than lettuce, they seem to have excellent nutritional value to *Biomphalaria*, if their cell walls and other defences are disrupted. One wonders how successful snails in experiment A would have been ingesting completely undisrupted macrophyte leaves, without freshly cut edges. Thomas suspects that under normal circumstances, *Biomphalaria* prefer epiphytic algae or dead plant tissue over the tougher living macrophyte.

As previously mentioned, much of what is known about chemorecep-tion in freshwater gastropods has been derived from experiments on the behaviour of *Biomphalaria* when presented with various plant extracts. One cannot be certain that a snail will, in fact, eat a plant to which it seems to be attracted, much less that it can assimilate the material and grow. But such evidence is suggestive. Bousfield (1979) showed that *Biomphalaria* was significantly attracted to *Groenlandia densa*, *Potamogeton crispus*, *Callitriche obtusangula*, *Ranunculus trichophyllus* and *Elodea canaden-sis* in circular arenas. Extracts from *Apium nodiflorum* and *Rorippa nastur-tium-aquaticum* actually repelled *Biomphalaria*, presumably with some sort of chemical defence. *Biomphalaria*, *Bulinus*, and *Helisoma* may differ in their attraction to different food types (Madsen 1992), a subject to which we shall return in Chapter 9.

Large data sets regarding the response of snails to varying chemicals have been gathered by J. D. Thomas and his colleagues using 'diffusion olfactometers', batteries of small rectangular central chambers with two end chambers, circular in cross-section. A filter paper soaked in some test substance is placed in one chamber, a control in the other. Individual snails are then introduced into the central chambers and observed at intervals. A substance is deemed to be an 'arrestant' if snails tend to rest on the test disc itself, and an 'attractant' if snails tend to spend more time on the test side generally.

Eleven different aquatic macrophytes were homogenized and used in laboratory olfactometer experiments by Sterry et al. (1983). They found that *Lemna paucicostata* and *Apium nodiflorum* strongly attracted *Biomphalaria*, and that *Callitriche obtusangulata* and *Oenanthe aquatica* were mildly attractive. Of the remaining macrophytes, four (including *Nasturtium officionale*) were strongly repelling and three (including *Elodea canadensis*) did not elicit any response. One could have hoped for better agreement among the results of these studies. Sterry and colleagues found a known foodstuff, *Nasturtium*, repellant to *Biomphalaria*. Bousfield found *Apium* strongly repellant, while Sterry found the same species strongly attractive. Foliage age, stage, and previous challenge by herbivores prob-ably all contribute to the attractiveness of any sample of macrophyte tissue.

As early as 1959, Pimentel and White observed that laboratory cul-tures of *Biomphalaria* apparently could not eat the pondweed *Najas* when fresh, but would eat the leaves if they were decaying. Thus it is interest-ing that Sterry and colleagues found that their snails were even more attracted to decaying *Lemna paucicostata* than to fresh. And in general, the molecules that attract *Biomphalaria* most strongly are those, such as maltose, more characteristic of decaying vegetation than living (Thomas 1986). Thus both Sterry *et al.* (1983) and Thomas (1986) concluded that the chemosensory apparatus of *Biomphalaria* may be best interpreted as an adaptation to detritivory.

There is also some evidence that adults and juveniles of individual species may ingest different resources. Thomas and colleagues (1985) placed 15 adult *Biomphalaria* and 15 juveniles in a single 6-litre aquarium furnished with macrophytes, green algae, diatoms, and detritus in such a way as to mimic the natural environment. Gut contents were analysed several days later, both by number of individual snails containing each category and by mean number of cells using a Sedgewick–Rafter counter. They found that juveniles seemed to rely more heavily on detritus and the small diatom *Achnanthes*, rarely ingesting living macrophyte tissue or certain larger diatom taxa. But here it is hard to distinguish between selective ingestion, as might be ascribed to the difference in radular size, and differential microhabitat choice in the aquarium.

The ultimate challege is to culture an organism axenically, in a chemically defined medium. Chernin (1957) described a technique to obtain biologically sterile *Biomphalaria* by washing individual embryos with sodium hypochlorite. Young snails so obtained have been reared to adulthood on a diet of autoclaved brewer's yeast and formalin-killed *E. coli* (Chernin and Schork 1959), and through several generations on a formula of powdered whole milk, wheat germ, dried lettuce, salt, starch, and yeast extract (Vieira 1967). Vieira demonstrated in the process that vitamin E was necessary for reproduction. It is interesting to note that a microbial 'gut fauna' is not apparently required, at least on these diets, nor are sand grains in the gut. It would be interesting from an ecological standpoint to see if these results could be repeated on a variety of more natural diets.

Laboratory studies on the diet of medically important planorbids are interesting, but a critical part of the puzzle is generally missing – the actual availability of these myriad potential foodstuffs in the wild. For this reason, among others, the work of Calow (1973b, 1974, 1975, Calow and Fletcher 1972) on the refreshingly unimportant *Planorbis contortus* has taken considerable prominence. As clearly as Chernin and Vieira have demonstrated that *Biomphalaria* do not require bacteria, Calow has demonstrated that bacteria constitute the normal diet of *P. contortus*.

Setting aside other habitats, Calow noticed that on the rocky shores of

Malham Tarn in Yorkshire, England, *P. contortus* was limited to the areas 'devoid of algae but which are covered with a thin chocolate-brown film of detritus.' Gut content analysis confirmed that the snail generally ingested detritus, apparently eschewing algae and fungi. In a series of feeding-choice experiments, Calow showed that *P. contortus* preferred detritus over algae, and more specifically detritus infected with live bacteria over sterilized detritus. He even found some evidence of preferences among bacterial types. Using an interesting radio-labelling technique, Calow showed about 75% assimilation efficiency for two preferred types of bacteria, 10–20% efficiency for two algal foodstuffs, and only 1–5% assimilation efficiency for lignin and cellulose. This is as strong as evidence on gastropod diet gets.

An interesting postscript to these studies was added by Polunin (1982). He showed that the aerobic microbial decomposition of reed litter was markedly accelerated by the presence of *Planorbis carinatus*. Two mechanisms seemed to be involved: some sort of physical agitation or stirring and the enhancement of nutrient turnover. Thus it would appear that *Planorbis* may, in a sense, 'farm' its own bacteria.

Physidae

Although unimportant from a medical standpoint, snails of this family are very common and widespread in the parts of the world most frequented by freshwater biologists. Thus there is little information on physid culture, but a fairly rich literature on their diet and habitat in the wild.

Crowl and Schnell (1990) reported that Oklahoma populations of *Physa virgata* reached maximum size and abundance in the reaches of streams with reduced flow and fine substrate. Bickel (1965) accomplished the difficult trick of sampling quantitatively (or at least semi-quantitatively) three different substrates over one year in the Ohio River near Louisville, Kentucky: mud bottom, macrophytes, and wood/dock. *Physa integra*, by far the most common snail, was found almost entirely on plants and other solid substrate in the summer, strongly preferring *Potamogeton pectinatus*. Snails may have been excluded from the bottom by low dissolved oxygen. As the macrophytes died in the autumn and winter (and levels of dissolved oxygen increased), *Physa* appeared on the mud. But Bickel suggested that spring floods would tend to carry such individuals away, so that the population may be maintained by individuals overwintering on docks and submerged wood.

The fondness of *P. integra* for solid substrate was confirmed by Clampitt (1970) in an interesting comparative study of several adjoining north Iowa lakes and ponds. Clampitt surveyed the region in a thorough but non-quantitative fashion. He observed that *P. integra* was found in lakes on solid substrates, either on rocks nearshore or on offshore vegetation. *Physa gyrina* was never found offshore, in contrast, but was common in the shallows on mud and sand bottom as well as rock. Unlike *P. integra*, *P. gyrina* was also often found in nearby shallow ponds. *Aplexa hypnorum* is similar in some respects to *P. gyrina* in that it commonly inhabits shallow ponds and ditches subject to seasonal drying. Den Hertog (1963) surveyed 29 Dutch populations of *A. hypnorum*, finding a striking positive relationship with certain sandy- clay soil types. The adaptation of populations of such species as *P. gyrina* and *A. hypnorum* to marginal and transient habitats will comprise a major theme of discussion in Chapter 4.

Clampitt (1970) qualitatively analysed the gut contents of individuals from both *P. integra* and *P. gyrina* populations collected on rocky shores where they co-occurred. No interspecific differences were apparent. Detritus was always the most common item in the guts of *Physa*, followed by algae of all sorts, and some sort of gelatinous material (mucus?), with animal and vascular plant remains of minor importance. The only striking differences were between lakes, not between species: a preponderance of diatoms in the gut of snails from one lake, colonial blue-green algae from another. Among the animal remains were rotifers, ostracods, amphipods, dipteran larvae and the chaetae of oligochaetes.

Nor was any difference apparent between *P. gyrina* and *P. integra* in food-choice experiments. Clampitt found that both species accumulated on algae-covered stones or dead *Carex* stems rather than on clean stone controls, although there was no apparent preference between algae and dead plant material. (Clampitt reared his laboratory populations on a mixed diet of lettuce and dead maple leaves.) Neither *Physa* species was attracted to living *Ceratophyllum demersum* or *Myriophyllum exalbescens*, two common offshore macrophytes upon which *Physa* seemed to accumulate. This agrees well with the rarity of vascular plant remains in the guts of wild-caught snails.

Clampitt also performed several laboratory studies of comparative physiology. He concluded that the difference in the habitats of the two species is due to a higher temperature tolerance and drought tolerance of *P. gyrina*, which allow it to colonize small ponds, and a lower metabolic rate of *P. integra*, which allows it to survive in deeper water without surface breathing. I would speculate that their dietary overlap makes competition quite likely on those occasions where they co-occur.

Although not typically found in lotic environments, *Physa gyrina* was the primary grazer in a set of laboratory streams established by Kehde and Wilhm (1972). Only six genera of algae developed in the streams, including greens and blue-greens but apparently no diatoms, possibly due to artificially high temperatures. In any case, Kehde and Wilhm found that *Physa* reduced the standing crop of algae, without noticably affecting productivity or diversity. Patrick (1970) mentioned a study showing that *Physa heterostropha* will feed on most species of diatoms, seeming to have difficulty only with the tightly cemented *Cocconeis*. *Physa* sp. was mentioned by Hambrook and Sheath (1987) in their list of freshwater grazers known to ingest red algae. Coffman (1971) examined the guts of 75 macroinvertebrates inhabiting a small Pennsylvania woodland stream, recognizing three categories of contents: animal remains, detritus, and algae. Expressed in calories (an unusual approach that has much to recommend it), '*Physa* sp.' was a herbivore, with algae comprising about 75% of its gut content, even in the autumn, when detritus would be expected to be plentiful.

Carman and Guckert (1994) made an effort to assess the relative assimilation efficiencies displayed by *Physa virgata* grazing on microbiological and algal components of the periphyton. Their study was complicated by the fact that one of their radiotracers (^{3}H-amino acids) tended to adsorb abiotically. Carmen and Guckert calculated that *Physa* assimilated adsorbed organic carbon (such as their label) at 78% efficiency, bacterial carbon at 50% efficiency, and algal carbon at only 32% efficiency. This suggests substantial adaptation for detritivory in *Physa*.

Lowe and Hunter (1988) allowed algae to colonize glass microscope slides held in screened enclosures for 26 days. They then introduced *Physa integra* for 21 days at either two or ten snails per enclosure, holding one set of enclosures as controls. *Physa* reduced overall biovolume from a mean of 3.4×10^4 $\mu m^3/mm^2$ in the control to 1.8×10^3 at low density and 9.5×10^2 at high density. *Physa* also caused a significant reduction in diversity (Figure 3.2). Declines in relative abundance were suffered by six taxa of diatoms, six taxa of green algae, three taxa of blue-greens, and three taxa of golden-browns. The relative increases came to the green *Oedogonium*, which Lowe and Hunter attribute to a tightly adhering holdfast, and the prostrate diatom *Eunotia*. Swamikannu and Hoagland (1989) reported a similar effect from high grazing pressure by *Physa*, but found evidence that low-level grazing increases periphyton diversity.

Among the most remarkable studies of macrophytes and freshwater snails published to date has been that of Sheldon (1987) on *Physa gyrina*

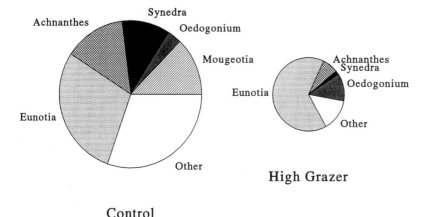

High Grazer

Control

Figure 3.2. Relative abundances of selected algal taxa on glass slides held in enclosures with ten *Physa* ('High Grazer') and no snails ('Control'). Algal biovolume on grazed slides was only 3% of that on the control. (Data of Lowe and Hunter 1988.)

in Christmas Lake, Minnesota. Her experimental plot was an extensive bed of submerged macrophytes on a seemingly uniform plain 5.7 m deep. She surveyed the area with SCUBA and found 14 common species of macrophytes, with from 6% to 13% of total leaf area lost to herbivores. The only herbivores identified by Sheldon were the snails: *Physa gyrina*, *Amnicola limosa*, *Promenetus exacuous*, and *Gyraulus sp.*, in order of abundance. (Recalling Clampitt's observations, one might wonder if the population of physids 5.7 m deep in a northern lake is indeed *P. gyrina*, but no matter.) She also reported that a number of predatory fish were common, including pumpkinseed sunfish, bluegill sunfish, black crappie, and northern pike, that all these were observed eating snails, and that shells were found in the stomachs of the fish. We shall return to the subject of predation in Chapter 7.

Sheldon performed an extensive series of tests to determine the dietary preferences of *Physa*, offering every pair of the 14 macrophyte species to individual snails in Petri dishes. She found a striking hierarchy of preferences, ranging from *Potamogeton richardsonii*, which was always preferred over all other offerings, to *Ceratophyllum* (with tough, needle-like leaves) which was never preferred. In general, Sheldon found that the fastest growing plants tended to be most favoured by the snails. She hypothesized that because of these preferences, an unchecked population of snails could rapidly reduce the diversity of macrophytic vegetation in Christmas Lake.

The heart of Sheldon's study was her exclosure experiment. She felt that any mesh small enough to include or exclude snails would cast an intolerable amount of shade. So she used a 5 mm mesh, quite passable by snails but too fine for their fish predators. Each cage was a 2 m tall cylinder, embedded in the bottom (at 5.7 m) and held upright with floats. The 12 experimental cages were completely meshed around, while the 12 control cages had large slices cut from the mesh to allow entrance for fish. After a five-month growing season, over five times more snails were found in the experimental exclosures than apparently inhabited the controls or the lake itself. Both macrophyte species richness and diversity seemed to be greatly reduced by the snails, from an average of 8–9 species per plot down to 2. The few species remaining in experimental exclosures tended to be those found least palatable to *Physa* in the laboratory choice experiments.

One might harbour the minor reservation that snails may not be the only herbivores in Christmas Lake. So to tie the whole project up with a red ribbon, Sheldon surveyed the snail biota and macrophyte diversity in eight other Minnesota lakes nearby. There was a fair amount of similarity in the species of plants encountered, while the snail communities varied considerably. Yet the negative correlation between macrophyte species richness and snail density was striking. Four lakes with snail densities over 65/kg of plant matter were all dominated by *Ceratophyllum*, while four lakes with low snail densities had higher macrophyte diversities, similar to that of Christmas Lake.

It may be recalled that *Ceratophyllum* was the macrophyte upon which Clampitt found his *Physa* populations accumulating in Iowa. There is an important general message here. Pulliam (1988) has pointed out the dangers of assuming that the environment is most favourable where population densities are greatest. In fact, the best habitats ('sources') may have low population densities, and the worst habitats ('sinks') high densities. Here we have been able to gather enough information to deduce that floating rafts of *Ceratophyllum* may be a 'sink' for *Physa*. Throughout most of this chapter (and elsewhere in the present work), such information is rarely available. We are forced to assume that population density and habitat favourability are positively related, keeping Pulliam's concerns at the fore.

Lymnaeidae

The richness of the literature on lymnaeid culture is due both to the medical and veterinary importance of lymnaeids and to an unexplainable interest of basic researchers in the biology of unimportant species. Noland and Carriker (1946) list 40 studies of lymnaeids involving at least some culture work, starting in 1854. Among the successful diets have been dried leaves, algae, and the higher plants *Chara*, *Elodea*, *Lemna*, and *Cabomba*. More recent workers with the *Fasciola* host *Lymnaea natalensis* have raised snails in shallow aquaria with mud bottoms, into which cultures of *Oscillatoria* have been innoculated (Kendall and Parfitt 1965). The algae serve as food for younger snails, until they have grown to a size at which an artificial diet can be substituted. An algal diet has also been found suitable for the culture of *L. truncatula* under similar conditions (Taylor and Mozley 1948, Kendall 1953). Boray (1964a,b) reported that cultures of lymnaeids, in this case *Lymnaea tormentosa* (the host of *Fasciola hepatica* in Australia), can grow and reproduce fed only filter paper or cellulose powder. Boray also reported snail growth on diets of softer plants, such as the watercress (*Nasturtium*) and duckweed (*Lemna*), but felt that most macrophytes presented cell walls too difficult for the snail to penetrate.

Among the most complete culture studies of freshwater snails of which I am aware is the work of Skoog (1978) on *L. peregra*. The author monitored 14-day shell growth on 24 juveniles fed four different diets, and 40-day growth and egg production on 10 individuals fed five diets. Skoog found diatoms to be best for juvenile growth, followed by mixed blue-green algae (dominated by *Calothrix*), then the green alga *Cladophora*. But mixed blue-green algae were significantly better than any other diet (diatoms, spinach, *Rivularia*, or *Cladophora*) for promoting 40-day growth in adult *L. peregra*. And the egg production promoted by spinach was significantly better than all others.

Bronmark (1985) presented laboratory evidence that grazing by *L. pereger* may stimulate growth in *Ceratophyllum*. He placed terminal, unbranched portions of the plant in 28 aquaria, half of which received two *L. pereger* and two *Planorbis planorbis*. After nine weeks the *Ceratophyllum* in the aquaria with snails had grown significantly longer than controls. Bronmark attributed this phenomenon to the removal of epiphytes by the snails, reducing shading and competition for nutrients. Olfactometer tests showed that *L. pereger* was significantly attracted to *Ceratophyllum* but could not distinguish grazed and ungrazed plants, and that epiphytes had no attraction by themselves. Bronmark made the surprising suggestion that macrophytes may excrete chemical cues as an adaptation to attract snails. But we have noted that *Ceratophyllum* is among the most unpalatable of all macrophytes. *Lymnaea pereger* can in fact grow and reproduce quite well on a diet of other sorts of macrophytes exclusively. For his study of the importance of mineral particles in *L. pereger* guts, Storey (1970) maintained healthy cultures over periods of months on a variety of rations. He found good long-term growth on the starwort *Callitriche*, unspecified coccoid algae, and even filter paper. Surprisingly, growth was not as good on a diet of the filamentous alga *Cladophora*. Storey noted some growth on a diet of bark and rotting leaves.

Before turning to more field-oriented studies, some laboratory evidence of bacterial feeding in lymnaeids should be noted. Underwood and Thomas (1990) placed growing shoots of previously grazed *Ceratophyllum* in beakers with either four *Planorbis planorbis* or four *L. pereger* (a total of eight replications) for 35 days. (It may be recalled that Calow, 1974, found excellent evidence of bacterial feeding in *P. contortus*, a snail quite similar to *P. planorbis*.) Underwood and Thomas sampled plant tissues at the beginning, middle, and end of the experiment, applied a stain, and examined for bacteria and algae. They observed that *L. pereger* made as large an impact on attached bacteria as did *Planorbis*. All of this effect could be due to mechanical disturbance, and the authors emphasized that the standing crops of epiphytic bacteria are low and unlikely to constitute a significant food source. Nevertheless, the possibility of bacterial feeding in *Lymnaea* is certainly not ruled out.

Turning now to observations in the field, Bovbjerg (1965, 1968, 1975) noted that the natural diets of four local species, *Lymnaea stagnalis*, *Stagnicola exilis*, *S. elodes*, and *S. reflexa*, most visibly included a variety of filamentous algae and macrophytes (for example, *Utricularia*, *Ranunculus*, *Lemna*, and *Potamogeton*). Gut contents contained the usual collection of 'organic debris', 'substratum materials', vascular plant materials, algal strands, blue-green cells, and miscellaneous small animal fragments. He placed 30 *S. reflexa* in a tank with alternating patches of the filamentous green *Spirogyra* and any one of several macrophytes, and found after 12 hours that the snails were always more common on the algae. Bovbjerg recorded an average of 80% of the snails on algae over the despised *Ceratophyllum*, 77% on algae over *Myriophyllum*, and 70% on algae over *Ranunculus*. But all other dietary preferences paled when compared with carrion. Prompted by an observation of several snails feeding on a dead insect, Bovbjerg performed a series of chemoreception experiments in a large Y-shaped chamber, 30 cm + 2(15) cm long. At these distances, none of the lymnaeids showed any significant ability to orient toward chopped *Potamogeton*, but 76–86% of the snails could find chopped crayfish.

Calow (1970) failed to find any higher plant material in the crop contents of 350 *L. pereger* collected from *Elodea*. Rather, he identified about 70% diatoms, 25% green filamentous algae, and 5% other algae, an almost exact match to the epiphytic flora of *Elodea*. Choice experiments showed that *L. pereger* was highly and equally attracted to *Elodea* extract and filamentous green algae, followed by normal *Elodea* (with attached epiphytes), then diatoms, then clean-scrubbed *Elodea*. (This result does not match that of Bronmark especially well.) The attraction of the *Elodea* extract may seem unexpected, but it may materialize that crushed vascular plant material is extremely strong but mildly attractive, while such material as diatom cells, although highly attractive, may emit almost no chemical cue.

The experiments of Lowe and Hunter (1988) on grazing by *Physa* were in some measure patterned on earlier experiments by Hunter (1980) involving *Lymnaea elodes*. The 1980 work was performed in a shallow Michigan lake where about 82% of the snails were *L. elodes*, the remainder being *Physa gyrina* and *Helisoma trivolvis*. Hunter suspended a series of enclosed slide racks in the lake, half of the enclosures without snails and half with *L. elodes* at natural proportions and densities. (There was no initial period for periphyton colonization as in the 1988 work.) Results were very similar, however: the snails reduced algal standing crop, productivity, and diversity, removing all filamentous algae and fungal hyphae, and nearly all diatoms, especially the larger ones. Remaining but in reduced numbers were smaller diatoms such as *Gomphonema*, and again, apparently unaffected by grazing was the adpressed diatom *Cocconeis*.

Cuker (1983a) also examined grazing in a population of *L. elodes*, this inhabiting a lake in arctic Alaska. He placed rocks covered with algae in 10 dark plastic tubs of lake water, maintaining all at lake temperature with a continuous bath. Cuker introduced snails at nine densities (keeping an ungrazed control) and sampled epilithic algae periodically for the month of July. Once again algal standing crop and primary productivity were severely reduced by the snails. In this particular algal flora, the diatoms and filaments were reduced by the grazing while the absolute densities of coccoid green and blue-green algae, said to have mucus sheaths, were unaffected. Using a radiolabelling technique, Cuker estimated that the snails actually assimilated 68% of algal carbon grazed, incorporating 25% into the tissues and respiring 43%.

Cuker (1983b) performed a second set of experiments to assess the impact of *L. elodes* on other grazers in his arctic lake, especially chironomids and microcrustaceans. Algal-coated rocks and snails at several densities were again placed in chambers, these suspended in the lake for one month. Tube-dwelling chironomids were reduced an order of magnitude at the highest snail densities, free-living scrapers reduced 85%, and predatory chironomids by 66%. Part of this effect may be due to mechanical disturbance by the snails, but competition for periphyton must also play a role. Among the other organisms, ostracods and cladocerans were reduced by high snail densities, while copepods, trichopterans, and mites were unaffected. This is the first occasion we have mentioned that the guild to which freshwater snails belong may not be exclusive. Others will follow in Chapter 9.

Knecht and Walter (1977) and Walter (1980) performed a thorough comparison of diet and habitat in two Lake Zurich lymnaeid populations, *L. auricularia* and *L. pereger*. Walter took 256 samples of 1 m² from the bottom at depths of 3–9 m using SCUBA and a suction device. If food was abundant, the snails tended to aggregate on stones, while if food was less abundant, they aggregated where the organic content of the mud was greatest. Walter's impression was that the total density of snails was more constant than the densities of the two species considered separately, implying that they replaced each other. The only minor difference between the species was that *L. pereger* was more common on stones.

Knecht and Walter analysed gut contents of individuals from the two species collected from submerged structures, comparing them with each other and with substrate samples. They detected no significant difference between the diets of *L. auricularia* and *L. pereger*. By volume, the 20 most common items identified from the guts (amounting to 97% combined) included 47% green algae (8 taxa), 4% diatoms (5 taxa), 3% blue-green algae (2 taxa), and a remarkable 43% protozoans (4 taxa of flagellates, 1 ciliate). There were significant differences between the gut contents of both snails and the substrate upon which they grazed. But the absence of any material differences in the diet or habitat of these two snail species suggested to the authors that food was not limiting and that population sizes were perhaps being regulated by fish predation.

We now turn from the lymnaeids of deep lakes to those spending the greatest portion of their lives in the open air, on mud. Two such amphibious lymnaeids of veterinary importance are common in New Zealand marshes, the native *Lymnaea tormentosa* and the North American *Lymnaea* (*Pseudosuccinea*) *columella*. Harris and Charleston (1977) took 489 quadrat samples from sites where one or the other snail occurred. (Interestingly, they do not seem to occur together.) They counted snails and collected a variety of environmental data, including proportion of substratum

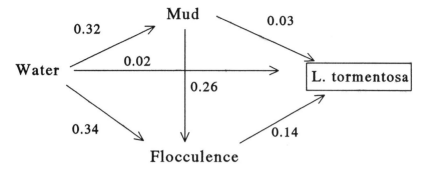

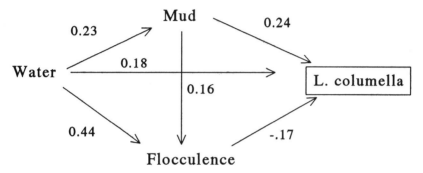

Figure 3.3. Path diagrams showing the hypothesized effects of quadrat proportion under water, quadrat proportion exposed mud, and flocculence on population density of *Lymnaea tormentosa* and *L. columella* in New Zealand. (Harris and Charleston 1977.)

under water (three categories: 0, 0–25%, >25%), proportion of exposed mud (same three categories), and flocculence (yes or no, measured by dropping a ruler to see if it stuck). Harris and Charleston introduced path analysis, a technique first developed for population genetics, as a method for analyzing these data. The technique is an extension of multiple regression, where a model of the relationships among variables is supplied and tested with 'path coefficients' based on regression coefficients. Path analysis is especially useful in cases such as we commonly encounter in ecology, where there is a good deal of intercorrelation among variables.

Figure 3.3 shows that Harris and Charleston proposed identical, completely connected models for the effects of the environment on *L. tormentosa* and *L. columella*. The authors would not venture to estimate the significance of the path coefficients, but a comparison is nonetheless valid. The coefficients show that the distribution of *L. tormentosa* seems to be positively influenced by flocculence and unaffected by exposed mud, while *L. columella* is influenced negatively by flocculence and positively by mud. I like this technique and would encourage further experimentation with it, applied to variables distributed in a more normal fashion.

Soil type is especially important to the amphibious lymnaeids of marshes, for reasons of refuge in times of cold or drought. Foster (1973) used an auger to collect 456 soil samples (348 cm³ each) from Benton County, Oregon. He sampled only in places that were wet or that seemed to have been inundated previous to his July and August survey. He placed the samples in water, examined them over a week's time, and observed *Lymnaea bulimoides* emerging. Foster located his sample stations on a U.S. Soil Conservation Service map to determine the soil 'series'. His samples included 16 of the 42 soil series recognized by the SCS in Benton County, but snails were found in only 5 series. Four of the series are described as 'dark brown silt loam' (over 50% silt, less than 20% clay, 1–9% sand, 0.4–0.6% fine gravel). The remaining series was also over 50% silt, but 20–30% clay, and the remainder sand. All were 'medium acid (pH 5.6–5.8), friable, slightly plastic, gritty and sticky' with 'moderately slow permeability, slow surface runoff and slight erosion hazard'. This is a remarkably vivid portrait of the environment to which a lymnaeid might repair to pass the heat of summer.

Lymnaeids occasionally inhabit lotic environments. In 1985, Eichenberger and colleagues recounted over six years of observations on vegetational succession in a large (250 m long, 45 cm deep) artificial river constructed in Switzerland. The populations of *Lymnaea pereger* and *Gammarus pulex* built up to such levels by years four and five as to eliminate the 'second vegetation cycle' (moss, *Vaucheria*, *Cladophora*, *Spirogyra*) and initiate a third stage: 'an epilithic crust composed of unicellular and very small filamentous blue-green and green algae and diatoms'. The artificial river was still in this 'third stage' at the close of the experiment, a stage which reminds me very much of what I generally see when collecting snails in riffle areas.

Acroloxidae and Ancylidae

Among the pioneering works of physiological ecology was that of K. Berg (1952, 1953) involving the European limpets *Ancylus fluviatilis* and *Acroloxus lacustris*. I quote Berg (1952) to set the stage. '*Ancylus fluviatilis* . . . occurs mainly in running water, it is rheophilous. It is nearly always found on a stony bottom, sometimes in a swift current. *Acroloxus lacus-*

tris is found in stagnant water – it is limnophilous. It is especially seen on stems of big aquatic plants.' Berg noted that aquatic insect larvae and several other types of macroinvertebrates inhabiting swift streams have generally been found to consume several times the oxygen of lentic water populations, that the oxygen consumption of lotic water macroinverte-brates usually falls much more dramatically in response to reduced oxygen tension, and that lotic macroinvertebrates are less tolerent of heat stress. Berg performed a series of experiments to see if such relationships held for the two limpets, perhaps explaining their different habitat choice.

They do not. The two limpets were much more similar in their res-piratory rates than workers with other aquatic invertebrates seem to find, and did not differ in their responses to thermal shock. In fact *A. fluviati-lis* held more constant oxygen consumption in the face of reducing dis-solved oxygen than *A. lacustris*, entirely the opposite of expectation. *A. fluviatilis* showed poorer survival under anaerobic conditions, in accor-dance with expectation, but this could hardly explain their field distribu-tion. Berg concluded 'in ecology it is dangerous to draw conclusions as to the requirements of the animals exclusively from the characteristics of the environment'.

An explanation for habitat choice in these two limpets came somewhat later from England's Lake Windermere, where they co-occur. Geldiay (1956) set out trays of rocks, without limpets but previously colonized by algae (I believe), to depths of 12 m. He observed some colonization by *A. fluviatilis* down to 5 m, where the water must be very calm. Thus *A. fluviatilis* is not really 'rheophilous' as much as 'lithophilous'. Geldiay observed no limpet colonization at 12 m, and a population he trans-planted to that depth died, suggesting that light was insufficient to support algae. He never reported any *A. lacustris* in his rock trays, no matter how calm. That species was restricted to the stems of *Schoenoplectus, Typha, Phragmites*, and water lily leaves. Thus *A. lacustris*, too, is not so much affected by current as by substrate, as we saw in Clampitt's *Physa* studies and perhaps also in Walter's data on *Lymnaea*.

Figure 3.4 shows the results of Geldiay's comparison of the algae in 30 *A. fluviatilis* stomachs with their substrate. He also noted that large volumes of sand were found in the guts, but does not mention detritus or other dietary components. The figure shows broad agreement between gut and substrate, although the large diatom *Synedra* does not seems to have been collected from the substrate, nor was the filamentous *Ulothrix* found in the guts. Geldiay concluded that the limpets browse

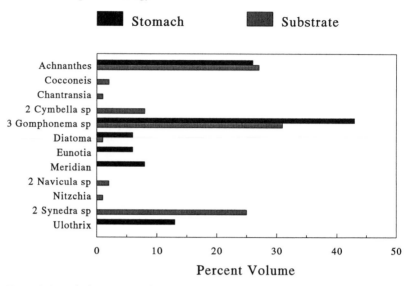

Figure 3.4. Algal contents of 30 individual *Ancylus fluviatilis* guts, compared with the Lake Windermere substrate from which they were collected. (Data of Geldiay 1956.)

'indiscriminately on the algal felt that covers the rocks and stones where they live', although to my eye, some of the differences look significant.

The North American ancylid *Ferrissia rivularis*, very similar ecologically to *A. fluviatilis*, was among the most herbivorous of all 75 macroinvertebrates in Coffman's (1971) Pennsylvania stream study. Less than 10% of the calories in its gut contents were due to detritus or animal material. Streit (1975) found that *A. fluviatilis* assimilated the diatom *Nitzschia* and the green *Scenedesmus* equally well, on the order of 40–60% efficiency. Interestingly, Streit found that *A. fluviatilis* could not assimilate the blue-green *Anabaena*, a result that dovetails well with the findings of Calow (1973b).

One of the nice aspects of Calow's (1973b) work on the diet of *Planorbis contortus* (previously discussed) was that he simultaneously examined *A. fluviatilis* using similar techniques. For example, as Calow observed that *P. contortus* was limited to stone surfaces with detritus but without algae, he noticed that *A. fluviatilis* occurred only 'on the stone sides where there is an abundance of epilithic algae'. Algae were by far the most common items in the gut content of Malham Tarn limpets, followed by detritus, with fungi a distant third. In the same chambers where *P. contortus* aggregated on detritus-covered filter paper and ignored algae,

A. fluviatilis aggregated on algae and ignored detritus. Previous work by another author (and his personal observations) led Calow to offer the limpet a choice between algae and the lichen *Verrucaria*. *Ancylus* did eat the lichen, but still preferred algae.

In other studies, Calow (1973a) concentrated only upon the algal component of the diet of *A. fluviatilis*. He compared the gut contents of limpets from Malham Tarn with the algal flora of grazed stones, finding broad agreement while noting some evidence of selection for diatoms and against blue-green algae. He then performed a series of food choice tests, discovering significant differences in every comparison. *Ancylus* prefers mixed diets of diatoms over all other offerings. The next most preferred is the filamentous green algae (*Cladophora*), then unicellular greens (*Scenedesmus*), and finally the blue-greens (*Rivularia*). Among the diatoms, Calow found a significant preference for *Gomphonema* over *Navicula* and *Achnanthes*. One might speculate that such strong preference among the herbivores might affect the algal flora of Malham Tarn. But Calow found little evidence of this using the (admittedly weak) method of comparing stones with snails with stones without snails. Among the complicating variables here are the population size of *A. fluviatilis*, which may not be large enough to produce an effect on the Malham Tarn flora, and Calow's observation that the limpets generally manifest dietary preferences best when satiated, a rare condition in the wild.

Pulmonate reproduction

Outcrossing

The freshwater pulmonates are hermaphroditic, both sperm and egg being manufactured in a single regionally differentiated gonad, the ovotestis. Spermatogenesis seems to precede oogenesis by at least a few weeks in *Physa fontinalis* (Duncan 1959), *Lymnaea stagnalis* (Duncan 1975), *Laevapex fuscus* (Russell-Hunter and McMahon 1976), *Bulinus globosus* (Rudolph 1983) and *Biomphalaria glabrata* (Trigwell and Dussart 1998). In contrast, Richards (1962) reported that female fertility precedes male fertility in *Gyraulus parvus*. Judging by reproductive function (rather than histological section), Wethington and Dillon (1993) reported that male maturity preceded female maturity in 46 of 50 *Physa heterostropha*, female preceded male in one case, and three individuals matured in both functions during the same seven-day period.

The genital systems of the several families have been described, compared and figured by Duncan (1960, 1975), Geraerts and Joosse (1984), and Visser (1981, 1988). Eggs and sperm are passed together through a narrow 'hermaphrodite duct', with a 'seminal vesicle' region, where the sperm may mature, be stored, or ultimately degraded (Hodgson 1992). Egg and sperm then enter a rather elaborate region, usually called the 'carrefour'. Opening into the carrefour are fertilization pocket(s), an albumen gland, a female duct and a male duct. Most freshwater pulmonates rarely self-fertilize. They seem to have mechanisms that favour fertilization by allosperm (donated by a partner) over autosperm (produced endogenously). But the physiological and/or anatomical adaptations of the hermaphrodite duct and carrefour to separate eggs from autosperm in inseminated individuals are at present poorly understood. The male and female reproductive systems are separate from the carrefour onward in all freshwater pulmonates except the Acroloxidae.

Autosperm is generally transported through the male duct into the vas deferens, along a prostate gland, to the penis. The penis is held in a sheath behind the head and, at mating, everted through a sack called the praeputium. A muscular 'preputial organ' may serve as a holdfast during copulation. Pulmonate copulation has been well described (Rudolph 1979a,b, van Duivenboden and ter Maat 1988, Wethington and Dillon 1996, Trigwell et al. 1997). There is no evidence of 'courtship' behaviour; snails seeking to mate as male do not seem to behave in any special manner to increase the receptivity of a prospective partner. Nor is there evidence that pulmonates have any behaviours or chemical cues to induce insemination from a prospective partner who might otherwise be unwilling. (It might seem that inductive behaviours would be especially adaptive for snails whose reproduction has been delayed for want of an allosperm donor, but such is not the case.) Rather, the likelihood that mating is initiated between a pair of snails encountering each other is a function of the maximum autosperm store (sometimes called the 'masculinity') of either (Figure 3.5). The energetic cost of mating as a male is not negligible (DeVisser et al. 1994). So if both snails have recently mated as males, they generally pass each other by, even though neither may have ever mated as female.

A snail carrying autosperm in storage (a 'male'), upon encountering a second individual, typically mounts the shell of that individual (the 'female') and crawls across her shell to orient himself in parallel fashion. The male positions himself along the edge of the female's shell aperture, everts his penis, and inserts it through the female's vagina a considerable distance into the oviduct. Sperm may apparently be conducted all the

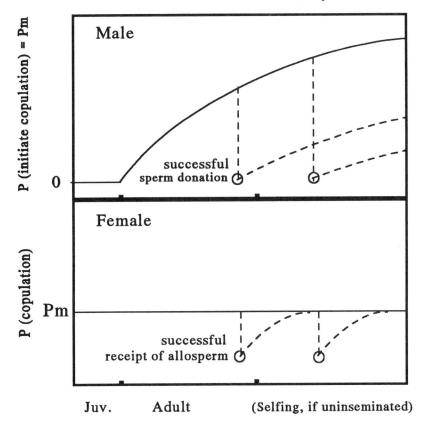

Figure 3.5. Model showing the probability of mating in pulmonate snails as a function of life history stage. The abscissa is divided into juvenile, reproductively mature but not self-fertilizing adult, and (only if uninseminated) self-fertilizing adult stages. Solid lines trace mating probabilities if no prior copulation has taken place, as male and as female separately. Note that the initial likelihood of female copulation is Pm, the likelihood that a male initiates copulation. Dashed lines show mating probabilities subsequent to a successful copulation. Based on studies with *Physa heterostropha* (Wethington and Dillon 1996).

way to the carrefour region during copulation, which typically requires 15–30 minutes, but which may go on much longer. The male also deposits a rather soft and temporary 'copulatory plug' in the female's oviduct. This 'plug' seems to constitute little barrier to reinsemination, however, and may have other functions.

But copulation is by no means assured to the snail who mounts a partner's shell and attempts to evert his penis. Snails mounted as females can display a great variety of rejective behaviours, including shell shaking or jerking to dislodge the prospective suitor, clamping the shell tightly to the substrate, lifting the shell high from the gonopore, expanding the foot to fill the aperture, or butting (Rudolph 1979b, DeWitt 1991). Suitors smaller than the snail they are mounting are more likely to gain a successful copulation (DeWitt 1995). Snails that have been inseminated recently are more likely to display rejective behaviour (Figure 3.5). And interestingly, snails who have not themselves mated recently seem to display increased frequency of rejective behaviours when mounted as female, even though they may carry uninseminated eggs. Wethington and Dillon (1996) have interpreted such behaviour as 'gender conflict' – fighting among hermaphrodites to assume the male role. Although manifesting itself in the same sort of shell-shaking and struggling that characterizes female rejection of unwanted males (usually termed 'sexual conflict'), gender conflict has different origins.

Pulmonate snails, although simultaneously hermaphroditic, generally cannot mate with a single partner in both roles simultaneously (*Helisoma* appears to be an exception (Abdel-Malek 1952, Pace 1971) and very rarely *Biomphalaria* (Trigwell *et al.* 1997)). At least initially, in most cases, one snail must serve as the male and the other the female. Chain copulation, involving many individuals mating in both roles simultaneously, has been commonly observed, especially in dense populations warmed in the spring. But in ordinary circumstances, if (only) one snail of a pair carries substantial autosperm stores, that snail (the 'male') will mount the female and attempt mating. And to the extent that he is not put off by rejective behaviours arising from sexual conflict, a unilateral mating will result. If both snails in the encounter have substantial autosperm stores, gender conflict results, as both vie to assume the male role. After one individual prevails, and insemination has taken place, the (previously male-acting) individual loses his tendency to mate as male. No further gender conflict exists, and the previously female-acting individual generally gains copulation as male. Bilateral mating, or 'swapping', is the result.

Some of the allosperm received from a partner in copulation may be diverted (hydrolysed and absorbed?) in the recipient's bursa copulatrix. A fraction is transported through the female reproductive system (the oviduct) to the fertilization pocket(s) mentioned previously, and to the seminal vesicles. It is not entirely clear that fertilization in fact occurs in fertilization pockets; either the fertilization pocket or the seminal vesicles may be sites for fertilization and/or sperm storage.

The sperm storage capacity of freshwater pulmonates seems to be quite large. Using genetic markers of various sorts (albinism, allozymes), workers have typically reported storage of allosperm for 60–100 days, and several thousand fertilizations (Cain 1956, Richards 1973, Rudolph 1983, Rollinson and Wright 1984, Rollinson 1986, Vianey-Liaud et al. 1987). Rudolph and Bailey (1985) demonstrated that *Bulinus* can store sperm over seven weeks of starvation, eight weeks of low temperature, or four weeks of desiccation. Evidence for multiple paternity has been gathered on many occasions (Mulvey and Vrijenhoek 1981b, Rollinson et al. 1989, Wethington and Dillon 1991). The subject of sperm competition in molluscs generally has been reviewed by Baur (1998). There is one report (Monteiro et al. 1984) that *Biomphalaria* may use stored allosperm (received from a previous partner) rather than its own autosperm to inseminate a partner. This phenomenon, termed 'sperm sharing', was not confirmed in *Bulinus* (Rollinson et al. 1989).

Fertilized eggs are enclosed in perivitelline fluid from the albumen gland and coated by a perivitelline membrane. The term 'albumen gland' may be a bit misleading, because the perivitelline fluid is primarily galactogen, with smaller amounts of proteins, calcium, and other components. Eggs are then passed down through the oviduct where they are encapsulated with jelly-like mucopolysaccharides and mucoproteins, from various secretory tissues. Planorbids and ancylids typically lay eggs in rather tough, flat, ovoid masses, while the spawn of physids and lymnaeids is more loosely packed, irregular, and convex. Egg laying is generally nocturnal and may be cued by changes in diet, temperature, and water quality (Van der Steen 1967). Egg masses may be cemented to any sort of firm substrate, typically near the water's surface. The amphibious *Lymnaea truncatula* oviposits on moist soil. The process of oviposition itself seems to require only a few minutes, but it may be preceded by some surface preparation and followed by some anchorage behaviour (Rudolph and White 1979). The number and size of the eggs, the timing of the reproductive cycle, and various other aspects of life history will be discussed in Chapter 4.

Selfing

All freshwater pulmonates seem to be able to reproduce successfully by self- fertilization. Usually there are adverse fitness consequences, both to the selfing parent directly (termed self-fertilization depression) and to the selfed offspring (inbreeding depression, in its strict sense.) But as pointed out by Doums et al. (1994), early workers (who tended to report an advantage for selfing) biased their estimates by comparing the fitness of isolated snails with the mean fitness of mass-cultured (and almost certainly crowded) outcrossers. DeWitt and Sloan (1958, 1959) were the first to compare the fitness of isolates with uncrowded pairs, showing clear overall advantages to outcrossing in *Pseudosuccinea columella* and *Physa heterostropha*. Reductions in both the lifetime fecundity and the offspring viability of isolated snails have been documented in populations of *Lymnaea peregra* from Lake Geneva (Jarne and Delay 1990), *Bulinus globosus* from Niger (Jarne *et al*. 1991, Njiokou *et al*. 1992), *Physa heterostropha* from South Carolina (Wethington and Dillon 1997) and *Lymnaea peregra* from France (Coutellec-Vreto *et al*. 1998).

Copulation certainly does, however, have some negative effects on fitness (van Duivenboden *et al*. 1985). It interferes with egg laying, is energetically expensive, and exposes the copulants to predation. Njiokou and his colleagues (1992) reported that *B. globosus* from an Ivory Coast population showed improved fertility and offspring survivorship when reared in isolation. And in fact, that population does seem to reproduce at least partly by self-fertilization in nature, to judge from the poor fits to Hardy–Weinberg expectation observed at allozyme loci. Jarne (1995) has reviewed allozyme data collected from 24 studies (16 species, 260 populations) of pulmonates worldwide. Self-fertilization seems to be the 'predominant' mating system in *Biomphalaria pfeifferi* (2 studies), *Bulinus forskalii* (2 studies), *Bulinus senegalensis* (1 study), and *Bulinus truncatus* (3 studies). To this list could be added *Ancylus fluviatilis* (Stadler *et al*. 1995). *Bulinus truncatus* and *A. fluviatilis* are allotetraploids, although the chromosomal situation is not abnormal in the other species, as far as is known.

Bulinus truncatus populations often contain aphallic individuals, lacking the terminal regions of their male reproductive system. Phally polymorphism seems to have both genetic and environmental components. Development in a warm climate tends to promote aphally (Schrag and Read 1992, Schrag *et al*. 1994). But the breeding experiments of Doums *et al*. (1996a) demonstrated significant correlations in aphally both between parent and offspring and within sibships. Aphally may be an adaptation to lessen the consequences of trematode parasitism (Schrag and Rollinson 1994). Regardless of the origin of the phenomenon, it seems clear that self-fertilization should be more likely in populations where aphally occurs in high frequency. And in fact, inbreeding depression in *B. truncatus* generally seems negligible (Doums *et al*. 1996b).

Jarne (1995) characterized 11 of the 16 pulmonate species in his tabulation 'predominantly' cross-fertilizing. This does not mean that the fits to Hardy–Weinberg expectation have been uniformly good, even in this subset. In fact, heterozygote deficits are very commonly reported even among outcrossing species. One likely factor must certainly be the incidence of outcross sterility (but self-fertility) noted in *P. heterostropha* by Wethington and Dillon (1993). Our large laboratory populations of *P. heterostropha* showed roughly 6% male outcross sterility, 2% female outcross sterility, 2% double-outcross sterility, and 19% self-sterility. Another factor is likely to be 'mixed mating', the production of selfed progeny along with outcrossed progeny that has occasionally been detected even in snails demonstrably holding stores of allosperm. Wethington and Dillon (1997) monitored the lifetime reproductive output of 26 *P. heterostropha* known to have been inseminated, and characterized 10 as 'mixed-maters', producing offspring by self-fertilization at a frequency greater than 1%. The absence of any fitness differences between the mixed maters and the more purely outcrossing individuals suggested that the phenomenon may be accidental, resulting from imperfect suppression of autosperm.

Prosobranch diet and habitat

Once again I feel moved to introduce a section by quoting Boycott (1936), possibly because this practice allows me to generalize with relative impunity. 'The 10 operculates (of England) live almost exclusively in running water; being gill breathers and unable to come to the surface and gulp in air as the pulmonates do, they presumably need water which is fairly well oxygenated and also free from particles which might choke their gills.' One could immediately note a number of prominent exceptions, such as the numerous lentic water taxa in the superfamily Viviparoidea and the amphibious groups in the superfamily Rissoidea. But again, if we understand the nature of generalizations, I consider Boycott's statement valuable.

Neritidae

Work on north European *Theodoxus fluviatilis* provides a useful comparison with the pulmonates. Skoog (1978) performed a series of laboratory growth studies with Norwegian *Theodoxus* along with the *Lymnaea peregra* work we have discussed previously. His experimental design was similar: 10 adults per food item, maintained individually in filtered water, with food and water refreshed every other day. Skoog carried his *Theodoxus* experiment 80 days, rather than the 40 days of data he gathered for adult *Lymnaea*. *Lymnaea* grew much faster, averaging 11.7 μm of shell growth per day on their preferred diet, as opposed to a maximum of 0.09 μm per day for *Theodoxus*. *Lymnaea* also generally laid eggs in culture, while *Theodoxus* did not, but there was some mortality among the former, while no *Theodoxus* died over a period twice as long.

Skoog observed some growth on all four diets offered *Theodoxus*. A mixed culture of diatoms gave the best growth, followed by mixed blue-green algae, *Cladophora* with diatoms, and an animal diet (crushed *Jaera*). But the differences in growth rate were not significant. (It may be recalled that mixed blue-green algae was significantly better than any other diet for adult *Lymnaea* growth.) Skoog noted that his results agreed well with those of Neumann (1961), who concluded that *Theodoxus* is not selective. *Theodoxus* seems in fact to have even less dietary specificity than an extreme generalist among the pulmonates. In a second set of experiments, however, Skoog found that the animal diet was significantly worse in promoting juvenile growth in *Theodoxus*.

One much more commonly encounters studies performed in lotic waters when surveying the prosobranch literature. Jacoby (1985) anchored 12 aluminium-framed chambers in the rocky outfall of Swedish Lake Erken, using vaseline-coated plastic strips as snail barriers. She placed 50 uniformly sized rocks from a local quarry in each, and allowed two weeks for colonization by periphyton. She then allowed *Theodoxus* to graze at 5 per chamber, 40 per chamber (about their natural density of $450/m^2$) and 70 per chamber. After another 2 weeks, Jacoby found significantly lower periphyton biomass (as chlorophyll a) in the low density chambers, and much greater abundances of the green filamentous *Cladophora*. *Theodoxus* guts contained 65% diatoms, 30% detritus and bacteria, and 5% algal filaments, by volume. We shall have more to say about prosobranch grazing in riffles in the next section.

Pleuroceridae

Pleurocerids are common and conspicuous in quite a few of the little North American streams that freshwater benthic ecologists have found convenient for study. For snails in such an environment, current and substrate are the critical habitat variables. In eddys and pools, one may find snails on any substrate where food is available, but in moving water, clearly snails will require a solid substrate upon which to hold. Houp (1970) mentioned this phenomenon in a population of *Pleurocera acuta* from a small Kentucky stream, and Krieger and Burbanck (1976) make similar observations in the *Goniobasis suturalis* population of a larger Georgia river. *Goniobasis livescens* seems to require a solid substrate in exposed regions of Lake Erie, but may be found on softer substrates where protected (Krecker 1924).

Big Sandy Creek, Alabama, is typical of many streams in the piedmont and coastal regions of the United States where pleurocerids find solid substrate less than common. The creek is inhabited by a large population of *Goniobasis cahawbensis*. As it passes Big Sandy Spring, a population of *G. carinifera* is also introduced, and the two species co-occur for about 1 km before *G. carinifera* disappears. The temperature and dissolved oxygen are much more constant in the spring water; temperatures are generally cooler and dissolved oxygen generally lower. Thus it seemed possible to Hawkins and Ultsch (1979) and Ross and Ultsch (1980) that the distribution of these two pleurocerids might have a physiological basis.

They measured oxygen consumption at several temperatures, 'heat coma temperature', 'critical thermal maximum', 'maximal exposure temperature', and 'ultimate lethal temperature' for upstream *G. cahawbensis*, downstream *G. cahawbensis*, and *G. carinifera*. The results were quite reminiscent of those obtained by Berg using limpets 20 years earlier: very few differences. *G. carinifera* had lower lethal temperatures, but these were still much in excess of field temperatures. Temperature preference tests showed that both species seemed to prefer water somewhat warmer than ambient in the wild. The authors therefore turned their attention away from physiology and toward depth, current, and substrate.

Ross and Ultsch sampled 18 transects from 2.5 km above the spring to 3.3 km downstream, recording snail occurrence in two categories of depth, two categories of current, and 17 substrate types. The data only indirectly address the question of *G. carinifera's* restricted range, and the absolute preferences of the two species cannot be determined. But the relative preferences of *G. cahawbensis* and *G. carinifera* in sympatry are summarized in Figure 3.6. The spring-dwelling *G. carinifera* significantly prefers water shallower and faster than *G. cahawbensis*, and hence is found hanging onto logs and vegetation at streamside significantly more often. *Goniobasis cahawbensis* was less common in the shallow rapids, but seemed to prefer the other three habitat types, shallow/slow, deep/slow, and deep/fast, equally well. (The authors measured current speed at the surface, so the 'deep/fast' category is problematic.) *Goniobasis cahawbensis* is more common on sand. The authors speculated that feeding prefer-ences may play a role in the differing habitat choice. I would add that if Big Sandy Creek becomes progressively slower and sandier downstream, such habitat choice might explain the disappearance of *G. carinifera* 1 km from the spring.

Depth and Flow

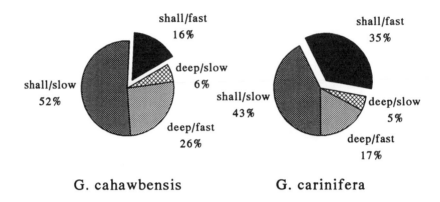

G. cahawbensis   G. carinifera

Substrate

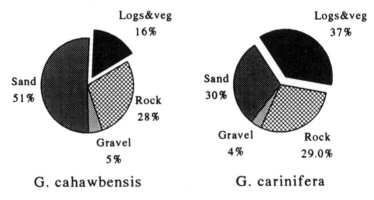

G. cahawbensis   G. carinifera

Figure 3.6. Habitat selection in two species of *Goniobasis* sympatric in Big Sandy Creek, Alabama. (Data of Ross and Ultsch 1980.)

Richardson and Scheiring (1994) performed a study similar in many respects to that of Ross and Ultsch. They confirmed that G. *cahawbensis* seems to be a generalist with regard to substrate and flow. The second species inhabiting Little Schultz Creek, Alabama, was *Goniobasis clara*, apparently found in greater proportion in riffles.

Johnson and Brown (1997) reported that a population of G. *semicarin-*

ata showed maximum adult abundance in slow-flowing, sunny regions of a Kentucky stream. Hawkins and colleagues (1982) found that the distribution and abundance of the pleurocerid *Juga plicifera* in several Oregon streams seemed to be influenced by at least four factors: gradient, current, substrate, and canopy. Hawkins and Furnish (1987) performed a series of experiments in a pair of 50 m channels fed by a small stream. In the bottom they placed trays filled with one of four grades of stones (1, 2, 5, and 10 cm diameter), and overhead they arranged 0%, 75%, or 95% shading. Invertebrates were allowed to colonize for 18 weeks, then trays were removed and inventoried. Hawkins and Furnish found *Juga siliculata* significantly more common in the reaches with 0% shade, and significantly less common on substrate 1 cm in diameter.

Hawkins and Furnish had previously obtained field data suggesting that in shaded areas, where primary productivity may be reduced, negative correlations may exist between the abundance of *Juga* and that of both detritivorous and herbivorous macroinvertebrates. Thus the primary objective of this particular research was not to determine habitat preferences, but rather to assess the affect of *Juga* itself on the remainder of the macroinvertebrate community. So in one of the parallel channels they stocked *Juga* at natural densities, while from the other they attempted to exclude as much snail immigration as possible. Early in the experiment, attached algae (especially *Melosira*) was strikingly more common in the removal channel than in the control, with chlorophyll a 4–9 times higher. But by the end of the 18 weeks, there was no significant difference in chlorophyll between the two treatments. The differences were in the abundances of other macroinvertebrates. Six taxa of insect larvae were significantly more common in the channel from which *Juga* had been removed, primarily detritivorous and herbivorous chironomids. Here is yet more evidence that in some circumstances, competition may occur for even the lowliest particles. And snails are not the only animals involved.

Juga from the Pacific northwest has been a 'white rat' for many laboratory studies of periphyton grazing (Sumner and McIntire 1982, Gregory 1983, Lamberti *et al.* 1987, 1989, DeNicola *et al.* 1990, DeNicola and McIntire 1991). Such studies consistently find that snails reduce chlorophyll a and periphyton biomass markedly (in the range of 30–50%), but effects on primary productivity have been variable. There has been a little evidence from these studies suggesting differential grazing in *Juga*. Lamberti's data seem to show that *Scenedesmus* may be more difficult to ingest than the blue-green *Phormidium*, diverse filamentous greens, or three common diatom taxa, but the effect does not show in DeNicola's experiments. DeNicola followed the relative abundances of *Scenedesmus* and three diatoms of the genus *Navicula*, one large species and two small adnate species, in laboratory streams colonized at intervals with *Juga*. In three of his four experiments, I cannot see any difference between the relative abundances of the algal taxa in grazed and control treatments.

I intend to sidestep the issue of grazer effects on periphyton communities here – it seems to be a complicated function of a lot of things. For example, it stands to reason that the algae left on a slide or tile after some weeks of exposure to grazers should be a function of both the grazing pressure and the growth rate of the algae. Hence McCormick has manipulated both nutrient levels and densities of *Goniobasis curryana* in experiments in and about an ephemeral Kentucky stream, often with surprising results (McCormick and Stevenson 1989, McCormick 1990). This creek is quite unusual for a pleurocerid habit in that it dries up into a series of pools each summer. Thus McCormick feels comfortable modelling 'stream' processes with static-water containers.

McCormick has found that the addition of *G. curryana* at low to moderate densities positively affects the abundance of the dominant greens *Ankistrodesmus*, *Kirchneriella* and *Mougeotia*, and the diatom *Achnanthes minutissima* in wading pools set at creekside. The same effects are noted upon the addition of crayfish as well, and seem to be due at least partly to measurably higher nitrogen levels in pools inhabited by animals. Adversely affected, however, are the diatoms *Nitzschia palea* and *Synedra tenera*. In smaller containers with higher grazer densities and lower nutrient levels, however, McCormick's observations are more familiar: significant reduction of all algae except the adnate *Cocconeis*. In pools of the stream itself, *Goniobasis* removed the diatom 'overstorey', leaving behind the resistant green filament *Stigeoclonium* (McCormick and Stevenson 1991).

High grazer density would seem to be the case in Walker Branch, Tennessee, the home of a large population of *Goniobasis clavaeformis*. Elwood and Nelson (1972) released radiolabelled phosphate into the stream on several occasions during the course of a year, sampling water, periphyton, and *Goniobasis*. Their estimated rates of streamwide primary production were insufficient to account for the estimated food consumption rates of the *Goniobasis* population. Elwood and Nelson concluded that the snails must be consuming significant amounts of allochthonous detritus. They were in fact able to maintain laboratory populations of *Goniobasis* for months on diets of dead leaves only.

Over the last 20 years, the Walker Branch population of *Goniobasis clavaeformis* has found itself surrounded by rings of ecosystems ecologists (Malone and Nelson 1969, Elwood and Goldstein 1975, Elwood *et al.* 1981, Mulholland *et al.* 1983, 1985, 1991). The snails have been found to be very effective shredders and processors of leaf matter. Mulholland and his colleagues (1985) stocked four fibreglass channels with white oak leaves and three densities of *Goniobasis*, keeping a control without snails. Samples at 30 weeks showed much finer particulate matter in channels with snails. The observation that bacterial abundance was significantly lower on the organic particles remaining in channels with snails constitutes some evidence of bactivory, although variable levels of nutrients confound this interpretation somewhat. Regarding algae, workers have found that the *Goniobasis* population reduces algal biomass and productivity in Walker Branch, and thus the levels of phosphorus uptake. What one gets from artificially enriching a stream such as this is not more algae, but more snails. As reported by other workers, *Goniobasis* grazing seems to select against diatoms (especially *Achnanthes*) and for filamentous greens (*Stigeoclonium*) and reds (*Batrachospermum*).

Guys Run is a rapidly flowing stream in the mountains of Virginia with which I have some personal experience. By way of summary for the pleurocerids, I quote from Miller (1985): 'Contrary to predictions of the River Continuum Concept, the shredder functional group in the Guys Run drainage and in other temperate woodland streams was found to be a minor part of total macroinvertebrate standing biomass.' Why? The pleurocerid *Leptoxis carinata* constitutes from 33% to 81% of the macroinvertebrate biomass in Guy's Run, and Miller has classified this snail as a 'grazer', not a 'shredder'. He is correct in that the majority of the *L. carinata* population is almost certainly grazing the majority of the time. But *Leptoxis* is also quite capable of shredding dead leaves, and certainly will if it can. In Guys Run, however, leaves are swept into 'leaf packs' by the rapidly flowing water, or deposited in loose piles at pool bottoms, and in neither case can *Leptoxis* grab a hold. Whether snails are herbivores, carnivores, bacteriovores, grazers, shredders, filter feeders (coming soon) or whatever, depends on 'what they can get'. And in Guy's Run, 'what they can get' is a function of substrate and current. We'll have more to say about this at the chapter summary.

Pomatiopsidae

Although they have gills, as do all prosobranchs, many pomatiopsid species are truly amphibious, or possibly even terrestrial. A well-studied example is *Oncomelania*, the intermediate host of *Schistosoma japonicum*. The young and newly hatched *Oncomelania* are generally aquatic, but adults spend the majority of their time on moist soil, and can survive 129 days of desiccation (Komiya 1961). Komiya lists typical habitats in Japan as 'on the soil in river beds', 'above the edge of the water in irrigation ditches', and 'uncultivated marshy areas'. Several analyses have detected little effect of soil chemistry or texture (Pesigan *et al.* 1958), but moist organic soil is clearly the preferred site for feeding and egg laying. One might think that an amphibious habit could provide some insulation from the effects of pollution and low dissolved oxygen, but Garcia (1972) has found otherwise.

Van der Schalie and Davis (1968) reviewed 11 reports on *Oncomelania* culture from 1933 to 1965, and notable subsequent contributions have come from Davis (1971), French (1974), and Iwanaga (1980). It became clear very early that these snails require a mud substrate in culture, usually formed into banks or islands, and that, under some conditions, enough algae, diatoms, detritus and/or bacteria may be present in the mud to make additional food supplements unnecessary. Davis (1971), however, found that innoculating the substrate with a 5% suspension of the diatom *Navicula luzonensis* greatly increased average fecundity. Following Liang (1974), French (1974) showed that the addition of small pieces of algal mat (*Fisherella, Nostoc, Schizothrix*, etc.) resulted in higher average fecundities than the artificial dietary supplements commonly offered *Oncomelania*. These techniques recall some of the methods used to culture the lymnaeid hosts of *Fasciola*.

Based on a detailed analysis of stomach and faecal contents from five snails collected at each of 18 sites in the Philippines, Dazo and Moreno (1962) concluded that *Oncomelania* was a herbivore. The most common identifiable food items included 5 genera of green algae, 9 genera of diatoms, 5 of desmids, 2 of euglenoids, and 4 of blue-green algae. The authors reported a correlation between population density and the abundance of green algae and diatoms, but offered no statistics, and the relationship is less than striking judging from the raw data. Protozoans and nematodes were fairly common among the gut contents. And actually, the most common item in the guts at all 18 sites was 'vegetative matter'.

Among the most remarkable results obtained from the study of any gastropod diet are those of E. D. Wagner and colleagues in the mid-1950s (Ritchie 1955, Wagner and Wong 1956, Winkler and Wagner 1959). It was discovered that *Oncomelania* survival was greater on a diet of ordinary filter paper than on any of nine other items, including plain soil, fish foods, liver extract, or leaves. Using radioactively labelled cellulose, Winkler and Wagner demonstrated fairly convincingly that the cellulase originates in the snail's crystalline style. In summary, 50 years of research has shown that the diet of *Oncomelania* may include just about anything one could find on mud.

The North American pomatiopsid genus *Pomatiopsis* has received some attention due, at least partly, to its biological similarity with *Oncomelania*. *Pomatiopsis lapidaria* can on occasion be found in moist environments metres from any permanent water body. *Pomatiopsis cincinnatiensis* is often restricted to fairly narrow strips of muddy river banks, more after the fashion of *Oncomelania*. Van der Schalie and Getz (1963) have shown that the two *Pomatiopsis* species are more tolerant of subzero temperatures and less tolerant of high temperatures than the four species of *Oncomelania*. The substrate moisture preference and resistance to desiccation of *P. lapidaria* are similar to that of *Oncomelania*, while *P. cincinnatiensis* seems better adapted to drier conditions.

Van der Schalie and Getz (1962) noticed a relationship between soil moisture and the abundance of *P. cincinnatiensis* within sites, the snails apparently preferring 75–92% saturation. The authors felt that they might be systematically overestimating soil moisture in the field, however, since their sampler dug several centimetres into the substrate. So they constructed a test chamber with compartments containing a uniform soil over a range of moisture contents, placed 20 snails in each compartment, and allowed them to move about in darkness for 24 hours. The preference for 70% saturation was striking, while moisture ranges of 0–40% and 90–100% seemed unacceptable to the snails. Since the authors detected little variance in temperature and soil texture at their several study sites, they concluded that moisture was the primary determinant of *Pomatiopsis* microdistribution.

Van der Schalie and Getz also compared the soil textures at 24 sites inhabited by *P. cincinnatiensis* in Ohio and Michigan with 54 sites from which the snail was absent, and could find no differences in the relative proportions of sand, silt, and clay. Their laboratory results on substrate choice were quite interesting, however. The authors placed 25–50 snails in the centre of a dish which had been equally divided into zones of (rather unnatural) sand and (quite familiar) loam, and noticed that after 24 hours, the snails significantly preferred loam, by greater than 2:1. As *P. cincinnatiensis* is not generally found on sand, this was not surprising. They then repeated their experiment comparing loam with pure clay, another substrate upon which the snail is not typically found. The snails preferred clay 10:1. Van der Schalie and Getz do not offer an explanation for this odd result, but I would speculate that the apparent preference for clay was a temporary one, and would have changed as soon as the snails got hungry.

Pomatiopsids of the Asian subfamily Triculinae are much more aquatic, often living on rocks, sticks, and mud in streams with rapid current. *Tricula aperta*, the intermediate host of *Schistosoma mekongi*, has been cultured by several workers (Liang and van der Schalie 1975, Liang and Kitikoon 1980, Lohachit *et al.* 1980, Kitikoon 1981a). The number and quality of the experiments does not approach that available for other medically important snails, for various reasons not under the control of the investigators. *Tricula* is generally cultured in Petri dishes with at least some mud, after the fashion of *Oncomelania*, although the mud is not vital. (Mud may be more important as a refuge for food algae.) The life cycle can be completed on provisions of the blue-green algae *Nostoc muscorum* or the diatoms *Navicula* and *Achnanthes*, and *Tricula* will also ingest both maple leaves and filter paper. Kitikoon found over an order of magnitude improvement in the fecundity of *Tricula* in an artificial stream, compared with the more traditional Petri-dish method. It seems to be a general result that animals from lotic waters are difficult to rear in the laboratory, for reasons of temperature, flow, dissolved oxygen, or all three.

Hydrobiidae

The best studied hydrobiid is the European *Potamopyrgus*, whose habitat Boycott (1936) described as 'Running waters, rivers, canals, brooks, ditches, *running* ponds [Boycott's emphasis], often in the meanest little trickles'. The species also commonly inhabits brackish water, and can survive for months in full sea water, although apparently reproduction is affected (Todd 1964, Duncan 1966).

Much unlike the outwardly similar pomatiopsids *Oncomelania* and *Pomatiopsis*, *Potamopyrgus* may be found on rocks, mud, or macrophytic vegetation, with little sign of substrate specificity. Heywood and Edwards (1962) estimated that about 30% of the individuals in one reach of the River Ivel, England, were on macrophytes and the remainder on the muddy river bed, but it is difficult to know if this represents any preference without estimates of total surface area available.

In a circular arena, Haynes and Taylor (1984) found that *Potamopyrgus* was attracted to homogenized, filtered extracts of the macrophyte *Apium nodiflorum*, fresh or decayed, the large alga *Ulva*, miscellaneous microscopic algae, and crushed amphipod. The snail was indifferent to an extract of fresh *Elodea*, but attracted by decayed *Elodea*, and was repelled by an extract of the watercress *Nasturtium*. In paired choice experiments, *Potamopyrgus* preferred (in this order) tissue extracts from (1) the amphipod *Gammarus pulex*, (2) the snail *Lymnaea peregra* and well-decayed *Elodea* tied, (3) decayed *Apium*, (4) miscellaneous green algae, (5) partly decayed *Elodea*, and (6) diatoms. When interpreting the results of this sort of experiment, it is important to remember that a particular foodstuff may appear to be highly desirable while entirely unavailable to the snails (e.g. crushed amphipod). It should also be kept in mind that two variables are being tested simultaneously: the actual preference for a foodstuff and the ability to locate it. It is quite possible, for example, that diatoms are the optimum diet for *Potamopyrgus* in the wild, but that diatoms excrete little chemical cue when compared with rotten macrophyte.

Haynes and Taylor did not offer their snails any terrestrial leaf litter, but such a potential food resource could be quite important, especially in the 'mean little trickles'. The experiments of Hanlon (1981) fill this gap admirably, including both food preference and measurements of growth rate. The author placed 200 *Potamopyrgus* in tanks with two 900 mm² squares of a fresh-fallen beech leaf standard diametrically opposite two 900 mm² squares of one of 11 other tree species, left these in the dark for 24 hours, and counted the individual snails on each leaf. Hanlon performed five replicates and then repeated his experiments with a lime-leaf standard. His results (Figure 3.7) show a striking preference for the 'softer' leaves (poplar, willow, and elm) over the leaves with harder cuticles, such as beech and oak. The order of preference with the lime-leaf standard was similar.

Note in Figure 3.7 that mechanically abraded beech leaves were significantly preferred over unabraded controls. Hanlon performed some choice experiments comparing freshly fallen leaves with those that had been submerged in a stream for three months. *Potamopyrgus* significantly preferred the partly decayed leaves over the fresh leaves, perhaps both because they were softer and because such leaves have certainly been colonized by bacteria and fungi.

Hanlon placed 3 g of each leaf species in a litre of filtered pond water, added 60 *Potamopyrgus*, and allowed the snails to feed for 10 weeks at constant temperature. The top axis of Figure 3.7 shows the growth in shell height of a sample from each of these 13 experiments. Clearly *Potamopyrgus* can assimilate dead leaf litter and grow. Moreover, the correlation between the two halves of the figure is gratifying. It is commonly taken for granted, especially by researchers who do not have small

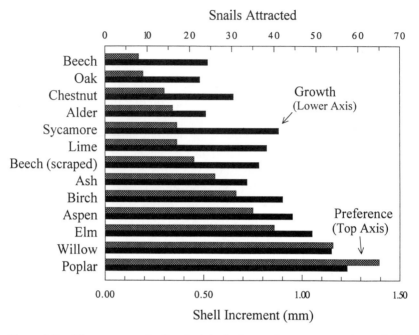

Figure 3.7. The upper axis (hatched bar) shows the mean number of 200 *Potamopyrgus* attracted to 13 types of leaves (5 trials of 24 hours), and the lower axis (solid bar) shows the mean ten-week increase in shell height for 30 snails fed on each sort of leaf. (Data of Hanlon 1981.)

children, that animals tend to prefer the diets most beneficial to them. This is the main rationale for choice experiments. But to my knowledge, Hanlon's data are the only demonstration that freshwater snails actually grow better on diets they tend to prefer.

Two nicely complementary experiments have been performed with lake populations of the North American hydrobiid *Amnicola*. The experiments of Kesler (1981) were directed specifically to the population of *Amnicola limosa* inhabiting Nonquit Pond, Rhode Island, and its effect on the periphyton. Kesler exposed glass slides for one week at four sampling periods to allow periphyton to colonize, then placed them for a second week either in an enclosure containing *A. limosa* at normal density (45–107 snails/m^2) or an exclosure containing no snails. Generally diatoms were the only algae colonizing these slides at densities sufficient to report. At all periods, Kesler found that *A. limosa* grazing reduced the abundances of all diatoms except our friend, the adnate *Cocconeis*.

Cattaneo and Kalff (1986) anchored 10 cubical frames in mesotrophic

Lake Memphremagog, on the Quebec/Vermont border, containing artificial macrophytes colonized by periphyton. They screened two each with mesh sizes 0.1 mm, 0.25 mm, and 0.5 mm (effectively excluding snails), and 2.0 mm (admitting *Amnicola*). Two frames were left unscreened as controls. Between 31 July and 29 August there was a marked decline in diatom biomass in the 2 mm and unscreened containers, and an increase in blue-green algae, particularly striking when compared with more finely meshed chambers. The author's inventory of all grazers in all chambers convinced me that *Amnicola* almost exclusively accounted for this effect. Small diatoms (<100 μm^3) such as *Achnanthes minutissima* suffered especially significant declines, while the abundance of the blue-green algae *Gleotricha pisium* (colonies up to 2 mm diameter) showed a strong increase. Diatoms over 100 μm^3 and the green algae (really very few colonies smaller than 1000 μm^3) were unaffected by *Amnicola*. It is interesting to note that total grazer biomass remained fairly uniform in all enclosures, with oligochaetes, chironomids, and cladocerans replacing snails behind fine mesh sizes. None of these taxa seemed to have an effect on the periphyton community as strong as *Amnicola*, however. Thus in Lake Memphremagog, snails have some effect on other grazers as well as on the grazed.

Ampullariidae

The capacity of these snails to enclose air within their mantle cavities confers buoyancy unique among the prosobranchs. Their proboscis has been modified into a pair of palps used to manipulate bulky items, such as leaves. Thus in spite of their large size and apparently ponderous shells, many species of ampullariids are characteristically observed in floating mats of vegetation and among emergent macrophytes. We left our discussion of macrophytes and pulmonates with the impression that the match was, more or less, a draw. In some places the pulmonates seemed to be nibbling meekly about the edges of a plant population too large to notice, while in others they seemed entirely frustrated by tough cell walls. But if macrophytes have nightmares, it would be of ampullariids. Some taxa may be restricted to the tropics and subtropics at least partly because the growth rates of temperate macrophyte populations are insufficient to support viable population sizes.

Ampullariids have attracted some attention as biological control agents for both aquatic weeds and medically important pulmonates, and are not difficult to culture. Santos and colleagues (1987), for example, fed *Elodea* to their experimental population of *Pomacea lineata*, the 'staple food in the pond' in Brazil. Meenakshi (1954) noted that Indian populations of *Pila virens* generally eat aquatic vegetation such as *Vallisnaria* and *Hydrilla*, but consumed filter paper 'readily and in large quantities'. Experimental populations of the Indian *Pila globosa* grow quite well on pure diets of *Ceratophyllum demersum*, which as we have seen is nearly invulnerable to attack by pulmonates (Haniffa and Pandian 1974, Haniffa 1982).

Salvinia is a large, aquatic fern that has become a pest in many tropical regions. Thomas (1975) found that although *Lymnaea* and *Indoplanorbis* could consume the tender leaves of young plants, the hispid texture of the older fronds discouraged herbivory. But he reported that Indian *Pila globosa* consumed *Salvinia* with no apparent difficulty, 20 g snails eating an average of 2 g of plant tissue per day. The most extensive experiments in ampullariid weed control have taken place with *Marisa cornuarietis* in the United States. Seaman and Porterfield (1964) performed a series of large-scale experiments, involving hundreds of snails and pounds of macrophytes. Of seven Florida aquatic weeds, they found that *Marisa* totally ingested first *Najas*, followed by *Potamogeton illinoensis*, *Salvinia*, and *Ceratophyllum* over 25 days. The snails ingested somewhat more *Alternanthera* tissue than was produced, retarded *Pistia* growth somewhat, and had no effect on water hyacinth. Seaman and Porterfield attribute the resistance of *Pistia* and water hyacinth to their habit of floating on the surface, with leaf surfaces aerial.

The food-choice experiments of Estebenet (1995) suggested that *Pomacea canaliculata* prefers macrophytes in the following order: *Zannichellia* > *Myriophyllum* > *Chara* > *Rorippa* = *Potamogeton* > *Elodea*. Growth recorded over 120 days on these diets matched the preference data exactly. It is thus quite clear that ampullariids can process and eat large food items, such as macrophytes. They may even attack other snails, as Chapter 7 will document. But nothing I have written here should be construed as evidence that they cannot graze on small particles as well. In fact, African populations of *Pila* and *Lanistes* may be abundant where no macrophytes are available, on the stony shores or sandy bottoms of lakes.

Ampullariids have evolved an unusual method for harvesting particles from the surface film as well. In calm waters such as may prevail in laboratory aquaria, *Pomacea* may climb to the surface, fold its foot into a funnel, and initiate a current with its pedal cilia (Johnson 1952, Cheesman 1956, McClary 1964). A very shallow layer of surface water is drawn down this funnel, from which the snail collects food particles with its foot. Periodically the snail turns its head into the funnel and ingests the entrapped particles. Both Johnson and McClary describe this behaviour as primarily a response to the placement of food on the water surface, but Cheesman suggested that the snails may be collecting and ingesting the proteinaceous surface film itself.

Viviparidae

The most striking modification of the basic gastropod feeding method in fresh water has been the evolution of filter feeding. The gills of *Viviparus* are characterized by unusually large, triangular lamellae whose tips hang over a ciliated gutter or 'food groove' running across the floor of the mantle cavity (Cook 1949). Cilia propel water currents through the mantle cavity of all prosobranchs, but in *Viviparus* they also direct mucus and entrapped particles to the food groove. Particles collected on the gill filaments are carried to the tip, where they also fall into the groove. A food/mucus string forms, which is carried forward and collected into a ball or 'sausage'. Periodically the snail will turn its head and eat the collected food.

The evolution of a new feeding apparatus does not necessarily imply the forfiture of the old. Just as was the case with ampullariids, viviparids have retained their ability to graze. In fact, *Campeloma* are sometimes baited with carrion. But viviparids are more benthic than the closely related ampullariids. Duch (1976) collected *V. contectoides* from a New York river and placed 20 individuals in a levelled, darkened aquarium, half with a sterile rock bottom and half with sterile silt-mud. The snails initially moved about randomly, but after six hours all 20 had collected on the mud side. Gut analysis of 50 wild-caught snails showed primarily diatoms, some filamentous green algae, and very little blue-green algae. Duch placed mixed cultures of diatoms in random areas of a sterile silt substrate, and once again released 20 *V. contectoides*. The animals again appeared to move about randomly until encountering a patch of diatoms, until at one point 80% of them were aggregated on diatom patches.

Studier and Pace (1978) also noticed *V. contectoides* most commonly burrowing in fine silt. They compared oxygen consumption in 16 snails (half male, half female) enclosed for 2 hours in 250 ml Erlenmeyer flasks containing sterilized natural substrate, fine sand, coarse sand, aquarium gravel, or no substrate. The authors found no significant differences in oxygen consumption among all these trials, concluding with the suggestion that *Viviparus* distribution may be more a function of the food in the substrate than its texture.

The dominant snail among the reeds fringing Japan's Lake Biwa is the viviparid *Sinotaia historica*, reaching summer densities of about three individuals per 10 cm of stem. Higashi and colleagues (1981) enclosed small PVC rods previously colonized by algae with 5 mm mesh and introduced *Sinotaia* into half of them, leaving the other half as controls. The results were familiar: snails significantly reduced algal biomass and primary productivity. Diatoms seem to have constituted the bulk of the algae on the rods. From a comparison of grazed and ungrazed rods, it appears that the *Navicula* population suffered more from grazing pressure than did either *Synedra* or *Cocconeis*, although no estimates of significance were furnished.

Bithyniidae

Filter feeding has also evolved in *Bithynia*, in a manner remarkably similar to, but entirely independent of, its occurrence in the Viviparidae (Tsikhon-Lukanina, cited in Monakov 1972, Meier-Brook and Kim 1977, Tashiro and Colman 1982). The apparent facility by which snails of this family switch from grazing to filter feeding should have attracted much more attention for them from ecologists than it has. Lilly (1953) generally observed English *B. tentaculata* grazing actively over macrophytes or hard substrates during the growing season, and buried in the mud during winter. She did not notice any filter feeding, identifying the structure generally called a 'food groove' as a channel to clear silt from the gills. But Tsikhon-Lukanina cultured laboratory populations of this species on suspensions of the green algae *Scenedesmus*, and considered Soviet populations of this species primarily filter feeders.

Bithynia siamensis is the first intermediate host of a human liver fluke in Thailand. Kruatrachue and colleagues (1982) have made an effort at laboratory culture, using small round containers with a thin layer of mud on the bottom. (It is not certain that the mud is required.) They offered *B. siamensis* three diets: a complex artificial one, a suspension of unspecified diatoms, and diatoms + artificial. Survival was poor on diatoms alone, but addition of diatoms to the artificial diet significantly improved fecundity. Brendleberger and Jurgens (1993) reported no difference in rates at which *B. tentaculata* filtered green algal cells of three taxa: the 6 μm (flagellated) *Chlamydomonas*, 6 μm (non-motile) *Chlorella*, and the 10 × 20 μm (flagellated) *Chlorogonium*. Although *Bithynia* could grow on a diet of *Chlamydomonas* alone, the laboratory populations of Brendleberger (1995) seemed to require solid food (lettuce) for reproduction.

The most thorough studies of the biology of *Bithynia* to date have been those of Tashiro on the population of Oneida Lake, New York (Tashiro 1982, Tashiro and Colman 1982). Here the snails seem to graze only during spring blooms of attached algae and sometimes during the autumn, when the concentration of suspended particles is much reduced. At all other times, they remain stationary under rocks and on pilings, filter feeding. In the laboratory, Tashiro showed good assimilation both for suspensions of *Chlorella* and for collections of natural 'aufwuchs' (unspecified) offered on slides.

Valvatidae

This small group of small snails has attracted small attention. North American species are primarily inhabitants of lentic waters, occuring at greater depths and colder temperatures than most other gastropods. But the European *V. piscinalis* is common in streams. Tsikhon-Lukanina has also reported filter feeding in *Valvata*, and has reared these little snails on a diet of suspended algae. *Valvata* do often live in very turbid environments, and their bodies are covered with cilia which constantly cleanse them of silt. But Cleland (1954) and Fretter and Graham (1962) have not reported any special devices for collecting particles. Instead, *Valvata* has a long, heavily ciliated extension of the mantle, the pallial tentacle, which these authors believe directs silt (as well as faeces) away from the body. *Valvata*'s gill is not made of triangular filaments, but is bipectinate, and often projects at least partly out of the mantle cavity, open to the water. Thus both Cleland and Fretter and Graham believe that *Valvata* is strictly a grazer.

Prosobranch reproduction

Gonochorism

The first reports of sex chromosomes in molluscs were made by Jacob (1959a,b) working with the Indian *Thiara crenulata* and *Paludomus tanschaurica* (a pleurocerid). Based on the older technique of paraffin sectioning, Jacob's observations have subsequently been questioned (Patterson 1973). More reliable documentation of sex chromosomes was obtained by Burch (1960) and Patterson (1963) from North American *Pomatiopsis*. The largest pair of chromosomes in the complement of male *P. cincinnatiensis* are unmatched with regard to centromere position, but such is not the case in the females. The largest chromosome in the complement of male *P. lapidaria* seems to have no homologue at all ($2N = 33$), although again this is not the case in females ($2N=34$). Thus *P. cincinnatiensis* seems to show an XY system of sex determination, while *P. lapidaria* shows XO. Patterson (1969) also reported a large, dimorphic pair of chromosomes peculiar to males of the North American viviparid *Tulotoma angulata*.

The above constitutes almost the entirety of our current knowledge regarding sex determination in putatively gonochoristic molluscs. (Marine species are even less known than the freshwater fauna.) In light of their oft-skewed sex ratios, and the unexplored potentials of sequential hermaphroditism, parthenogenesis, and environmental influence, much additional research is needed.

Although the gonochoristic prosobranchs of fresh water are diverse in their evolutionary origin, some generalization regarding their mode of reproduction is possible. The males generally possess a large, muscular penis that is not retractable. It arises from the head behind the right tentacle, and lies curled under the mantle when not in use in *Theodoxus* (Fretter and Graham 1962), bithyniids (Lilly 1953), pomatiopsids (Dundee 1957, van der Schalie and Getz 1962, Davis *et al.* 1976), and most hydrobiids. The penis of the ampullariids arises from the edge of their mantle, and that of the viviparids is a modified right (copulatory) tentacle. The thiarids, melanopsids, and pleurocerids have no penis at all, nor any external clue to their gender. Thus it would seem that sexual reproduction has arisen at least four separate times in prosobranch groups that have exploited fresh waters.

Female ampullariids of most species retain a vestigial penis, leading Keawjam (1987) to speculate that Thai *Pila* may be protandrous hermaphrodites. But careful study has uncovered no evidence of hermaphroditism throughout the life history of *Pomacea* (Andrews 1964, Lum-Kong and Kenny 1989, Estebenet and Cazzaniga 1998). The absence of a penis in male pleurocerids and thiarids also seems to indicate potential hermaphroditism, although again, several complete studies of their maturation have detected no evidence (Jewell 1931, Magruder 1935, Dazo 1965).

Male prosobranchs generally achieve copulation by mounting the shell of a female, orienting in the same direction as their mate, and extending their penis into a vaginal duct or opening on the female's right. Very little courtship or special behaviour prior to mating has been documented for any freshwater prosobranch. Mating cues may be chemical. Many populations of freshwater prosobranchs manifest great genetic and morphological similarity with one another, yet co-occur with little or no hybridization. It is also possible that prosobranchs do show mating behaviours which we have yet to observe. Opportunity for detailed observation of fleeting behaviours simply may not present itself in the wild, and unlike the pulmonates, most prosobranchs are not easily kept in culture. There is considerable indirect evidence that female prosobranchs can store sperm or fertilized eggs for months before egg deposition.

Typically, prosobranchs cement small (1 mm) eggs singly or in small clusters on solid substrates and cover them with sand grains, mud, or faeces. Female ampullariids, pleurocerids, and melanopsids are endowed with egg-laying grooves or 'ovipositors' on the right, dorsal side of their foot for this purpose. Eggs hatch directly into crawling juveniles after several weeks, depending on temperature. A minor, but nonetheless striking exception is provided by the tropical neritid *Neritina*, whose populations typically inhabit small rocky streams flowing directly into the sea. *Neritina* spawns planktonic larvae, which are carried to the ocean and develop into marine juveniles. Juveniles migrate back into fresh waters.

Ampullariid reproduction differs in quite a few respects from the general prosobranch pattern outlined above. In the more amphibious genera of ampullariids (including *Pila* and *Pomacea*), the male achieves copulation by gripping the margin of the female's shell and rolling her over so that her aperture faces his. The penis is then inserted down the length of the female's pallial oviduct. In some circumstances, both snails may withdraw until their opercula are almost closed, while in other circumstances, the female may continue actively crawling (Andrews 1964, Albrecht *et al.* 1996). In any case, copulating pairs remain nearly inseparable for many hours. Female ampullariids may emit some chemical or behavioural cue to indicate receptivity to insemination. In her laboratory colony of *Pomacea canaliculata*, Andrews (1964) observed 'Several males are usually attracted to the female at the same time, and two or more may be attached to her, with the sheath in her mantle cavity, but only one inserts his penis into the oviduct. Even after copulation has begun other males remain in the vicinity.' It would be most interesting to learn how the successful male is determined.

Ampullariid eggs are exceptionally large, packed with albumen and ultimately measuring 3–5 mm in diameter. In most ampullariids, the egg is covered with a calcareous shell and laid above the water level on vegetation (*Pomacea*) or in excavated pits (*Pila*). *Pomacea urceus* inhabits neotropical marshes and savannahs liable to de-water entirely during the dry season. As the rains end, it burrows into the mud and deposits very large eggs (10 mm diameter) just inside the aperture of its shell. It then withdraws deeply inside its shell and aestivates. The young hatch during the dry season, but remain under their mother's shell until the water rises (Burky 1974, Lum-Kong and Kenny 1989). *Marisa cornuarietis* is more strictly aquatic than *Pomacea* or *Pila*. *Marisa* and *Lanistes* lay eggs (2–3 mm diameter) on vegetation or other solid support below water level, unprotected by a calcareous shell.

Pleurocerids and melanopsids also differ from the general prosobranch pattern in that no behaviour distinguishable even as mating has ever been reported. But spermatophores (2–4 mm) have been reported in their culture water as eggs are being deposited. Pleurocerids lay eggs of the ordinary prosobranch type, while melanopsids produce gellatinous spawn containing several hundred eggs (Bilgin 1973). Dazo (1965) observed that the pleurocerid pallial oviduct is saclike, lined with epithelial cells 'into which spermatozoa, that have entered the female tract, insert their heads and remain embedded for an indefinite time'. Both Dazo and I (Dillon 1986) have observed the production of fertile eggs by North American pleurocerids isolated in the winter. Thus it seems likely that insemination may occur as early as the late summer or autumn, and that females may store sperm over winter. Further, I have some indirect evidence that female *Goniobasis proxima* mate but a single time in their entire 3–4 year lifespan (Dillon 1986, 1988).

Ovoviviparity is demonstrated by the viviparids, the thiarids, the pleurocerid *Semisulcospira*, and the hydrobiid *Potamopyrgus*. Fertilized ova are brooded in a modified pallial oviduct, sometimes called a 'uterus' in viviparids (Vail 1977), or a 'brood pouch' in *Semisulcospira* or *Potamopyrgus* (Fretter and Graham 1962, Davis 1969). The thiarids have a special 'cephalic brood pouch' in the back of their heads (Berry and Kadri 1974, Muley 1977). At least one species of thiarid, the Indian *Melania crenulata*, seems to release young from its brood pouch while they are still at the free-swimming veliger stage (Seshaiya 1940). Many of the ovoviviparous groups reproduce parthenogenically, as will be developed in the next section.

In some cases, sex ratios in prosobranch populations seem to be remarkably balanced. For example, on 13 June 1987 I hand-collected all the *Goniobasis proxima* from a small rock ledge (about 1 m² area) in the Elk River of western North Carolina (population *Elkp* of Dillon 1984a). The site, submerged under about 8 cm of clear, gently flowing water, was especially chosen to yield a large sample of snails unbiased with regard to size. The 487 individuals collected were cracked with pliers and sexed by direct examination of gonad under low magnification. I counted 62 (generally large) parasitically neutered snails, making no attempt to estab-

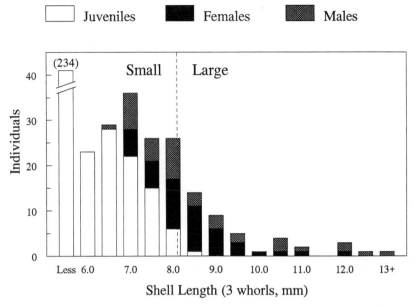

Figure 3.8. *Goniobasis proxima* sampled from a single square metre of rock substrate in the Elk River, western North Carolina, in June of 1987. The shell length is of the last three whorls, due to apical erosion (Dillon 1984b). Size categories are minima.

lish their sex. (Sexing is nearly impossible in the absence of healthy gonads.) The 425 individuals remaining are plotted by their shell lengths in Figure 3.8.

Goniobasis proxima is a perennial, laying its first eggs in the early spring at age two, and capable of living four to five years. Thus my June sample primarily included young-of-the-year juveniles, hatched perhaps three months previously and remaining less than 6 mm in shell length. The next-largest fraction were aged about 15 months in June, ranging from approximately 6 mm shell length to 9 mm in my sample. These snails seemed to be maturing, but had not as yet laid eggs. Egg laying was accomplished by the fraction of the population aged 27 months or older, larger than about 9 mm in my June sample.

The sex ratio at my population appeared to be balanced. The mature, healthy fraction of the sample included 38 males and 46 females, not significantly different from 1:1 by a chi-square test. Further, Figure 3.8 contains no evidence that either sex matures before the other, grows more quickly, or survives better. Dividing the mature snails at their approximate median size (the dashed line on Figure 3.8) yields roughly identical sex ratios in the 'large' and 'small' fractions, not different from the combined ratio overall.

Such is not the case in other prosobranch populations, however. Female ampullariids and viviparids may be larger and more long-lived, although males may record higher rates of ingestion and metabolism (Aldridge *et al.* 1986, Estebenet and Cazzaniga 1998). *Marisa* may display additional dimorphism in body pigmentation and in aperture shape (Demian and Ibrahim 1972). Perhaps at least partly because of differences in the growth rates and/or survivorships of the sexes, published sex ratios often vary tremendously among prosobranch taxa, among conspecific populations, and over time. And even if such factors are analytically removed, significant sex ratio biases sometimes seem to remain.

For example, van der Schalie and Getz (1962) used a square frame to obtain quantitative samples of the *Pomatiopsis cincinnatiensis* population inhabiting a mud bank at Tecumseh Station, on Michigan's River Raisin. They sampled on 11 dates, from August to October of 1959 and April to August of 1960, measuring and sexing each snail collected. (Males become distinguishable from females by the presence of a penis at 1.5 mm, shortly after hatching.) Example data from three dates are shown in Figure 3.9.

It is quite clear that the sexes are dimorphic with respect to size; the maximum shell length of the male reaches 3.8 mm while that of the female is 4.8 mm. Van der Schalie and Getz also offered evidence that males grow more quickly than females, and enjoy better survivorship. So there is certainly ample opportunity for sampling bias to creep into an estimate of sex ratio. But focusing on just the cohort of *P. cincinnatiensis* hatched in the summer of 1959, the authors concluded that the sex ratio was never balanced, rising from 20% male near birth on 10 August 1959 (the fraction smaller than about 3 mm in Figure 3.9) to 34% male at reproduction 11 August 1960. The authors noted an identical phenomenon, significant bias in the sex ratio even after adjusting for differential growth and survival, at a second site on the River Raisin near Clinton, Michigan. The ratio at Clinton was 34% just after birth, rising to 40% male at reproduction.

What is going on here? *P. cincinnatiensis* is one of the (just three) freshwater molluscs for which sex chromosomes have been described. Might males have different habitat preferences? Might males suffer reduced embryonic survival? Van der Schalie and Getz offered no explanation, nor can I. But in any case, sex ratio bias is widely reported in the gonochoristic freshwater prosobranchs. In Table 3.1 I have collected the most

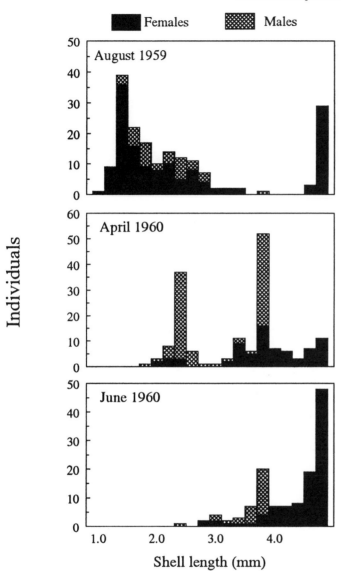

Figure 3.9. Size distribution of *Pomatiopsis cincinnatiensis* from the Tecumseh Station on the River Raisen, Michigan. (Data of van der Schalie and Getz 1962.)

Table 3.1. *Sex ratios (as proportion male) for populations of gonochoristic freshwater prosobranchs*

Author	Population	Sample size	Sex ratio
van der Schalie and Getz (1962) (Michigan)	*Pomatiopsis cincinnatiensis*, Tecumseh	135	0.20[a]
	Clinton	83	0.34[a]
Liang and Kitikoon (1980) (lab-reared)	*Tricula aperta*, alpha race	272	0.50
	beta race	138	0.49
Demian and Ibrahim (1972) (Egypt)	*Marisa cornuarietis*, wild	418	0.41[a]
	lab-reared	396	0.45[a]
Dazo (1965) (lab-reared)	*Pleurocera acuta*	74	0.32[a]
	Goniobasis livescens	318	0.13[a]
Mancini (1978) (Indiana)	*Goniobasis semicarinata*	374	0.39[a]
Dillon (this work) (North Carolina)	*Goniobasis proxima*	84	0.45
Brown (1988) (west Africa)	*Sierraia leonensis*	56	0.48
	S. expansilabrum	50	0.50
	S. outambensis	30	0.47
Calow and Calow (1983) (Hong Kong)	*Sinotaia quadrata*	271	0.50

Notes:
[a] indicates a significant difference from 1:1, by goodness-of-fit χ^2.

reliable estimates of sex ratio available for a variety of populations from this group. I have excluded studies where the sampling technique was unspecified, the sample size unknown, or the methods were likely to yield size bias. Excluded also are studies where sex ratios have been calculated across generations without first verifying that such a practice does not introduce bias. (This criterion is more likely to exclude studies reporting skewed sex ratios than those reporting equality, e.g. Calow and Calow 1983.) Where sex ratio is demonstrated to vary by age (as in *P. cincinnatiensis*) I have recorded only the estimate at the earliest age.

The techniques used by the various authors contributing to Table 3.1 have been quite diverse. Demian and Ibrahim (1972) used external shell and body pigmentation characters to sex large samples of *Marisa* just as they reached maturity (25–30 mm) from both field and laboratory colonies. The west African bithyniid *Sierraia* sets a thickened aperture lip upon reaching sexual maturity, a feature that facilitates survey of its sex ratio. D. Brown (1988, personal communication) sexed samples of *Sierraia* evenly selected to represent the size range displayed by adults from each of three species. Lack of a penis makes pleurocerids especially difficult to sex. Since both males and juveniles are missing an egg-laying groove, the only reliable method of sexing (especially small) pleurocerids is direct examination of their gonads. But Dazo's (1965) data on laboratory-reared cohorts, together with Mancini's (1978) data on 'mature' *G. semicarinata* from the field (summed over 13 months) convince me that the excess of females observed by these authors is a genuine phenomenon.

In the end, the data of Table 3.1 remain enigmatic. I suspect that the environment may play a role in the determination of prosobranch sex, although again, established theory on the evolution of environmentally determined sex does not seem to fit the particulars of prosobranch natural history (Charnov and Bull 1977, Bull 1983, Head *et al.* 1987). Answers await the day when clever and patient workers, at last, reliably rear a variety of prosobranchs under appropriately controlled conditions.

Parthenogenesis

Parthenogenesis is known to have arisen only four times in the phylum Mollusca, three of these in the freshwater prosobranchs. Honours for the initial discovery go to my hero, A. E. Boycott (1919) working with the hydrobiid *Potamopyrgus* 'jenkinsi'. (It is now generally accepted that European populations described under this nomen in the late nineteenth century are *Potamopyrgus antipodarum*, described earlier in New Zealand.) Parthenogenesis was subsequently recognized in the North American viviparid *Campeloma rufum* (van Cleave and Altringer 1937) and in Indian thiarids (Seshaiya 1936, Jacob 1957a,b). It now seems clear that the phenomenon is quite widespread in the Thiaridae and *Campeloma*, although it has not subsequently been documented in other hydrobiids. In any case, the biological parallels in these three situations are most striking.

Parthenogenesis is apomictic in all three groups, as far as can be told. Parthenogenetic females produce ova of the same ploidy as themselves, without a reduction division. Males may be entirely absent from some populations, rare in others, and common in yet others. Sexually reproducing females do not seem to have been distinguished within bisexual populations of thiarids, *Potamopyrgus*, or *Campeloma*, although there is good evidence that sexual reproduction does take place under some circumstances in all three groups. In no case has the female anatomy lost its capacity to receive sperm. (The sperm channel, bursa copulatrix, and receptaculum seminalis all remain.) Parthenogenesis does not seem to originate from interspecific hybridization in any of these taxa. In neither *Potamopyrgus*, nor *Campeloma*, nor the thiarids have putative ancestors been identified from among the sexually reproducing mollusc fauna currently extant. Rather, allozyme studies seem to suggest multiple origins of parthenogenesis from within sexually reproducing conspecific populations (Johnson 1992, Dybdahl and Lively 1995).

Wallace (1992) has made some observations suggesting that parthenogenetic *Potamopyrgus* are polyploid, perhaps autotriploid, but a consistent chromosome count remains elusive. *Potamopyrgus* males seem to be diploid. Jacob's (1957a) discovery of several $2N=32$ populations of *Melanoides tuberculata* prompted him to suggest polyploid origins for *M. lineatus* (71–73 chromosomes), *M. scabra* (76–78 chromosomes) and most other Indian populations of *M. tuberculata* (90–94 chromosomes). The karyotypic situation in *Campeloma* is unclear.

Heaven knows that more work is needed in this area; I have already emphasized that we are frighteningly ignorant regarding sex determination in gonochorists, and the situation in parthenogens must be complicated yet further. However, it is tempting to speculate that the karyotype of parthenogenetic females is unbalanced, yielding viable ova by mitosis but generally inviable meiotic products. Males (and sexually fertile females?) may arise from the rare reduction divisions that may be fortuitously viable. The incidence of viable diploid progeny in each population would, of course, depend on the genetic background of the clone(s) present. So too will the reproductive success of the sexuals. Sexual males would seem more likely to contribute to the next generation if they can find diploid, sexual females (not as yet humanly distinguishable), rather than relying upon parthenogenetic females to furnish fertilizable gametes.

Wallace has gathered extensive data on the sex ratio in *Potamopyrgus* from both its Australasian source of origin and its currently impressive European range extension. Her (1992) review of 207 populations from outside New Zealand found males to be absent or very rare (less than 1%) in 183, and present at frequencies of 1–9% in 22 additional. Only in two Welsh populations did the frequency of males exceed 10%. The situation is quite different back home in New Zealand, however, where Wallace's survey of 58 populations found 18 with less than 1% males but 24 populations with 10% males or greater. In fact, some New Zealand populations of *Potamopyrgus* approach 50% male. Wallace recognized 'no obvious relationship between sex ratio and habitat' in her New Zealand populations, although to my eye, her 27 'stream/drain' populations seem to hold especially high proportions of males. Sex ratio seemed fairly stable in three New Zealand populations that Wallace monitored over a 15-year period, although apparently significant trends were noted in two others.

Male *Potamopyrgus* are smaller than females, with a shell perhaps not quite as elongated (Wallace 1979). Hermaphroditism has been described (Wallace 1985). Although copulation has not been observed to my knowledge, females from high-male populations 'regularly' hold sperm in their receptaculum seminalis (Wallace 1992). Mother–offspring analysis of allozyme inheritance has demonstrated at least occasional outcrossing in several bisexual *Potamopyrgus* populations from New Zealand (Phillips and Lambert 1989). Wallace (1979) designed a rather heroic 'breeding trial' involving several hundred sets of one to several female *Potamopyrgus* with or without males, held together in meshed containers over four years. While the potential for sperm storage rendered some of his data ambiguous, Wallace's results with a 'semi-albino' line convince me that outcrossing did, in fact, occur in her experiment. She reported that females from high-male populations were much less likely than females from low-male populations to reproduce successfully without male contact. Given the reproductive success of the former group of females when males were subsequently added, Wallace's data convince me that *Potamopyrgus* populations indeed vary in the degree to which they reproduce sexually.

Data on sex ratios in *Campeloma* are neither as common nor as reliable, both because of sexual dimorphism in size and body, and because of collecting difficulties. It is clear, however, that both unisexual and bisexual populations are found in the southern United States, while the north seems to be inhabited only by unisexual populations (Karlin *et al.* 1980, Johnson 1992). Again it appears that parthenogenesis is associated with a range extension, in this case the natural colonization of higher latitudes as the glaciers receded. Karlin and his colleagues performed a mother–offspring analysis of allozyme genotypes in *Campeloma* from both unisexual and bisexual populations. The offspring being brooded by 11 mothers from two bisexual populations did occasionally show outcrossed genotypes, although results were far from Mendelian expectation. But seven mothers from each of two unisexual populations, each heterozygous at a minimum of one allozyme locus, bore entirely heterozygous offspring, identical to themselves.

The incidence of sexual reproduction in thiarids parallels that of *Potamopyrgus* and *Campeloma* in many ways. Both entirely female populations and bisexual populations of *M. tuberculata* are common in Israel, and the frequency of males can be rather high (Livshits *et al.* 1984). Livshits and Fishelson (1983) found that the esterase and acid phosphatase zymograms shown by the offspring of individual *M. tuberculata* from some of these bisexual populations often did not match those of their mothers. This constitutes considerable evidence of sexual reproduction. But the *Melanoides tuberculata* populations introduced to Malaysia and Hong Kong are entirely female (Berry and Kadri 1974, Dudgeon 1986), as is the endemic thiarid fauna of Australia (Stoddart 1985). Males are quite rare among the Indian thiarids, where they appear chromosomally sterile (Jacob 1957b). Again, as was the case for *Potamopyrgus* and *Campeloma*, some thiarids seem to be exceptionally well adapted for colonization (Madsen and Frandsen 1989).

One hypothesis to account for parthenogenesis in *Potamopyrgus* has commonly been offered in reverse fashion, that sexual reproduction may be an adaptation to a worsened environment, such as an increased level of trematode parasitism. But sexual reproduction is generally believed to have evolved with the Phylum Mollusca itself, in Precambrian times, and it seems most likely that all three groups of freshwater parthenogens evolved from sexual ancestors. Thus more recently this hypothesis has been rephrased to the effect that parthenogenesis has evolved as a response to environmental improvement. In either case, Lively (1989, 1992) has reported that in New Zealand bisexual populations of *Potamopyrgus* tend to be found where parasite loads are especially high. Lively noted that his data are consistent with the 'Red Queen' hypothesis that sex remains adaptive to promote genetic variation as an escape from parasites, while parthenogenesis evolves because of the twofold advantage of asexual reproduction where parasite pressure is relaxed. (That *Potamopyrgus* does in fact incur a 'cost of sex' has been verified by Jokela *et al.*, 1997.) Lively has, at times, offered data which suggest to him that the 'temporal variation', 'spatial variation', and 'reproductive assurance' hypotheses for the origin of reproductive mode do not fit his data as well as 'Red Queen'. More recently, however, it has become apparent that parthenogenesis has not led to a reduction in genetic diversity in *Potamopyrgus* (Phillips and Lambert 1989, Dybdahl and Lively 1995). And no relationship is apparent between parasitism and the frequency of males in the *M. tuberculata* populations of Israel (Heller and Farstey 1990).

If parasitism has been the driving force in the evolution of reproductive mode in *Potamopyrgus*, what is the role of other life history traits? Is it a coincidence that the only parthenogenetic hydrobiid is the only hydrobiid that broods its young? Is it a coincidence that the only parthenogenetic hydrobiid is the only hydrobiid that has successfully colonized a new continent in historic times? It should be hoped that any hypothesis to account for the origin of parthenogenesis in freshwater molluscs would offer some explanation for ovoviviparity and all the other life history phenomena that seem so closely associated with it.

Hermaphroditism

The valvatids are hermaphrodites; a condition very unusual among prosobranch snails. The male system seems to develop slightly before the female, but mature *Valvata* manufacture egg and sperm simultaneously in a single 'hermaphrodite gland' (Cleland 1954). The large penis (folded behind the head) and the female opening are both located to the right side of the body, as is normal for prosobranchs. Copulation has not been described.

As there is only a single duct leading from the hermaphrodite gland, self-fertilization seems quite possible in *Valvata*. There are no data on this, as far as I know. Cleland never found ova in any region but the hermaphrodite gland, and suggested that copulation may be necessary to stimulate egg release, as in other prosobranchs. Hence the opportunity for self-fertilization may be reduced. Some *Valvata* species seem to group their eggs into larger masses than is typical for freshwater prosobranchs (Heard 1963). Masses are cemented onto vegetation or other hard surfaces. This may necessitate periodic migration, since *Valvata* populations often inhabit deeper bottoms characterized by soft mud substrate.

Summary

The trophic machinery mounted by typical gastropods seems somewhat more selective than that of the bivalves. Snails are better able to select the surface onto which they place their mouths, and their radular ribbon and associated musculature seem finely adapted. Yet the overall impression one takes from a review of gastropod diet is that the animals will stuff anything into their mouths they can rip off the substrate.

The gastropod literature contains many studies of artificial culture showing growth and reproduction on a bewildering variety of organics, from fish meal to filter paper. The most interesting of these studies compare growth and reproduction on varieties of natural diets, and here we are not surprised to see differences both between species and between age categories within species. There is no question that snails may be attracted by chemical cues emanating from food items, and that feeding may be stimulated by these cues. And there is no question that snails of different age, size, and species vary in their preferences. I simply stipulate that as they feed upon a dead worm they have travelled two hours to search out, they are just as likely to eat the chaetae as the flesh.

Pulmonate snails are primarily adapted for calmer waters, while the more heavily bodied prosobranchs are more commonly encountered in lotic environments. Planorbids and ampullariids seem especially adapted for softer substrates and floating vegetation by virtue of their shells, which are rendered light by the air bubbles they enclose. Ancylids and neritids prefer firm substrates, by virtue of their limpet forms, and lymnaeids and pomatiopsids seem especially amphibious. Gastropod diets may include any small organic particle: macrophytic vegetation (living or dead), litter from the bank of the river or pond, algal filaments, fungal hyphae, detritus, bacteria, and even carrion. Larger ampullariids, lymnaeids, and planorbids are especially liable to consume tougher foodstuffs. Filter feeding has evolved in the Viviparidae and the Bithyniidae.

Of all the aspects of freshwater mollusc biology, the reproductive function of pulmonates is certainly among the best known. Perhaps as a consequence of this attention, we can say for certain that the phenomenon is bewilderingly complex. The snails are best considered simultaneously hermaphroditic, but usually the male function matures first, and sometimes the reverse is true. Given that a pair of pulmonates meet, the gender role(s) they may assume is a function of prior reproductive history, among other variables. And pulmonates also have the ability to self-fertilize, which is preferred in some populations, and engage in 'mixed mating', the production of outcrossed and selfed progeny simultaneously. All this reproductive diversity strikes me as 'weedy'. The pulmonate reproductive system seems adapted for colonization and spread into disturbed or newly opened habitat, or recolonization after a population crash. A single colonizer can carry stored allosperm for thousands of fertilizations, or self-fertilize if necessary. Then as the population grows and expands, its members can cross-fertilize to preserve the genetic diversity remaining.

Most prosobranchs are gonochoristic; the particulars of their reproductive biology being as diverse as their origins. Sex chromosomes have been reported in several taxa. Sexes may be dimorphic with respect to the speed and size at which they mature, yet even factoring out such considerations, prosobranch sex ratios often seem highly skewed in the wild. Hermaphroditism has evolved once in freshwater prosobranchs (the valvatids) and parthenogenesis at least three separate times. One striking parallel between the three parthenogenetic groups is that all are ovoviviparous. While parthenogenesis seems to be a common feature of reproduction in freshwater molluscs that brood their young, no molluscs lay parthenogenetic eggs. I would not be so bold as to predict that an oviparous parthenogen will never be discovered; on the contrary, I think that unrecognized parthenogenesis is quite likely among the prosobranchs. However, males seem to be at least moderately common in all the populations of oviparous gonochorists thus far examined. Williams (1975) has also observed that the offspring of parthenogens generally tend to be large compared with the size of their mothers, continuously produced, liberated into an environment essentially the same as that experienced by their mothers, and subjected mainly to intraspecific competition. All these generalizations seem to hold quite nicely for the thiarids, *Campeloma*, and *Potamopyrgus*, as will be seen in Chapters 4 and 5.

A second striking parallel among the three parthenogenetic groups is that all have recently expanded their ranges by (sometimes spectacularly successful) colonization. I view parthenogenesis as one of many life history traits that have evolved in a set of freshwater molluscs I will term 'R-selected' (in a specially narrowed sense, to be detailed in the next chapter). Individuals from R-selected populations may produce exceptionally large numbers of offspring, or offspring of exceptionally large size, as is typical of such ovoviviparous groups as *Potamopyrgus*, *Campeloma*, and the thiarids. R-selection also favours asexual reproduction, including parthenogenesis and self-fertilization, as may be demonstrated by the pulmonates. Thus I would class my own theory regarding the evolution of hermaphroditism and parthenogenesis in freshwater molluscs as 'Reproductive Assurance', that asexual reproduction of any type facilitates colonization by increasing the likelihood of reproduction in a small number of colonists (Heath 1977). It further seems possible to me that ovoviviparity may increase the physiological likelihood of parthenogenesis, especially as a consequence of polyploidy, independent of its adaptive value. It is to the larger issues of life history evolution that we now turn.

4 · Life history

If natural selection works to maximize the total offspring an individual leaves behind, how can it happen that one perfectly successful population of freshwater bivalves produces, on the average, 10 offspring per parent lifetime, while another produces 10^6? Life history studies address variation in fundamental demographic parameters such as birth rate (including age at reproduction, clutch size, and developmental time) and survivorship (lifespan, semelparity/iteroparity), as well as the relationship between individual age and size, from propagule to adult. Because the energy available to an organism is finite, life history studies are ultimately concerned with what have come to be called 'trade-offs' between parental growth, maintenance, and reproduction, semelparity and iteroparity, propagule size and number, and many other factors. Trade-offs are not inevitable, however, as we shall see.

The efficiency of natural selection on a trait is dependent on its heritability, that portion of the total phenotypic variance that is additively genetic. But the heritability of life history traits generally seems to be less than that observed for morphological, physiological, or even behavioural traits (Price and Schluter 1991). Because selection is expected to act most efficiently on traits as directly tied to fitness as survivorship and reproduction, the numerator of the heritability may be small. And because the expression of such traits is unusually sensitive to the environment, the denominator of the heritability will be large. Thus this chapter begins with a cautionary section, exploring the difficulty in distinguishing (additive) genetic variance (available for selection to act upon directly) from what has come to be called in the life history literature 'phenotypic plasticity'.

In the early 1980s, researchers began to focus their attention on the importance of body size on life history evolution (Peters 1983, Stearns 1984). At much the same time, data began to accumulate supporting the theory that body size at maturity might be a function of age-specific mortality schedules (Charlesworth 1980, Reznick 1985). These and other related observations were combined by Charnov (1991) into what has been called 'the first comprehensive model of life history evolution' (Harvey and Nee 1991). Although Charnov developed his model specifically for mammals, I would judge its principles of such broad applicability as to provide a useful framework for freshwater molluscs. Charnov finds no evidence of a 'spent salmon' phenomenon in mammals. Thus his model involves no trade-off between current reproduction and future survival; mortality is derived from extrinsic sources only. The mortality of newborns and young juveniles is density dependent. A key component of Charnov's model is that survivorship − not neonatal or early, but about the age of adulthood − determines size at reproduction. If survivorship in this window is good, the animal will mature early, and if survivorship in this window is poor, maturity will be delayed to a larger size. Size at maturity then determines reproductive effort. Charnov explained the tendency for larger mammals to wean larger offspring as an adaptation to speed the offspring through the (presumably extra-perilous) later juvenile stages.

Following Charnov, in the second section of this chapter it is shown that adult size seems to be the primary determinant of reproductive effort in freshwater molluscs, although some important exceptions are noted. The partition of reproductive effort into many small or a few large offspring is explored in the third section, and the determinants of adult age and size examined in the fourth section. The chapter concludes with an overview of the timing of reproduction through life under the heading of 'life cycle pattern'.

Genetics, environment, and demography

At this point it might be useful to clarify some confusions which occasionally appear in the life history literature regarding the term 'phenotypic plasticity'. Occasionally this term is used as a synonym for what is perhaps more clearly designated 'total phenotypic variance' by quantitative geneticists (Brown 1985b). But more usually the term 'phenotypic plasticity' is used to describe a component of (or 'a possible explanation for') total phenotypic variance. A dichotomy is made between variance in some trait of interest (for example, size at reproduction) that is either 'genetic adaptation under specific environmental selective pressures' and variance that is 'non-genetic, environmentally induced phenotypic plasticity' (Byrne et al. 1989).

But total phenotypic variance may be partitioned into three categories: genetic, environmental, and genotype/environment interaction. 'Phenotypic plasticity' (as that term is used by Byrne above) is likely not strictly 'non-genetic', but rather due to the interaction of genotype and environment. This distinction is not trivial. A large literature rests on the assumption that life history traits are (additively) genetic, and thus may respond to selection pressure. To the extent that variance in such traits is due to genotype/environment interaction, they will not respond to selection directly. But it is certainly possible to imagine selection for genes that, for example, trigger maturation in response to a variable environmental cue. If, however, variance in life history traits is controlled entirely by the environment, there can be no evolutionary alteration in characters that must be critical to fitness. This would amount to a major blow to Darwinism.

In this chapter I will adopt what I take to be conventional meanings for two key terms. I will use the term 'genetic' to describe variance that is available for selection to act upon directly, meaning additive (as opposed to dominance or epistatic) genetic variance. The term 'phenotypic plasticity' I will use to describe variance that is certainly not (additively) genetic, but most likely due to genetic/environmental interaction.

Environment

Between 1969 and 1972, van der Schalie and Berry (1973) performed an extensive series of experiments upon the effects of temperature on freshwater snails. Their apparatus enabled them to assess the effects of differences as small as 2°C simultaneously over large numbers of snail populations held in vessels connected by a flow-through system, minimizing variation due to water chemistry. Thirty newborn snails were introduced into each vessel and reared for about 50 days beyond earliest maturity, holding food and photoperiod constant. Example results are shown in Figure 4.1.

First, it is clear that small differences in ambient temperature may have profound effects on snails in culture. As little as 4°C may translate into an order-of-magnitude difference in 50-day production of viable eggs. Note that variation in total production of young was variously attributable to differing survivorships, ages at first reproduction, fecundity, and egg viability. For example, although 30 juveniles were placed in all vessels, in the 14°C *L. stagnalis* experiment first egg laying occurred at day 92, and only 12 individuals were alive when the experiment was terminated at day 111. Thus the (standardized) 334 eggs counted by van der Schalie

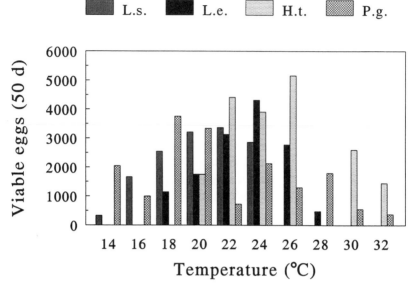

Figure 4.1. Viable eggs laid by four species of pulmonate snails, *Lymnaea emargin-ata, L. stagnalis, Helisoma trivolvis,* and *Physa gyrina,* as a function of temperature (van der Schalie and Berry 1973). Data are standardized to 50 days of reproduction. $N = 30$ individuals to start, although mortality occurred in all vessels.

and Berry, corrected by their 0.987 viability, reflect an average fecundity of (at most) 1.5 eggs/adult/day. In the 20°C *L. stagnalis* experiment, egg laying began at day 64 and 24 snails survived to day 111. In this vessel the 3217 eggs, corrected by 0.998 viability, are attributable to an average fecundity of (at most) 2.8 viable eggs/adult/day.

So if a matter of only a few degrees of temperature can be of such consequence to the biology of snail populations in culture, is it reasonable to expect that demographic data collected from the field will reflect anything other than the immediate environment? To what extent could such data as fecundity, survivorship, and population size collected from natural populations of freshwater molluscs be repeatable?

More than temperature contributes to the wide variance seen within-species in Figure 4.1. It is clear from the rather jagged tracings that considerable within-treatment variance is involved as well. Each datum in Figure 4.1 is a single measurement; it would perhaps be unreasonable to expect replications in a study already so large. But in the course of 'ranging' these experiments – determining the range of temperatures over which to perform detailed tests – individual measurements were occasionally replicated. A second estimate of 18 °C *L. stagnalis* was 1458 viable eggs/30 starting snails/50 reproductive days, compared with the 2458 shown in Figure 4.1. For 24 °C *H. trivolvis*, van der Schalie and Berry's second estimate was 889 viable eggs/30 snails/50 reproductive days (compared with 3905 in Figure 4.1) and for 24 °C *P. gyrina* it was 178 (compared with 2129). What would account for such disparate values (of both survivorship and fecundity) under such controlled circumstances? My suspicion is that there is a great deal of genetic variance for fitness traits, as well as for everything else, hidden in any sample of 30 one-week-old gastropods. In any case, one is again driven to question the repeatability of life history measurements taken from natural populations of freshwater molluscs.

Some consolation may be taken, however, from comparisons between the four species. For even though the variance is great within species, Figure 4.1 shows that between-species variance is still marked. *L. stagnalis* seems adapted to the lowest temperatures, *H. trivolvis* seems to be the most fecund by a substantial margin, and *P. gyrina* seems to have the broadest tolerances. It should also be noted that there was striking variation in age and size at first egg laying, ranging from 21 days and 10 mm in *P. gyrina* to 64 days and 21 mm in *L. stagnalis*. So while one might be tempted to attach little significance to variation in population or life history traits within an order of magnitude, differences much greater than an order of magnitude are commonly recorded.

But although it seems clear that some of the variance shown in Figure 4.1 is interspecific and hence probably has a genetic component, its relevance to natural populations is by no means certain. Van der Schalie and Berry found that three other species of snails, *Helisoma anceps*, *H. campanulatum*, and *Amnicola limosa*, could not be cultured reliably in their apparatus at any temperature. One might speculate that diet, substrate, and/or water chemistry were unsuitable. Then perhaps all the variability shown between the four species in Figure 4.1 is a function of their relative abilities to assimilate lettuce or lay eggs on glass. In the final analysis, the results of van der Schalie and Berry might leave one discouraged about the prospect for meaningful generalization from laboratory studies regarding the population biology of freshwater molluscs.

Population studies

One might expect that the levels of phenotypic plasticity typically observed in laboratory studies of life history traits, when combined with the environmental variability and genetic diversity typical of freshwater mollusc populations, would result in an immense variety of life history pattern. Kilgour and Mackie (1991) sampled 17 Ontario populations of the cosmopolitan pill clam *Pisidium casertanum*, visiting each population 12 times during the course of one year. They recorded broods per year (BPY), proportion of the population with shelled larvae (PPSL), adult length at which shelled larvae first appeared (LSL), number of shelled larvae per reproductive adult of standard size (NSL), lifespan (LS), maximum adult length (MAL), and weight (MAW).

The authors observed two-fold or three-fold ranges in most variables: BPY 1–2, LSL 1.7–2.6 mm, NSL 2.2–7.1, LS 12.5–36 months, and MAL 2.4–5.0 mm. Greater variation was observed in PPSL (2.6–16.7%) and especially maximum adult weight (0.9–16.0 mm). The author's principal component analyses suggested that the populations of warmer environments tended both to reproduce earlier and die earlier. Highly significant negative univariate correlations were obtained between temperature and LSL, MAL, and MAW. Both MAL and MAW showed high positive correlations with the particle size of the substrate, particularly the coarse organic fraction. There was also evidence of a relationship between water quality (calcium concentration, conductivity) and demography (MAW, PPSL), a topic which will be pursued in Chapter 8.

What fraction of such life history variation among populations of freshwater molluscs can be attributed to additive genetic variance? One approach, pioneered by Forbes and Crampton (1942), is to compare populations in a constant environment. Forbes and Crampton noticed that, cultured separately in quart jars, individual *Lymnaea elodes* from a New York population matured much earlier and at a smaller size than individuals from a Connecticut population. The New York population laid significantly fewer eggs, over a significantly shorter time, and had a significantly shorter life span. Since these differences persisted over three generations in a uniform laboratory environment, Forbes and Crampton were justified in concluding that 'their genetic causation seems probable'.

Calow (1981) noticed that the *Lymnaea peregra* populations of sheltered habitats (a canal and a small pond) matured at a substantially larger size than snails from exposed habitats (a lakeshore and a stream). He reared snails from both habitat types in the laboratory at constant temperature, ration, and water quality for three generations. Although no differences were noticed in growth rate, snails from sheltered environments consistently delayed reproduction until a larger size was reached. Their delay in reproduction was compensated for, however, by an increased number of eggs per individual per week once maturity was reached. We will return to further experiments with *L. peregra* along these lines later in this chapter.

Another approach occasionally taken by those interested in the origin of life history variation is reciprocal transplantation. McMahon (1975) found considerable evidence of a heritable component to variation in growth and fecundity comparing two New York *Laevapex fuscus* populations, while Geldiay (1956) could only distinguish environmental variance in growth among *Ancylus fluviatilis* from different sites in Lake Windermere. Hornbach and colleagues (1991) were impressed by the non-genetic component of life cycle variation among their native and transplanted populations of *Musculium partumeium*. The experiments of Hunter (1975) provide a cautionary note in this regard. He compared the growth and fecundity of *Lymnaea elodes* (= *L. palustris*) in an abandoned feeder canal ('MAR') and a swamp ('FAY') in upstate New York. His reciprocal transplant experiments showed the FAY environment to be much richer. The fecundity of MAR animals improved from 34.9 eggs/adult caged in their home environment to 255.7 eggs/adult caged in FAY, while that of the FAY animals decreased from 371.3 eggs/adult caged in their home environment to only 1.4 eggs/adult caged in MAR. Hunter interpreted his findings to demonstrate that *L. elodes* 'does not occur as a series of populations each tactically suited to specific freshwater habitats.' But I should note that each population remained strikingly superior when caged in its home environment, and might argue that the opposite is the case.

In sum, variance for life history traits will not be treated in this chapter as though it arises *either* from phenotypic plasticity *or* from genetic adaptation. Traits as complex as those controlling demography will doubtless have genetic, environmental, *and* interaction components. Experiments to date suggest that some component of freshwater mollusc life history variation will indeed generally be available for selection to act upon.

Reproductive effort

I am aware of several definitions of this term, falling into three categories. Sometimes reproductive effort is taken to mean the clutch size or the total number of newborn offspring produced, or their value in calories or grams of carbon. On some occasions, the yolk volume or proportion of carbohydrate, lipid and protein contained in the eggs or newborn is estimated. Defined in this general fashion, reproductive effort is expected to increase as a function of adult size. By a second category of definition, an attempt is made to remove the effect of adult size. Reproductive effort is estimated as the ratio clutch weight:body weight, carbon channelled to offspring:carbon contained in the female, egg volume:parent volume, or yolk volume:adult growth. By the third category of definition, reproductive effort is based on rates, as for example the percentage of non-respired, assimilated energy allocated to reproduction or the more elaborate cost indicies of Calow (1979) or Hughes and Roberts (1980). But a problem with these last two categories of definition is that linear relationships are assumed between adult and offspring weight or between rate of assimilation and allocation to reproduction. A non-linear relationship between whatever is in a numerator and whatever is in a denominator will result in a statistic that may be misleading. So for my purpose here, I will define the term 'reproductive effort' simply to mean the carbon content of the offspring or eggs produced by a female.

Pisidium

Because of the difficulty of collecting eggs in the wild, more detailed data are generally available regarding natural life history variation in brooding molluscs than in oviparous ones. The pisidiid clams have attracted considerable attention in this regard.

In 1986, Holopainen and Hanski published an extensive review of life history variation in *Pisidium*. Their Table 1 summarized data on 17 variables in 31 populations of *Pisidium* (9 species), including body size, clutch, life span, and timing of reproduction. In almost all respects, *Pisidium* populations display the range of life history variation we have come to expect. For example, it will be recalled from Kilgour and Mackie's (1991) survey of 17 Ontario populations of *P. casertanum* that all combinations of 1 to 2 broods per year and lifespans of 1 to 3 years were observed. In the Holopainen and Hanski tabulation, all 10 *P. casertanum* populations produced one clutch per year, but the northern European populations seem to be much longer lived, up to five years. In fact, over the entire data set, only one population produced two generations per year – the *P. subtruncatum* population of the River Tarrent, England.

Although populations were sampled over two continents, from lotic and lentic environments, temporary or permanent, Holopainen and Hanski were struck by at least one constant. It appears that across all populations, embryo size is rather uniformly about 1.0 mm shell length at birth. Exceptions only include the unusually large *P. amnicum* (about 1.8 mm) and one or two unusually tiny species (e.g. *P. moitessierianum*, 0.6

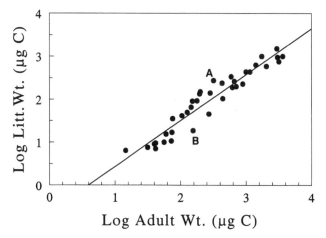

Figure 4.2. Mean annual litter weight in 13 iteroparous populations of *Pisidium* (7 species) as a function of adult weight. (Data collected by Holopainen and Hanski 1986.) Data 'A' and 'B' show the largest residual variation from the regression $\log_{10}$ (litter weight) = 1.071 (adult wt.) − 0.637.

mm at birth). Thus the rather large variation in total reproductive output displayed by populations of *Pisidium* seems to be due primarily to variable litter size. The authors suggested quite reasonably that the maximum number of embryos produced by a parent might be modelled as a power function of parent size.

As mentioned previously, the ratio of litter weight/parent weight has occasionally been used as a measure of reproductive effort. So in their Figure 3, Holopainen and Hanski graphed L/P as a function of log parent ash-free dry weight in 13 populations of 7 species. All of these populations were iteroparous, with from 2 to 4 broods in their lifetime. Recording one datum per brood, the authors plotted 39 points on their Figure 3 and noted a great deal of variation.

From the Holopainen and Hanski Figure 3 I read parent weight and, by multiplying the values on abscissa and ordinate, litter weight. I standardized the data to annual litter weight by summing the 2 broods produced by Tarrant *P. subtruncatum* in a single year and averaging the parent weight. I multiplied all data by 0.5 to convert (roughly) from AFDW to carbon weight (Russell-Hunter *et al.* 1968). The rationale for this conversion will become apparent presently. The (now 37) data are plotted on a log/log scale in Figure 4.2.

Rarely have I seen biological data taken outside a laboratory fit a regression so cleanly. The correlation coefficient (r) is 0.96, the regression coefficient is 1.07 ± 0.05 (s.e.), and the intercept is -0.637 ± 0.13. Converting back out of the log scale we obtain:

$$A = 0.23 \ B^{1.07} \tag{1}$$

where A is the average weight of all offspring produced in a year and B is the average weight of the parent, in μg carbon. It appears that over 13 populations, 7 species and 2 continents, annual reproductive effort in *Pisidium* is a power function of adult weight. Since the size of individual newborns does not generally vary among populations, and since almost all of these populations produce one brood per year, a rough approximation might be that *Pisidium* produce broods to about a quarter of their body weight annually.

As one would ordinarily expect for $N = 37$ data, inspection of 'studentized' residuals shows two samples outside the 95% confidence limits for the regression. The point labelled 'A' in Figure 4.2 corresponds to the second (and final) annual litter produced by *P. henslowanum* at 2 m in Paajarvi, Finland. The observed mean litter size here is unexpectedly large given the average size of the parents, 274 μg C rather than the expected 109 μg C. The point labelled 'B' corresponds to the first (of as many as 4) annual litters produced by *P. casertanum* at 20 m in Lake Esrom, Denmark. Here total annual litter weight is 18.5 μg C, rather than the expected 51.1 μg C. It is tempting to speculate that the *P. casertanum* population in Lake Esrom may be allocating a larger fraction of its energy to growth than to reproduction in its first reproductive year, and that in its last year, the Paajarvi *P. henslowanum* population may sacrifice growth or maintenance to increase allocation to reproduction. But the overall fit between size and reproductive effort is so impressive that small differences may be overemphasized. Each datum is, after all, a complex estimate with a great deal of experimental error attendant.

General survey

Pisidium are quite homogeneous reproductively, and range in adult size a little over two orders of magnitude. One wonders to what extent the relationship in equation (1) might apply across the broad diversity of freshwater molluscs. Browne and Russell-Hunter (1978) have reviewed the subject of reproductive effort in molluscs generally, including in their literature survey one population each of eight freshwater species. Among other attributes, the authors totalled the mean number of young per average female, the average weight of young, and the weight of the average female (in g C, hence my earlier conversion of Holopainen and

Hanski's data from AFDW). Included separately were data on the two annual generations of the limpets *Ferrissia rivularis* and *Laevapex fuscus*, plus first- and second-year data on the iteroparous *Corbicula* and three year's data on *Viviparus georgianus*, for a total of 13 observations, seven semelparous and six iteroparous.

Browne and Russell-Hunter measured reproductive effort using two coefficients: percentage non-respired assimilation annually diverted to reproduction, and the ratio of carbon channelled into reproduction:carbon in the average female. Their analysis of the entire data set, including marine as well as freshwater species, suggested that reproductive effort is higher in semelparous than in iteroparous populations, and that in iteroparous populations, reproductive effort increases with successive breeding seasons. Oviparous species seemed to allot more energy to reproduction than viviparous species.

From reports generally subsequent to 1978 I have collected data on reproductive effort in five additional species, amounting to 12 observations beyond those of Browne and Russell-Hunter (Table 4.1). For the pleurocerid *Leptoxis carinata*, May (pre-reproductive) mean adult weight (mg C) was obtained for three New York populations from Aldridge (1982, table 1). Adult weight was then substituted into population-specific regressions to obtain total egg carbon (Aldridge's table 3, obtained from cage experiments). Data relating mean individual size to fecundity were also obtained from cage experiments for three New York populations of the pleurocerid *Goniobasis livescens* from Payne (1979, appendix C). Adult weights (mg C) were calculated from Payne's population-specific length/weight regressions and average egg weight read from Payne's appendix D. Reproduction was primarily semelparous in all six of these pleurocerid populations.

For *Musculium partumeium*, Burky and colleagues (1985b) provided selection ratios (average number of juveniles born per average parent) for the spring generation of population AM (total), autumn generation of AM, and population DW, as well as average weights of newborns. I estimated average parent weight by back-calculating from stated reproductive efficiencies. Reproduction in the permanent pond (AM) was iteroparous, while in the temporary DW reproduction was semelparous.

Haukioja and Hakala (1978, table 1 and equation 1) provided equations to estimate dry tissue weight from the shell length of reproductive *Anodonta anatina* (= *A. piscinalis*) individuals from 13 populations in Finland. They also provided a general equation (2) to estimate the dry weight of glochidia as a function of shell length. Into these equations I

Table 4.1. *Statistics on reproductive effort in freshwater molluscs*

	Mean adult wt. (mg C)	Mean total annual offspring wt. (mg C)
Leptoxis carinata, New York (Aldridge 1982)		
River TIO	7.30	1.151
River SUS	11.20	3.820
River UNA	9.76	3.260
Goniobasis livescens, New York (Payne 1979)		
Lake GRL	153.0	0.278
Reservoir DER	104.0	0.238
Reservoir JAS	89.0	0.278
Musculium partumeium, Ohio (Burky *et al.* 1985b)		
Pond AM, spring gen.	9.406	3.292
Pond AM, autumn gen.	4.876	0.829
Pond DW	5.139	0.925
Anodonta anatina (= *piscinalis*), Finland (Haukioja and Hakala 1978)		
50 mm, population 95	200.0	28.84
75 mm, population 380	1044	220.4
Anodonta anatina, England (Negus 1966)		
River Thames	290.0	12.50

substituted two values selected to represent the range naturally occurring: a small individual (50 mm) from the population with the slowest growth rate (number 95) and a large individual (75 mm) from the population with the highest growth rate (number 380). The two values of dry glochidial weight obtained, as well as the two corresponding adult dry weights, were multiplied by 0.4 to approximate weight as carbon.

Negus (1966, table 5) estimated that 177 446 individual *Anodonta anatina* inhabited a 250 m reach of the River Thames at Reading. From the observed age frequency distribution (Negus' fig. 14b) and age-specific weights (Negus' table 2) it can be calculated that the 154 378 reproductive adults in this sample had an average wet weight of 5.8 g. Comparing the weights of gravid and non-gravid individuals of the same age, Negus estimated the total glochidial production for *A. anatina* inhabiting this reach to be 38.4 kg wet weight, or 0.25 g glochidia/reproductive adult. Multiplication by 0.05 to convert wet weight to weight as

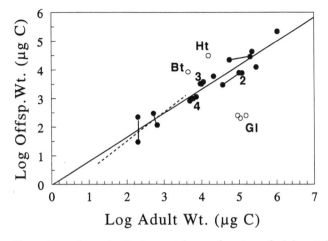

Figure 4.3. Annual offspring weight as a function of adult weight in 25 samples of freshwater molluscs (13 species). Samples of different generations from the same population are joined by line segments. Numbers 4, 3, and 2 count tightly overlapping data points. Open circles locate the outliers Bt (*Bithynia tentaculata*), Ht (*Helisoma trivolvis*) and Gl (*Goniobasis livescens*), not included in the final regression $\log_{10}$ (litter weight) = 0.839 $\log_{10}$ (adult wt) − 0.037. The dashed line locates the regression based on *Pisidium* data shown in Figure 4.2.

carbon (Russell-Hunter *et al.* 1968) gives the figures reported in Table 4.1.

In Figure 4.3 I have plotted all 12 of the data in Table 4.1 together with the 13 data collected previously by Browne and Russell-Hunter on a log–log scale. With $N = 25$ (14 semelparous and 11 iteroparous observations), the regression was highly significant (coefficient = 0.632 with s.e. = 0.145, $r = 0.67$). But judging from studentized residuals, the five data points shown as open circles in Figure 4.3 were well outside 95% confidence limits for the regression. The populations of both *Bithynia tentaculata* and *Helisoma trivolvis* allot an order of magnitude more carbon to reproduction than would be expected from their adult weights, and the three populations of *Goniobasis livescens* over an order of magnitude less. Setting aside these five data, the regression improved from excellent to near-perfect (coefficient = 0.84 with s.e. = 0.063, $r = 0.95$). Once again it appears that reproductive effort in freshwater molluscs may be modelled as a power function of adult weight. Converting the regression of Figure 4.3 from the log–log scale, we obtain:

$$A = 0.92 \, B^{0.84} \tag{2}$$

where A is the average total weight of offspring produced annually and B is the average weight of parents, in μg C.

One cannot fail to notice the similarity between this result and that obtained from the $N = 37$ *Pisidium* data analysed previously (Figure 4.2). The dashed segment in Figure 4.3 locates the position of the Figure 4.2 data, well within the Figure 4.3 confidence limits. With a few exceptions, apparently the power function relationship between parent and offspring weight holds over bivalves and gastropods, ranging in weight over five orders of magnitude.

Mode of reproduction (oviparous or viviparous, semelparous or iteroparous) does not seem to influence reproductive effort. This finding is somewhat at odds with the conclusions of Browne and Russell-Hunter (1978), based in large part on a subset of these same data. Perhaps, however, the non-linearity of the relationship between parent and offspring weight confounded at least some the analysis of Browne and Russell-Hunter. For example, the authors calculated the ratio '$C_R : C_{AF}$', carbon channelled into reproduction compared with carbon contained in the average female. From the regression of Figure 4.3, an adult with log weight $= 2$ (100 μg C) is expected to produce log weight $= 1.64$ (44 μg C) offspring, while an adult of log weight $= 6$ (1 g C) will produce log weight $= 5.00$ (100 mg C) offspring annually. Then reproductive effort estimated as $C_R : C_{AF}$ decreases from 44% to 10% as parent size increases. Since the larger species in Browne and Russell-Hunter's data set (*Corbicula*, *Viviparus*) were both iteroparous and viviparous, one might be tempted to conclude from $C_R : C_{AF}$ ratios that semelparous or oviparous species tend to direct more resource to reproduction.

The observation that 9 semelparous populations allocate no more effort to reproduction than 11 iteroparous ones challenges a good deal of thought on 'life history strategy' in freshwater molluscs. Previously it has been felt that molluscs reproducing semelparously were 'trading off' growth and maintenance for reproduction, i.e. they sacrificed future survivorship in favour of present offspring, much as the salmon dies after spawning. Most of the data in Figures 4.2 and 4.3 suggest otherwise.

We opened this chapter with the observation that in mammals there seems to be no trade-off between survivorship and reproduction (Charnov 1991). Support for the no-trade-off model in molluscs comes from the careful and detailed laboratory work of Rollo and Hawryluk (1988) involving Canadian *Lymnaea* (*Stagnicola*) *elodes* and *Physa gyrina*. Rollo and Hawryluk sampled 30 individuals of each species from a natural population in September and October, after they had reached adult size but some months before egg laying. They reared their animals at 22 °C on a constant excess of three diets: a 'high quality' control diet of fish food (immobilized in agar pellets) and a 50% and a 75% dilution with cellulose. For 35 days they monitored shell growth, wet weight, clutches of eggs, eggs per clutch, and clutch weight, as well as dry weight of food remaining and faeces. Their most interesting finding was that the consumption rates and the assimilation efficiencies of both *Physa* and *Stagnicola* are not ordinarily at a maximum. Both groups of snails can consume more food and assimilate more energy if challenged, and both groups of snails can continue to reproduce, regardless. The allocation to reproduction by *Stagnicola* was not strongly affected by a 50% dietary dilution, although with a 75% dilution, reproduction fell to 25% of control. Reproduction by *Physa* fell 50% in the 50% dietary dilution, but no further in the 75% dilution. Rollo and Hawryluk concluded that freshwater snails 'ordinarily operate at submaximal rates that are homeostatically maintained.' Reduction of available resources does not necessarily engender a reduction in growth, maintenance, or reproduction. And reproduction, I should add, would not necessarily engender a cost in future survival.

The USR model

Figure 4.3 does show, however, that the mean reproductive effort of occasional populations of freshwater molluscs may differ from expectation by an order of magnitude or more. Here I introduce a new 'USR' model of life history variation in freshwater molluscs, inspired by the 'CSR' model for plant strategies of Grime (1979, 1985, 1988).

Grime has suggested that a list of at least 20 attributes of plant morphology, physiology, and life history may be modelled as adaptations to varying levels of competition for resources, environmental stress, and disturbance. **C**-selected (**C**ompetitive) plants are characteristic of productive habitats and crowded vegetation. They are flexibly able to grow tall, produce dense canopies, and/or produce large root surface areas. **S**-selected (**S**tress tolerant) plants are adapted to predictably cold, arid, shaded, or otherwise unproductive habitats. They grow more slowly, reproduce less frequently, and are less (morphologically) flexible than **C**-selected plants, but are longer lived and more (physiologically) able to respond to transient resource availability. **R**- selected (**R**uderal) plants are adapted to rich, productive habitats subject to unselective disturbance, such as grazing, fire, inundation, or sporadic climate fluctuation. They grow rapidly, flower early, and reproduce profusely, but are generally annuals, or in any case short lived.

Although biologically disparate in many respects, both *Helisoma trivolvis* and *Bithynia tentaculata* are 'weeds'. *H. trivolvis* is common throughout North America from Canada to Florida, 'usually found in eutrophic waters: lakes, ponds, streams, and man-made impoundments and ditches' (Eversole 1978). The population from which Eversole obtained the Figure 4.3 datum inhabited the 'mildly polluted and eutrophic' outlet of Owasco Lake, New York. The population was annual and semelparous; eggs were laid from May to September (depressed in mid-summer) and no adults survived a second winter. Interestingly, Eversole also monitored two other New York *H. trivolvis* populations, although not in detail sufficient to estimate reproductive effort, and reported that one in a more mesotrophic environment was iteroparous, adults often surviving through two winters to lay a second batch of eggs.

Data on the population of *B. tentaculata* graphed in Figure 4.3 came from Mattice (1972). This is the rather famous population inhabiting 'richly eutrophic and somewhat polluted' Oneida Lake, New York. *Bithynia* seems to have been introduced from Europe into the Great Lakes during the latter half of the nineteenth century, and in 1876 was first recorded at the Lake Erie mouth of the Oswego River, the river draining Oneida Lake and the 'Finger Lakes' of central New York (Harman 1968). In Baker's (1918) benthic samples of Oneida Lake, to be described in Chapter 9, *Bithynia* was the fifth most abundant of about 23 gastropod species. But by Mattice's day most the Oneida Lake gastropod species had disappeared, and *Bithynia* had become dominant (Harman and Forney 1970). Mattice reported that this population is annual and semelparous, although this is not certain (Tashiro and Colman 1982); populations of *Bithynia* elsewhere display iteroparity (Fretter and Graham 1962).

I suggest that the *H. trivolvis* population of the Owasco Lake outlet and the *B. tentaculata* population of Oneida Lake tend to allot unusually high proportions of available energetic resource to reproduction, such that growth and/or maintenance of normal metabolic processes is noticeably reduced ('reproductive recklessness', Calow 1979). Following Grime, I will call such populations '**R**-adapted', here for **R**eproduction. I suggest that **R** populations will generally be widespread and/or 'weedy' and will tend to be found in eutrophic and perhaps temporary environments. They display mean reproductive efforts over one order of magnitude greater than expected from their body size. Semelparity is likely, but not required. Their large reproductive outputs may manifest themselves as exceptionally large numbers of small offspring, or as smaller numbers of exceptionally large offspring.

As mentioned previously, Figure 4.3 also indicates that the three populations of *Goniobasis livescens* examined by Payne (1979) seem to allot an unexpectedly small amount of energy to reproduction, judging from average adult weight. Pleurocerids are generally considered to be 'clean water organisms' (Dazo 1965). Perhaps as a consequence, *G. livescens* has disappeared from Oneida Lake since 1918, even as *B. tentaculata* has flourished (Harman 1968, Harman and Forney 1970).

The three populations graphed inhabit small but permanent mesotrophic lakes and reservoirs in central New York, with water of low to moderate hardness. Because pleurocerids are generally more southerly in their distribution and more characteristic of lotic than lentic habitats, Payne considered all three of his *G. livescens* populations to be stressed. Population densities nonetheless seem to have been fairly high, and productivities within the ordinary range for freshwater snail populations. Females of one population attained reproductive maturity at age 2 years, while those of the other two populations did not mature until age 3 years. Payne suggested that all were primarily semelparous; adults failing to survive the winter after reproduction. Some evidence was produced, however, that reproduction might occur iteroparously, in both years 2 and 3, as is normal for pleurocerids.

I suggest that such populations as these three of *G. livescens* in New York be considered **S**- adapted, for **S**tress-tolerant. Their habitat is poor (cold and oligotrophic), but permanent and predictable. Individuals from **S** populations will tend to allot an unusually large fraction of available energy to maintenance, such that their somatic growth rates are markedly reduced and/or their reproductive efforts depressed much below expectation for their adult sizes. But high habitat predictability ensures that lower fecundities do not necessarily translate to low population sizes or population growth rates. They may be semelparous or iteroparous.

Note that individuals from **S** populations may show reduced allocation to somatic growth rather than reproduction. The famously long-lived *Margaritifera margaritifera* populations of Holarctic upland streams come to mind in this connection, although I am unaware of any estimates of their reproductive effort. It is thus possible that **S** populations are hiding among the data fitting the regression so nicely in Figure 4.3. A regression of adult weight on age might identify latent **S** populations. In any case, reproductive effort an order of magnitude lower than expected from adult weight will be sufficient (but not necessary) to identify a population as **S**-adapted.

Among the most detailed studies of life history variation in freshwater molluscs available to date is that of the *Viviparus ater* populations of Lake Zurich, Switzerland, and Lake Maggiore, Italy (Ribi and Gebhardt 1986, Ribi *et al.* 1986, Gebhardt and Ribi 1987). In both populations, snails first mature in year 2 and may live up to 10 years, reproducing annually. But individual growth rates and age-specific survivorships average much higher in the harder waters of Lake Zurich (11.5 mg/l $CaCO_3$) than in Lake Maggiore (3.5 mg/l $CaCO_3$). Since the shells of Maggiore animals are pitted and dissolved, while Zurich shells appear normal, it seems reasonable to suggest that the Maggiore environment is abnormally stressful.

From cage experiments, Gebhardt and Ribi (1987, table 1) have obtained data on the reproductive effort of each of these two populations at ages 2–6 years. Their results are expressed in grams live weight, including shell, and as such are difficult to compare with the data of Figure 4.3. At first reproduction, age 2 Zurich females average 7.56 g and produce 2.32 g offspring, and by age 6, females average 8.29 g and produce 2.50 g offspring. The slope of the relationship between the average live weight of adult females and the average live weight of their offspring each of the 5 years does appear to fit equation (2), giving a very approximate conversion of 3×10^{-6} g C/g living *Viviparus*.

Lake Maggiore reproductive efforts would fall below expectation, assuming this very approximate conversion. Two-year-old Maggiore snails average only 5.7 g, with an average of only 1.37 g offspring, somewhat over 1 g less than expected from their size. Average reproductive effort increases with the size of the adult at age 3, but then remarkably, declines steadily through ages 4, 5, and 6. In Lake Maggiore, six-year-old snails average 6.75 g live weight (less than Zurich two-year-olds), and bear only 0.97 g offspring, perhaps over 2 g too low. It seems fairly clear that the Maggiore population of *Viviparus ater* is responding to stress with reduced growth, reproduction, and perhaps (judging from survivorship) maintenance as well. But is the population 'S-adapted', or simply stressed? An accurate conversion from grams live weight to grams carbon for both *V. ater* adults and newborns would be required to tell.

The 20 data fitting the regression line in Figure 4.3 (as well as all 37 data on Figure 4.2) suggest U-populations, Undifferentiated with respect to reproductive effort. Their allocation to reproduction is a power function of body size, as derived previously. They are not expected to exploit temporary or unpredictable habitats, nor are they expected in habitats that are predictably poor. But because U populations inhabit environments that are stably favourable, their energy allocation to maintenance of normal metabolic function is comparatively small. They may channel resources to reproduction while still small, or postpone reproduction and channel resources to growth instead. Any combination of life history attributes is possible; reproduction may be early or late, iteroparous or semelparous.

It is in rich, stable environments that competition would be most likely to occur. But here my model departs from that of Grime. Because competition for light, water and nutrients can be so obvious and pervasive in the plant communities of rich, stable environments, Grime's interpretation of such vegetation as 'Competition-adapted' was justified. But the evidence for competition in freshwater mollusc communities is not as strong as one sees in a climax forest. Although it is tempting, for example, to interpret the great diversity of life history observed in the 23 gastropod populations of 1918 Oneida Lake as an adaptation to minimize interspecific competition, other factors (predation, phylogenetic constraint) have certainly been involved. We will pursue this subject further in Chapter 9. Meanwhile, it should be borne in mind that the USR model has been proposed here to account for varying levels of reproductive effort, not interspecific interactions, and that U-populations are by definition undifferentiated in this regard.

Such predictions regarding other life history variables as emerge from USR, however, are collected in Table 4.2. It will be recalled from Chapter 2 that the sporadic occurrence of hermaphroditism and self-fertilization in the unionoids, as well as the predominance of these phenomena in the corbiculoids, seems associated with colonizing ability. Adaptation for colonization also seems to have played a role in the reproductive diversity displayed by the hermaphroditic pulmonates and the parthenogenesis displayed by certain prosobranch groups (Chapter 3). A line noting the apparent relationship between R-adaptation and asexual reproduction has been added to Table 4.2. The role of R-adaptation and S-toleration in the evolution of maturation size will be explored later in the present chapter.

Size at birth

In the preceding section we focused on the (rather unsurprising) fact that the total weight of young produced annually is a function of adult body size, in most cases. Our interest now turns to questions regarding the allocation of a parent's reproductive effort to juveniles. In particular, the trade-off called 'a few large vs. many small young' has been identified by Sterns (1976) as one of central interest.

Table 4.2. *The USR model of life history variation*

	R	S	U
Reproductive effort, relative to body size	High	Low	Normal
Asexual reproduction (cf. Chapters 2 and 3)	More likely	(–)	(–)
Growth and/or maturation (cf. section on maturity)	Rapid	Slow	(–)
Lifespan	Short	Long	(–)
Survivorship after reproduction	Often semelparous	(–)	(–)
Common in rich but unpredictible environments	Yes	No	No
Common in poor but stable environments	No	Yes	No
Adaptation to competition (cf. Chapter 5)	Intraspecific	Intraspecific	Interspecific

Note:
(–) indicates that a trait is undifferentiated, i.e., that no special prediction is made.

Pisidiids

We have previously touched on the observation of Holopainen and Hanski (1986) that the size of *Pisidium* larvae at birth seems to be rather fixed at about 1.0 mm shell length, over 31 populations of 10 species. Although this is generally true, the smallest species (*P. moitessierianum*) does give birth to the smallest juvenile (0.6 mm), while the unusually large *P. amnicum* bears the largest juvenile (up to 2.2 mm). From the 19 data graphed as circles on Figure 4.4, Holopainen and Hanski suggested that to a limited extent, size at birth might be constrained by parent size. Pill clams of the genus *Musculium* tend to mature slightly larger than *Pisidium*, perhaps 5–14 mm. Heard (1977) included data on the maximum length of shelled larvae for populations of *Musculium lacustre*, *M. partumeium*, *M. securis*, and *M. transversum*. This would be roughly equivalent to Holopainen and Hanski's 'birth size'. I have graphed these data as triangles in Figure 4.4, taking typical adult sizes from the scaled illustrations in Burch (1975). It appears that *Musculium* birth sizes are also rather invariant, here around 2.0 mm. Heard also included data on the largest shelled larvae found in 10 species of *Sphaerium*, ranging in adult

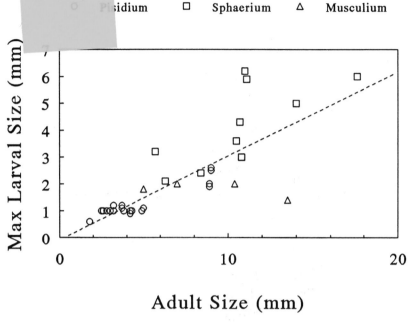

Adult Size (mm)

Figure 4.4. Shell size of the largest shelled larva as a function of the shell size of a typical adult in the three genera of pill clams. The regression ($r=0.80$) is (larval size) $=0.325$(adult size) -0.063. (Data from Heard 1977 and Holopainen and Hanski 1986.)

size from 6mm to 18 mm in shell length (Burch 1975). Their larvae seem more variable, ranging from 2mm to 6 mm, and a relationship between birth size and adult size seems fairly clear.

Summing over all 33 pisidiid species, the relationship between the size of a typical adult and the maximum size of the larvae it bears goes from 'fairly clear' to 'quite convincing'. The simple linear regression:

$$y=0.325x-0.063 \tag{3}$$

fits the data with a value of $r=0.80$. Evidently larger pill clams tend to have larger offspring. But two points need to be emphasized. First, the entire range of size at birth shown in Figure 4.4 is 0.6–6.2 mm, one order of magnitude, while the range in reproductive effort, just for the genus *Pisidium* alone, is three orders of magnitude (Figure 4.2). Clearly variance in offspring size is not as important as offspring number in accounting for variance in pisidiid reproductive effort. Second, much of the variation in offspring size seems to be related to higher levels of phylogeny. *Pisidium*

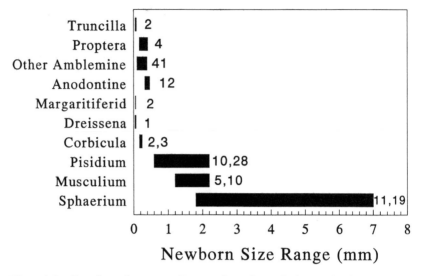

Figure 4.5. Bars show the range of size at release for 10 freshwater bivalve taxa. The number of species surveyed for each taxon is given at right, followed by the number of populations, if different. Size was measured across shell at maximum dimension.

clearly bears the smallest young, *Musculium* intermediate, and *Sphaerium* the largest.

General survey

As brooders, one might not be surprised to learn that the three pisidiid genera give birth to the largest offspring among the freshwater bivalves. In unionaceans, for example, glochidia are released at sizes an order of magnitude smaller. From Chapter 2 it may be recalled that glochidia are classified into three types: the 'hooked' type characteristic of the subfamily Anodontinae, the 'hookless' type of most Ambleminae, and the 'axe-head' type of the North American genus *Proptera*. I have surveyed the works of Lefevre and Curtis (1912), Coker *et al.* (1921) and Kat (1984) and obtained glochidial sizes (maximum shell dimension) for 61 taxa of freshwater mussels, primarily from scale drawings. Figure 4.5 shows that hooked (anodontine) and axe-head (*Proptera*) glochidia tend to be released at larger sizes, up to 0.47 mm and 0.41 mm, respectively. Among amblemine genera, *Truncilla* is distinguished by an unusually small adult and juvenile size, the two species surveyed releasing glochidia of 0.05 mm and 0.08 mm. Mussels of the family Margaritiferidae also produce exceptionally small glochidia. Thus a large phylogenetic component to offspring size is as apparent in unionaceans as it was in pill clams. But

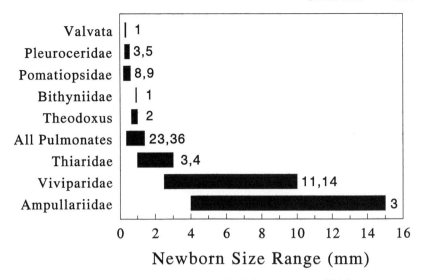

Figure 4.6. Bars show the range of size at birth for nine taxa of freshwater gastropods. The number of species surveyed for each taxon is given at right, followed by the number of populations if different. Size was measured as maximum shell dimension.

combining all 61 mussel species surveyed, glochidial sizes range only from 0.05 mm to 0.47 mm, not quite an order of magnitude.

Figure 4.5 also shows data on size at release for *Dreissena*, *Corbicula*, and enlarged samples of *Pisidium*, *Musculium*, and *Sphaerium*. Again, most of the *Pisidium* data, expanded from 19 to 28 populations, come from Holopainen and Hanski (1986), while *Sphaerium* data (expanded from 10 to 19 populations) are primarily from Heard (1977). The effect of enlarged data sets are negligible, however. Judged generally by largest shelled larva, size at birth still ranges only from 0.6 to 2.2 mm for *Pisidium* populations, while increasing slightly to 1.8–7.0 mm for *Sphaerium* and 1.2–2.2 mm for *Musculium*.

Figure 4.6 shows the result of a fairly thorough literature search for data on shell size at first release in freshwater gastropod taxa. For inclusion in this data set, I required direct observation of eggs and hatchlings (or liveborn) in the laboratory. About 55 references were ultimately discovered containing sufficiently detailed information. Size was taken as the maximum shell dimension at hatch, or release in the case of the brooders (*Potamopyrgus*, Thiaridae, Viviparidae). Note that in some cases, juvenile gastropods have shells shaped so differently from those of adults that their maximum dimension will be measured in a different plane.

Over all 74 freshwater gastropod populations, the shell size of juveniles at release seems to range not quite two orders of magnitude, from 0.3 mm in *Valvata piscinalis* to 15 mm in the ampullariid *Pomacea urceus*. But just as was the case with the bivalves, Figure 4.6 shows that most of the variance is attributable to levels of classification higher than species. In no family does offspring size vary more than fourfold. Especially striking in their constancy are the pulmonates, where hatching size in 36 populations, taken about evenly from four families, ranges only from 0.36 mm to 1.4 mm.

There seems to be some weak tendency for taxa with large-bodied adults (ampullariids, viviparids) to produce large offspring, and small taxa (*Valvata*, pomatiopsids) to produce smaller offspring. But pleurocerids have rather large adults and small juveniles. One can also detect a positive relationship within taxa (such as pulmonates) between adult size and size at hatching, although perhaps not as marked as that shown in Figure 4.6. The pulmonate with the smallest adult size surveyed (the planorbid *Armiger crista*) had the smallest hatchling size, while the largest adults (*Lymnaea stagnalis*) have among the largest hatchlings (1.3 mm).

So in summary, over all freshwater molluscs, shell size at release ranges from 0.04 mm to 15 mm. The great majority of this variance seems to be attributable to modes of reproduction that differ strikingly among diverse higher taxa. One would not expect *Dreissena*, with planktonic larvae, to produce young comparable in size to parasitic glochidia, much less egg-laying pulmonates or brooding thiarids. Within higher taxa, variance is generally much less than an order of magnitude, and seems to be positively related to adult size. The 'many small vs. few large young' trade-off does not seem as important in freshwater molluscs as it may be in other groups.

Maturity

We have now seen that both reproductive effort (as total weight of young) and individual offspring size (setting aside a phylogenetic component) are functions of the size of the mother in freshwater mollusc populations. What determines the size at which a snail or bivalve matures? Our observations regarding offspring size are reminiscent of those for the much better studied mammals (Charnov 1991). In Charnov's view, the positive correlation observed between parent size and newborn size in mammals may be an adaptation to speed juveniles through an especially perilous childhood. And a view prevailing in the literature of life history evolution holds that delayed maturation may also be an adaptation to similar selection pressures (Reznick *et al.* 1990). If the survivorship schedule is deeply concave, selection will favour those individuals allotting all their energetic resource to growth through the most hazardous age classes, deferring reproduction. But if early survivorship is high relative to late survivorship, selection will act to promote reproduction at the earliest possible date.

Somatic growth rates in freshwater molluscs are commonly shown to be sensitive functions of temperature, diet, water quality, and other environmental variables. Thus size, rather than age, is generally found to be a better predictor of reproductive maturity. The heritable component of size at maturity may be large. In the laboratory, individual *Biomphalaria glabrata* vary in shell diameter at onset of (selfed) egg laying from 5 mm to 16 mm. Richards and Merritt (1975) produced an outcrossed *B. glabrata* line and subjected it to bidirectional selection. In four generations the mean size at onset of selfed-egg production for the low-selected line was 8.6 mm (range 7–11 mm), and for the high it was 12.1 mm (range 11–16 mm), the trait apparently showing almost no overlap. F_1 hybrids between snails of these two lines were intermediate in shell diameter at onset of egg laying, although nearer the large-diameter parent. Although Richards and Merritt did not monitor age, they stated that there was 'no indication that age at onset of egg laying was consistently correlated with diameter.'

Wethington and Dillon (1993) have also documented significant differences in size at first reproduction between isofemale lines of *Physa* reared in a constant, laboratory environment, although we have not performed any selection experiments thus far. As in *Biomphalaria*, age at first onset of selfed egg laying in *Physa* seems to be a function of ambient temperature, among other things. But size at onset of egg laying seems to be genetically controlled, and I have an extensive data set showing growth rate in my laboratory populations to be highly heritable.

Lam and Calow (1989a) quantitatively sampled the *Lymnaea peregra* populations at sites on three rivers near Sheffield, England (Sheaf, Rivelin, and Don) from January 1985 to December 1986, noting several quite striking life history differences. Survivorship over the first few months of life was similar in all three populations, but much greater for the Don population during the autumn, winter, and early spring. (The Don site was protected from strong current, and artificially warmed by pipes from an adjacent steel factory.) That population began breeding earlier than the Sheaf and Rivelin populations, at a much smaller mean shell length (6.2 mm in the 1985 cohort, as compared with 10.9 mm and 11.6 mm). Maximum lifespan for individuals of all populations was about one year. Data on survivorship and shell growth at two of Lam and Calow's populations are shown in Figure 4.7.

The authors considered the possibility that Don snails may have been induced to breed earlier in the season by the warmer water temperatures they enjoyed. But they noted that the mean growth rate in the cooler Sheaf waters actually exceeded that observed in the Don. In late March 1985, 24 weeks after hatching, the Don snails were 6.2 mm in shell length and laying eggs, while Sheaf snails were 7.6 mm and not. Why did Sheaf snails continue to grow, postponing reproduction until May? It does appear that Sheaf snails ultimately produced more eggs, on average, as would be expected from their larger sizes at reproduction. Although there were no differences between the two 1985 cohorts in mean egg volume or percentage hatchability of the eggs they laid, 43 egg capsules collected from the Sheaf site contained on average 57.2 eggs, while 40 from the Don showed a mean of only 15.9. Lam and Calow indirectly estimated a mean of 521 eggs per individual Sheaf snail, and only 50 eggs per snail in the Don. Thus in spite of the lower survivorship to reproduction at Sheaf, snail population density there in the summer of 1985 far exceeded that observed in the Don.

In June, 1985, Lam and Calow (1989b) brought 50 adult snails from each of their three study sites into the laboratory, collected a sample of their newborn, and initiated both controlled-breeding and mass–culture experiments. In the former, F_1 individuals were reared in small containers at 12 h light:12 h dark at 20 °C for 100 days, paired randomly with another of its own population until onset of breeding, then re-isolated. F_2 individuals were collected and reared in a similar fashion. In the latter experiments, 50 F_1 individuals from each of the three populations were reared in large, continuously flowing tanks at natural light and ambient temperatures (mean $= 16$ °C, s.d. $= 4$ °C). A great variety of life history variables were measured on all snails, including growth rate, shell length at reproduction, egg volume, and total number of eggs produced.

From the controlled-breeding experiments, Lam and Calow observed significant intrapopulation variance for most life history traits. Between populations, however, results were equivocal. The authors did observe a highly significant difference in shell length at onset of breeding in the mass-culture experiments, consistent with field observations. The Don culture began laying eggs at a mean shell length of 7.5 mm (10 weeks), while the Sheaf postponed to a mean shell length of 10.2 mm (13 weeks).

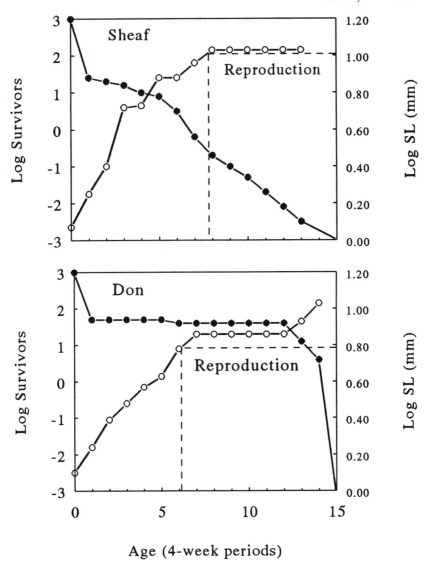

Figure 4.7. Growth and survivorship data of Lam and Calow (1989a) on 1985 cohorts of *Lymnaea peregra* from the Rivers Sheaf and Don, England. Time is measured in four-week intervals from the May date of maximum juvenile density. The left axis (and closed circles) show age-specific survivorship (l_x, standardized to 1000 individuals at start), while the right axis (and open circles) show mean shell length (SL). The dashed lines mark the size and approximate date of the onset of egg laying.

There was no difference in the total number of eggs produced, perhaps because the Don ultimately bred for 26 weeks, Sheaf only 22. Significant among-population variance in shell length at maturity was not observed in the controlled-breeding experiments, either at F_1 or at F_2 generations. My reading of these results would be that the Sheaf and Don populations do, in fact, show significant genetic divergence in their mean sizes at reproduction as shown in Figure 4.7, but that this effect may be extinguished if reared at a constant temperature of 20 °C. And I would suggest that the selection agent has been differential survivorship in the vicinity of reproductive age, quite possibly in the form of a steel mill.

The contrast between the *L. peregra* populations of England and the *Bulinus globosus* populations of Zimbabwe studied by Shiff (1964c) is in some respects jarring. Zimbabwean *Bulinus* mature in about 10 weeks and reproduce year-round, except in temporary ponds during the dry season. But one can see precisely the same relationships between survivorship and age at reproduction within a single *Bulinus* population as were manifest between populations of English *Lymnaea* or Oklahoma *Physa*.

The absence of readily distinguishable cohorts introduces an added challenge to the study of *Bulinus* demography. Shiff (1964b) preceded his field studies with careful observations on the relationships between shell growth, age, and temperature in the lab. He estimated *lx* in the field by establishing arbitrary 'cohorts' as groups of snails newborn on each of five dates during the year. He followed the survivorship of each 'cohort' from month to month by counting the number of snails in their expected size range. Shiff's (1964c) 'Table II' reported cohort ages in 12-day ranges (since 12 days are required for egg hatch) and included estimates of egg fertility. For my Figure 4.8, I standardized viable embryos (not total eggs) to 1000 and age to the last of the 12 days for each period.

Shiff provided a thorough description of both his study population and the climate of Zimbabwe, doubtless with those of us at home in temperate regions of the northern hemisphere in mind. His *B. globosus* population inhabited two small, interconnected ponds (area 448 m², maximum depth 80 cm). Rains came in mid-November, 1962, and on November 22 the habitat started to fill. Shiff took his first population sample on November 27, by which time the (really very few) *Bulinus* that had survived the dry season had already begun to lay eggs. Water temperatures continued to warm through the next several months, with flooding (and much reduced survivorship) in January and February. The upper half of Figure 4.8 shows that the November 27 'cohort' reproduced in February

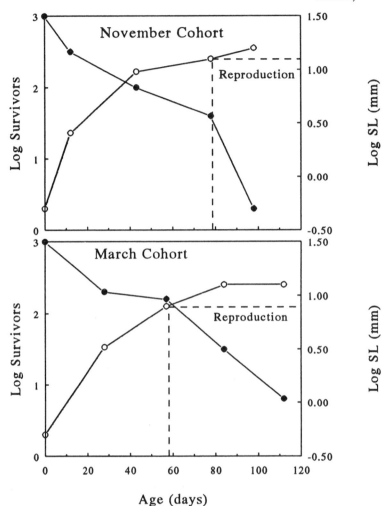

Figure 4.8. Growth and survivorship data of Shiff (1964c) on two cohorts of *Bulinus globosus* from a small pond in Zimbabwe. Time is marked in days from hatching; remainder of the details are as in Figure 4.7.

and March at about 13 mm shell length. The 'cohort' born March 6 (Figure 4.8 at bottom) experienced environmental conditions much different from those of its parents. Temperatures cooled, the rains ceased, and survivorship (at least through ages 20–60 days) was much higher. Maturity was achieved significantly earlier in the March cohort than in the November, at about 8 mm shell length. The pond continued to shrink, however, to dryness in August–October.

Thus again we see two snail samples of different survivorship schedules and sizes at reproduction. Rather than from different populations, however, the samples come from different cohorts of a single population in a fluctuating environment. Heavy rain and flooding around age 60 days seems to have inflicted increased mortality in the November cohort and prompted a delay in maturation.

As *B. globosus* is one of the more famous intermediate hosts of human schistosomiasis, much interest in its biology is ultimately parasitological. One might imagine that the levels of parasitic castration might fluctuate as dramatically as other features of the Zimbabwean environment. Unfortunately, Shiff collected no data on parasitism in his *B. globosus* population. I am aware of only one study in which parasitism has been monitored well enough to adjust age-specific survivorships: the work of Brown *et al.* (1988) on *Lymnaea elodes* populations in three northern Indiana ponds.

Brown's ponds A and B were quite vernal, reliably drying in mid- to late-July annually, while pond F was more permanent, drying only in exceptional years. He and his colleagues took monthly quantitative samples of snails during field seasons (April–August) with an Ekman dredge, examining a subsample of each for trematode infection. Since cohorts were discrete, survivorships could be estimated from declines in (size-specific) abundance. Individual snails were reared in containers in each pond to estimate age-specific fecundity. The populations inhabiting ponds A and B did not reproduce sufficiently to replace themselves, while the population in pond F did.

Figure 4.9 shows that survivorship to months 9 and 11 was much greater in pond F than in pond A. But adult *L. elodes* of pond F reproduced semelparously and did not survive beyond month 13, while in spite of the harsher environment they experienced, some individuals of the pond A population survived a second year to reproduce iteroparously. Parasitism increased from 5% to 25% over months 9 to 13 in pond F and 5% to 35% from months 9 to 24 in pond A. Although certainly not of negligible consequence to the snail populations, these levels of parasitism did not affect the overall contrast between the surivorship schedules in ponds A and F, nor between the life histories displayed.

One might have predicted from survivorship schedules that pond F individuals would mature at a smaller size than pond A individuals. This does not seem to have been the case; snails from both populations reached adulthood at a mean shell length of about 14 mm. But somatic growth

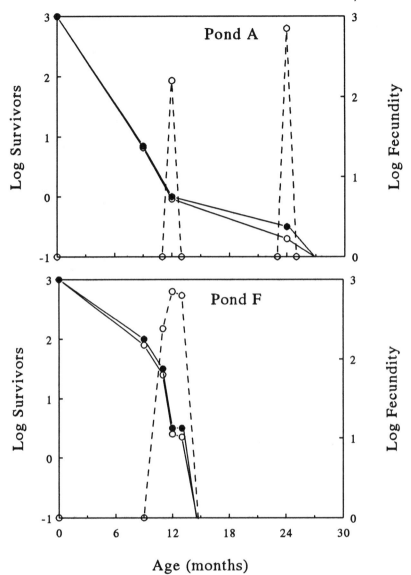

Figure 4.9. Life history data on two populations of *Lymnaea elodes* from ponds in northern Indiana (Brown *et al.* 1988). Survivorship (left axis, solid lines) is standardized to 1000 individuals at date of maximum egg production. The upper survivorship curve shows uncorrected 'head count' while the lower curve is corrected for parasitic castration. Mean fecundity (right axis, dashed lines) is given in eggs per individual per month, corrected for parasitism.

rates were greater in pond F, so that this minimum size was reached in 11 months rather than 12.

By far the more spectacular life history difference in the two populations was their reproductive effort. Although adults were of comparable sizes, individual population F snails laid on average $242 + 715 + 637 = 1594$ eggs (corrected for parasitism), over three months of reproduction, then died. Brown and colleagues estimated that pond A snails produced a corrected average of just 163 eggs in the single month of reproduction available to them in their first year, before dry-out. At least some, however, survived to breed again.

It would appear that the pond F *L. elodes* population studied by Brown *et al.* (1988) displays the attributes of **R**-selection (Table 4.2). The environment seems to be very rich, promoting good growth and high survivorship (9.8% survive to age 9 months, as compared with 0.7% at pond A). But the environment is also unpredictable. (Brown reported that pond F did dry occasionally, although not in the years of his study.) In response, an annual, semelparous life cycle seems to have evolved, population F individuals producing an order of magnitude more offspring than the iteroparous population A. But how much of this phenomenon is additively genetic?

Brown (1985a) used transfer experiments of two different designs to answer this question. All snails were held in small, floating chambers with screened openings, either two or four per chamber. In his 'reciprocal transfer' experiment, Brown reared juvenile snails in all combinations of two ponds, two populations, and two densities for 50 days, until pond A dried. He recorded age and shell length at maturity, shell growth increment, clutch size, and total fecundity. For his 'parent–offspring' experiment, Brown collected eggs from pond F chambers laid by parents of both populations. He held eggs and newborns in the lab at 5 °C over winter, then returned the juveniles to pond F chambers until the last died in mid-November. He recorded the same five life history traits as in the reciprocal transfer experiment, plus age and shell length at death.

Brown clearly demonstrated significant genetic divergence in life history traits between populations A and F of a remarkable (and, to me, unexpected) nature. The tendency for population F snails to grow faster is heritable. Reciprocal transplant experiments demonstrated that, reared in either environment at either density, population F snails showed a significantly greater growth increment. Their ages at maturity were not different from population A snails, but of 45 values of the *F*-statistic, the highest was a population difference in shell length at maturity ($F = 78.7$, with 1,88 d.f.). Reared at either density in pond A, population F snails reached maturity at a mean shell length of 20 mm while population A snails were reaching maturity at the (more normal) 15 mm.

Brown's parent–offspring experiments confirmed the genetic nature of the population difference in growth rate. The six other values of F from his reciprocal transfer experiment that I would judge significant at the (Bonferroni-corrected) 0.05 level included pond (i.e. environmental) effects on shell length at maturity and shell growth increment, as well as age at maturity and total fecundity. This is unsurprising, since these two ponds were initially selected for study because of their striking environmental differences.

At first these results would seem to directly contradict the thesis I have advanced in this section, that differences in survivorship schedules such as displayed in these two populations (Figure 4.9) should promote maturation at a smaller size in population F, not larger. But the key is to note that reproduction may be accelerated in two fashions: lowering the size at maturity or raising the growth rate. Even though their size at maturation may be larger, population F *L. elodes* mature a month earlier than pond A snails. Although maturity is best considered a function of size in freshwater molluscs, survivorships generally seem more time specific than size specific. Thus age at maturity may be more critical, in an evolutionary sense, than size at maturity.

Lymnaea elodes, it may be recalled, was one of the two snail species in the dietary dilution studies of Rollo and Hawryluk (1988). Individuals from their Canadian population of *L. elodes* reared in the lab were not growing or reproducing at maximal rates, but rather at submaximal rates homeostatically maintained. I suggest that the Canadian population of *L. elodes* studied by Rollo and Hawryluk was, like most mollusc populations, Undifferentiated with respect to life history. I further suggest that R-selection may have worked to bring the growth rate displayed by Brown's pond F *L. elodes* much closer to maximal.

In Table 4.2 I have included a line labelled 'growth and/or maturation', indicating my prediction that growth rate will be more rapid in R-populations, and/or that maturity will come early. This I would take to be an adaptation to environmental conditions which are generally good, rich in nutrients and safe, but which may unpredictably turn hostile. S-populations inhabit poor environments, where juvenile survivorship may be predictably low, and hence maturation delayed.

Life cycle pattern

All of the factors we have discussed thus far, survivorship, age and size at reproduction, and semelparity vs. iteroparity, can be summarized as a 'life cycle pattern', the timing of reproduction through life. It is difficult to imagine an experimental design that could test the heritable component of overall 'life cycle pattern'. Brown's (1985a) study populations of *Lymnaea elodes*, it may be recalled, included both an annual, semelparous population F and a perennial, iteroparous population A. Reproduction ordinarily spanned three months in population F and 12 months in population A; clearly Brown's 50 day reciprocal transplant experiments could not hope to provide a fair estimate of effort. A single-season comparison of fecundity for both populations (in environment F) might have been obtained from his parent–offspring experiment, but his experimental means were only in the 600–800 egg range, half the 1594 observed from pond F in the wild. This seems likely to have been a cage effect. And clearly, Brown could not hope to measure semelparity/iteroparity, when fewer than 0.1% of population A snails were expected to survive to a second breeding season.

But there is no reason to suspect that variance in overall life cycle pattern is any more than the sum of the variances of other attributes (such as growth rate or size at maturity) whose heritable component has, at least occasionally, been established. Thus it would seem fair to proceed.

Unionoids

All unionoid populations are iteroparous and perennial, displaying life cycle patterns classifiable into a fairly manageable number of discrete categories. Sterki (1895) recognized two groups: short term ('tachytictic') breeders gravid only in the summer, and long term ('bradytictic') breeders remaining gravid through the winter for glochidial release the following spring. 'Breeding' may not be the best term for holding glochidia in the marsupium, but its use has historical precedent. Ortmann (1909) counted 13 tachytictic and 36 bradytictic species among the fauna of Pennsylvania. The tabulation of Coker *et al.* (1921) included (to my eye) 36 tachytictic North American species and 18 bradytictic, with data on another 12 species too fragmentary to permit classification.

Spawning time did not enter into the original tachytictic/bradytictic system used by the early workers to classify unionoid life history. But with the recognition that some populations may spawn in the autumn came the realization that animals gravid in the winter might be short term

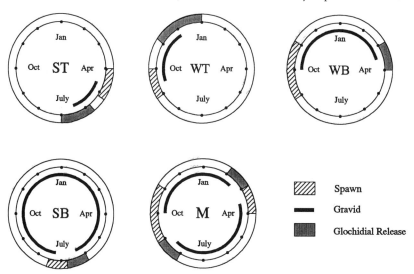

Figure 4.10. Example life cycles for the unionoids, illustrated by northern hemisphere examples given in the text. *ST*, summer tachytictic, *WT*, winter tachytictic, *WB*, winter bradytictic, *SB*, summer bradytictic, *M*, multicyclic.

(tachytictic) 'breeders.' So a third category, 'winter tachytictic', was distinguished from bradytictic and (the more ordinary) summer tachytictic life histories by Howard (1915). By extension, I would add a fourth category. Bradytictic populations might spawn in either spring or autumn, just as is the case for tachytictic populations. At least four life history categories would seem distinguishable, given a single reproductive period per year: summer tachytictic (spawning on rising temperatures and releasing glochidia before winter), winter tachytictic (spawning on falling temperatures and releasing glochidia in the autumn or winter), summer bradytictic (spawning on rising temperatures and holding glochidia through the winter), and winter bradytictic (spawning on falling temperatures and holding glochidia through the winter). Examples of these life cycles are shown in Figure 4.10.

The *Pleurobema cordatum* population of the Tennessee River (in Alabama) displays a summer tachytictic (*ST*) life history. Yokley (1972) collected adult mussels monthly, noting both gametogenic stage and gravidity. Gametogenesis occurs in the autumn and winter, followed by a fairly synchronous spawning in April and early May. Glochidia appear in late April, and gravid females may be found in the population until early July, with June the peak month for glochidial release.

Kondo's (1987) survey of gravidity in seven Japanese unionids extended for over two years, with 10–40 adults collected for each species monthly. Six of these populations were summer tachytictic as above, spawning as the water temperatures increased, albeit at differing trigger temperatures. But Kondo's population of *Inversidens yanagawensis* spawned in September on both years, reached peak gravidity in October or November, and released glochidia in November and December. Pale chubs carried *I. yanagawensis* glochidia November through March. This nicely illustrates a winter tachytictic (*WT*) life cycle.

A summer bradytictic (*SB*) life cycle is illustrated by the *Lampsilis radiata siliquoidea* population of Shannon Lake, Minnesota. Trdan (1981) sampled about 40–70 individuals from this population each season over a period of three years, varying the actual month of collection, noting glochidial development. Trdan observed that egg production and fertilization occurred in late June or early July, such that by mid-July, embryos were in late stages of gastrulation. Glochidia matured by August, but overwintered in their marsupia, embedded in a gelatinous matrix. Glochidial release occurred during the first three weeks of June.

The *Anodonta cygnea* population of Lake Trasimeno, central Italy, is winter bradytictic (*WB*). Giusti and colleagues (1975) sampled ten adults from this population monthly from July 1970 to May 1972, examining both gametogenesis and gravidity. Gametes were produced year round, although at a greater rate from July through September. Spawning occurred in September and October, on falling water temperatures, and females were gravid through the winter, from October to March.

It has long been clear that if gravidity is brief, there may be opportunity for more than one complete reproductive cycle in a single year. Although his data were fragmentary, Allen (1924) offered some evidence that such was the case in an Iowa population of *Anodonta imbecillis*. The terms 'tachytictic' and 'bradytictic', predicated as they are upon a single reproductive cycle per year, do not seem especially appropriate to this situation. The term 'multicyclic' (*M*) seems the most descriptive. Heard (1975), for example, reported two consecutive breeding cycles per year in the *Anodonta peggyae* population of Florida's Lake Talquin. He made 12 monthly samples (102 individuals) noting both gametogenesis and gravidity. While oogenesis took place throughout the year, spermatogenesis occurred March through April and July through September. The population also appeared gravid year round, but marsupia tended to contain embryos in March, September and October, and early larvae in April and November. Based on these observations, Heard's conclusion

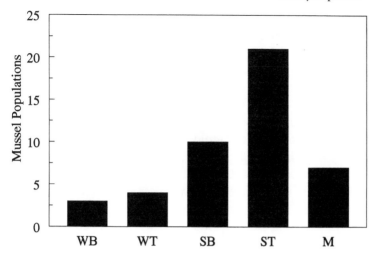

Figure 4.11. A sample of 45 unionoid populations categorized by life cycle.

that this population of *A. peggyae* produces two generations per year seems inescapable. 'Multicyclic' is included as a fifth category of unionoid life history in Figure 4.10.

Over the last hundred years, a large volume of information has been published regarding the life cycles of unionoids. Early reports tended to be fragmentary; based on small samples or short term observations. I have resurveyed the literature to extract only those studies involving samples over periods of a year or more. Neither the number of samples, nor the number of individuals per sample, could be small. Data on gravidity were required; data on gametogenesis desirable but not necessary. The 45 populations (22 studies) for which adequate information is available are graphed by their life history category in Figure 4.11.

It appears safe to conclude that most unionoid populations spawn on rising, rather than falling, temperatures. The summer tachytictic cycle is by far the most commonly reported life history in the literature, followed by the summer bradytictic. The seven populations thus far known to spawn on falling temperatures include five *Anodonta*: two from Florida (*WT*, Heard 1975), *A. piscinalis* from Poland (*WB*, Tudorancea 1972), and *A. cygnea* from Portugal (*WB*, Galhano and Da Silva 1983), in addition to the Italian *A. cygnea* mentioned previously. Other than the Japanese *Inversidens* used to illustrate *WT*, the only non-*Anodonta* known to spawn in cooling water is the hyriid *Cucumerunio* from New South Wales (*WT*, Jones *et al.* 1986).

Multicyclic life histories are not commonly reported. They seem to occur more frequently in lower latitudes, including in Louisiana (Parker *et al.* 1984) and Zimbabwe (Kenmuir 1981), as well as New South Wales (Jones *et al.* 1986). Their comparative rarity may reflect both the difficulty of recognizing *M* life cycles in the field, and the obscurity of tropical faunas in general. It also seems clear from Figure 4.11 that tachytictic cycles are more commonly reported than bradytictic. I can see little evidence of either a systematic or a latitudinal trend, however. Both *SB* and *ST* populations seem equally common in Wisconsin, Minnesota, Alabama and Florida.

At least some life history diversity may reflect adaptation for the partitioning of fish hosts. The examples illustrated in Figure 4.10 include a *WB* population releasing glochidia in the spring, an *ST* releasing in the summer, and a *WT* releasing in the autumn. Available fish hosts over mixed mussel beds may well differ among seasons, or in any case, the same individual fish would seem less likely to be multiply infected. Host fish partitioning may, at least in some cases, be even finer. As may be recalled from Chapter 2, *Villosa nebulosa*, *V. vanuxemi*, and *Medionidus conradicus* are very similar unionids co-occurring in Big Moccasin Creek, Virginia. Zale and Neves (1982b) reported that all three show *SB* life cycles, spawning on the rising water temperatures. But their 'set points' seem to differ, such that in 1979 *M. conradicus* spawned July 8–16, *V. vanuxemi* spawned July 23–30, and *V. nebulosa* spawned August 13–20. This may be a coincidence. It may reflect an adaptation to minimize hybridization. Or it may, in later months, translate to slightly differing windows for glochidial release, yielding in turn the preferential infection of three different glochidial hosts, a darter, a sculpin, and a bass, respectively. In any case, I am intrigued.

Corbiculoids and *Dreissena*

As a consequence of their brooding habit, *Pisidium* life cycles share some similarity to those of the unionoids. For example, egg laying occurs in the spring in the *P. casertanum* population of Lake Paajarvi, Finland, with summer parturition (Holopainen 1979). But Holopainen reported a strikingly different cycle in the *P. moitessierianum* population of the same lake, eggs being laid in the autumn and brooded over winter for release the following summer. Thus applying the terminology developed for unionaceans, *P. casertanum* might be said to display an *ST* life cycle, and *P. moitessierianum* an *SB* life cycle. A glance at the *Pisidium* life history review compiled by Holopainen and Hanski (1986) suggests that *P.*

amnicum (three populations) tends to be bradytictic but that most other species (13 populations of *P. casertanum*, 2 populations of *P. subtruncatum*, 1 population each of *P. lilljeborgii*, *P. hibernicum*, *P. variabile*, and *P. compressum*) tend to be tachytictic, as far as can be told from the data tabulated. Some populations, for example the *P. casertanum* population from a temporary pool in Ontario (Mackie 1979) and Hong Kong *Musculium lacustre* (Morton 1985), may be multicyclic (*M*).

Rather in contrast to the situation in unionoids, however, some pisiidids may lay eggs and brood young all year round, regardless of season. The marsupia of *Sphaerium*, *Musculium*, and the deep water *Pisidium conventus* are generally found to contain sacs of developing young, in various stages, throughout the year. Births seem to occur in all the warm months, with spring and autumn peaks apparent in some populations.

The corbiculoid life span is also much shorter than that displayed by unionoids. In their 1986 review, Holopainen and Hanski found no *Pisidium* populations displaying maximum individual lifespans beyond five years. Annual life cycles are commonly reported in *Pisidium* populations. For example, Mackie (1979) found the *P. casertanum* population inhabiting one roadside pond near Guelph, Ontario, to be annual and semelparous: embryos and larvae were brooded through the winter months, released in July and August, and all adults expired after parturition. Way and Wissing (1982) reported that the *P. variabile* population of a small Ohio pond was annual and iteroparous: April embryos were released in June, and a second batch of September embryos were released in November, after which the adults expired. As has been pointed out by Burky (1983) such life cycles are more similar to those typically displayed by gastropods than bivalves. General questions regarding such brief and semelparous life cycles will be pursued in the next section.

There seems to be a great deal of life history variation among populations of the pest bivalves *Corbicula* and *Dreissena*. This is related to the wide range of climates and habitats into which populations of these species have spread (whether cause or effect is an open question). This is also clearly related to the large numbers of studies devoted to the biology of these animals. Hornbach (1992) reviewed 23 studies of *Corbicula* life history, concluding that in general, maturity is reached in one year with maximum life span in the range of 2–4 years. Some populations have a single reproductive period in the summer, others show bimodal reproduction. Nichols (1996) has reviewed *Dreissena* life history. Although zebra mussels usually mature in one year, in some environments two years may be required. Each mussel typically spawns over a period of 6–8 weeks; in some populations this is synchronized, in other populations spawning extends through the year. The time required for larvae to develop to shelled juveniles can range from 8 to 240 days. Typical life spans extend for several years.

Gastropods

The first formal classification of freshwater gastropod life cycles was that of Russell-Hunter (1961b, 1964, 1978), based primarily on the number of generations per year and survival after egg laying. He reviewed several dozen populations of snails, mostly European pulmonates, and recognized seven life cycles: (1) one generation per year, reproducing semelparously in the spring, (2) one generation per year, reproducing semelparously in the late summer, (3) two generations per year, the spring-born iteroparous and the summer-born semelparous, (4) two generations per year, both semelparous, (5) three semelparous generations per year, (6) one generation per year, reproducing iteroparously, and (7) a two-year generation time.

Russell-Hunter documented great interpopulation variability in freshwater gastropod life cycles. British populations of *Lymnaea peregra*, for example, may display five of the seven categories of pattern (1–4 and 6), while *Ferrissia rivularis* and *Physa fontinalis* may display any of four. He stated (1961b:165) 'In most cases, these interpopulation variations could be determined phenotypically by trophic or temperature differences in the environment.' But he clearly did not mean to imply no role for genetics. Regarding number of generations per year, he documented life cycle (5) only in the physically small *Lymnaea truncatula* and life cycle (7) in some populations of the large *Lymnaea stagnalis*. And regarding semelparity/iteroparity, he stated 'The selective advantages of an annual cycle, in which the adults do not survive long after breeding to compete for food with their offspring, are obvious.' Russell-Hunter's primary conclusion in 1961 was that 'selection has produced genotypes which can show phenotypic plasticity.'

Russell-Hunter's classification was expanded and modified by Calow (1978). Calow combined (1) and (2) into category *A*, and called categories (3), (4), and (5) *B*, *C*, and *F*, respectively. He added two rather rare life cycle types: (*D*) three generations per year, spring and summer generations iteroparous, autumn generation semelparous, and (*E*) three generations per year, spring generation semelparous, summer iteroparous, autumn semelparous. Calow combined Russell-Hunter's categories (6) and (7) into a single category *G*.

Calow then surveyed the literature and gathered life cycle data on 52 populations (or population groups) of pulmonates and 14 populations of prosobranchs, representing 37 species combined. He de-emphasized the non-genetic component of the variation catalogued, focusing instead on its adaptive significance. Semelparity, in Calow's view, might not be an adaptation to minimize parent/offspring competition, as suggested by Russell-Hunter. Instead he suggested that it may represent 'reproductive recklessness', an adaptation to maximize reproductive effort at the expense of adult survival.

In reviewing data collected during the last 15–20 years on the life cycles actually displayed by freshwater snail populations, it has become clear to me that the classification system of Russell-Hunter and Calow requires some clarification regarding semelparity and iteroparity. Note, for example, that if two generations per year are the norm, both the first (usually spring) generation and the second (summer or autumn) generation could be semelparous or iteroparous. That would imply four types of cycles: abbreviated by their initials *ss*, *si*, *is*, and *ii*. But Russell-Hunter and later Calow described and figured only two types of such cycles: *B* showing *is* and *C* showing *ss*. What is one to do with the Indian *Bellamya* population described by Khan and Chaudhuri (1984), where the animals reach reproductive age in half a year but typically reproduce at age 1 and age 1½ years as well? I believe it is in the spirit of the currently-existing classification system to recognize types *si*, *is*, and *ii* as subtypes of of *B*, reserving *C* for complete semelparity. This system is illustrated in Figure 4.12.

Similarly, given three generations per year, there are eight possible permutations of *s* and *i*. Russell-Hunter recognized *sss* in the wild, which became Calow's *F*. Calow's *D* would be *iis*, and his *E* would be *sis*. Since the distinction between *D* and *E* is in the reproduction of the first (spring) generation, it would seem consistent to also include *iis*, *isi*, and *iii* as subtypes of *D* and *ssi* and *sii* as subtypes of *E*, again reserving *F* for complete semelparity. This system is illustrated in Figure 4.13.

I agree with Russell-Hunter that it is important to distinguish between snails that mature in one year and those that require two years to mature. Figure 4.12 shows as category *G* only those populations that mature in 12–23 months and reproduce iteroparously. New category *Hi* includes populations that require 24–35 months to mature and reproduce iteroparously, with new category *Hs* describing those reproducing semelparously at age 24–35 months. The iteroparous categories *G* and *Hi* include those populations that reproduce twice, surviving a spring reproduction

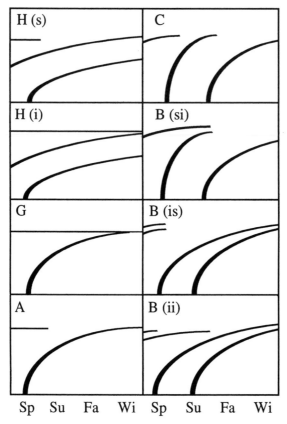

Sp   Su   Fa   Wi   Sp   Su   Fa   Wi

Figure 4.12. Schematic diagrams to illustrate eight types of freshwater snail life cycles involving two generations per year (*B* and *C*), one generation per year (*A* and *G*) and a two-year generation time (*H*). The ordinate for each graph is a measure of size. Each curling line represents a cohort, born at size 0 and growing with time until maturity. Cohorts are semelparous when they are mature for a single period of reproduction, and iteroparous when they are mature over two or more periods of reproduction.

but not a second, summer or autumn reproduction, and hence not surviving to an additional year.

Note that the diagrams shown in Figures 4.12 and 4.13 are intended only as examples, not as absolute patterns. Each of the iteroparous generations for cycle types *B* and *D–F* is shown surviving to contribute to exactly two cohorts, but could in reality contribute to any number greater than one. Life cycle types *G* and *H* are shown with one, spring-born cohort per year, but additional cohorts per year are certainly pos-

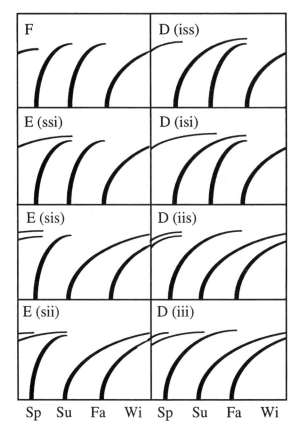

Figure 4.13. Schematic diagrams to illustrate eight types of freshwater snail life cycles involving three generations per year. See Figure 4.12 for further details.

sible. For example, the Hong Kong thiariid *Brotia hainanensis* ordinarily requires 2½ years to mature, reproduces twice a year, and lives 3½ years (Dudgeon 1989). By definition this is an *Hi* cycle, but is far from matching anything in Figure 4.12.

Since 1978, several populations of freshwater snails have been reported to display three-year generation times. Two such populations of *Goniobasis livescens* were discovered by Payne (1979), and one each of *Theodoxus fluviatilis* and *Viviparus georgianus* were reported by McMahon (1980) and Jokinen *et al.* (1982), respectively. In addition, McKillop (1985) suggested that Canadian populations of *Bulimnea megasoma* might be triennial, although Gilbertson *et al.* (1978) reported type **A** in Minnesota. Thus I suggest a new (ninth) category of freshwater snail life

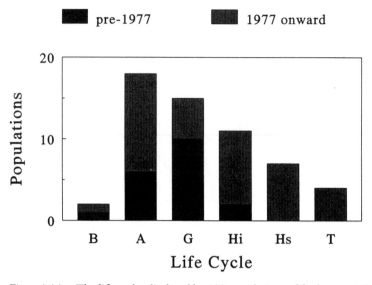

Figure 4.14. The life cycles displayed by 120 populations of freshwater pulmonates, combining Calow's pre-1977 data set with 16 studies published subsequently. See Figures 4.12 and 4.13 for an explanation of life cycle types.

cycle, type *T*, to describe populations where individuals require 36–47 months to mature. All *T* populations for which there are adequate data seem to show semelparity, but the category could be subdivided into types *Ts* and *Ti* should this become necessary.

The data collected by Calow are shown as 'pre-1977' in Figures 4.14 and 4.15. I have re-examined all references to type-**G** life cycle in Calow's data set, dividing the populations into *G* (narrower definition), *Hi* and *Hs*. In those situations where Calow combined populations (generally of a single species surveyed in a single work), I have reseparated them. To Calow's data set I have added 33 post-1977 references to 104 populations of snails, expanding the species represented from 37 to 70. Especially useful were McKillop's (1985) descriptions of 14 Canadian populations and Jokinen's (1985) study of 13 populations co-occurring in a single Connecticut lake.

Along with Figures 4.14 and 4.15 I feel obliged to issue two categories of disclaimer. First it must be emphasized that the cohort data from which freshwater snail life cycles are inferred ordinarily leave wide latitude for interpretation. As shell growth rates slow at reproduction, cohorts become increasingly difficult to distinguish. Thus on such questions as whether a particular cohort has reproduced semelparously or

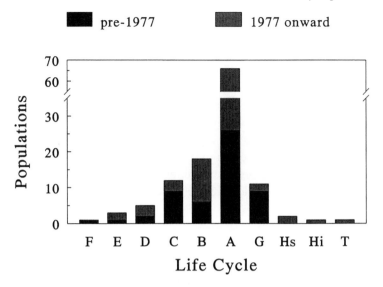

Figure 4.15. The life cycles displayed by 57 populations of prosobranch snails, combining Calow's pre-1977 data set with 21 studies published subsequently. See Figure 4.12 for an explanation of life cycle types.

simply become absorbed by a more recently born cohort, one must rely on the judgement of the individual researchers involved.

Second, even if the real life history of all snail populations were known, considerable judgement would be called for in sorting them into neat categories. For example, Brown *et al.* (1989) concluded that most individuals of a Louisiana population of *Viviparus subpurpureus* reproduced semelparously at age two years, although some were reproductive at age one, and that although most individuals of a co-occurring population of *Campeloma decisum* were semelparous, some survived to reproduce a second time at age three. Both these populations appear as *Hs* data in my Figure 4.15. The Connecticut population of *Viviparus georgianus* that I have mentioned earlier among the *T* data comprises males that first mate at age two years and die, while females reproduce semelparously at age three. The large range of life cycles reported for *V. georgianus* (*G* – Browne 1978, *Hi* – Buckley 1986, both from New York) is interesting to note in passing.

Calow's (1978) observations are generally supported by the enlarged data sets of Figures 4.14 and 4.15. There is a very striking difference between the life cycles displayed by pulmonates and prosobranchs. Life cycle *A* is now the most common type in both groups, not just pulmonates. But while 32% of pulmonate populations display generation times of less than one year, only two such prosobranch populations are known (one of the Canadian *Bithynia tentaculata* populations studied by Pinel-Alloul and Magnin 1971, and the Indian *Bellamya bengalensis* of Khan and Chaudhuri 1984). While 39% of prosobranch populations have generation times of 24 months or more, the phenomenon is rare in pulmonates. McKillop (1985) reported two-year generation times for *Lymnaea stagnalis* and *L. palustris*, while Morris and Boag (1982) obtained the same result for *Helisoma trivolvis*. Since pulmonates and prosobranchs often co-occur, and since there is no systematic difference in their sizes at reproduction, one might infer a phylogenetic component to growth rates, with prosobranchs slower.

Semelparity seems to be more common among pulmonates. Of the 77 pulmonate populations that reproduce first at age one year, 66 do so and die (type *A*). I have collected records for 33 prosbranch populations maturing at age one, of which only 18 show life cycle type *A* and 15 show *G*, that is, living to reproduce again. Brown and colleagues (1998) have observed that such broad differences between pulmonates and prosobranchs may extend to their competitive abilities, dispersal capabilities, and risk from predation as well. Such clear-cut differences along phylogenetic lines comprise important evidence that variation in life cycle and autecology, although perhaps not always available for selection to act upon, are nevertheless not entirely the result of phenotypic plasticity.

Modelling life cycles

Can the variation shown in Figures 4.14 and 4.15 be interpreted as direct adaptation? Throughout this chapter we have repeatedly been impressed by the primacy of body size as a determinant of life history variation. Perhaps populations displaying longer generation times are simply composed of animals reaching reproduction at a larger size for some other reason (predation, for example). And in much of Russell-Hunter's work (and occasionally implicit in our discussion of life cycles) one finds evidence that semelparity or iteroparity may be influenced by environmental harshness. Populations may appear to reproduce semelparously in an environment where they suffer high mortality, when otherwise adults would survive to reproduce again. It is tempting to speculate that all the variation shown in Figure 4.14, for example, could be explained by two variables: individual size and the general background level of mortality specific to a habitat. Perhaps life cycle patterns are not generally adaptations themselves, but rather consequences. I have designed a model to explore this notion.

Because both survivorship and somatic growth are clearly expected to vary as a function of ambient temperature, I divided model years into 12 months and specified a temperature factor, f, for each month. I surveyed the literature and found mean monthly water temperature data for two representative environments: one quite variable and the other less so. Brown (1979) measured water temperatures at four Iowa habitats containing *Physa* and *Lymnaea*: a slough, a pond and two river sites. During the calendar year 1974, the four habitats varied together from 0 °C in December to 30 °C in July. By contrast, Lam and Calow (1989a) found mean monthly temperatures at three English rivers inhabited by *Lymnaea peregra* to range only from about 5 °C to 17 °C over two years.

For my model I assumed that both growth and survivorship were greatest in the month with the highest mean temperature observed, and that growth and survivorship halved with each drop of 10 °C from optimum. Then for each month t, I calculated a temperature factor:

$$f_t = \tfrac{1}{2}^{0.1(Topt - Tobs)} \tag{4}$$

where *Topt* is the optimum temperature (30 °C in Iowa, 17 °C in England) and *Tobs* is the mean temperature observed for each month. My selection of the warmest month as optimal was arbitrary. Any month might be selected for this purpose, without affecting the overall forms of the model life cycles generated.

In the wild, survivorship is expected to be influenced both by age and month, and probably by the interaction of the two as well. But as a simplifying assumption, I modelled monthly survivorship as:

$$N_{t+1} = mf_{t+1}N_t \tag{5}$$

where N_t is the cohort size at month t, N_{t+1} and f_{t+1} are the new cohort size and temperature factor, and m is a constant proportion surviving, independent of age. Values of N_t less than 1.0 were set to 0.

The earliest effort at modelling body (or more precisely, shell) growth in freshwater snails of which I am aware is that of Baily (1931). Baily measured the aperture length of 16 individual *Lymnaea* (*Pseudosuccinea*) *columella* every 3 days, beginning about two weeks after their birth and proceeding to the death of the last individual at day 139. Egg production was recorded in all nine individuals that survived to maturity.

Baily fitted individual growth curves to each of his nine data sets, and calculated a 'composite' by averaging the parameters of the equations, rather than the raw data. The result was a simple logisitic model:

$$y = 10.1 / 1 + e^{2.45 - 0.066x} \tag{6}$$

where y is the length of the aperture in mm and x is age in days. He experimented with a more elaborate, cubic parabolic model for the asymptotic value of y, but the simple logisitic model above, with asymptote at 10.1 mm, provided a nearly perfect fit.

I converted the 0.066 coefficient from days to months by assuming 1 month / 30 d. All Baily's observation were made at room temperature. So to introduce an effect for fluctuating environmental temperature, I included f as a second coefficient of x. When $f = 0.5$, for example, twice the time will be required to reach asymptotic size. Then a more general model of somatic growth in pulmonate snails would be:

$$y = k / 1 + e^{2.45 - 1.97fx} \tag{7}$$

where y is size in mm, x is age expressed in months (but see below), k is the asymptotic value of y, and f is the temperature factor.

I am not entirely convinced of the generality of this equation as a model for somatic growth in freshwater snails. *Pseudosuccinea columella* is an average-sized snail; somatic growth in 3 mm or 30 mm animals may better be described by different equations, not different values of k. Further, comparison of Figures 4.14 and 4.15 suggests that pulmonate and prosobranch growth rates may differ substantially. Thus I would not offer this model for application to prosobranchs.

Figure 4.6 suggests that size-at-birth (y_0) might typically take the value of 1.0 mm for pulmonates. Then the calculation of y_1, with $x = 1$ and reference to month-of-birth for f_1, is straightforward from equation (7). But since f is month-specific, its inclusion into equation (7) necessitates monthly recalculation of growth rate. So to calculate a second month's growth, for example, I inserted y_1 into equation (7) with f_2 and solved for x_1', an *apparent* age in months entering month 2. Then y_2 was obtained by setting $x_2 = x_1' + 1$. Although size plotted as a monthly datum increases regularly when calculated in this fashion, growth within months may be visualized as a 'sawtooth' affair, the arc on each cusp proportional both to the suitability of the temperature (f) and the closeness of y to k.

Eight of Baily's snails lived to produce 100 or more eggs. For 6 of these 8, the size at which this milestone was reached was about 9.0 mm, or 90% of the asymptotic size k. Thus for the purpose of the present general model, each snail surviving to size $0.9k$ produced 100 offspring monthly, ambient temperature permitting. A commonly reported temperature for the onset of egg laying in field populations is about 10°C (DeWitt 1955, Geldiay 1956, Burky 1971, Heppleston 1972, Eversole 1978). Thus egg laying was allowable from April to October in England and from May to October in Iowa.

In overview, then, the major parameters of this model are k, the asymptote for the logistic model of somatic growth, m, the general background survivorship and thus a measure of environmental harshness, and a schedule of 12 values of f, derived from monthly temperature readings at snail-containing habitats in Iowa and England. The initial population size was $N_0 = 10^6$ juveniles, size $y_0 = 1$ mm in April. This cohort began to decrease in number according to equation (5) but to increase in somatic size according to equation (7). For each month t, when $N_t > 1$ and $y_t > 0.9k$ with mean ambient temperature greater than 10°C, the cohort produced $100N_t$ newborn. These then were subjected to the same monthly survivorship, somatic growth, and reproduction schedules as their parents. I allowed three years for stabilization, and examined the life cycle manifested by the model population in the fourth year. Ultimately I performed about 30 runs of this model for each of the two habitats, varying k from 4 mm upward and m from 1.0 downward until the extinction of each model population. Results are shown schematically in Figure 4.16.

I found that values of m could be set down to 0.55 in the milder and less variable English environment without extinction, but only down to 0.9 in Iowa. And increasing maximum size k above 5.5 mm in England or 7.0 mm in Iowa did not affect results. But in just these two environments, over a rather small range of parameters, model populations displayed 5 of the $A–H$ general types of life cycles documented in pulmonates. As one might predict, environmental harshness may be manipulated to induce semelparity or iteroparity, while the number of generations per year seems quite sensitive both to size at maturation and growing season.

In particular, only a single generation will pass in the six months of growing season available in Iowa if k is greater than about 7.0 mm, but two generations will pass if less. Complete semelparity is expected if m is less than 0.95; some iteroparity is expected otherwise. Brown (1979) in fact reported that the 30–35 mm *Lymnaea stagnalis* and the 20–25 mm *L. palustris* displayed type A life cycles in the ponds upon which this environment was modelled, and that the 7–10 mm *Physa gyrina* displayed a

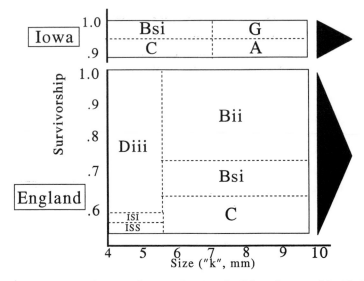

Figure 4.16. Schematic diagram showing the life cycle types (classified according to Figures 4.12 and 4.13) displayed by model populations varying two parameters: survivorship (*m*) and maximum size (*k*, in mm). Two levels of environmental fluctuation were examined, high (Iowa) and low (England). Varying *k* above 10 mm did not change the outcome; values of *m* less than those shown resulted in the extinction of the model population.

life cycle type *C*, both consistent with $0.9 < m < 0.95$. But Brown reported three generations per year (*sis?*) for a fourth species in these ponds, *Physa integra*, a result my model was unable to predict.

Figure 4.16 shows that the mild, seven-month growing season in England seems to promote the passage of at least two generations regardless of size at maturation. Two generations are expected in England if *k* is greater than 5.5 mm and three generations if less, the latter generally matching a type *D* life cycle. The observations of Lam and Calow (1989a) on the *Lymnaea peregra* inhabiting the environments upon which this model was based, however, included two type *A* populations and one *B(is)*, none of which would have been predicted by the model.

In an overall comparison of Figures 4.14 to 4.16 one is struck by several contrasts. First, generation times greater than one year (life cycle types *H* and *T*) will not emerge from my model even when *k* is set very large, although type *H* is definitely (if rarely) observed in natural populations of pulmonates and type *T* may be. Life cycle types *E* and *F*, also rare in nature, were similarly absent from my model. Both *E* and *F* require that the first of three generations reproduce semelparously, dying in mid-summer at the peak of the growing season. My intuition suggests to me that no amount of adjustment to the parameters of my model will result in such behaviour. And perhaps most strikingly, given the frequency of type *A* life cycles in the wild, one might be surprised to find that *A* seems to be predicted only under rather special circumstances in my model.

So although I feel that much of the variance in life cycle catalogued in Figures 4.14 and 4.15 may be accounted for by size at maturity and environmental harshness, it would appear that more direct adaptation may also play a role. I suggest that the observed frequency of populations with type *H* life cycles may constitute evidence of **S**-adaptation in the wild. Such populations, I have predicted, will live in unproductive but stable environments, allocate an unusually large fraction of their energetic resource to maintenance, and delay their reproduction. I also note that data from wild populations seem to reflect a greater incidence of semelparity than expectation from modelling efforts. Thus I suggest that life cycle types *E*, *F*, and (in its unusually high frequency) *A* may constitute evidence of **R**-adaptation in the wild. Some populations with these life cycles may in fact allocate greater than expected energetic resource to reproduction, at the expense of individual maintenance and ultimately survival.

Throughout this chapter a recurring theme has been the great variation in life history attributes often observed among conspecific populations of freshwater molluscs inhabiting different environments. In the observations of Jokinen (1985) on 14 gastropod populations inhabiting a single small lake in Connecticut, one can place this theme in perspective. The largest snail in the lake was *Campeloma decisum*, apparently showing a type *G* or *H* life cycle, although not present in numbers sufficient to sample. Jokinen found juveniles throughout the growing season, from spring to autumn, for three species: *Fossaria modicella*, *Physa ancillaria*, and *Gyraulus circumstriatus*. Thus for these three she was unable to discern any cohorts – probably a common and underreported phenomenon. She identified type *A* life cycles for the larger *Helisoma campanulatum* (adults about 13 mm) and *H. anceps* (9 mm), plus the 5 mm limpet *Laevapex fuscus* and the only common prosobranch in the lake, *Amnicola limosa* (5 mm). Type *B(is)* cycles were identified for the larger *Pseudosuccinea columella* (11 mm), as well as the 4–5 mm planorbids *Gyraulus deflectus* and *Promenetus exacuous*. Another 5 mm planorbid, *Planorbula armigera*, showed a type *C* life cycle. Jokinen found that the two smallest animals had three generations per year: *D(iss)* for the 2.5 mm planorbid *Micromenetus dilatatus* and *E(sis)* for the 3 mm limpet *Ferrissia fragilis*.

It would appear here that life cycle is a function of phylogeny, with prosobranchs manifesting types *A* and *G*, regardless of size. Life cycle also seems to be controlled by size at maturity, with type *A* generally displayed by the larger snails and types *D* and *E* the smallest. But *P. columella* is comparatively large and shows two generations per year (*B*), while 5 mm pulmonates showed three different types of life cycles, including *Laevapex* with only a single generation per year. Neither phylogeny, nor size, nor environmental harshness (which would be fairly uniform in this example) seems likely to account for the diversity of life cycles observed by Jokinen. At least some of this variance would seem to be direct adaptation. There is considerable merit in Jokinen's suggestion that in some situations, selection may act upon demographic parameters of snail populations to minimize competition. To this subject we next turn.

Summary

Fundamental aspects of the demography of freshwater snail populations are very sensitive to environmental fluctuation. Varying ambient temperature over a few degrees, for example, may result in order of magnitude differences in survivorship and reproduction. Little wonder, then, that comparisons of the life histories displayed by allopatric populations, which generally inhabit different environments and display some genetic divergence, uncover bewildering diversity. Much of this variance seems due to phenotypic plasticity. But a heritable component of interpopulation life history variation can usually be distinguished, especially with laboratory studies. And life history variation between freshwater mollusc species, which often ranges much greater than an order of magnitude for any character, clearly has a genetic basis.

In freshwater mollusc populations, reproductive effort is a power function of adult weight. In *Pisidium*, the annual weight of offspring seems to be determined by adult weight alone, such that individuals surviving to a second year reproduce according to their new, larger size. There is no evidence that *Pisidium* populations sacrifice survivorship for reproduction; they do not apparently allocate extra energetic resource to reproduction and die as a consequence. The same could be said for freshwater molluscs generally.

Exceptionally, however, populations of freshwater snails have been described whose adults, on the average, have reproductive efforts an order of magnitude greater than expected for their body size. These populations inhabit rich environments but reproduce semelparously. Also exceptionally, populations have been described with an order of magnitude lower reproductive effort than would be expected from body size, inhabiting poor but stable environments. Inspired by the CSR model of plant strategy, I here propose the 'USR' model of life history variation in freshwater molluscs. Mollusc populations of the former type are here designated **R**eproduction-adapted, those of the latter type **S**tress-adapted, and all other populations are considered **U**ndifferentiated.

Juvenile size at hatch or parental release varies from about 0.4 mm shell size up to 15 mm in freshwater molluscs. The great majority of this variance is attributable to phylogeny – genus, family, or higher level classification. Within taxon, variation in juvenile size is generally much less than an order of magnitude, and positively correlated with parent size. Thus it seems that the 'many small vs. few big offspring' trade-off is not a major variable in the life history strategies of freshwater molluscs.

Given, then, that both newborn size and overall reproductive effort generally seem to be functions of adult body size, the determinants of age and size at reproduction take special importance in freshwater molluscs. This seems to be largely a function of age-specific survivorship. Setting aside (rather immediate) newborn mortality (following Charnov 1991), we note that if early survivorship is relatively high, freshwater snails mature rapidly, while if early survivorship is low, reproductive maturity will be delayed. Size at reproduction usually seems to be the critical variable, although I am aware of at least one case where a (apparently **R**-selected) population achieved early maturity by improved growth rate.

Interpopulation differences in size and age at reproduction, and in survivorship after reproduction, manifest themselves strikingly in the variable life cycle patterns displayed by freshwater mollusc populations. Five distinct life cycle patterns can be distinguished in unionid mussel populations. Although previous authors have recognized seven categories of freshwater snail life cycles, here I recommend an expanded and clarified system involving nine types, some with subtypes. Among both pulmonate and prosobranch populations, the annual semelparous type *A* life cycle is most common. Pulmonates often show two generations per year and rarely three generations per year or two years per generation. Prosobranchs very rarely show two generations per year and much more commonly show generation times of two or even three years.

A modelling effort suggests that at least some of the observed variation in pulmonate life cycle pattern may be explained as a consequence of varying size at reproduction and environmental harshness. Life cycle pattern may often best not be viewed as directly adaptive. Subjectively, however, the observed distribution of pulmonate life cycles does seem to include both more semelparity and more examples of lengthened generation time than my model would suggest. So once again, we find some evidence of both **R**- and **S**-adaptation in data taken from the wild.

5 · *Population dynamics and competition*

An understanding of population regulation is central to ecology. Murdoch (1994) observed that 'Persistent species must consist of regulated single populations or regulated collections of subpopulations. Regulation should therefore be ubiquitous in natural populations, provided we look for it at the appropriate spatial (and temporal!) scale.' Since most of us, I would imagine, think of extinction as a rare thing, most of us consider that 'density dependent' controls must act at some population size. Populations must be repressed by factors which 'intensify as the population density increases and relax as the density falls', the definition of Huffaker and Messenger (1964). But density dependence turns out to be more difficult to demonstrate in natural populations than one might expect. Frustration may have driven some ecologists away from the study of population regulation entirely (Krebs 1991, 1992).

The reduced emphasis I may seem to place on density independent factors in the present chapter may surprise many of my colleagues. Molluscs are among the most conspicuously successful colonizers of unstable freshwater habitats, and unquestionably show great volatility in local population size, linked to vagaries of their environment. The densities of the *Bulinus globosus* and *Biomphalaria pfeifferi* populations inhabiting small rivers on the Zimbabwean highveld may fluctuate to near zero as a function of temperature and rainfall (Woolhouse and Chandiwana 1990a,b, Woolhouse 1992). But viewed at a larger scale, the densities of *B. glabrata* and *B. pfeifferi* in Zimbabwe are much more constant. Similarly, Eleutheriadis and Lazaridoudimitriadou (1995) reported that the monthly density of a *Bithynia graeca* population from northern Greece fluctuated dramatically over a 38 month sampling period, from less than 5 snails/25 cm^2 up to about 100 snails/25 cm^2. The authors linked these fluctuations to variation in water chemistry (especially phosphate concentration). But again, viewed at a larger (regional) scale, population densities were much less volatile. The March density of B. graeca varied only from about 3 to about 12 individuals/25 cm², over four Marches. To my eyes, data sets such as this show strong evidence of density dependence.

Previous reviews of regulation in freshwater mollusc populations have divided the topic into biotic factors (McCullough 1981, Burky 1983, chapter 11 of Brown 1994) or abiotic factors (Macan 1961, Appleton 1978, chapter 10 of Brown 1994). In the present chapter I address the former, leaving the latter for Chapter 8. I begin with descriptions of population growth, both under ideal laboratory conditions and in nature. I then turn to crowding effects (primarily studied in the laboratory) and to field studies of population regulation. The latter studies are of three kinds: experiments (usually involving enclosures), long-term censuses, and shorter-term observations regarding specific perturbations. I conclude this chapter with a survey of the evidence for interspecific competition in freshwater mollusc populations, focusing on several well-documented invasions and artificial introductions.

Population growth

Discussions of population dynamics conventionally begin with an examination of the form of exponential growth taken by populations unlimited by resources and unchallenged by predators, parasites, or pathogens. Such data are generally obtained in the laboratory. DeWitt (1954) seems to have been the first researcher to examine the growth of a freshwater mollusc population growing exponentially in a laboratory setting. Although the techniques in use today are not substantially different from those employed by DeWitt, the years have seen considerable variation in terminology and convention. For the purpose of this review, a population growing exponentially increases as:

$$N_t = N_0\, e^{rt}$$

where N_0 is the population size at start and N_t is the population size t units of time in the future. Curiously, the standard unit of time in laboratory studies of freshwater snail demography has come to be the fortnight, 14 days. The constant e is the base of the natural logarithms, and r is the intrinsic rate of increase, which will be the object of our interest. For convenience, let $e^r = \lambda$, the 'finite rate of increase', the factor by which a population increases per unit time.

Values of r are generally obtained by solving Euler's equation:

$$\sum_{X} l_x \, m_x \, e^{-rx} = 1$$

where l_x is the proportion of the population surviving to age class x and m_x is the mean number of viable offspring produced by animals of that age class. Values of m_x are conventionally expressed in units of 'female offspring per female parent', but since essentially all the freshwater molluscs for which such data have thus far been gathered have been hermaphrodites or parthenogens, this convention is here ignored. The total number of offspring produced by an average parent in its lifetime is the net reproductive rate $R = \sum l_x m_x$.

Laboratory studies

The experiments of Pointier and colleagues (1991) nicely illustrate the application of demographic methodology to laboratory populations of freshwater snails. The authors collected *Biomphalaria glabrata* from Celigny Pond, Guadeloupe, and removed 81 of their apparently viable eggs to a separate container for observation. Hatchlings were transferred to larger containers to minimize crowding, and four adults were ultimately cultured per 1 litre aquarium. Snails were held at 25 °C and fed dried lettuce, with weekly water change. Pointier and colleagues recorded the proportion surviving and the mean viable eggs they produced every two weeks ('fortnightly') for 21 fortnights, by which time 65% of the adults had died, and the mean fecundity of the survivors had decreased substantially ($m_{21} = 25.2$, $l_{21}m_{21} = 8.7$).

Pointier *et al.*'s results are shown graphically in Figure 5.1. Note that since the experiment was terminated before the population had exhausted its reproductive capacity (either by mortality or senescence), the value of $R = 1419$ obtained by the authors is a slight underestimate. Solving Euler's equation, the authors obtained a value of $r = 0.84$ per fortnight, and hence $\lambda = 2.32$. Considered objectively, a mean reproductive rate of 1419 viable offspring per adult per lifetime, and a population that increases by a factor of 2.32 every 14 days, are both quite remarkable statistics. But as we shall see, such statistics are not unusual for freshwater snails.

Pointier and colleagues performed similar analyses on a total of seven snail populations – two each of the medically important *Biomphalaria glabrata*, *B. alexandrina*, and *B. straminea*, plus one population of *Melanoides*

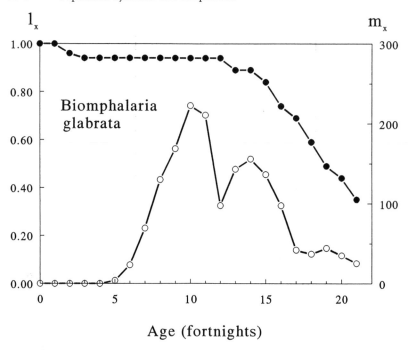

Figure 5.1. Proportion surviving (l_x, closed circles) and average per capita production of viable offspring (m_x, open circles) for a cohort of *Biomphalaria glabrata* (Pointier *et al.* 1991).

tuberculata, which has been nominated as a possible agent for the competitive control of *Biomphalaria*. The *Melanoides* experiment started with 40 juveniles newly liberated from adults collected on Guadeloupe. All snails surviving to reproductive age seem to have been females. Although raised under conditions identical to those previously described for *B. glabrata*, the life history of *Melanoides* proved strikingly different (Figure 5.2). Reproduction commenced at fortnight 9, but did not peak until the second year of Pointier *et al.'s* study, at fortnight 56, and was still ongoing when the experiment was terminated at fortnight 133. In the concluding fortnight, after over five years of observation, 25% of 40 *Melanoides* were still alive, bearing 15 offspring among them. Although the net reproductive rate of *Melanoides* ($R = 493$) does not seem as impressive as the $R = 1419$ of *B. glabrata*, newborn *Melanoides* are much larger than pulmonate hatchlings, and presumably show much greater survivorship. The intrinsic rate of increase $r = 0.24$ obtained by Pointier and colleagues is one of only two such values available for a non-pulmonate freshwater mollusc.

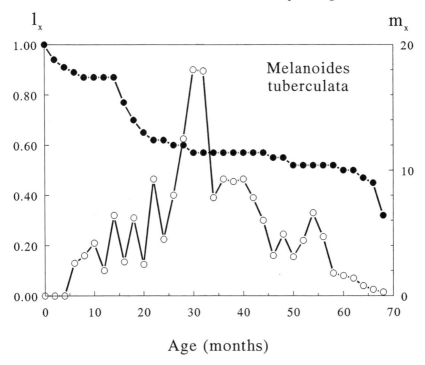

Figure 5.2. Proportion surviving (l_x, closed circles) and average per capita production of viable offspring (m_x, open circles) for a cohort of *Melanoides tuberculata*. These data are extracted from the much larger (fortnightly) data set of Pointier *et al.* (1991).

The metabolism of a small, slow, aquatic ectotherm is expected to depend strongly upon the temperature of its environment. The relationship between temperature and the intrinsic rate of increase displayed by pulmonate populations has been explored in at least ten separate investigations; data from a sample of these are shown in Figure 5.3. Striking temperature effects have in fact been demonstrated in *Bulinus tropicus* by Prinsloo and van Eeden (1969), and in *Biomphalaria pfeifferi* and *Helisoma duryi* by de Kock and Joubert (1991). The increase in the rate of increase for *Helisoma* is spectacular, from $r = 0.64$ at 15 °C ($\lambda = 1.89$) to $r = 2.99$ at 30 °C ($\lambda = 20.1$). Perhaps not surprisingly, the North American *H. duryi* is another of the snails receiving attention as a potential competitor against the medically important planorbids of the tropics.

However, the intrinsic rate of increase r seems rather insensitive to temperature in most of the populations shown in Figure 5.3. The Zimbabwean population of *B. pfeifferi* studied by Shiff and Garnett (1967) varied only from $r = 0.24$ at 18 °C to $r = 0.44$ at 27 °C, much in contrast

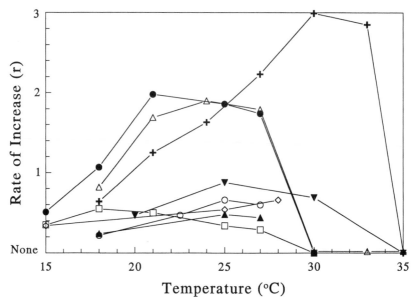

Figure 5.3. Intrinsic rate of increase (*r*, per fortnight) as a function of temperature for eight pulmonate populations. From de Kock and Joubert (1991): crosses, *Helisoma duryi*; open triangles, *Biomphalaria pfeifferi* (South Africa). From Shiff and Garnett (1967): upright closed triangles, *Biomphalaria pfeifferi* (Rhodesia). From Sturrock and Sturrock (1972): inverted closed triangles, *Biomphalaria glabrata*. From Prinsloo and van Eeden (1969): squares, *Lymnaea natalensis*; closed circles, *Bulinus tropicus*. From Shiff (1964a): open circles, *Bulinus globosus*. From Brackenbury and Appleton (1991a): diamonds, *Physa acuta.*

to the South African *B. pfeifferi* of de Kock and Joubert. An equally striking comparison may be made between Shiff's (1964a) results with Zimbabwean *Bulinus globosus* and Prinsloo and van Eeden's findings with South African *Bulinus tropicus*. Perhaps the five populations unaffected by temperature were in fact limited by something else under their particular experimental conditions – food, crowding, or their own physical constraints. But it is certainly also possible that no *Biomphalaria* population has the wiring and plumbing necessary to achieve the *r* = 2.99 posted by *H. duryi* under any circumstance, as a consequence of the sorts of life history adaptations we reviewed in Chapter 4.

Investigators have not reached consensus regarding some important aspects of the culture technique used to assess population growth in pulmonates. Most of the authors whose work is shown in Figure 5.3 offered their snails lettuce, which was the favoured snail food into the 1980s.

More recently most workers (including de Kock and Joubert, 1991, and Brackenbury and Appleton, 1991b) have shifted to various commercially available fish foods of complex recipe. The culture volume and rate of water change are also critical variables in studies of this sort, as we shall see presently. Shiff's (1964) snails seem to have been the most crowded of the lot on display in Figure 5.3, at six individuals per 'large' aquarium, changed every two months. Both Prinsloo and van Eeden (1969) and de Kock and Joubert (1991) used a continually circulating system with a charcoal-and-sand filter designed to keep their culture water fresh. Food, water or both may account for the impressive values of r displayed at all temperatures by *B. pfeifferi* in de Kock and Joubert's hands.

Prompted by observations of improved population growth in naturally fluctuating room temperature 'controls', de Kock and van Eeden (1986) designed a device to vary the water temperature delivered to their circulating system in a daily cycle. They reared cohorts of South African *B. pfeifferi* in three temperature regimes: a constant 23 °C, a 5 °C fluctuation about 23 °C, and a 10 °C fluctuation about 23 °C. The experiment was carried until the last of 3×25 snails died, 11 fortnights in total. Interestingly, it was the 10 °C daily fluctuation (18–28) that yielded the greatest population growth in *B. pfeifferi* ($r = 1.36$, $R = 987$), not substantially different from the 5 °C fluctuation ($r = 1.31$, $R = 309$) but rather greater than the constant 23 °C treatment ($r = 1.04$, $R = 322$).

Variation in culture technique should be kept firmly in mind while inspecting Table 5.1, a summary of demographic data from laboratory populations of freshwater snails. I have included separate entries where the populations of a single species are of different origin, or have been re-tested in a separately published paper. But in those circumstances where snails from a single population have been tested under varying conditions in a single investigation (such as the eight populations shown in Figure 5.3), the single best performance, as judged by r, is included in Table 5.1. For example, under the Prinsloo and van Eeden (1969) entry in Table 5.1, the single row for *Bulinus tropicus* comes from 21 °C, where $r = 1.98$ (Figure 5.3). If unspecified, the temperatures at which the Table 5.1 data were derived were 'room temperature', or 23–25 °C. Note that room temperature may not be the optimum, as data from *Lymnaea natalensis* (Figure 5.3) will attest. Values of R may be difficult to interpret, since investigators sometimes terminate their experiments before the end of the reproductive lives of their animals. Thus Table 5.1 includes data on the length of each experiment (x_{max}) and reproduction ($l_x m_x$) at last count, if available. The smaller the value of the final $l_x m_x$, the more

Table 5.1. *Demographic data summarized from laboratory studies of 36 populations of freshwater snails*

Reference	Population	r	R	x_{max}	$l_x m_x$ at x_{max}	Culture density, waterchange	Notes
Ahmed *et al.* 1986	*Bulinus abyssinicus* Somalia	1.41	623	7	140	4/640 ml, daily	
Brackenbury and Appleton 1991b	*Bulinus tropicus* S. Africa	0.44	43	9.5	0	51/20 L, biweekly	FF
	Physa acuta S. Africa	0.66	49	6	0	51/20 L, biweekly	FF, 28°C (opt.)
De Kock and Joubert 1991	*Biomphalaria pfeifferi* S. Africa	1.90	182	12.5	0	Circulating system	
	Helisoma duryi S. Africa	2.99	5428	14.5	$l_{14.5} = 0.8$	Circulating system	30°C (opt.)
De Kock and van Eeden 1981	*Biomphalaria pfeifferi* S. Africa	2.37	2696				
De Kock and van Eeden 1985	*Bulinus tropicus* S. Africa	3.37	1656	8.5	$l_{8.5} = 0.31$	Circulating system	FF
	Lymnaea natalensis S. Africa	1.81	1372	9.5	$l_{9.5} = 0.09$	Circulating system	FF
De Kock and Van Eeden 1986	*Biomphalaria pfeifferi* S. Africa	1.36	987	11	0	Circulating system	FF, 18–28°C daily
De Kock *et al.* 1986	*Bulinus tropicus* S. Africa	3.20	715	7.5	$l_{7.5} = 0.2$	Circulating system	FF, 28°C, Gen.#7
DeWitt 1954	*Physa gyrina* Scio, Michigan	0.06	418	17.6	24	1–8/1–2 L, unchanged	Selfers and outcrossers combined
	Physa gyrina Ann Arbor, Michigan	0.03	492	23.3	29	1–8/1–2 L, unchanged	Selfers and outcrossers combined

Reference	Species, Location						
DeWitt and Sloan 1958	Pseudosuccinea columella Florida	1.02	360	19.5	0	?	Calcium alginate
Doums et al. 1994	Bulinus truncatus Algeria	5.24	196	3.5	45	4/640 ml, 3/wk.	Aphallic-selfing
Estebenet and Cazzaniga 1992	Pomacea canaliculata Argentina	0.09	20.1	30	0	40/10 L	
Joubert and de Kock 1990	Biomphalaria pfeifferi S. Africa	1.70	337	12.5	$l_{12.5} = 0.64$	50 L aquaria, weekly	FF
	Helisoma duryi S. Africa	2.4	6691	12.5	$l_{12.5} = 1.0$	50 L aquaria, weekly	FF
Loreau and Baluku 1987a	Biomphalaria pfeifferi Zaire	0.47	217	24.5	0	30/3.5 L, weekly	Rorippa leaves
Parashar et al. 1986	Indoplanorbis exustus India	1.44	4360	11.5	0	50/20 L, weekly	Rat feed, 30°C
Pointier et al. 1991	Biomphalaria glabrata Celigny, Guadeloupe	0.84	1419	20.5	8.7	4/200 ml, weekly	
	Biomphalaria glabrata Dubelloy, Guadeloupe	0.86	2144	?	?	4/200 ml, weekly	
	Biomphalaria alexandrina Kalyub, Egypt	0.78	807	?	?	4/200 ml, weekly	
	Biomphalaria alexandrina Kafr al Hamza, Egypt	0.70	711	18.5	10.8	4/200 ml, weekly	
	Biomphalaria straminea Epinette, Martinique	0.88	462	14.5	6.5	4/200 ml, weekly	
	Biomphalaria straminea Madame, Martinique	1.01	256	?	?	4/200 ml, weekly	
	Melanoides tuberculata Guadeloupe	0.24	492	133	0.37	4/200 ml, weekly	

Table 5.1 (*cont.*)

Reference	Population	r	R	x_{max}	$\sum_x m_x$ at x_{max}	Culture density, waterchange	Notes
Prinsloo and van Eeden 1969	*Bulinus tropicus* S. Africa	1.98	890	15	0	Circulating system	21°C (opt.)
	Lymnaea natalensis S. Africa	0.55	382	15	0	Circulating system	18°C (opt.)
Rankin and Harrison 1979	*Physa marmorata* St Vincent Sta. 71	1.76	458	?	?	2/200 ml, 5/week	Dasheen leaves
	Physa marmorata St Vincent Sta. 51	1.52	425	?	?	2/200 ml, 5/week	Dasheen leaves
	Physa marmorata St Vincent Sta. 77	2.70	1712	?	?	2/200 ml, 5/week	Dasheen leaves
	Physa marmorata St Vincent Sta. 69	0.80	36	?	?	2/200 ml, 5/week	Dasheen leaves
Shiff 1964a	*Bulinus globosus* Zimbabwe	1.93	467	30.5	3.2	6/'large aquarium', 2 months	
Shiff and Garnett 1967	*Biomphalaria pfeifferi* Zimbabwe	0.48	111	17.5	11.2	2/L	
Sturrock 1966a	*Biomphalaria pfeifferi* Tanzania	0.86	182	15	0	10/10 L, 2 months	
Sturrock and Sturrock 1972	*Biomphalaria glabrata* St Lucia	0.47	324	15.5	63.7	9–12/10 L, biweekly	
Wethington and Dillon 1997	*Physa heterostropha* South Carolina	1.41	808	30	0	1/200 ml, weekly	FF

Notes:
Values of r and study duration (x_{max}) are expressed in (14-day) 'fortnights.' All studies at optimum temperature (if known) or 'room temperature' (23–25°C) unless specified. Food was lettuce unless specified otherwise; 'FF' designates commercial fish food.

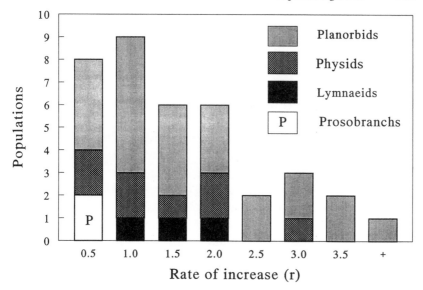

Figure 5.4. Values for the intrinsic rate of increase (*r*, per fortnight) from 36 laboratory populations of freshwater snails. The data are from Table 5.1.

confident one can be about the value of *R*. Initial time contributes another source of variation to Table 5.1; some authors have started their experiments at date of egg laying, others at date of hatch.

Most authors seem to have collected snails from the wild and derived demographic data from a cohort of their (wild-conceived) F_1 progeny. Occasionally laboratory lines have been used (Parashar *et al.* 1986) or the source of the snails has been unspecified. But such levels of inbreeding as may occur under typical laboratory culture regimes do not seem to impact freshwater snail demography substantially. De Kock *et al.* (1986) monitored survivorship and fecundity in a population of *B. tropicus* from the wild through ten generations of culture. The population size was set at 20 individuals per generation, mass-cultured in a circulating system at 28 °C and fed a mixture of commercial fish foods. Values of *r* were initially about 2.5, dropped to 1.5, then rose to the 3.2 posted at generation 7 (Table 5.1). Survivorship seemed to suffer through generations F_8 and F_9, but the population rebounded at F_{10} to finish at a strong $r = 1.9$. De Kock and his colleagues concluded that laboratory breeding exerts 'unpredictable influence on the population dynamics of *B. tropicus*', but one might wonder whether such wide fluctuation constitutes evidence of any influence of laboratory breeding at all.

It is quite clear, however, that mating system may strongly influence population demographics. Forced to self-fertilize in isolation, the value of r for (naturally outcrossing) *Physa heterostropha* halves from the 1.41 posted in Table 5.1 to 0.78 (Wethington and Dillon 1997). But interestingly, the highest value of r recorded in Table 5.1 (r = 5.24) was obtained by Doums *et al.* (1994) with (one of four) entirely aphallic populations of the famous self-fertilizer, *B. truncatus*.

The 36 values of r are plotted in Figure 5.4. Most of the highest figures have been recorded for South African planorbids, generally by researchers using systems for circulating filtered culture water. The two lowest values are those of the pioneer DeWitt (1954), who did not specify a food type and may not have changed the water in his one or two litre vessels over the entire 246–326 days of his experiments. The difference among representatives from the three pulmonate families is not striking, but the lowest values (setting aside DeWitt's data) belong to the two non- pulmonates studied thus far, *Melanoides* and *Pomacea*. Additional studies of other non-pulmonate populations would be welcome.

Field studies

The quantity r is a parameter of the logistic growth equation, and as such has primarily found use to describe populations under ideal circumstances, perhaps those newly colonizing a rich habitat or growing unlimited in an artificial environment. A stable age distribution is assumed. But r has also been calculated as a statistic to describe the growth (or decrease) of a natural population or cohort with some (inevitably fluctuating) schedule of survivorship and fertility (Marti 1986, Woolhouse and Chandiwana 1990a,b). The assumptions involved have been reviewed by Loreau and Baluku (1987b).

Especially where reproduction is continuous, as it may be in tropical populations of medically important planorbids, construction of complete life tables is a daunting task. O'Keeffe (1985) sampled populations of *Bulinus globosus* from two Kenyan reservoirs weekly over the course of two years using 'snail traps': plastic sheeting framed in wire, dangling from a float. He estimated size-specific fecundity from snails reared in large, floating plastic boxes with mesh windows. O'Keeffe was able to relate age and size in wild-caught snails using a series of cohort growth curves specific to month of birth, which he derived from growth in his boxes (for juveniles) and from mark–recapture studies (for larger snails). His aim was to assign all snails in every sample to a week of birth and a fecundity.

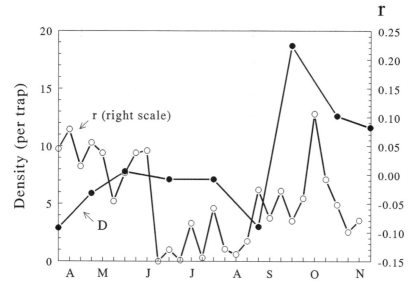

Figure 5.5. Monthly mean number of *Bulinus globosus* per trap (closed circles, left axis) in Benes reservoir, Kenya, compared with the intrinsic rate of increase (per week) ultimately posted by cohorts hatched on each date (open circles, right axis). (Data are from O'Keeffe 1985.)

Ultimately O'Keeffe recognized 29 weekly cohorts over a single year at his Benes Reservoir site, and 29 + 14 cohorts in two years at his Nguni Reservoir site. Figure 5.5 shows *r* (per week) achieved by the 29 Benes cohorts as a function of hatch date, compared to O'Keeffe's estimate of overall population density. Natural values of the *r* statistic bear little resemblance to such values derived from laboratory studies. Those few *Bulinus* cohorts that hatched in April matured in the winter (of the southern hemisphere) to post positive values of *r*, while most other cohorts failed to replace themselves. O'Keeffe performed a multiple regression analysis with *r* as dependent variable and water temperature, snail density, rainfall, conductivity, oxygen, pH, and relative water depth as independent variables. He found a very strong negative correlation between *r* and mean water temperature three months post-hatch. Across both sites and both years, cohorts spending their third month in water less than about 29 °C generally replaced themselves, while those maturing in warmer water did not.

It is possible to estimate *r* from natural populations of freshwater snails with considerably fewer field data than those marshalled by O'Keeffe, if one is willing to accept more assumptions. Growth and fecundity may be approximated from laboratory studies. In fact, some authors have estimated age-specific fecundity from the abundance of juveniles in the field, circumventing the collection of egg data entirely. Dazo *et al.* (1966) obtained an estimated maximum $r = 0.67$ per fortnight for natural Egyptian populations of *Bulinus truncatus*, and $r = 0.62$ for Egyptian *Biomphalaria alexandrina*. Saint Lucian *Biomphalaria glabrata* populations show a maximum $r = 0.59$ per fortnight (Sturrock 1973a,b). A population of *Biomphalaria pfeifferi* inhabiting a stream in eastern Zaire showed a maximum $r = 0.06$ per fortnight (Loreau and Baluku 1987b). Workers in the field typically cite temperature and rainfall (or current speed) as the important determinants of population growth in medically important planorbids.

Regarding the *Bulinus globosus* population of a pair of small, interconnected ponds in Zimbabwe, the picture painted by Shiff (1964c) was one of catastrophe followed by catastrophe. As may be recalled from Chapter 4, Shiff's study ponds dried out entirely August–November, 1962, cutting the resident snail population from almost 100 000 to about 1000. Rains fell and conditions improved to bring the population back up to about 100 000 by January, but these were almost entirely swept away by excessive flooding in February and March, 1963. Shiff estimated $r = 0.12$ per fortnight for a cohort of *Bulinus* maturing between these two calamities (lower Figure 4.8), and $r = -0.26$ for a second cohort caught by the floods (upper Figure 4.8).

In temperate climates, where generations are discrete and more easily followed, the researcher wishing to compile life tables and estimate r in the field finds his or her task considerably simplified. Egg laying occurs only in early summer in the *Lymnaea elodes* populations of small Indiana ponds. It may again be recalled from Chapter 4 that Brown *et al.* (1988) followed survivorships in cohorts of *L. elodes* over four years from three sample ponds, estimating their fecundities from field enclosures. I estimated l_x and m_x for an average cohort of all three populations, both corrected for parasitism and not, and calculated the six values of R shown in Table 5.2. I then estimated a value of r for each from population growth rates projected to stable age distribution using spreadsheet software. Brown's population 'F' seems to have been growing nicely during his four-year study period, while populations 'A' and 'B' suffered losses. The effects of parasitic castration in each population, although not negligible, were not strong enough to reverse fortunes, a subject to which we shall turn in the next chapter.

The compilation of life history tables from natural populations of

Table 5.2. *Original estimates of demographic parameters from annual cohorts of freshwater molluscs in the wild*

Data source	Population	R	r	x_{max} (months)
Browne 1978	*Viviparus georgianus,* New York			
	Pop. JAM	2.61	0.57[b]	36
	Pop. SON	2.17	0.47[b]	36
	Pop. TUL	1.38	0.20[b]	24
	Pop. CAZ	0.34	−0.68[b]	24
Brown *et al.* 1988	*Lymnaea elodes,* Indiana			
	Pop. A (parasitized/not)	0.19/0.23	−0.71/−0.62[b]	24
	Pop. B	0.61/0.70	−0.24/−0.021[b]	36
	Pop. F	10.2/11.8	1.33/1.36[a]	13
Avolizi 1971	*Sphaerium,* New York			
	S. striatinum	0.78	−0.10[a]	6
	S. simile	0.24	−0.90[a]	6

Notes:
[a] per month. [b] per year.

freshwater molluscs may be accomplished with maximum ease where young are brooded, and released synchronously over some short time period. Mortality can then be estimated by following discrete cohorts, and fertility estimated by dissecting a sample of adults just prior to reproduction. That the literature does not include more studies of temperate pisidiids and viviparids using this approach comes as something of a surprise.

Browne (1978) sampled populations of *Viviparus georgianus* from four lakes in central New York over the course of a year. Cohorts were unambiguous, such that all snails sampled could easily be assigned to a birth year. Maximum lifespan appeared to be three years at populations 'JAM' and 'SON' but two years at populations 'TUL' and 'CAZ'. Sex ratios approximated 1:1. Uterine dissection showed a gestation time of nine months, embryos appearing in June to be born the following spring. I obtained l_x from Browne's table 2 and m_x from his figure 4, showing mean midwinter fecundities at the four populations (halved to yield females-per-female). Table 5.2 shows that the four populations varied substantially in their r estimated over Browne's sample year, from 0.57 in Jamesville Reservoir (JAM) to −0.68 in Cazenovia Lake (CAZ). Browne noted that the *Viviparus* population density at Cazenovia Lake was substantially lower than those observed at the other three lakes, and appeared 'to be on a steep decline', but could offer no explanation.

Cazenovia Lake was also the site of Avolizi's (1971) cage experiments involving *Sphaerium striatinum* and *S. simile*. The experimental units were (apparently small) transparent plastic boxes with nylon mesh windows, contained in angle-iron racks. A total of 76 *S. striatinum* ranging in size from 3.6 mm to 13.2 mm were distributed in five such cages in May, 1969. Another five cages received 49 *S. simile*, ranging from 5.8 mm to 17.6 mm. Avolizi sampled these cages monthly from July through November, monitoring survivorship and counting the number of newborn spat. My analysis of his results (Table 5.2) did not indicate that either population was reproducing at a rate sufficient to replace itself. Moreover, these values of *r* are systematically overestimated, assuming that all newborn spat survive to the (rather advanced) sizes of their parents. Reproduction in Avolizi's cages seemed sporadic, with no spat born to most individuals over the entire study period. Many animals may not have reached maturity. In addition, the eye catches some evidence of an inverse relationship between the initial stocking densities of each cage and its offspring yield. Thus we are brought to the subject of crowding.

Crowding

In the previous sections of this chapter we have seen that the reproductive potential of freshwater molluscs is generally quite high when grown under conditions as benign as can be engineered by human hand, but that in the wild, population growth rates are often zero or less. One explanation for this phenomenon might be intraspecific competition.

It has long been evident that the effects of crowding on freshwater mollusc populations are many and complex. In addition to the physical competition for food and other resources one would ordinarily expect as density increases, and beyond the general degradation of the environment by waste products and oxygen depletion, freshwater molluscs interact chemically in ways that we humans only dimly understand. From his observations that crowding effects in *Bulinus forskalii* cultures could be at least partly extinguished by activated charcoal, Wright (1960) proposed that crowded snails produce a 'growth-inhibiting pheromone'. This was but one year after P. Karlson and A. Butenandt coined the term 'pheromone' for use in insects. The 1970s and 1980s saw an explosion of interest in chemical interactions among pulmonates, driven largely by the

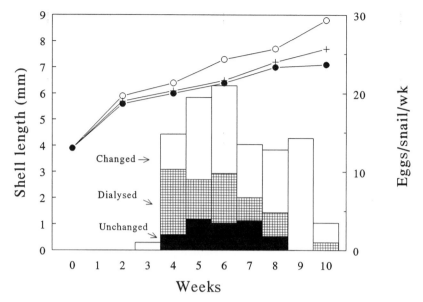

Figure 5.6. Mean length (lines, left axis) and weekly fecundity (bars, right axis) of 15 *Bulinus tropicus* raised in three water treatments: unchanged (closed bars and symbols), dialysed (crossed bars and symbols) and changed weekly (open bars and symbols). (Data of Chaudhry and Morgan 1987.)

search for a new compound that might be useful against the intermediate hosts of trematode parasites. Inhibitory pheromone(s) do seem to exist, and stimulatory ones as well, but their use against economically important snail populations has not materialized.

Chaudhry and Morgan (1987) reared three groups of 15 juvenile *Bulinus tropicus* in vessels containing 900 ml of gently aerated water, previously 'conditioned' in a fish tank, with chalk initially added as a source of calcium carbonate. Snails were fed scalded lettuce 'in excess', and any individual dying was replaced with a snail of similar size from a stock tank. One group of snails received a weekly water change, the second group was held entirely without exchange of medium, and in the third group, the culture was dialysed against fresh (fish conditioned, carbonate enriched) water.

Chaudhry and Morgan's results (Figure 5.6) well illustrate the crowding phenomenon in freshwater pulmonates. Even at a (fairly low) constant density, fed in excess, the *Bulinus* seem to be most adversely affected by the absence of water exchange. Water analysis showed a dramatic decline in the concentration of Ca^{+2} over 10 weeks in the unchanged tank (from 0.5 to 0.1 mmole/l), an increase in ammonia from 0.09 ppm to 4.1 ppm, and a drop in pH from 7.2 to 6.5. There were small increases in the concentrations of Na^+, K^+, and Mg^{+2}. These various water quality problems were corrected in the dialysed treatment, and apparently as a consequence, both growth and reproduction showed significant improvement.

The 15 *Bulinus* living in dialysed water nevertheless grew more slowly than the 15 individuals whose water was entirely refreshed weekly. Some material unable to pass through a dialysis membrane would seem to be responsible. Levy *et al.* (1973) began the characterization of such a substance, from water in which stocks of *Lymnaea cubensis* had been cultured for one month. They evaporated the culture water almost to dryness, dialysed it against distilled water, then gathered its protein component with ammonium sulphate precipitation. Levy and his co-workers called the precipitate (redissolved, with ammonium sulphate removed) 'inhibiting substance' (IS), and demonstrated its striking negative effects on growth and egg laying in both *L. cubensis* and *Biomphalaria glabrata*. Inhibiting substance had no effect on planarians or fish, however. The activity of IS appeared stable for some weeks at room temperature, but was abolished by heating in solution at 85 °C or trypsin digestion.

Thomas and Benjamin (1974) examined growth and reproduction in *B. glabrata* over an unusually complete experimental design. They manufactured their own 'standard snail water' ('SSW') from distilled, deionized water and five reagent-quality salts, treated with pressurized air and CO_2. Into volumes of 25, 50, 100, 200, and 400 ml of SSW they placed 1, 2, 4, 8, or 16 young snails of uniform size in 15 combinations, generally 10 replicates of each combination. Water was changed every three days, growth was monitored over 12 days, and reproduction (egg masses) were totalled over 15–21 days. A clear reduction was apparent in both growth and reproduction in the five highest (0.04) density cultures: 1/25, 2/50, 4/200, and 16/400. A proteinaceous inhibiting substance may have been at least partly responsible. Thomas and Benjamin preferred to hypothesize, however, that some chemical factors (particularly Ca^{+2}) were being rapidly depleted at densities of 0.04. The further experiments of Thomas and his colleagues on Ca^{+2} dynamics in pulmonate cultures will be featured in Chapter 8.

In addition to growth inhibition at high densities, Thomas and Benjamin also felt that they saw some evidence of growth inhibition at very low densities (1/400 to 2/200 snails per ml). They hypothesized that the growth of (especially) juvenile snails might be promoted by some chemical factor secreted by adults. In follow-up studies, Thomas *et al.* (1975) found evidence of such growth-promoting factors in SSW conditioned for three days with moderately high densities of adult *B. glabrata* (40 snails per 4 litres), fed lettuce. They called this 'heterotypically conditioned medium', HetCM. Thomas and colleagues showed significant improvement in the growth of juvenile (20 mg) *B. glabrata* reared for 12 days in HetCM, over that observed in plain, unconditioned SSW. They fractionated HetCM using a series of membrane filters, and were able to attribute improved growth to several substances of greater than 500 molecular weight units. Interestingly, some of the authors' substance(s) also seemed to inhibit growth at some concentrations.

Vernon (1995) has found evidence that substances secreted by one *B. glabrata* individual may augment the hatching success of eggs laid by another, suggesting that the absence of such 'social facilitation' may at least partially explain the poor reproductive performance of isolated snails. One gets the impression that humankind's chemical ignorance is profound.

In an effort to remove food competition as a variable among treatments, the authors of the crowding studies we have thus far discussed sought to provide lettuce 'in excess of requirements' in all cultures. Ensuring that food is always present does not, however, seem necessarily to forestall competition for it. Mooij-Vogelaar and van der Steen (1973) monitored two-week growth in sets of *Lymnaea stagnalis* cultures with equal densities (two adults per 500 ml) but different quantities of lettuce. The authors measured lettuce offered and consumed in units equal to 9 mm^2, as judged on a transparent, gridded surface. Their 'Experiment V' returned an average growth increment of 3.22 mm at 551 units of lettuce per snail per day, approximately the minimum ration totally consumed. At four times minimum ration (2204 units daily), snails consumed an average of 814 units (leaving uneaten 1390 units) and grew 6.73 mm. And at 20 times minimum ration, snails consumed an average of 1069 units daily, left uneaten 9951 units, and grew 9.57 mm. It would appear that large snails feeding on a single lettuce leaf cannot avoid interfering with one another. Larger food volumes seem to result in less jostling, and hence more feeding. A similar phenomenon was reported by Brown and his colleagues (1994), rearing *Physa virgata* at various densities on periphyton-covered tiles. The authors attributed reductions in grazing rates to increased behavioural interference in crowded cultures.

It will be recalled that pulmonate snails have large capacities for sperm storage, a single insemination typically furnishing allosperm sufficient to fertilize weeks of egg laying. There is evidence that the frequency at which mating occurs in crowded cultures is far from adaptive, constituting rather a substantial interference to growth and reproduction. Van Duivenboden *et al.* (1985) showed that oviposition rates in sets of 12 (adult, previously inseminated) *Lymnaea stagnalis* declined when grouped and rose when isolated. Oviposition rates remained uniformly high, however, when surgery was performed to cut the vas deferens of 12 snails before grouping. Controls subjected to 'sham surgery' – operated upon but vas deferens left intact – continued to display reductions in fecundity associated with grouping. Van Duivenboden and his colleagues offered some evidence suggesting that excessive copulation as female, rather than as male, was more responsible for the reduction in observed oviposition rates. Bohlken *et al.* (1987) and Bayomy and Joosse (1987) have also documented rapid increases in oviposition rate when adult pulmonates are removed to isolation, and likewise attributed the phenomenon to a reduction in copulation interference.

Crowding effects are not unique to pulmonates. They have been documented in most of the freshwater molluscs successfully cultured to date, including the prosobranch *Tricula* (Liang and Kitikoon 1980) and the pisidiid *Musculium* (Mackie 1976b). Paterson (1983) introduced sets of 1, 3, 5, 7, 9, and 12 *Elliptio complanata* into a respirometer of 20 litre volume, monitoring respiratory rates over 12 hours. All animals exhibited elevated respiratory rates initially, then equilibrated long before dissolved oxygen was reduced to any level that might cause an adverse impact. The average equilibrium rate was, however, significantly higher for single mussels than for groups of three, and significantly higher for groups of three mussels than for groups of five. Groups of 7, 9 or 12 mussels all equilibrated at a uniformly low respiratory rate (about 0.2–0.3 mg $O_2/g/hr$) that Paterson characterized as 'comfortable'. And when single mussels were placed in chambers containing water previously conditioned by higher numbers of individuals, they equilibrated at the low, 'comfortable' rate. Paterson hypothesized that *Elliptio* release 'some detectable substance' which may influence the metabolism of their neighbours.

Paterson preceded his physiological studies with an Ekman grab survey of the natural distribution of his animals in Morice Lake, New Brunswick. They were, in fact, aggregated; they were absent from about 59% of his (522.6 cm²) grabs but present in numbers as great as 13 in the remainder. Paterson translated natural density to a 'standard lower density' of about 3.4 mussels per his respirometer chamber. Hence he considered that the lower respiratory rates he measured in crowded

Elliptio a reflection of 'comfort', and the higher rates shown by isolated individuals some indication of stress. But Paterson noted that just the reverse might well be true; higher respiratory rates may indicate health. The assumption that natural unionid aggregation is an adaptive response, rather than the potentially deleterious consequence of microhabitat preferences that must inevitably overlap, warrants further study.

We have thus far discussed crowding effects as though they were primarily phenomena of artificial culture. But while crowding certainly is most easily recognized and studied in the laboratory environment, it is clear that natural populations of freshwater molluscs can, on occasion, become crowded to such levels that growth and reproduction are strikingly impacted. Berrie (1968) monitored growth rates and population dynamics in a population of *Biomphalaria sudanica tanganyicensis* inhabiting a small, closed pool near Uganda's Lake Victoria. No more than two metres across, the pool suffered a dramatic volume reduction in May when local drainage operations seem to have lowered the water table. Berrie estimated that its resident 1500 *Biomphalaria* were reduced to a volume of 150 litres for five months, during which time there was virtually no shell growth or reproduction. He extracted a large sample of this water with activated charcoal, stripped the charcoal with various solvents, and via bioassay attributed growth inhibition to a chloroform-soluble 'toxin', perhaps an ester of molecular weight 360 (Berrie and Visser 1963). In mid-October, Berrie (1968) removed 1490 snails from the pool, returning but 53 individuals. Rain in the amount of 25 cm fell during the next three weeks, and when Berrie resampled in early November, he found that his much-reduced population of *Biomphalaria* had posted substantial shell growth. Reproduction was well under way by December.

Population regulation

Berrie's studies of *Biomphalaria* (initiated in 1960, although not published until 1968) were among the first in which experimental manipulations were used to establish what factors (if any) may regulate the size of natural freshwater mollusc populations. Subsequent investigators have taken a variety of approaches: experimental manipulation of population density or food supply *in situ* or in facilities designed to duplicate the natural environment. Our discussion here leads from such experimental studies to more analytical treatments, where the question of population regulation is examined through long-term observation. We will discuss the results of experiments especially designed to measure predation effects in Chapter 7.

Experimental approaches

Researchers have found unionid clams to be especially suitable candidates for direct manipulation of population density in the field. We have previously mentioned the studies of Hanson *et al.* (1988) involving *Anodonta grandis* caged at various depths in Narrow Lake, Alberta, over the course of a growing season. Although depth does seem to affect growth rate, Hanson *et al.* could see no effect of stocking densities that varied an order of magnitude, over the range of depths naturally occupied by the animal. Kat (1982a) established square-metre plots on the bottom of Norwich Creek, Maryland, and placed marked, measured *Elliptio complanata* at densities ranging from 25% to four times normal. His recovery rates were quite good a year later, generally in the 70–80% range. Analysis was hindered by small sample sizes, especially given the complication that mussels of different ages are expected to grow at different rates. A series of separate one-way analyses of variance (rather than analysis of covariance) nevertheless suggested that growth rates were inversely related to stocking density.

Snail population densities are generally manipulated using cages. Aldridge (1982) could not see an effect of adult density on the fertility of *Leptoxis carinata* caged in a New York river, although artificial removal of all *Leptoxis* eggs prior to their hatch did stimulate increased egg laying in both low and high density treatments. Aloi and Bronmark (1991) monitored three-week shell growth in *Lymnaea emarginata* held at four densities in flow-through chambers located by Trout Lake, Wisconsin. The chambers were opened to colonization by periphyton two weeks prior to the beginning of the experiment. Both periphyton biomass and primary productivity decreased rapidly upon the introduction of the snails, to their effective detection limits. The growth rates ultimately achieved by the snails appeared to be a direct function of their stocking densities. A limitation of both the studies of Aldridge and Aloi and Bronmark is that it is difficult to relate their experimental densities to the density of snails actually observed in the field.

Turning to experiments involving both density manipulation and resource augmentation, we come to one of the few studies involving freshwater molluscs that may truly be considered influential to the study of ecology as a whole. Eisenberg (1966) constructed large (4.23 m²) pens on the exposed, frozen soil surrounding a small Michigan pond. He

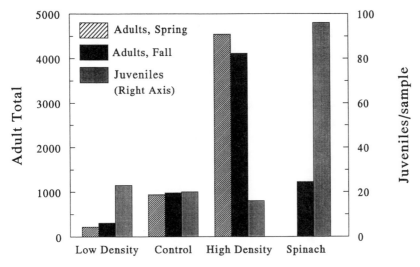

Figure 5.7. Demographic data from the cage experiments of Eisenberg (1966). The left axis is the average density (per 4.23 m² pen) of adult *Lymnaea elodes* in the spring (several June estimates) and in the final autumn census. The right axis shows the average number of young-of-the-year juveniles (per 80 cm² sampler) counted over 2–3 July weeks.

enclosed his pens in May, by which time pond levels had risen and a large, resident population of *Lymnaea elodes* had emerged from aestivation. Using several techniques in tandem, he estimated the numbers of snails enclosed in each of 12 pens, and in June, adjusted densities such that four pens held approximately five times their normal density, and four pens held only 25% normal density. To two other pens he added packages of frozen spinach at three or four-day intervals. (No initial estimates of density were made in these two pens.) He estimated the numbers of young-of-the-year snails in all 14 pairs on several dates in July, and re-estimated the densities of old adults in September.

Eisenberg found very little mortality in his enclosed populations of adult snails from May to September, even at high density (Figure 5.7). But regardless of the size of the resident population of adults, he found roughly the same numbers of juveniles born in all 12 pens of his density experiments. Average adults were thus much less fecund at high densities than they were at low. Fecundities showed further improvement in the pens to which Eisenberg added spinach. It would appear that the size of the *Lymnaea elodes* population in this small pond is regulated by food, acting through fecundity. In a series of follow-up experiments, Eisenberg (1970) confirmed his earlier results, adding observations on somatic growth. He reported that snails living at low density grow more rapidly than those at high density, and those whose normal diet is augmented with spinach grow more rapidly still.

Gilbert *et al.* (1986) also varied both density and feed in their cage study of the *Bithynia tentaculata* population inhabiting the St Lawrence River, Quebec. Their experiments involved three starting densities (10, 30, and 90 adults per cage) and two levels of food – unsupplemented and supplemented to excess with a mixture of lettuce, spinach, and chicory leaves. Assessed at the end of a three-month season, the authors could distinguish no effect of food level on the numbers of offspring produced per parent. The 10 adult treatment yielded high numbers of relatively large offsping, the 30 adult treatment yielded lower numbers of large offspring, and 90 adults yielded high numbers of small offspring. These results seem to reflect crowding in both parental and offspring generations.

'Cage effects' are a general problem with studies of this kind. In spite of the best efforts of researchers, the environment inside a cage may bear little resemblance to the environment outside. It is odd, for example, that adult mortality seemed to have been nearly negligible inside Eisenberg's cages over four months, while outside the cages it seems to have been over 90% (to judge from Eisenberg's figure 1). Cages certainly provide unusual protection from predation, as we shall see in Chapter 7, but Eisenberg's own efforts to assess this effect experimentally proved inconclusive. Thus the resource augmentation study of Osenberg (1989), involving both caged and uncaged controls, provides an especially welcome insight.

Lawrence Lake, Michigan, is home to eight species of freshwater snails: three populations of small, numerically dominant prosobranchs (*Amnicola limosa, Marstonia lustrica* and *Valvata tricarinata*) and five populations of pulmonates (two *Gyraulus*, two *Helisoma*, and a *Physa*). Osenberg (1989) established eight 2.5 m² stations at about one metre depth and applied four treatments: caged (surrounded by a wood frame supporting 1 mm mesh), fertilized (with resin-coated phosphate pellets on wooden dowels), caged-and-fertilized, and control. His experiment ran from mid-August to mid-September, after which Osenberg quantitatively sampled both snails and macrophytes. Although the experiment may have been too brief to return significant density effects for most snail populations, fertilized treatments did show greater numbers of *Gyraulus parvus* and newborn *Physa*. Snails of most populations also seemed to grow more rapidly where nutrients were applied. Cage effects generally were not significant. But the highest total snail biomasses (and macrophyte biomasses) were recorded in the caged, fertilized treatments, an effect Osenberg attributed to higher nutrient retention where water circulation was reduced. Here it should be kept in mind that macrophytes may serve as habitat, cover, and substrate for egg laying as well as food; treatments that benefit macrophytes benefit snails in more ways than one.

Among the most effective methods of eliminating cage effects (but not problems of interpretation) is not to use cages. Repeated drainage and refilling of a fish pond in northeastern Tanzania had complex and unpredictable influences on its snail fauna (Pringle and Msangi 1961, Pringle and Raybould 1965). Hershey (1992) reported the effects of five years of artificial fertilization upon the benthos of an entire (1.8 ha) lake in arctic Alaska. The four snails present, *Lymnaea elodes*, *Valvata lewisi*, a *Physa* and a *Gyraulus*, typically 'live for several years and appear to take 2–3 yr to reach adult size' in this extreme environment. In late June, 1985, a fibre-glass-reinforced plastic curtain was used to divide the lake in half. Then from 1985 to 1990, the western sector was continuously fertilized (with both N and P) using a solar-powered pump, to about five times its normal annual nutrient loading. Midsummer *Valvata* densities were estimated from quantitative samples taken at 4–5 m on both sides of the partition, and adult *Lymnaea* were counted by divers swimming measured transects. Results are shown in Figure 5.8.

One must not expect rapid responses from arctic molluscs. By 1987–88, however, the density of adult *Lymnaea* was significantly greater in the fertilized half of the lake than in the unfertilized control. Data on water quality and productivity are not at hand, but it appears that fertilization effects were building; divers were unable to swim transects in 1989 because of reduced visibility due to an algal bloom. Although adult *Lymnaea* densities had dropped in the fertilized half of the lake by the summer of 1990, Hershey began to notice increased densities of juveniles. The *Valvata* population did not post a significant increase until 1989, as the *Lymnaea* population seemed to be falling. Although artificial fertilization of their habitat will clearly affect freshwater snails in many ways, both populations certainly reacted as though they were food-limited. Hershey noted that the density response of *Valvata* 'is consistent with the cascade concept', as increased resources are first expected to benefit the *Lymnaea* grazing the shallows, then later the detritivorous *Valvata* of deeper water.

It is certainly not true, however, that the addition of fertilizer to any freshwater habitat will necessarily benefit resident molluscs. Daldorph

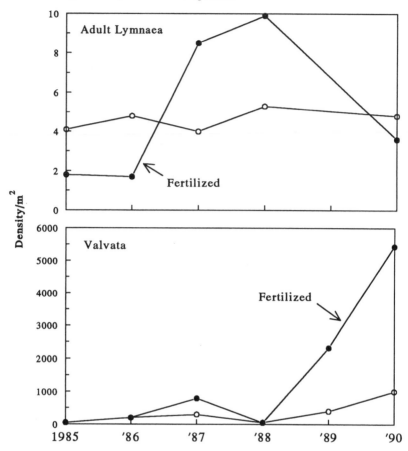

Figure 5.8. Densities of *Valvata* and adult *Lymnaea* on the fertilized side of an arctic lake (closed circles), compared with densities on the unfertilized control side (open circles). (Data of Hershey 1992.)

and Thomas (1991) constructed cylindrical enclosures (0.44 m²) in a shallow, narrow channel (of negligible summer flow) draining reclaimed wetland in Sussex, U.K. The channel, spring-fed from neighbouring chalk downland and the recipient of agricultural and domestic run-off, was naturally rich in both macrophytes and molluscs, including *Bithynia tentaculata*, *Planorbis planorbis*, *Physa fontinalis*, and *Lymnaea peregra*, in order of abundance. To their enclosures of this already-rich soup, Daldorph and Thomas added two levels of nitrate and two levels of phosphate, each treatment (and controls) replicated seven times. Increased phosphate loading clearly promoted the growth of floating and non-rooted macrophytes, while high nitrates promoted phytoplankton blooms, to the general detriment of both macrophytes and periphyton. At the end of a (five month) growing season, the authors could find no statistically significant treatment effects for any individual snail species. But setting aside the amphibious *L. peregra*, and combining across the three more aquatic species, Daldorph and Thomas could see an inverse relationship between snail density and phytoplankton levels. This suggests that artificial fertilization had proven broadly detrimental to resident snail populations in this circumstance.

With an eye toward the war against medically important pulmonates, Thomas and Daldorph (1991) took a second approach somewhat different from that of previous workers. Rather than augmenting the resources inside a second set of 20 large (7 m²) enclosures constructed in their Sussex drainage channel, the authors undertook to remove the resources available to the resident snail populations, in a conscious effort to effect their demise. They monitored population densities of *Bithynia*, *Planorbis*, *Physa* and *Lymnaea* over five months in control enclosures and in enclosures subjected to three treatments: mechanical harvest of macrophytes, light exclusion by black polythene cover, and light attenuation with the floating weed *Hydrocharis*. But only the black polythene treatment had a significant effect on snail densities (deleterious to all except *Planorbis*), and this only after 24 weeks had elapsed.

Although the actual mechanism by which opaque plastic sheeting adversely affects populations of *Bithynia*, *Physa*, and *Lymnaea* is of little importance from a bioengineering standpoint, it is of great relevance to us here. In particular, we have been prospecting through the literature for evidence that freshwater mollusc populations are regulated by food. And since the snail populations of the Sussex drainage channel did not suffer when their macrophyte resource was entirely removed by hand, nor benefit when at least some elements of it were stimulated by phosphates, it would appear that the macrophytes were not themselves being eaten. The significant depression in dissolved oxygen Thomas and Daldorph recorded in the black polythene treatment during the summer months, perhaps due to macrophytes rotting below, may have been responsible for the lower snail densities they observed.

It is very difficult to find any such quibbles with the thorough experiments of Hill and his colleagues (1992, 1995) involving multiple sorts of resource augmentation, as well as density manipulation. Upper White Oak Creek, in eastern Tennessee, is about as different from Eisenberg's boggy Michigan pond as one can imagine a freshwater habitat. And

Goniobasis clavaeformis is about as different from *Lymnaea elodes* as one can imagine a freshwater snail. But the densities of *Goniobasis* on the bare rocks of little mountain streams can be every bit as striking as the densities *Lymnaea elodes* often achieves on the brown bogs of northern America.

Hill *et al.* (1992) enclosed *Goniobasis* at natural densities (about $900/m^2$!) in 12 one-litre chambers of circulating water (changed daily), immersed in a bath regulated to natural temperature. Four treatments were replicated three times: low biomass periphyton ('LBP') with dead leaves, LBP without leaves, high biomass periphyton ('HBP') with leaves, and HBP without leaves. Low biomass periphyton was supplied on natural rocks from White Oak Creek, changed daily, while rocks with high biomass periphyton were brought from nearby East Fork Poplar Creek, a richer stream with no resident *Goniobasis* population. Hill measured two-month growth (live mass) on five small and five large snails from each of the 12 chambers. Snail growth in both LBP treatments was quite similar to that observed *in situ* in White Oak Creek. But in HBP treatments, small snails grew two times faster, and large snails grew five times faster. (Addition of leaves had no effect.) Hill and his colleagues concluded that 'food limitation appears to be a chronic condition for grazers in streams like White Oak Creek, where grazer biomass is high and primary productivity is low.'

To test the hypothesis that *Goniobasis* are limited by periphyton, and periphyton by sunlight, Hill *et al.* (1995) designed a blocked, split-plot experiment including two levels of snail density and two levels or irradiance, replicated four times. They cleared trees and branches away from four sites on White Oak Creek, leaving adjacent areas shaded, and placed into the stream bed Plexiglas channels, divided lengthwise, screened with rather coarse mesh at each end. Natural cobble was added, as well as juvenile blacknose dace to prey upon the aquatic insects (which must have been included) but not snails. Then large, adult *Goniobasis* were stocked at natural densities on one side of each channel, and at 0.05 natural density on the other. From late June to early August, snail growth (wet mass increment) was strongly influenced by both population density and light regime. The average growth in marked, uncaged controls was about 190 µg/snail/day, intermediate between the values posted in the normal-density-shaded and normal-density-unshaded treatments. Mean growth in the low-density-unshaded channels was 832 µg/snail/day, almost ten times that observed in normal-density-shaded. Interestingly, periphyton biomass showed significant responses to snail density, but not to light regime, as ravenous snails at high density easily stayed ahead of any gains that the hapless algal cells might have enjoyed in the sunlight. Rosemond and colleagues (1993) have confirmed Hill's results in the G. *clavaeformis* population of nearby Walker Branch, where nutrient additions stimulated both algal growth and weight gain in the snail population.

The general impression one takes from the experiments of Berrie, Eisenberg, Osenberg, Hershey, Hill, and Rosemond is that populations of freshwater molluscs may be food limited, and hence subject to density-dependent control. And in at least the well-studied populations of *Goniobasis* in Tennessee streams, somatic growth itself may be regulated in density–dependent fashion.

Perturbations

Over against whatever evidence might be marshalled for density dependence in mollusc populations, the literature contains numerous and sometimes spectacular examples of density-independent effects.

We noted in Chapter 3, for example, that the relationship between certain snail populations and macrophytes seems to be an especially close one. Lodge and Kelly (1985) documented a 'summerkill' event in the small (1 ha) Radley Pond, associated with warm weather and anoxic conditions. The submerged macrophyte population suffered greatly, as did populations of two snails living upon them; 99% of the *Lymnaea peregra* population and 35% of the *Valvata piscinalis* population died. Heywood and Edwards (1962) were on hand when macrophytes seem to have been eliminated from England's River Ivel. They attributed a 78% drop in the summer density of *Potamopyrgus jenkinsi* to this calamity. Drought is another widely cited perturbation. During 1975–76, only about half the normal precipitation fell on the River Roding catchment in Essex, England. Flows were reduced to negligible at some sites, insufficient to mix and dilute urban sewage. Extence (1981) reported a significant decrease in the mean annual density of the Roding *Potamopyrgus* population, which he attributed both to stranding and to organic pollution. Effects upon other species were mixed; pisidiids (apparently more than one species) seem to have significantly benefited from the drought.

All too often, it is simply impossible to separate fluctuations in population size due to natural exigencies from those due to the interference of man. Most of the long-term declines in populations of freshwater molluscs documented to date have been attributed to human alteration of the environment (e.g. Covich 1976, Eckblad and Lehtinen 1991, Lewandowski 1991, Nalepa *et al.* 1991). Shaw and Mackie (1989, 1990) reported that the decline (or extinction) of *Amnicola limosa* in a variety of Canadian lakes seemed to be a function of pH. They found some populations inhabiting lakes so close to their physiological tolerance limits for acidity that adult size (and hence fecundity) seemed to be affected. But whether the apparent extinctions are natural, or due to the phenomenon of acid rain, cannot be determined.

Payne and Miller (1989) offer an example of a perturbation (of uncertain nature) in a freshwater mollusc population that gives every appearance of being natural. Using S C U B A, the authors took 17–24 samples of 0.25 m² each from a gravelly shoal on the lower Ohio River. Although the 256 *Fusconaia ebena* they collected in 1983 ranged from 1 mm to 90 mm in standard shell length, 71% of all individuals were clustered between 12.8 mm and 19.5 mm. The authors identified this as a single cohort born in 1981, distinguishing much smaller cohorts born in 1979 (6%) and 1982 (3%). They were easily able to follow the 1981 cohort in samples from similar dates in subsequent years. It again comprised 71% of 269 mussels sampled in 1985, and 74% of 219 mussels sampled in 1987, due to lack of strong subsequent recruitment. Payne and Miller offered no explanation for the exceptional reproductive success of their *Fusconaia* population in 1981. Perhaps such dynamics are not unusual in populations of individuals where the average longevity is 11–18 years.

Chief among perturbations must be the new colonization of a previously unoccupied habitat. On a number of famous occasions, such 'colonizations' have come at the hand of man, and the consequences have been unfortunate from our perspective. Worldwide zebra mussel invasions have been reviewed by Kerney and Morton (1970), Stanczykowska (1977), Mills *et al.* (1996), Ram and McMahon (1996) and Karatayev *et al.* (1997), *Corbicula* invasion by McMahon (1983b) and Counts (1986), *Potamopyrgus* by Bondesen and Kaiser (1949) and Ponder (1988a), and the invasions of other snails by Madsen and Frandsen (1989). For the present purposes, the artificially mediated colonizations of such pests as *Dreissena* and *Corbicula* may serve as models of processes which must, at some lower frequency, occur naturally.

The people at Glen Lyn Power Plant, on the New River at the Virginia/West Virginia border, were not surprised to discover that *Corbicula* had ascended the Ohio and Kanawha Rivers to threaten their water works. Their first quantitative samples (October, 1976) contained no individuals greater than 10 mm (standard shell length), clearly all young-of-the-year (Rodgers *et al.* 1977). Population densities of 18–29/m² suggested to the authors that (unsampled) parents may have

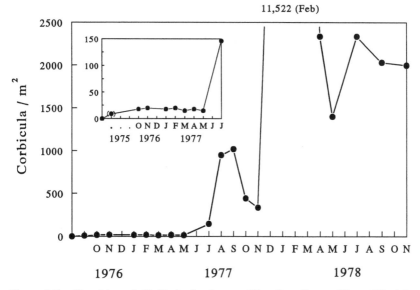

Figure 5.9. Densities of *Corbicula fluminea* at Glen Lyn Power Plant, Virginia (station 4, thermally influenced). (Data of Graney *et al.* 1980.)

arrived, undetected, in 1975. Graney *et al.* (1980) established a programme of regular sampling at six stations, collecting benthos from five $0.25 \, m^2$ quadrats at each station monthly or bimonthly. Figure 5.9 shows that *Corbicula* densities peaked at $11\,522/m^2$ in February 1978, after which they seemed to stabilize around $2000/m^2$.

It may be recalled from Chapter 2 that *Corbicula* densities of about $1000/m^2$ seem to have been sufficient to impose a substantial reduction in the phytoplankton concentration of the (very large) Potomac River in Maryland (Cohen *et al.* 1984). Calculations based on per capita filtration rates suggested an even larger relative impact for the $350/m^2$ population of North Carolina's smaller Chowan River (Lauritsen 1986b). From a comparison of laboratory diet to the known concentrations of particulate matter *in situ*, Foe and Knight (1985) suggested that the population densities of *Corbicula* in the Sacramento–San Joaquin Delta of California might be approaching their carrying capacity. Might the widespread declines that have subsequently been reported in North American *Corbicula* populations (Sickel 1986) represent some measure of density-dependent adjustment?

The spread of *Dreissena* through most of Europe predated the modern science of ecology. There have been some reports of boom-and-bust phenomena (Walz 1974), but most populations seem, by now, to have acquired a permanence. Stanczykowska *et al.* (1975) compared 1962 *Dreissena* densities in 13 Polish lakes with the densities observed in 1972. She found five lakes in which densities had increased substantially, four in which densities were substantially reduced, and four lakes in which *Dreissena* densities appeared unchanged. The introduction of *Dreissena* to North America is much more recent. Although *Dreissena* densities may have reached local equilibrium a few years after its arrival in some regions, viewed at an appropriate scale (the Great Lakes, the Mississippi drainage) the population continued to give the appearance of log-phase growth after five years (Griffiths *et al.* 1991).

Long-term trends

It should be apparent from the previous discussion that freshwater mollusc populations are often subject to density-independent perturbations, and that the existence of such perturbations does not disprove density-dependent control. The key is to examine population sizes over sufficient scales of time and space.

The best available record of freshwater mollusc population density comes from the *Pisidium* fauna of Denmark's large (17.3 km^2), temperate, eutrophic Lake Esrom. Beginning in 1953, and extending irregularly to 1973, P. M. Jonasson collected 5 or 10 bottom hauls (250 cm^2 each) using an Ekman grab in the deepest spot in the lake (22 m – the 'Endrup' station). An additional set of samples was taken in 1977, and an account of the entire data set was offered by Holopainen and Jonasson (1983). As one might expect, the environment at 22 m seems to be fairly constant among years, although within years there are regular thermal cycles (about 2–15 °C) and cycles of summer oxygen depletion. Yet Holopainen and Jonasson reported that the annual abundances of the three *Pisidium* species at Endrup seemed to vary substantially – *P. subtruncatum* and *P. casertanum* swapped dominance twice as *P. henslowanum* was reduced to (local) extinction.

Figure 5.10 shows summer densities for the three *Pisidium* species at the Endrup station. To construct this figure I averaged the (1–3) records collected for June, July, or August from the larger data set kindly provided by I. J. Holopainen. Summer data were not available for 1953 or 1968. Holopainen and Jonasson could offer no explanation for the large population explosions in 1955, nor for the 1956–1959 declines. The only substantial perturbation identified by the authors over the entire 24-year

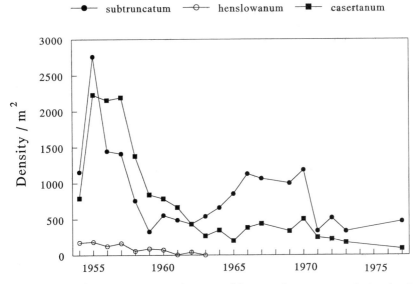

Figure 5.10. The average summer densities of three *Pisidium* species at the 'Endrup' station of Lake Esrom, Denmark. (Data of Holopainen and Jonasson 1983.)

record was the construction of one additional municipal sewage outflow in 1961.

Additional, shorter records of the Esrom *Pisidium* fauna are available from other sites on the lake. Irregularly between 1954 and 1961, Jonasson made five hauls of his (250 cm²) Ekman grab at each of three depths on the 'Tumlinghus transect' (Holopainen and Jonasson 1989b). His shallowest samples (11 m) contained six *Pisidium* species: the three Endrup species as well as *P. nitidum*, *P. hibernicum*, and *P. milium*. *Pisidium casertanum* and *P. nitidum* were difficult to distinguish when young, and often tallied together. But again through the courtesy of I. J. Holopainen, I was able to extract average summer densities from the original data for the remaining four species. Both *P. subtruncatum* and *P. henslowanum* showed notable peaks in 1955–57, matching their deeper-water brethren, while populations of the rarer *P. milium* and *P. hibernicum* seemed more stable through the seven-year record.

Dennis and Taper (1994) have offered the parametric bootstrap likelihood ratio (PBLR) test as a robust and powerful approach for the detection of density dependence in time series data. If populations are regulated in a density dependent fashion, one would expect that years of unusually low population density (caused, for example, by unusually

Table 5.3. *Statistics on the long-term densities of freshwater mollusc populations*

	Years	Mean (/m²)	s.d.	t (observed)	t (α = 0.05)
Endrup, 22 m					
P. casertanum	1954–73	749	668	−0.60	−2.84
P. henslowanum	1954–62	102	60.8	−0.95	−3.08
P. subtruncatum	1954–73	899	552	−1.77	−2.72
Tumlinghus, 11 m					
P. henslowanum	1954–60	416	267	−2.26	−3.17
P. hibernicum	1954–60	42	27	−3.44★	−3.27
P. milium	1954–60	79	38	−1.64	−2.93
P. subtruncatum	1954–60	456	250	−4.03★	−2.99
Scotland					
Ancylus fluviatilis	1950–58	263	126	−6.69★★	−3.00

Notes:
★ $P < 0.05$
★★ $P < 0.01$
Sources: The values of t are from PBLR tests for density dependence (Dennis and Taper 1994). The 'Endrup' and 'Tumlinghus' data were taken in the summer at two different stations in Denmark's Lake Esrom (Holopainen and Jonasson 1983, 1989b). *Ancylus* data are from Russell-Hunter (1961a).

severe weather) would be followed in the next year by large population growth, and similarly that years of exceptionally high population density would be followed by especially large population declines. There should be a negative correlation between the density at time t (n_t) and the change to the next time (n_{t+1}). To perform the PBLR test, one regresses log (n_{t+1}/n_t) on n_t, comparing the t-statistic from the regression with critical values obtained by bootstrapping.

I performed PBLR tests on the three *Pisidium* data sets from the Endrup station (Figure 5.10) and the four shorter sets from Tumlinghus, testing my values of t using a brief SAS program kindly provided by B. Dennis. The 1977 observations were not included in the analyses of *P. subtruncatum* and *P. casertanum*, but values for 1968 were extrapolated linearly. Results are shown in Table 5.3. Although none of the three Endrup data sets showed evidence of density dependence, two of the four Tumlinghus records did return values of t significant at the 95% confidence level: the most common *P. subtruncatum* and the most scarce *P.*

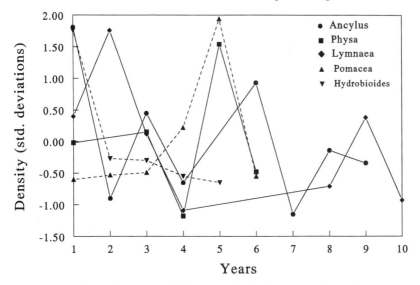

Figure 5.11. Annual estimates of density in five freshwater snail populations, standardized by statistics cited in the text. *Pomacea paludosa* data are from Kushlan (1975), and *Hydrobioides manchouricus* data are from Chung *et al.* (1980). *Lymnaea peregra*, *Physa fontinalis* and *Ancylus fluviatilis* data are from Russell-Hunter (1961a).

hibernicum. It is difficult to imagine any substantial biological distinction among these two species and *P. henslowanum* or *P. milium*, or any environmental distinction among their habitats. One might hypothesize that all four populations were regulated in density dependent fashion, but that density dependence was undetectable in two of the populations due to the brevity of the data record.

The only long-term data set regarding the sizes of European *Dreissena* populations of which I am aware is rather spotty. Stanczykowska and Lewandowski (1993) reported 14 estimates of *Dreissena* density in Poland's Mikolajskie Lake, taken irregularly over the 31 years between 1959 and 1989. The population fell dramatically from over 2000/m² in 1959 to less than 100/m² 1960–66, gradually grew back to about 2000/m² in 1976, then again dropped below 100/m² in 1977–89. Lewandowski (1982) viewed these large fluctuations as density dependent cycles of growth, overgrowth, and crash.

Given the relative ease by which typical populations of freshwater snails may be sampled, I am surprised and, I suppose, disappointed that more long-term census data are not available. I am aware of but three studies involving censuses of five years or better: two of prosobranchs and one of Scottish pulmonates. Oddly, there seems to be no long-term study of any economically important pulmonate.

Kushlan (1975) gathered six years of data on *Pomacea paludosa* in the southern Everglades of Florida, from 1966–67 to 1971–72. He offered averages from a set of 4.5 m² traps, surveyed monthly. The author noted very little annual variation, except during the single year 1970–71, which saw a population spike that Kushlan attributed to prolonged high water. I calculated his grand mean to be a (rather low) $0.11/m^2$ (s.d. $= 0.16$), standardized his data, and plotted it in Figure 5.11. The other prosobranch data set is that of Chung *et al.* (1980) on the bithyniid *Hydrobioides* (or *Parafossarulus*) *manchouricus*, an intermediate host for medically important trematodes in Korea. The authors monitored population density over five years, from 1975 to 1979, noting a large drop and then relative stability. Their average June data are pictured in Figure 5.11, standardized by the grand mean of $41.8/m^2$ (s.d. $= 42.0$).

Russell-Hunter (1961a) contributed the three remaining data sets to Figure 5.11. The *Physa fontinalis* population and the *Lymnaea peregra* population were sampled from the rocky littoral zone of Loch Lomond, Scotland, and the *Ancylus fluviatilis* population was sampled in a small highland stream north of Glasgow. Russell-Hunter carefully lifted stones and counted all snails attached, estimating surface area by tracing the outline of each stone on a piece of paper. Samples were taken in late May of each year, eight times from 1950 to 1958 for *Ancylus*, five times from 1953 to 1958 for *Physa*, and seven times from 1949 to 1958 for *Lymnaea* (at the 'old' site). Grand means (and standard deviations) were *Ancylus* $260/m^2$ (142), *Physa* $77/m^2$ (30.2), and *Lymnaea* $93/m^2$ (49.3). Russell-Hunter was impressed with the intercorrelations among the densities of these three species, which he felt reflected a shared dependence upon the severity of the winters.

Although none of the data sets depicted in Figure 5.11 is especially lengthy, I am personally impressed by the stability generally on display. To my eye, the density of none of these populations appears to be a 'random walk'. I analysed Russell-Hunter's *Ancylus* record using the PBLR technique, linearly extrapolating a value of $280/m^2$ for the missing 1954 datum. The value of t I obtained was significant at the 0.01 level (Table 5.3). Thus the single record of freshwater snail density of sufficient length to test for density dependence does, in fact, show strong evidence of the phenomenon.

In summary, my survey of eight moderately long data sets on the density of freshwater molluscs returned three values significant at the putative 95% confidence level or better (Table 5.3). This result is quite comparable to those reported in populations of other groups (e.g. 2 of 16 insects, 17 of 92 birds, Dennis and Taper 1994, Murdoch 1994). In light of the generally recognized difficulty of detecting density dependence using analytical methodologies, I take my 3/8 success rate as some evidence of a signal.

Interspecific competition

In our review of the diet and habitat of freshwater molluscs through Chapters 2 and 3 we found both extraordinary diversity and remarkable uniformity. Certainly the habitats of planorbids and lymnaeids often differ; for example, many populations of the latter are much more amphibious. But their diets are quite general, and broadly overlap. And if sometimes lymnaeids and planorbids may be spied jostling each other on the same blade of grass, surely the same regulatory phenomena we have documented within species in the present chapter thus far ought to apply between species. In the following sections we will review the evidence of competition between populations of freshwater molluscs in natural situations, as well as where artificial introductions may have brought the phenomenon into more vivid focus.

Natural situations

Some pairs of closely related species seem to show strikingly discordant local distributions, as though they exclude one another. For example, *Lymnaea peregra* and *L. auricularia* have similar geographic ranges in Great Britain but are rarely found together. In those unusual habitats were they do co-occur, there seems to be an inverse relationship between their temporal abundances. Adam and Lewis (1992) reported a striking example of *L. auricularia* replacing (outcompeting?) *L. peregra* in an English gravel-pit lake.

It is perhaps not surprising that among the schemes for the biological control of medically important planorbids are various ideas for the promotion of interspecific competition. *Biomphalaria straminea*, for example, is not as effective a vector of schistosomiasis in tropical America as is *Biomphalaria glabrata*. The former species has replaced the latter on several occasions, both in long-term laboratory studies and in the wild (Barbosa

1973, Barbosa *et al.* 1984, Guyard *et al.* 1986). It is not clear whether the phenomenon is due to competition or hybridization, but the end result seems to be desirable.

The potential for competition between quite diverse freshwater molluscs is well illustrated by the relationship between the Alaskan populations of *Lymnaea elodes* and *Valvata lewisi* we have touched on previously. The tiny, delicate prosobranch, with a penchant for feeding on fine deposited detritus, reaches mean densities of $1000/m^2$ on the deeper sediments of Toolik Lake, but may also be found (at about $80/m^2$) in the lake shallows (Hershey 1990). The large pulmonate *Lymnaea*, with its strong jaws and formidable radula, grazes at about equal densities ($80/m^2$) throughout the lake. It will be recalled that the artificial fertilization study of Hershey (1992) suggested that populations of both these species seemed to be food limited (Figure 5.8). Hershey (1990) caged four *Valvata* either with or without a *Lymnaea* on both sides of her artifically fertilized lake. After 12 days, *Valvata* enclosed on the fertilized side had laid significantly more eggs than those enclosed on the unfertilized side, and those without *Lymnaea* significantly more than those with. The latter result is unsurprising, since the addition of a single *Lymnaea* probably more than doubled the snail biomass depending on the resources of an $800 cm^3$ cage. Hershey also conducted a more extensive set of laboratory experiments (at lake temperatures) confirming the negative impact of *Lymnaea* on *Valvata* fecundity, although again, aquaria with both species supported higher biomass than aquaria with *Valvata* alone. Interestingly, the effect of *Valvata* on *Lymnaea* fertility seemed to be positive. Hershey concluded by suggesting that trout predation may have greater impact upon the *Lymnaea* populations of arctic lakes than upon the *Valvata* populations, facilitating their continued coexistence.

Competition experiments are more easily performed if the species compared are of roughly equal size. Four larger pulmonates are commonly found in the ponds of the Crooked Lake Field Station in northern Indiana: *Helisoma trivolvis*, *Aplexa hypnorum*, *Physa gyrina*, and *Lymnaea elodes*. Brown (1982) calculated pairwise MacArthur–Levins niche overlap statistics in this community of four, based on the duration of their reproductive periods, their habitat preferences (quantitative samples of eight lakes varying in permanence and cover) and their diet (laboratory experiments). We will explore Brown's assumption that high habitat overlap implies increased competition in Chapter 9. In any case, his analysis suggested that the greatest likelihood of competition might be between *Physa gyrina* and *Lymnaea elodes*.

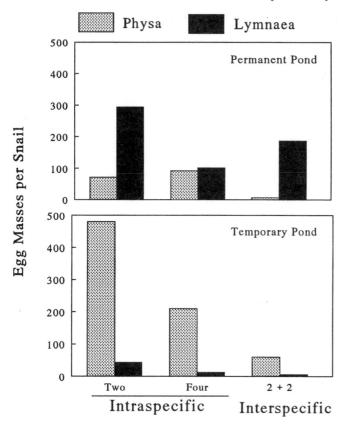

Figure 5.12. Nine-week production of egg masses for *Physa gyrina* and *Lymnaea elodes* in containers floating in a permanent pond ('F') and a temporary pond ('A'). The treatments within each pond (two of each, four of each, mixed two-and-two) are explained in the text. (Data of Brown 1982.)

Brown floated $15 \times 5 = 75$ one-litre, screened containers in each of three separate ponds: temporary ponds 'A' and 'B' and permanent pond 'F'. Fifteen containers in each pond received two *Physa*, four *Physa*, two *Lymnaea*, four *Lymnaea*, and two *Physa* plus two *Lymnaea*. Brown tended each container weekly from mid-May to mid-June, counting egg masses, adding food ('pond vegetation'), and replacing any mortality. His data on egg mass production are shown in Figure 5.12. Brown's ANOVA returned significant effects (and significant interactions) for all factors: species, pond, and density treatment. Assuming no systematic differences in the number of eggs per egg mass (probably fair within species but more problematic between), it appears that *Lymnaea* has the competitive advantage in the permanent pond, while *Physa* is superior in the temporary pond. For *Physa*, interspecific competition appears more severe than intraspecific competition. Even in habitats to which they are better adapted, a pair of *Physa* are always more adversely affected by a pair of *Lymnaea* than by another pair of *Physa*. For *Lymnaea*, interspecific competition is not as severe as intraspecific, at least in the permanent ponds they favour.

Brown's results highlight the importance of a variable introduced in Chapter 4, life history 'strategy'. Although all the members of a single population may be expected to mature, reproduce, and expire according to roughly similar schedules, this can by no means be said for populations of two different species. Through life history variation, interspecific competition may be mitigated, or aggravated, in ways not generally available in conspecific comparisons.

Bronmark and his colleagues (1991) designed an extensive set of cage experiments to test competition among the large, aquatic *Lymnaea stagnalis*, the smaller, more amphibious *Lymnaea peregra*, and aneuran tadpoles in a small Swedish pond. *Lymnaea peregra* typically shows annual life cycles, while *L. stagnalis* is perennial. Within a month of the initiation of Bronmark's experiment, essentially all of the *L. peregra* had laid eggs and expired. Bronmark could find no competition effects of either lymnaeid on the other, in growth or reproduction. Had his experiments continued, he would have needed to shift his attention to the 1 mm size classes.

Mackie et al. (1978) enclosed sets of 20 *Musculium* in small (about 50 cm^3) plastic vials with nylon mesh windows and placed these at 0.3 m on the bottom of Britannia Bay, on the Ottawa River. This density was chosen to mimic the natural density of *Musculium*, about 1700/m^2. He monitored total reproduction in three replicates for each of nine treatments: 20 *M. transversum*, 20 *M. securis*, and mixtures in the ratios 19:1, 18:2, 16:4, 10:10, 4:16, 2:18, and 1:19. Mackie concluded that *M. transversum* is 'dominant', averaging about six offspring per parent when present at high densities (80% or greater), holding *M. securis* to only about two offspring each. But in treatments where *M. securis* was present at densities of 80% or greater, *securis* averaged about four offspring per parent as *transversum* ceased reproduction entirely. As juveniles are of considerable size at release, Mackie felt that competition was primarily for space. It is this theme we shall pursue through the next two sections, before returning to the snails.

Corbicula and the North American bivalves

Since its introduction into the United States, there has been speculation that *Corbicula* might competitively inhibit native bivalves (Kraemer 1979).

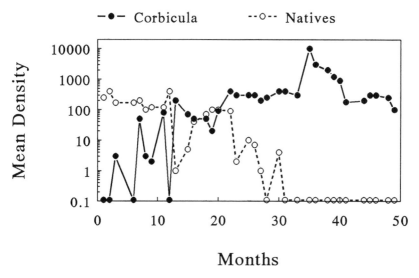

Figure 5.13. Densities of bivalves (N/m^2) in the Altamaha River, Georgia, October 1971 to October 1975 (Gardner *et al.* 1976).

Among the best observations are those of Boozer and Mirkes (1979) on the co-occurrence of *Corbicula* and *Musculium partuneium* in a sedimentation basin at a nuclear plant in South Carolina. They reported that a dramatic increase in the number of *Corbicula* on the floor of the basin was associated with a small decrease in the abundance of *Musculium*. At the same time, the population of *Musculium* attached by byssal thread to the wall of the basin increased dramatically. (*Corbicula* was very rarely found bysally attached to the wall.) Although the evidence is anecdotal, it is certainly consistent with displacement of *Musculium* by *Corbicula*.

We have the Georgia Power Company to thank for the wealth of information available on the bivalves of the middle Altamaha River. The quantitative unionid samples taken by Sickle in 1968 (but not published until 1980) will be discussed in Chapter 9. The first report of *Corbicula* in the region was made in 1971. Beginning in October of that year, and continuing for four years thereafter, Gardner *et al.* (1976) collected quantitative samples at six-week intervals from a number of sites with a Petersen dredge. At the start, densities of native bivalves (primarily pisidiids, but with at least 4 or 5 unionid species) were about $200/m^2$. But Figure 5.13 shows that the results of the 1972 and 1973 samples (months 3–27) were spectacular, even on a log scale. In the space of three years, *Corbicula* densities rose to about $10\,000/m^2$, while native bivalves disappeared. Gardner's descriptions of subsequent qualitative samples indicate that the disappearance of unionids, at least, is not a sampling artifact. The authors mention no environmental variation, such as pollution or development, that might have influenced these results. These data stand as, admittedly circumstantial, evidence of competition.

To my knowledge, however, there are no additional data or observations supporting competition between North American bivalves and *Corbicula* in the fifty-year history of the introduction. From measured filtration rates, water flow rates, and population size, Leff *et al.* (1990) estimated that *Corbicula* could significantly reduce the concentration of food particles suspended in a small South Carolina stream. This is very much in accordance with the earlier findings of Cohen *et al.* (1984), Foe and Knight (1985), and Lauritsen (1986b). But their extensive field survey failed to show any apparent affect of *Corbicula* on the native *Elliptio complanata* population, in distribution, abundance, or individual size. Miller and Payne (1994) found no significant correlations (negative or positive) between *Corbicula* density and the density of native unionids in 403 quadrat samples taken over several years in the gravelly shoals of the Ohio River. In general, it has been concluded that *Corbicula* is not a threat to native North American bivalves (McMahon 1983b).

Dreissena and the unionids

It has long been remarked that on otherwise soft, silty lake bottom, *Dreissena* tends to aggregate on whatever islands of solid substrate are available, often especially including the shells of living unionids. From the standpoint of an individual *Dreissena*, the advantages of life on the posterior margin of a unionid are easy to discern. The larger bivalve sets up a water current from which the smaller animals can thieve, and holds them above the silt more efficiently than an inanimate substrate (but see Nichols and Wilcox 1997). There is strong evidence of direct competition between zebra mussels and host unionids for algal cells and organic matter generally in the wild (Parker *et al.* 1998). *Dreissena* impacts go far beyond theft from their host's intake current, however. Mackie (1991) listed eight effects of zebra mussel colonization upon host unionids, including occlusion of shell margins so great as to interfere with shell opening or prevent closure, and burden so heavy (often greatly exceeding the weight of the unionid itself) as to overbalance or bury it entirely. Tucker (1994) noted the opposite effect on a gravel bar in the Mississippi River. Zebra mussels paved the substrate to such a degree as to prevent burial by unionids, resulting in their demise when water levels fell.

In Europe, competition between *Dreissena* and unionids has been the object of some interest since the observations of Sebestyen (1938).

Lewandowski (1976) found that 85% of the unionids in Poland's Mikolajskie Lake had mussels attached to them, mostly about the siphons, at an average density of 20 individuals per unionid. The *Anodonta piscinalis* population was more affected than the *Unio tumidus* population, because the latter mainly occurred at 0.5 m, where *Dreissena* is generally scarce. A sample of *Anodonta* severely overgrown with *Dreissena* showed lesser shell growth than a sample of unburdened individuals aged 3–4 years at time of collection, although the effect disappeared as *Anodonta* growth rate slowed with age.

The introduction of *Dreissena* to North America, apparently in the vicinity of Lake St. Clair, occasioned immediate concern for the fate of the diverse unionid fauna native to the Laurentian Lakes. In 1986, prior to the detection of zebra mussels, Nalepa and Gauvin (1988) found maximum unionid densities ($7.8/m^2$) in Lake St Clair at Puce, Ontario. But between June and September, 1990, mean unionid densities (at a depth of 4 m) dropped from about $3.3/m^2$ to $0.5/m^2$, as the incidence of dead unionid shells covered with zebra mussels increased (Gillis and Mackie 1994). A re-survey in the summer of 1991 found 0.06 unionids/m^2 at Puce, and by the summer of 1992, no living unionids remained at that site. Unionid densities did not decline over the same period in the northwestern portion of Lake St Clair, where zebra mussel infestation was lighter (Nalepa 1994).

Unionid populations were also entirely eliminated from some regions of Lake Erie during the period 1989–1991 (Schloesser and Nalepa 1994, Dermott and Kerec 1997). Haag *et al.* (1993) collected 80 *Lampsilis radiata* and 80 *Amblema plicata* from a bay on the western end of Lake Erie, scrubbed off all attached zebra mussels, and weighed, measured, and marked each unionid. They then covered half of each population with zebra mussels and allowed recolonization to about 200 *Dreissena*/unionid, the prevailing density in Lake Erie. Animals were returned to the lake for three months in large open baskets. Although no mortality was observed in *Amblema*, the mean glycogen content and cellulase activity of the encrusted animals was significantly lower than observed in controls. Control populations of *Lampsilis* suffered about 35% mortality, as did encrusted males, while the 83% mortality of encrusted females was significantly worse. Again, encrusted *Lampsilis* of both sexes showed significantly lower mean glycogen content and cellulase activity than unencrusted controls. The laboratory experiments of Baker and Hornbach (1997) confirmed both that zebra mussel infestation causes nutritional stress in unionids, and that the magnitude of such effects seems to depend on the mussel species being fouled.

Significant unionid mortalities have now been associated with the spread of zebra mussels into the Mississippi and Hudson Rivers (Schloesser *et al.* 1996, Strayer and Smith 1996). The prospects for the diverse unionid fauna of North America, already so threatened by impoundment, dredging, siltation and pollution, dim with each passing year.

Helisoma duryi

Interest in the potential of the large North American *Helisoma duryi* as a competitor against medically important planorbids dates to 1941. As may be gathered from Figure 5.3, *Helisoma* populations have greater intrinsic rates of increase than typical *Biomphalaria* populations at warm temperatures, at least under the favourable food and water conditions provided by De Kock and Joubert (1991). Joubert and De Kock (1990) recorded values of $r = 2.34$ for *H. duryi* and 1.69 for *B. pfeifferi* in large aquaria at $26\,°C$, fed and changed daily. The authors saw no competitive effects between the two species when reared together ($r = 2.12$ and $r = 1.60$, respectively), as long as water was changed constantly and eggs removed.

However, it has been well established that *H. duryi* has a strikingly negative effect on the growth and reproduction of African *Biomphalaria* and *Bulinus* when the environment is more restricted (Frandsen and Madsen 1979). Detailed experiments involving different densities of *Helisoma* and *Biomphalaria*, reared together, separately, in partitioned aquaria and in variously conditioned waters, have tended to rule out allelopathic chemicals (Madsen 1979a,b, 1982, 1984). Such chemical inhibition as has been described appears to be attributable to non-specific pollutants and perhaps competition for inorganic ions (Madsen 1987).

Nor in fact is the markedly negative effect of *H. duryi* due exclusively to competition for food. Lassen and Madsen (1986) reared *Helisoma* and *Bulinus* together and separately in small cages immersed in a single large aquarium, agitating and changing water regularly to remove any possible effect of fouling or inhibiting chemicals. Food (dog food plus lettuce) was supplied at three levels: less than the amount consumed daily, daily consumption, and in apparent excess of daily consumption. As expected, they found strikingly lower growth and fecundity in *Bulinus* caged together with *Helisoma* than in *Bulinus* caged with an equal number of conspecifics. In contrast, eight *Helisoma* showed improved growth and fertility caged with eight *Bulinus* than with eight other conspecifics. The negative effect of *Helisoma* on *Bulinus* remained striking even when excess food was left floating on the water all day. Excess food did seem to relieve a little of what appeared to be egg predation by *Helisoma*, although its negative effect on *Bulinus* was still significant. Further, confirmatory experiments of Meyer- Lassen and Madsen (1989) also pointed strongly to egg predation.

Such results prompted Madsen (1986) to perform a number of behavioural experiments comparing the two species. *Helisoma* seems generally to be more active, running into other snails more often and able to find food faster. But perhaps as importantly, *Bulinus* seem to be more adversely affected by collisions with *Helisoma* than with conspecifics, withdrawing into their shells rather than simply turning aside. It is possible that *Helisoma* is directly antagonistic.

El-Emam and Madsen (1982) found that laboratory populations of *H. duryi*, *B. truncatus*, and *Biomphalaria alexandrina* all showed optimal growth at 26–28 °C and optimal reproduction on the same artificial diet. They noticed only minor differences in tolerance of extreme temperatures, starvation, and darkness. Unfortunately, their habitat preference may not be especially similar in the field. In 1972, *H. duryi* was introduced into a sugar plantation in Tanzania, where *Biomphalaria*, *Bulinus*, and *Lymnaea* were widespread. But Madsen (1983) reported that it was subsequently restricted to a very small area where (perhaps not coincidentally) *Biomphalaria* was not present. Madsen noted that the canals are regularly treated with molluscicide, and one wonders how much of the problem may be *Helisoma*'s ability to disperse and colonize.

Joubert and his colleagues (1992) were unable to introduce viable populations of *H. duryi* into small ponds on the South African lowveld. A few wild populations of *Helisoma* have become established about Africa (De Kock and Joubert 1991, Brown 1994), but by comparison with the immediate and rapid spread through Africa shown by such exotics as *Physa acuta* and *Pseudosuccinea columella*, this is unimpressive. It is also important to note that *H. duryi* shares the New World with *Biomphalaria glabrata*, with little evidence of competition between the species. Perera et al. (1986) noted that Cuban populations of *H. duryi* do not seem to multiply as quickly as do *Biomphalaria*, nor do they seem to survive seasonal drying. Thus the prospects for *H. duryi* as a biological control agent do not seem promising.

The ampullariids

In Chapter 3 we noted that ampullariid populations can, on occasion, be prodigious consumers of macrophytes. The diet of *Biomphalaria* is more typical of freshwater snails generally: certainly including macrophytes but

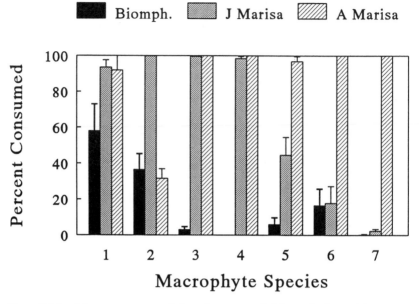

Figure 5.14. Mean percentage of plant tissue eaten (with standard error) in macrophyte choice tests involving ten each *Biomphalaria glabrata* ('Biomph.'), juvenile *Marisa cornuarietis* ('J Marisa'), and adult *M. cornuarietis* ('A Marisa'). See text for key to the seven macrophyte species. (Data of Cedeno-Leon and Thomas 1982.)

probably slanted toward detritus and algae. Cedeno-Leon and Thomas (1982) compared the macrophyte preferences of *Marisa cornuarietis* with that of *Biomphalaria glabrata*, placing pre-weighed standard sized disks or rings of the 10 plant species in randomly chosen quadrants of small aquaria. The plants offered were as follows: (1) *Rorippa*, (2) *Lemna*, (3) *Cabomba*, (4) *Elodea*, (5) *Ceratophyllum*, (6) *Potamogeton*, (7) *Pistia*, (8) *Myriophyllum*, (9) *Eichornia*, and (10) *Salvinia*. The authors then introduced an individual *Marisa* or *B. glabrata* previously 'experienced' with these types of food and allowed it to feed for 24 hours.

Cedeno-Leon and Thomas performed 10 replicates on each of a great many size classes of snails, and Figure 5.14 shows a small part of the data. *Myriophyllum* results were similar to those shown for *Ceratophyllum* (5), while *Eichornia* and *Salvinia* results resembled those for *Pistia* (7). There was a great deal of variance between replicate snails, as reflected by large standard errors. Nevertheless, it is clear that most of the macrophytes were almost untouched by adult *Biomphalaria*. Juvenile *Marisa*, half the size of adult *Biomphalaria*, in contrast, ate most of the macrophytic vegetation available to them in 24 hours. There were striking differences in preference. Among the favoured food of juvenile *Marisa* were *Elodea* and *Cabomba*, essentially uneaten by *Biomphalaria*, while *Potamogeton* seems to be much higher on *Biomphalaria*'s list. Large adult *Marisa* ate almost everything offered in 24 hours, with the exception of the little duckweed *Lemna*, among the favourite foods of the juveniles. Cedeno- Leon and Thomas speculated that adult *Marisa* find this small floating weed difficult to capture.

Large ampullariids may in fact ingest eggs, neonates, and even adult pulmonates under some circumstances. So in further experiments, Cedeno-Leon and Thomas offered *Marisa* some *Biomphalaria* egg masses, but these were not preferred over any sort of vegetation. We will continue our examination of the predatory proclivities of *Marisa* in Chapter 7.

The first reports that *Marisa* could have significant negative effects on *Biomphalaria* came from Puerto Rico (Oliver-Gonzalez *et al.* 1956), and were rapidly verified in the laboratory (Chernin *et al.* 1956a, Michelson and Augustine 1957). Chernin *et al.* noted that when placed together in a large aquarium with watercress (*Nasturtium*) as food, *Marisa* effectively eliminated *Biomphalaria glabrata*. Their most striking demonstration involved 25 juvenile (10 μm) *Marisa* placed with 50 adult *Biomphalaria* (comparable in size to the *Marisa*) and a huge excess of watercress. Chernin reported a total of 10 new *Biomphalaria* born in five weeks from this aquarium, compared with over 2000 from a control tank. Chernin demonstrated *Marisa* predation on both *Biomphalaria* eggs and juveniles, but felt that most of their negative effect was due to incidental disruption of eggs.

A chain of five ponds in Mayaguez, Puerto Rico, was the site of the first controlled introduction of *Marisa* into a natural setting (Radke *et al.* 1961). Each of these ponds was choked with foliage, primarily water lily (*Nymphaea*), and each was inhabited by a large population of *Biomphalaria glabrata*. In October, 1957, 200 adult *Marisa* were released into each pond. Every 6 weeks, Radke took 25 dips with a long-handled scoop at each of 10 sites. The results (Figure 5.15, upper) show a dramatic increase in the *Marisa* population after one year, and the disappearance of *Biomphalaria*. The snails had cleared all ponds of vegetation by week 75, and no *Biomphalaria* could be found alive.

Accompanying Radke's more controlled study was a larger effort to employ *Marisa* as a biological control agent against *Biomphalaria* in 111 irrigation ponds serving a large section of southern Puerto Rico (Ruiz-Tiben *et al.* 1969). The first ponds (average area 0.5 ha, all with established *Biomphalaria* populations) were 'seeded' with *Marisa* in 1956, and

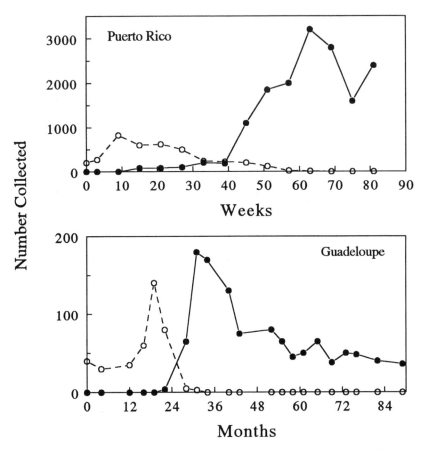

Figure 5.15. Abundance of *Biomphalaria* and (top) *Marisa* in a series of Puerto Rican ponds from 1957 to 1958 (Radke *et al.* 1961) and (bottom) *Ampullaria* at site on Grand Etang Lake, Guadeloupe, from 1979 to 1986 (Pointier *et al.* 1988). Number collected is per 10–20 minutes of sampling with a fine sieve at Grand Etang Lake, and per 25 dips of a long-handled scoop at ten sites in the Puerto Rico ponds.

as these populations grew, *Marisa* were harvested to spead to other ponds. Ponds were sampled annually, and *Marisa* reintroduced as necessary. The total effort (111 seedings, 229 reseedings, 179700 *Marisa* introduced) may be judged a rousing success; *Biomphalaria* seems to have been eliminated from 89 of the 97 ponds still in operation in 1965. It was the author's general impression that about four years were required for a complete replacement of *Biomphalaria* by *Marisa* in small ponds. A longer-term effort to eliminate *Biomphalaria* from 30 'major reservoirs' of Puerto Rico met with similar success. Over the period 1956–76, Jobin and colleagues (1977) reported the establishment of *Marisa* in 22 reservoirs, and elimination of *Biomphalaria* from 25.

The rate at which the replacement phenomenon takes place (and perhaps its ultimate success) seems to depend on the amount of vegetation a pond contains. Jobin and colleagues (1973) introduced *Marisa* into seven of nine similarly sized Puerto Rican farm ponds, at rates of 1–5 snails/m^2. After two years, *Biomphalaria* populations were eliminated from two ponds, substantially reduced in three others, and unaffected in the two controls and in the two ponds whose surfaces were completely covered with macrophytic vegetation. *Marisa* failed to become established in a pond which 'contained heavy mats of vegetation and very little water during 1969, frequently producing anaerobic conditions.' It is certainly possible that a *Biomphalaria* reduction might have been observed in the seventh pond, had study continued beyond two years.

Marisa is not the only ampullariid capable of interfering with populations of pulmonates. Grand Etang Lake is isolated in the rainforests of Guadeloupe, 3 km from the nearest human habitation. The flora and fauna are not extensive. Pointier *et al.* (1988) state that *Ampullaria glauca* was 'artificially introduced' in 1976, although this introduction could have been natural, judging from the paper. Well established in the lake at that time was a large population of *Biomphalaria glabrata*, associated with the floating macrophyte *Pistia stratiotes*. It took six years, but an *Ampullaria* population boom in 1982 had predictable results (Figure 5.15, lower). By 1986, *Ampullaria* had grazed the *Pistia* down from dominance to just a single patch only a few metres across, that patch reportedly the last refuge of *Biomphalaria* in Grand Etang. The more rapid response to *Marisa* in Puerto Rico was doubtless due to the large size of the release population.

The thiarids

Of all the biological agents offered for the control of medically important pulmonates, *Thiara granifera* and *Melanoides tuberculata* have in recent years displayed the most promise for lotic waters. The former species, a native of the Far East, began to be widely reported in Caribbean regions in the 1950s and 1960s. The latter species, initially an inhabitant of the milder climates of the Middle East and Africa, seems to have spread through the New World tropics about ten years later. As the two thiarids have spread, workers have noted striking declines and even extinctions of *Biomphalaria* and other pulmonates. Pointier and McCullough

(1989) have reviewed independent reports of this phenomenon from Puerto Rico, Venezuela, the Dominican Republic, Cuba, Martinique, Dominica, and Guadeloupe.

Pointier *et al.* (1994) provide a good set of example data from 29 small, rocky streams running north into the Gulf of Mexico on the coast of Venezuela. Prior to the arrival of *T. granifera* and *M. tuberculata* in 1975, *B. glabrata* inhabited seven of these rivers. But occasional, qualitative surveys conducted in 1975–90 have documented the spread of the thiarids to all 29 streams, and the disappearance of *Biomphalaria*. The complicating factor here (as throughout the Caribbean) is that all seven of the *Biomphalaria* populations have been treated with molluscicides.

By the 1980s Pointier *et al.* (1989) reported that *B. glabrata* had been all but extinguished from the island of Guadeloupe, coincident with the spread of *Biomphalaria straminea*. But a few populations remained in shallow, flowing waters managed for watercress culture. Pointier and his colleagues collected 60–70 quadrat samples of 0.05 m^2 each from each of two watercress beds, two or three times yearly from 1981 to 1987. Both *B. glabrata* and *B. straminea* were common, at least initially. So in January 1983, Pointier introduced 700 *Melanoides* into each pond, for a calculated initial density of just over $1/\text{m}^2$. Within two years, *Melanoides* had peaked at $9941/\text{m}^2$ in one pond and $13388/\text{m}^2$ in the other, and both *Biomphalaria* populations had disappeared. Figure 5.16 (upper) shows population density data from one of these watercress ponds. A larger set of follow-up studies confirmed the negative impact of *Melanoides* on *B. glabrata* in Guadeloupe watercress beds, but yielded no evidence of an effect in smaller cattle ponds or in *Colocasia* fields subject to irregular drying (Pointier *et al.* 1993).

At this point the reader might well thumb back to compare Figure 5.1 and Figure 5.2, note the scales on their abscissas, and marvel that a population with $r = 0.024$ might outcompete a population with $r = 0.84$. However, *Melanoides* seems to be better adapted to stable environments than *Biomphalaria*. The Guadeloupe watercress beds afforded constant water supply, constant food, and very small temperature fluctuations (3 °C annually). Pointier notes that *Melanoides* 'is able to maintain very high densities for a long time', that is, intraspecific competition does not seem especially severe. And as long as the water is well oxygenated, *Melanoides* shows remarkable resistance against pollution, such as one might expect in rich beds of watercress. Finally, it is my guess that at least occasionally, these beds of watercress were harvested, and that

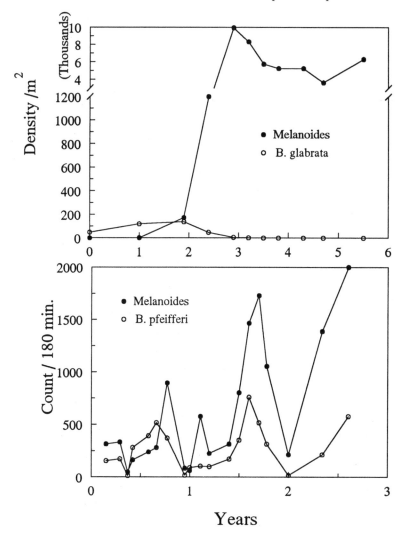

Figure 5.16. The upper graph shows the densities of *Melanoides tuberculata* and *Biomphalaria glabrata* in a watercress bed on Guadeloupe from the beginning of 1982 until mid-1987 (data of Pointier *et al.* 1989). The lower graph shows the relative abundance of *M. tuberculata* and *Biomphalaria pfeifferi* in Kenya's Matingani Stream from early 1987 until late 1989. (Data of Mkoji *et al.* 1992.)

Biomphalaria populations suffered greatly, but that the more benthic *Melanoides* tended to be left undisturbed.

I am aware of only one experiment matching a thiarid against *Biomphalaria* under controlled conditions. Gomez *et al.* (1990) designed an elaborate experiment with four controls and nine treatments, involving 3 gallon or 1.5 gallon aquaria, 5 or 10 snails, and *B. glabrata* and *Thiara granifera* either together or partitioned by a fine screen. The snails were fed powdered rat food, which must have rapidly settled and tended to foul the water. Oddly, the authors make no mention of aeration or water change over the 12-month duration of the experiment. I should assume that the tanks were aerated, and that water must have been changed occasionally, perhaps irregularly. Neither *Thiara* nor *Biomphalaria* reproduced during the first five months of the experiment, quite normal for the former but rather unusual for the latter. Juveniles remained in the aquaria, competing against their parents and each other. At the end of 12 months, populations had grown about tenfold, probably to the carrying capacity of the tanks. In all cases, even when the two populations were separated by screens, the densities of *Thiara* were strikingly greater than those of *Biomphalaria*. Gomez attributed the superiority of *Thiara* to 'chemical factor(s), possibly water-soluble pheromones.' A second interpretation would be that *Thiara* does better than *Biomphalaria* in aerated water fouled with rotten rat food.

Finally, it is important to note that *Melanoides* populations native to Africa peacefully coexist with an extensive pulmonate fauna, including *Biomphalaria pfeifferi*. In fact, Mkoji *et al.* (1992) reported positive associations between the abundances of *Melanoides* and *Biomphalaria* in the Sudan and Nigeria. They monitored snail population densities at four sites in Kenya over three years. At each site, Mkoji and his colleagues established six stations, at which three collectors searched for ten minutes. The densities shown in the lower half of Figure 5.16 are counts over $6 \times 10 \times 3 = 180$ person-minutes at one of these four sites, Matingani Stream. Their results at the other three sites were the same. There was no tendency for native *Melanoides* populations to replace or outcompete established pulmonate populations. In fact, *Melanoides* and *Biomphalaria* (as well as *Bulinus tropicus* and *Lymnaea natalensis*, not shown) seemed to fluctuate together. Mkoji pointed out that native African populations will have, by now, had the opportunity to adapt to one another, so that any competitive displacement will have occurred long ago. Also, the sites they studied were all what one would consider typical pulmonate habitat: subject to considerable fluctuation in volume and shallow only at narrow littoral zones. Pointier and McCullough (1989) offer the following summary of the prospects for biological control with thiarids: favourable in the neotropics if waters are permanent, shallow, well oxygenated, and free of floating plants.

Summary

The intrinsic rate of increase (r) has been estimated under benign laboratory conditions for a broad range of planorbids, physids, and lymnaeids. Although comparisons are complicated by differing details of culture, populations of all three pulmonate families typically demonstrate $r = 0.5$ to $r = 2.0$ per fortnight, with values ranging to $r = 3.0$ on occasion. This is to say that given suitable physical conditions, an excess of food, and protection from competition and predation, typical pulmonate populations may be expected to increase by factors of 1.6 to 7.4 every 14 days.

Observed intrinsic rates of increase show a good deal of variance. They seem very sensitive to temperature for some species, while other species seem unaffected over broad ranges. It is difficult to know whether the populations in this latter situation are reproducing at their physiological limit, or whether they are limited by some aspect of their environment unrelated to temperature. Although the (probably low) levels of inbreeding experienced under typical laboratory culture do not seem to influence values of r, an artificially imposed mating system (whether selfing or outcrossing) may have a profound effect.

I am aware of only two reports of the intrinsic rate of increase for prosobranch populations. Both these values are much lower than those typically posted by pulmonates. No such data are available for any bivalve.

Although r was originally conceived as a parameter of the logistic growth equation, it has found use as a statistic to describe the instantaneous status of natural populations under conditions neither constant nor benign. Tropical cohorts of pulmonates typically exhibit $r < 1.0$ per fortnight in the wild, and often show $r < 0.0$. Fewer data are available for temperate populations (surprisingly), but they seem comparable to values from tropical pulmonates. Among the many candidate explanations for the lower values of r observed in the wild, one must include the effects of competition.

Competition for food seems to occur as jostling and disturbance in crowded laboratory cultures, even when food is made available 'in excess'. Excessive mating may also constitute a measurable disturbance in dense snail cultures, while waste products build up and available calcium is rapidly depleted. In addition, workers have been most impressed with the evidence of chemical interactions among snails cultured at various densities. Crowded pulmonate snails seem to produce substances that inhibit growth and egg laying in conspecifics, and growth-promoting factors may be produced in uncrowded situations. Although best documented in pulmonates, there is evidence of chemical messages passing through crowded populations of prosobranchs and bivalves, in the wild as well as in culture.

Experiments on the regulation of freshwater mollusc populations in the wild have generally involved the use of pens or cages. Workers have manipulated population density, directly adjusted food availability, or indirectly augmented food by the use of fertilizers. Of approximately ten such studies reviewed in this chapter, about half returned at least some evidence of food limitation in freshwater molluscs. The famous work of Eisenberg (1966) on Michigan *Lymnaea*, followed by the large-scale fertilization study of Hershey (1992) on Alaskan *Lymnaea* and *Valvata* and the thorough food and density manipulations of Hill *et al.* (1992, 1995) involving Tennessee *Goniobasis*, leave little doubt that many natural populations of freshwater molluscs are under density-dependent control.

Workers also often report striking (presumably natural) fluctuations in mollusc population densities (Brown 1994: 487–95). These are often associated with identifiable perturbations of the environment, such as drought, spate, or vegetation summerkill, although sometimes the cause remains obscure. Among the most spectacular fluctuations have been demonstrated by invading populations of *Corbicula* and *Dreissena* in North America, which may follow initial logarithmic growth rates with population declines, and then perhaps stability.

The literature does not contain a great wealth of long-term data on the density of freshwater mollusc populations. An inspection of five snail population data records of greater than five years duration (2 prosobranchs, 3 pulmonates) leaves the impression of some stability, however. Density dependence can be distinguished in a series of annual censuses using parametric bootstrap likelihood ratio (PBLR) tests. I applied such tests to seven data sets from the *Pisidium* fauna of Denmark's Lake Esrom, two records of 20 years, one of 9 years, and four of 7 years. One of the gastropod records was also of sufficient length to analyse using the PBLR technique. The three significant results obtained from these eight tests constitute a success rate comparable to those posted by workers with much better bird or insect data sets. This again constitutes some evidence of density dependence.

If natural populations rarely grow to the carrying capacity of their environments, and if they are generally adapted to exploit widely varying resources, one might expect competition to be a rare phenomenon in the wild. But workers using the same approaches as used for intraspecific competition, including the design of enclosure experiments and the examination of long-term trends, often report evidence of the interspecific phenomenon. *In situ* cage experiments have returned evidence of competition between North American populations of *Musculium transversum* and *M. securis* (Mackie *et al.* 1978), *Lymnaea* and *Physa* (Brown 1982) and even *Lymnaea* and *Valvata* (Hershey 1990).

Competition is a widely reported consequence of invasion or artificial introduction. There is some evidence of competition between *Corbicula* and bivalves native to North America, the American pisidiids perhaps most adversely affected. Much stronger evidence has accumulated regarding the adverse affect of *Dreissena* upon unionids. Fouling by zebra mussels lowers the glycogen and lipid stores of host unionids, resulting in decreased growth rates and increased mortality. Large ampullariids, voracious consumers of macrophytes (and perhaps pulmonate eggs and neonates), have reliably and efficiently eliminated *Biomphalaria* from larger ponds and lakes in the neotropics. Thiarids have been similarly effective in better oxygenated and less weedy environments, by virtue of the high population densities they achieve, and their environmental tolerance. The failure of *Helisoma duryi* to outcompete *Biomphalaria* in the wild, given its promise in laboratory trials, warrants further examination.

Traditionally, invertebrate population density has been viewed as a function of the time elapsed since the last calamity. But here we have reviewed a not inconsequential store of evidence for density-dependent regulation in freshwater mollusc populations, and for both intraspecific and interspecific competition. Thus the questions become: under what circumstances may competition be expected, and what shall be the outcome.

In the previous chapter, it was suggested that freshwater mollusc populations may profitably be divided into categories according to their reproductive effort: **R**-adapted, **S**-adapted, and **U**ndifferentiated. Because they are adapted to special environments, I would hypothesize that intraspecific competition at least occasionally becomes an important factor in the regulation of **R** and **S** populations, but that interspecific competition would be less usual. One might thus expect adaptations to minimize intraspecific competition in **R** and **S** populations, as for example the broad generalization of diet and substrate use manifested by the *Physa* population studied by Brown (1982). But interspecific competition would seem more likely to impact **U** populations. Hence **U** populations may be more likely to evolve specializations to minimize ecological overlap with other species: the unusual size of the ampullariids, for example, or the epibenthic habit of *Dreissena*. It may be recalled from Chapter 4 that unusual life cycles may evolve in large communities, such as those reported by Jokinen (1985) among the 14 snail populations of her Connecticut pond. I have added a row indicating the predicted relative importance of intra- and interspecific competition in **U**, **S**, and **R** populations to Table 4.2.

Successful molluscan invaders almost certainly include many **R**-adapted populations in their number. I would speculate that the reproductive efforts of *Physa acuta*, *Pseudosuccinea columella*, and *Helisoma duryi* would rank them among the R-adapted, for example. But it seems unlikely to me that an **R**-adapted or an **S**-adapted population could competitively eliminate a reproductively **U**ndifferentiated population in the wild. Although **R** populations may replace **U** populations (such as *B. glabrata* appears) in restricted or artificial environments, in the wild typical **U** populations probably display sufficient adaptations and habitat specializations to coexist with any generalist. Thus *H. duryi* seems ineffective against *Biomphalaria* in field trials.

It may be recalled from Chapter 4 that the single population of *Corbicula* for which adequate data are available seems to be **U**ndifferentiated in its reproductive effort. I speculate that such successful invaders as *Dreissena*, *Corbicula*, certain thiarids, and *Potamopyrgus* may not be **R**-selected. They may all prove to be reproductively **U**ndifferentiated species that have evolved special adaptations to minimize the impact of intraspecific competition in environments that are neither unpredictable nor poor. When introduced (typically by human hand) into a new region, they may exploit previously unused resources, and/or outcompete resident populations. The interactions among the ecologically diverse **U** populations of predictably rich environments will be the focus of Chapter 9.

6 · Parasitism

The theme of this chapter is well summarized by the title of a collected volume edited by Toft and her colleagues (1991): 'Parasite–host associations: coexistence or conflict?'. The answer to that question, over the entirety of biology, has proven quite complex. Here I examine a small, more easily tractible aspect of the problem, taking the perspective of freshwater molluscs.

Freshwater molluscs are known to host a wide variety of parasites. The first report of a haplosporidian from North America was Barrow's (1961) discovery in Michigan *Lymnaea*, *Physa*, and *Helisoma*. Haplosporidian disease subsequently decimated the oyster population on the east coast of North America, ending a way of life for thousands of fisherman. The oligochaete *Chaetogaster* lives on or in a wide variety of freshwater snails and may be parasitic (Buse 1971), commensal (Young 1974), or possibly even of some benefit to its host (Khalil 1961). This chapter will emphasize the Digenea, however, by virtue of the wealth of literature available. I conclude with brief discussions of two groups of parasites most associated with unionacean mussels, the aspidogastrid trematodes and union-icolid mites. Readers with interests in any other class of parasite to which molluscan flesh may be heir are referred to the primary literature directly.

The Digenea is by far the largest order of trematodes, that entirely parasitic class of flatworms generally termed 'flukes'. Essentially all of the approximately 40 000 digenetic trematode species seem to require a molluscan host to complete their life cycles, plus at least one (and perhaps as many as three) additional hosts. Llewellyn (1965) suggested that the Digenea may have evolved from free-living (turbellarian) flatworms that first became commensal with molluscs, then parasitic (or predatory) upon them, and only subsequently added other (usually vertebrate) hosts. Thus the host in which the adult worm is now found, although often accorded central importance, may in an evolutionary sense be secondary. Esch and

Fernandez (1994) divide trematode life cycles into those that are auto-genic and those that are allogenic. Autogenic species find their ultimate hosts in aquatic or semi-aquatic vertebrates, and thus may complete their life cycles within the confines of a single habitat. Allogenic species complete their life cycles in temporary residents, such as humans and live-stock. Thus the supply of eggs from autogenic trematodes might be fairly uniform to freshwater molluscs, while that from allogenics would likely be more sporadic. My discussion here will (to a disproportionate degree) emphasize trematodes with allogenic life cycles, since they include among their number all the species of medical and veterinary impor-tance, and hence are better studied. Katsigianis and Harman (1974) sug-gested that the life cycles of the digenetic trematodes might be divided into three categories, depending upon the method by which the worm gains entrance into its ultimate vertebrate host: (I) direct ingestion, (II) larval penetration, and (III) ingestion of intermediate host. For our review here we draw heavily from the literature of the several species of *Schistosoma* (a type II group) infecting humans, and from the livestock fluke *Fasciola* (type I). Some special reference will be made to the (type III) echinostome flukes as well.

The chapter begins by establishing the widespread incidence of dige-netic trematode parasitism in the freshwater molluscs. This is followed by sections on the consequences of trematode infection on the molluscan host, and on the defences mounted by the molluscs themselves. I finish the section on Digenea with a brief review of evidence regarding the relationship between fluke infection rates and snail population sizes. Among the more useful general reviews of these subjects are those of Wright (1971), Frandsen (1979), Malek (1980), Loker (1983), Rollinson and Simpson (1987), and Brown (1994).

The digenetic trematodes

Schistosomiasis, an important public health problem in the tropics world-wide, may be caused by several species of the blood fluke *Schistosoma*. Eggs pass from the human host in urine (*S. haematobium*) or in faeces (other species) and hatch into a ciliated larval stage known as a miracid-ium, about 50×150 μm. Schistosome miracidia may swim actively for up to 12 hours at 6–7 m/hr, hunting snail hosts. They are sensitive to cues from gravity, light, and temperature, and seem to be attracted by fatty acids, amino acids, and other small organic molecules released by living molluscs. Schistosome miracidia may burrow into any exposed region of a snail; head-foot, tentacle, and mantle collar seem to be especially common. After penetration they lose their ciliated covering and transform into a sac-like 'mother sporocyst', usually very near the penetration site. It is at this first stage of attack that host specificity is most commonly noted. The tropical planorbid genera *Biomphalaria* and *Bulinus* are the principal hosts for *S. mansoni* and *S. haematobium*, respectively. The oriental schistosomes develop in pomatiopsids: *S. japonicum* in *Oncomelania* and *S. mekongi* in *Tricula* or *Neotricula*.

Undisturbed, mother sporocysts grow to about 400–500 μm in 2–3 weeks and begin to release smaller daughter sporocysts. Over the following two-week period, large numbers of daughter sporocysts migrate to the snail's digestive gland, or less frequently its ovotestis. Daughter sporocysts are also rather sac-like, absorbing host nutrients through their body surface and growing several millimetres within about three weeks of their arrival. At this stage the host generally becomes castrated. There is evidence that castration may result directly from worm-elaborated excretory/secretory products, rather than as an indirect consequence of nutrient re-allocation (Crews and Yoshino 1990, 1991).

Cercarial emergence begins about three to five weeks post-infection and usually seems to peak at about 10 weeks. Cercariae, swimming trematode larvae with well-developed tails, emerge from daughter sporocysts, migrate from the host digestive gland, and break free into the open water. Prior to the onset of cercarial emergence, a trematode infection is termed 'latent' or 'prepatent', and subsequently it is considered 'patent'. The cercariae emerging from an infected snail over its lifetime may number in the thousands. As each cercarium is about 500 μm in total length, the snail may suffer considerable tissue damage during patency. Cercariae of medically important flukes may burrow into the exposed skin of humans washing, bathing, cultivating rice, etc., and develop into adult worms. Back in the snail, daughter sporocysts may regenerate, so that cercarial production from a single snail may last up to eight months.

The two species of *Fasciola*, *F. hepatica* and *F. gigantica*, have long been recognized as important parasites of livestock worldwide. The first trematode life cycle to be completed experimentally was that of *F. hepatica*, a feat accomplished by both German and English researchers in 1881–83. Eggs are released in faeces, but must be separated from faecal material and at least moistened (preferably inundated) before hatching. The ciliated miracidium that emerges can live about 8 hours absent a lymnaeid snail host. *Fasciola* miracidia do not, however, seem to be as attracted by chemical cues as are those of schistosomes. And in fact, workers commonly report apparently random miracidial movement even in very close proximity to suitable hosts. Miracidia penetrate an exposed portion of the snails's epithelium, as in schistosomes, but shed their ciliated coat upon entry, rather than subsequently. They form mother sporocysts about 70 μm in diameter, which grow into sac-like structures 0.5–0.7 mm, rupture, and release rediae as early as five days post-infection. (Unlike the daughter sporocysts produced by schistosomes, rediae have functional mouths.) Rediae migrate to the host's digestive gland, where they grow to 2.0 mm and may produce daughter rediae or free-swimming cercariae. The first cercariae begin to emerge about four to five weeks post-infection, as in schistosomes. Much unlike schistosomes, however, *Fasciola* cercariae lose their tails and encyst as metacercariae, usually on aquatic vegetation. Ingested by mammals grazing in marshy pastures (or even, rarely, by humans eating watercress), they develop into adult flukes in the liver and complete the life cycle.

The cosmopolitan trematode family Echinostomatidae includes several hundred species ultimately parasitizing a wide variety of vertebrates, especially aquatic or semi-aquatic birds and mammals. Human echinostomiasis is mild, usually accompanied only by diarrhoea. Eggs passing in faecal matter hatch to ciliated miracidia that penetrate their snail host, leaving behind their coats as in *Fasciola*, and transform to mother sporocysts. Although echinostomes, as a family, have been reported to use the entire spectrum of freshwater gastropod taxa as first intermediate host, there is often considerable host specificity for individual echinostome species. Echinostome sporocysts seem to penetrate unusually deep into their hosts. They migrate to the heart region, where they give rise to mother rediae in the ventricle. Mother rediae migrate to the host's digestive gland/gonad area, giving rise to daughter rediae, which give rise to swimming cercariae four to six weeks post-infection. Cercariae swim to second intermediate hosts, which can be fish, tadpoles, snails, clams, or even planarians. In the laboratory, workers often find it convenient to use new individuals of the same snail species serving as first intermediate host to serve as second intermediate host. Cercariae usually swim up the nephridiopore of a second snail and form metacercarial cysts in its kidney. Adult worms develop in birds and mammals that may feed on the second intermediate host.

Prevalence

It is unfortunate that more is not known about the great majority of the trematodes of which a direct economic consequence is not immediately obvious. Mackie (1976a) tabulated data from 27 separate studies on digenetic trematode infections in 19 pisidiid species. Pisidiids are known to host 13 species from the family Allocreadiidae and 7 from the Gorgoderidae, almost all of which are autogenic, requiring a second intermediate host (usually an aquatic insect) before maturing in a fish or an amphibian. Data on prevalence for most of these are rare; data on consequence to the molluscan host nearly non-existent.

For three years Mackie took monthly quantitative samples from the *Musculium securis* population of Britannia Bay near Ottawa, Canada, ultimately examining 1764 individuals for trematode parasites. Observable parasitism was limited to adults greater than 3.5 mm in length during the months of July, August, and September. Mackie noted, however, that the 40–50 day intramolluscan development times normal for trematodes of this sort would imply that infection often occurs among the juveniles. Prevalence peaked at 27.8% in July 1970, 25.0% in August 1971, and 21.8% in August 1972. Although uninfected adult *Musculium* generally carried 3–6 brood sacs with 2–8 larvae per sac throughout the summer, parasitized individuals appeared to be entirely sterile.

Clearly trematode parasitism may be of considerable consequence to populations of freshwater bivalves (Taskinen *et al.*, 1997). The gorgoderid *Phyllodistomum* has drawn some attention for its potential to infect (and perhaps help control) the zebra mussel *Dreissena*. Martell and Trdan (1994) reported a 61% *Phyllodistomum* infection rate in a Michigan population of the unionid *Venustachoncha ellipsiformis*. Infected individuals appeared to be entirely castrated. The authors implicated the trematode in the very low replacement rate demonstrated by this mussel population.

Ismail and Arif (1993) reported infections of at least seven different cercarial types in a dense population of *Melanoides tuberculata* from the United Arab Emirates. The overall infection rate increased from about 50% to about 90% over a year of observation. The total of 3737 snails examined included several hundred double- and even triple-infections, for an overall prevalence of 74%. Such levels of parasitism cannot fail to have a profound impact on the dynamics of a snail population. How commonly may populations of freshwater molluscs be infected at these rates?

Studies of human schistosomiasis afford the richest source of data on trematode infection rates in natural populations of snails. Anderson and May (1979) tabulated such data from about 30 studies conducted worldwide, involving all three major hosts of human schistosomes (Figure 6.1). The authors concluded that average prevalence of infection 'tends to lie in the range 1–5%, irrespective of the species of snail or schistosome, or

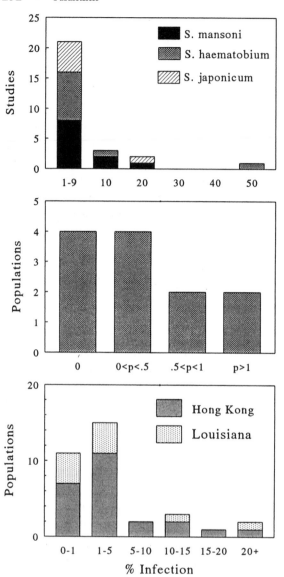

Figure 6.1. (Top) 27 studies of schistosome populations categorized by the percentage infection observed among their snail hosts (data collected by Anderson and May 1979). (Middle) Ten Iranian populations of *Bulinus* categorized in the same fashion (Chu *et al.* 1968). (Bottom) 'Hong Kong' data show percentage infection (all types of trematodes combined) for 24 populations of six different gastropod species (Tang 1985). 'Louisiana' data are percentage prevalence in ten populations of ancylid limpets from SE Louisiana (Turner and Corkum 1979).

of the geographical location'. Five of the studies to which Anderson and May referred were reports of wide-ranging infection rates over multiple populations, and as such could not be integrated into the present Figure 6.1. *Schistosoma mansoni* infection has been reported to vary in *Biomphalaria glabrata* populations from 5% to 25% in Puerto Rico, from 5.8% to 17.8% in St. Lucia, and 50% to 70% in Brazil. Liberian *Biomphalaria pfeifferi* populations show 0.8% to 44% infection rates for *S. mansoni*, and Philippine *Oncomelania quadrasi* show 0.9% to 17.4% rates for *S. japonicum*. But while a few studies demonstrating higher infection rates might be added to Figure 6.1, one's overall impression of the graph would remain unchanged. The prevalence of schistosome infection in populations of known hosts may be surprisingly low.

It is not clear how well Anderson and May's schistosome estimate may reflect trematode infection rates in mollusc populations generally. If other (medically unimportant) trematode species are present in these populations, 1–5% may underestimate the overall trematode impact. But most of the studies surveyed by Anderson and May were probably informed by the prior demonstration of schistosomiasis in the study area, and hence are as likely to represent overestimates as underestimates. A survey conducted by Chu et al. (1968), not among those reviewed by Anderson and May, was exceptional in that the authors tabulated monthly data on percentage prevalence from 17 Iranian *Bulinus truncatus* populations without apparent regard to local occurrence of human disease. Infections by both *S. haematobium* and the livestock parasite *S. bovis* were combined.

Estimates of rare phenomena, such as parasitic infection often seems to be, will be strongly biased (toward zero) by small sample sizes. At least 12 months of observations (uninterrupted by chemical treatment or habitat destruction) are available for 12 of the 17 populations monitored by Chu and his colleagues. Although data on many of these 12 extend to 20–28 months, populations were sparse in some cases. Sixteen months of collection yielded only 94 snails at Seyed Nur, and 17 months yielded 65 individuals at Boreh Hajat. All other sample sizes exceeded 200, and ranged to 28 309. The middle graph of Figure 6.1 shows that, averaged over 12 months or more, the prevalence of schistosome infection in Iranian *Bulinus* populations seems to be quite low. The modal infection rate would be less than 0.5%. The maximum observed was 2.6%, in 976 snails collected over 22 months at Farash Abad. So in the data of Chu et al. we find some evidence that even the 1–5% estimate of infection rate offered by Anderson and May may be high.

Tang (1985) surveyed 20 flooded furrows and irrigation ditches in the New Territories of Hong Kong, collecting 11 680 individual gastropods of 11 species. All were dissected and directly examined for evidence of parasitism. Tang discovered no trematodes in her 5633 individual *Biomphalaria straminea*, the recently introduced potential schistosome host that had originally motivated her study. But the calculated prevalence of parasitism in the remaining gastropod species would seem a fair estimate for natural populations. Among 6047 individuals, the total with observable trematode infections was 207, for an overall prevalence of 3.4%.

Tang tabulated the prevalence of parasitism separately for all 20 sites × 10 snail species × 12 distinguishable trematode species. As one might expect, her survey included quite a few small sample sizes of snails. I extracted from Tang's table only the 24 snail populations with $N > 50$, excluding *B. straminea*, and calculated percentage prevalence in each case combining all parasite types. Although this was not specified, her samples seem to have been of adult snails only. In the single case where Tang divided her sample into 'small' and 'big' individuals, I calculated prevalence for the 'big' category only.

The lowest graph in Figure 6.1 shows that across 24 Hong Kong snail populations for which we have an adequate (albeit single) sample, prevalence of parasitism is almost always less than 5%. One's eye may be drawn to more extreme cases, however. Tang found that a sample of 165 *Austropeplea ollula* (a lymnaeid) from Ngau Ha Tsuen included 41 individuals with obvious *Cortrema corti* infections, for a prevalence of 24.8%. Her sample of 132 *Radix auricularia* from the Lam Tsuen River included 11 with *Plagiorchis* infections and 13 with *Notocotylus* infections, for a combined prevalence of 18.2%. It would appear that in at least a few cases, parasitism may affect a substantial proportion of a snail population.

From the standpoints of both worm and mollusc, the most complete and unbiased parasitological survey of which I am aware is that of Turner and Corkum (1979). Their study organisms were the three ancylid limpet species of southeastern Louisiana: *Laevapex fuscus*, *Ferrissia fragilis*, and *Hebetancylus excentricus*. Limpets are more easily sampled quantitatively than most other freshwater gastropods – the authors simply lifted submerged and floating debris and vegetation, and removed all attached individuals with a scalpel blade. They visited five sites monthly over 14–18 months, ultimately gathering data on 12 limpet populations. Their data vividly illustrate several variables in parasite prevalence to which I have only alluded thus far: the diversity of trematode parasites and the temporal and spatial variation in their occurrence.

Table 6.1. *The number of months in which 14 taxa of trematodes were detected in 12 Louisiana populations of ancylid limpets*

	A	B	C	D	E	F	G	H	I	J	K	L	M	N
Ben Hur (18 months)														
Hebetancylus			6	10	1	12								
Ferrissia		1		2		2	2				1	1	1	
Beaver Pond Branch (17 months)														
Hebetancylus				2		5								
Ferrissia	1	1				3	1						1	3
Ramah (14 months)														
Laevapex	5		2			1		2		6				
Hebetancylus				1										
Ferrissia	1					1	2							
Sorrento (14 months)														
Laevapex	12	1	12	1	1			2						
Hebetancylus			2	1										
Ferrissia		4												
Head of Island Pond (15 months)														
Hebetancylus			2	1										
Ferrissia		1									1	1		

Notes:
The total number of months sampled is given by the name of each station. Letters A–N correspond to Turner and Corkum's taxa I–XVII, disregarding the rare types VIII, XII, XVI, XVIII, and XIX.
Source: Turner and Corkum (1979).

Turner and Corkum were able to distinguish 19 species of trematode larvae infecting their samples of limpets from southeastern Louisiana. Five of these species were single occurrences – detected in just one individual snail over 12 populations and 14 to 18 months – and as such may be considered 'fluke flukes'. The number of months in which each of the remaining 14 trematode species was detected in the 12 populations is shown in Table 6.1.

One's initial response to these data might be despair. Given the extremely patchy occurrence of the trematodes that infect mollusc populations, how can a fair estimate of the importance of parasitism be obtained? Part of the temporal variation is doubtless attributable to the population dynamics of the limpets, which seem to have generation times of six months or less. As limpet population densities fell, so too did

Turner and Corkum's sample sizes, and hence their ability to detect parasites present at very low frequencies. But understanding the nature of Turner and Corkum's sampling problem does not render it especially easy to solve.

Although the authors did not offer exact monthly sample sizes, they did record relative population densities. So to reduce bias, I calculated percentage prevalence (all trematodes summed) only for months in which the number of limpets collected per 15 person-minutes exceeded 10. This criterion excluded the *Hebetancylus* and *Ferrissia* populations of Sorrento entirely, and admitted only one or two months of data for *Ferrissia* at Ben Hur, *Ferrissia* at Head of Island Pond, and *Hebetancylus* at Ramah. Calculations on the remaining seven populations of limpets were based on five months of data or more. Percentage prevalence for the ten populations of limpets of southeastern Louisiana are added to the Hong Kong data in the lower graph of Figure 6.1.

In spite of the great differences in the climate, geography, snails, trematodes, and the sampling methods employed, the distributions of Tang's and Turner and Corkum's samples are strikingly similar. Both show strong medians less than 5% infection. But once again in Louisiana, as was the case in Hong Kong, an occasional population may be identified with a strikingly high parasitism rate.

Excellent data regarding temporal and spatial variation in parasitism rates within single snail populations have been gathered by Woolhouse and Chandiwana (1989). Their study took place in an 840 m stretch of the Kakwidibire River in Zimbabwe, passing through three villages. The *Biomphalaria pfeifferi* inhabiting this particular river host *Schistosoma mansoni*, while the *Bulinus globosus* host both *S. haematobium* and the cattle schistosome *S. mattheei*. Woolhouse and Chandiwana divided their study area into 22 sections, each 30–40 m long, and sampled all sections on seven dates over a one-year period using a 2 mm mesh scoop (2 scoops/metre of river bank/date). The climate in Zimbabwe features a warm rainy season from December to April, a cool-dry period from May to August, and a hot-dry period from September to November. Woolhouse and Chandiwana found that reproduction in both *Bulinus* and *Biomphalaria* peaked in the rainy season, declined markedly in cool-dry, and was negligible (for *Bulinus*, in any case) during the hot-dry. Interestingly, mortality was such that the density of catchable *Bulinus* seemed to remain approximately constant over the course of the study, at about 400 individuals per sample period.

Figure 6.2 compares *B. globosus* data for cool-dry August and hot-dry November. As one might expect, all 22 sections do not seem equally suitable for snails. The authors noted a significant positive correlation between *Bulinus* abundance and the abundance of several species of emergent macrophytes, noting that these areas seem especially suitable for egg laying. The seven individual snails found to harbour patent schistosome infections in August seemed to be randomly dispersed through the study area. In striking contrast, the November sample showed extremely high local infection rates centred at sites 1 and 2 and at sites 14 and 15, two regions of known human–water contact. Clearly the apparent prevalence rate of schistosome infection could vary from 0% to around 50%, depending on sampling date and place. Over all sections, Woolhouse and Chandiwana estimated the peak prevalence of *S. haematobium* in *Bulinus* to be 16% in the hot-dry season, falling to 2% by the subsequent cool-dry season.

We have not, to this point, discussed variation in parasite prevalence with the age of the snail host. Inset to Figure 6.2 are 'age–prevalence curves' corresponding to Woolhouse and Chandiwana's two sample dates, pooled over all sample sections. Not only does the overall likelihood of sampling an infected snail from the Kakwidibire River increase greatly in November, the relative prevalence of infection in young snails is much greater at that time. Distributions such as these, often called 'catalytic curves' have long been looked upon as a source of information on the 'force of infection', the per capita rate at which hosts become infected (Meunch 1959).

Quite a few catalytic curve analyses have been directed toward populations of schistosomes and snails. The primary difficulty has been that a realistic model requires estimates for a long list of parameters other than force of infection. Anderson and Crombie (1984) performed a series of laboratory studies exposing young cohorts of *Biomphalaria* to known numbers of *S. mansoni* miracidia, monitoring survivorship and prevalence over 40 weeks. Even under very simplified circumstances (constant environment, no reproduction), the authors could not fit their observed age–prevalence data to a four-parameter model (specifying average intrinsic snail death rate, death rate of infected snails, latency period, and force of infection). Anderson and Crombie found that adequate fits to their data could only be obtained by adding age dependency to their forces of infection and assuming an exponential rise in the death rates of shedding snails.

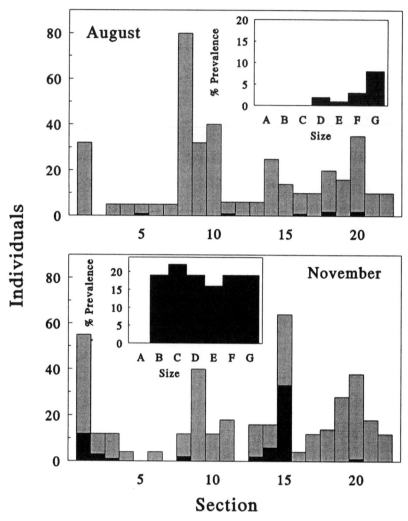

Figure 6.2. The *Bulinus globosus* collected in 22 contiguous 'sections' of the Kakwidibire River, Zimbabwe, each section 30–40 m long (Woolhouse and Chandiwana 1989). Two collectors sampled at an intensity of 2 scoops/metre of river bank. The proportion patently infected with schistosomes is blackened. Inset to each graph is the size distribution of infected snails (A, 4–5 mm; B, 6–7 mm; C, 8–9 mm; D, 10–11 mm; E, 12–13 mm; F, 14–15 mm; G, 16+ mm).

Woolhouse and Chandiwana (1990b) developed an extremely complete model to explain the variable prevalence of patent schistosome infection they observed in the *B. globosus* population of the Kakwidibire River. Their model called for estimates of nine parameters, most pertaining to the survivorship and fecundity of the snails, both infected and not. Then assuming the constant force of infection of 0.05 per individual snail per week fitted to their age–prevalence data, Woolhouse and Chandiwana suggested that annual variation in overall schistosome prevalence might be a function of temperature. That is, the striking differences between November and August shown in Figure 6.2 may not be due to population fluctuations in the worm or the input of miracidia. Rather, the authors propose that temperature dependence in sporocyst development rate may determine seasonality in the prevalence of schistosome infections. Infections may accumulate latently during the cold months, to 'mature en masse, together with new infections, after a few weeks at higher temperatures'.

Conclusions such as these are very much dependent on the impact of trematode parasitism on the survivorship and fecundity of the snail host. Such data are difficult to obtain from the wild. Gracio (1988) brought *B. globosus* collected from a large lake in Angola into the lab, divided them into patently infected and apparently uninfected groups, and monitored survivorship and reproduction for 101 days. The survivorship of the infected snails was about half that of the uninfected, and while uninfected animals averaged 74 egg masses laid over the interval, no viable eggs were laid by any infected snail. Similarly, Woolhouse (1989) brought *B. globosus* and *B. pfeifferi* from Kakwidibire River into his lab and compared the six-week survivorship of patently infected snails to size-matched individuals not shedding cercariae. He also estimated that (patent) schistosome infection approximately doubles the mortality rate of *Bulinus*, and triples that of *Biomphalaria*. But problems with studies of this sort include inability to determine initial snail age (complicated by parasitic gigantism) and inability to factor latent infections from control groups. A clearer understanding of the consequences of trematode infection is to be obtained by observation of individual snails infected in the lab.

Consequences

In the preceding section we were left with the impression that the overall prevalence of trematode infection in freshwater mollusc populations generally seems low, but that both temporal and spatial variation may be extraordinarily high. Some freshwater snail populations, at some times and places, may show very high percentage infection rates. What is the impact of trematode infection on a snail?

Holmes (1983) outlined 'three contrasting models of the way hosts and parasites may coevolve': mutual aggression, prudent parasite, and incipient mutualism. The key distinction among these models is the observed effect of the parasite on its normal host. Holmes' mutual aggression model (embracing the gene-for-gene and red queen hypotheses of other workers) predicts that parasites evolve to maximize their total reproductive output, even at the expense of their host's viability. His prudent parasite model predicts that parasites evolve to minimize adverse effects on their hosts, and his incipient mutualism model proposes that, at least occasionally, parasites may actually benefit their hosts. The theme of this section will be to demonstrate that trematodes infecting mollusc hosts are not prudent, much less mutualistic. They are always, insofar as I can determine, aggressive.

Schistosomes

Among the earliest and most thorough descriptions of the effects of trematode infection on snails was the work of Pan (1965) on *S. mansoni* and *B. glabrata*. Pan's experimental snails were 5.5–9.0 mm in shell diameter, approaching sexual maturity. He exposed 150 individuals to 'numerous' freshly hatched schistosome miracidia for three hours, keeping 100 snails as unexposed controls. The onset of egg laying occurred within the week in both treatments, but declined to zero among infected snails as the first cercariae began to emerge, about five weeks post-infection. Survivorship was also affected; mortality to week 18 was 77% in infected snails but only 11% in uninfected controls. Mean shell growth was significantly greater in the control population, giving no evidence of 'parasitic gigantism' in host snails. Pan continued his observations to week 30, by which time only two of his 150 infected snails survived, but did not see any recovery from infection. His *B. glabrata* had been entirely castrated by the parasites.

In his (apparently successful) effort to eliminate the confounding effects of snails exposed to schistosomes but not infected, Pan used an unspecified but doubtless high concentration of miracidia. Evidence that the actual number of miracidia infecting an individual snail may be of profound consequence is shown in Figure 6.3. Chu and his co-workers

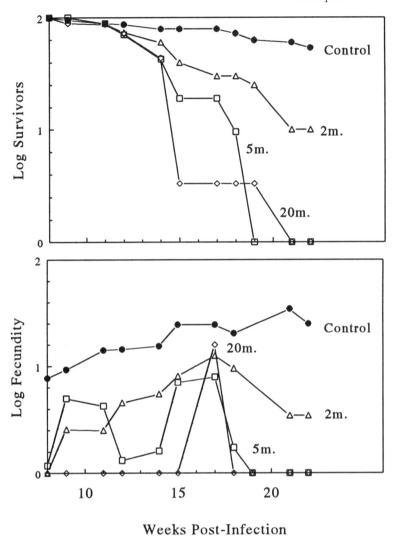

Figure 6.3. Survivorship and fecundity (eggs per survivor per week) in *Bulinus truncatus* infected with *Schistosoma haematobium* (data of Chu *et al.* 1966c). An uninfected control is compared with treatments of 2, 5 and 20 miracidia per snail. Survivorship is standardized to 100 at week 8 post-infection, the onset of cercarial shedding. (Actual *N* = 20–30.)

(1966c) exposed individual *Bulinus truncatus* to 1, 2, 5, 10, and 20 *S. hae-matobium* miracidia per snail (50 snails per treatment), keeping a sixth set of 50 snails uninfected as a control. The snails were approaching sexual maturity as were Pan's *Biomphalaria*, four weeks old at the beginning of the experiment. Snails began shedding cercaria on day 38 of the experiment, and on day 53 (roughly eight weeks post-exposure) the authors began to monitor survival and reproduction in a successfully infected subset.

Figure 6.3 is based on Chu *et al.*'s table 4, with survivorship standardized to 100, days (and fecundities) converted to weeks, and all data log transformed. Survivorship to week 22 was about 53% for the control, 10% for the two-miracidia treatment, and nil for five- and 20-miracidia. But while the effect of parasitism on reproduction was striking, only in the 20-miracidium/snail treatment did castration approach 100%. In the two-miracidium snails, fertility generally rose throughout most of the study, peaking at a (corrected) 76 eggs laid per six snails surviving to 17 weeks post-infection. It should be noted that the entire 76-egg effort probably did not come equally from all six animals. In the 20-miracidium treatment, the single survivor to 17 weeks quite suddenly laid 24 eggs, apparently affecting a 'self cure'. The occasional ability of host snails to out-last schistosome infection has also been noted in *B. glabrata* by Etges and Gresso (1965).

The age of the snail host is another important variable influencing the severity of schistosome attack. Sturrock (1966b) collected 140 *Biomphalaria pfeifferi* hatched on the same day and exposed groups of 30 individuals to five *S. mansoni* miracidia at ages one, three, five, and seven weeks, keeping the remaining 20 snails as an uninfected control. The top half of Figure 6.4 shows that considerable mortality is associated with schistosome attack, regardless of age. While the control snails suffered no mortality over 21 weeks of observation, only 18 of 30 snails exposed at age one week survived, including 8 uninfected (subtracted from Figure 6.4), and but 13/30 of the three-week treatment survived, including a single uninfected individual. Mortality was most severe in the seven-week treatment, where only a single individual survived to age 21 weeks.

Figure 6.4 also shows that parasitic castration was 100% only in the group infected at seven weeks, already mature when infected. Interestingly, these snails actually laid more eggs in weeks 9 and 10 while their infections were prepatent than uninfected controls. Similar phenomena have been reported by Minchella and Loverde (1981), Minchella

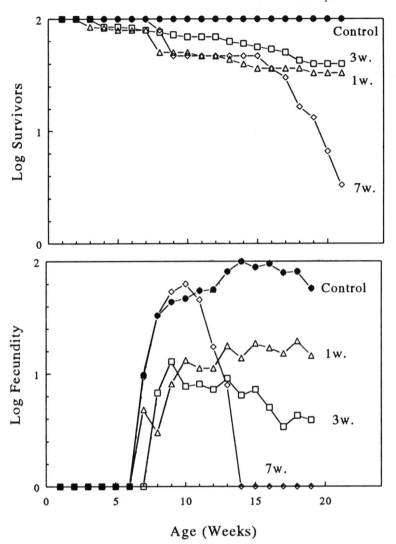

Figure 6.4. Survivorship and fecundity (eggs per survivor per week) in *Biomphalaria pfeifferi* infected with *Schistosoma mansoni* (data of Sturrock 1966b). An uninfected control is compared with snails infected at ages 1, 3, and 7 weeks. Survivorship is standardized to 100 at start (actual $N = 30$).

(1985), and Thornhill *et al.* (1986) in the *B. glabrata/S. mansoni* system, and have generally been interpreted as a compensatory response for expected future suppression of egg laying.

Although their fecundity was reduced, *Biomphalaria* challenged with five miracidia per individual at an early age often did not seem to lose their reproductive capabilities entirely. All the individuals included in the fecundity estimates shown in Figure 6.4 were actively shedding cercariae. Mean fecundities of snails infected at one week varied around 10–20 eggs/individual/week, while the three-week exposed snails averaged about 5–7. Sturrock did note a significantly high proportion of sterile and inviable eggs among those laid by infected snails. Such data from Chu and colleagues would have been welcome. However, Sturrock reported that by week 26, four of his one-week and three-week individuals had stopped shedding cercariae and raised their fecundities to about 80 eggs/week, apparently cured. One might speculate that, although five miracidia may attack and penetrate a one-week-old snail, not all five may develop. The effective parasite loads in these snails may have been lower.

All four infected groups of snails (including a five-week group I have neglected here thus far) showed accelerated shell growth during their pre-patent periods. Mean shell sizes at week 19 ranged from 10 mm to 12 mm, significantly greater than the 9 mm averaged by uninfected controls. In a second, larger and longer-term experiment involving three-week exposed snails only, Sturrock showed significantly greater mean size and greater size variance in parasitized *B. pfeifferi*. This constitutes considerable evidence for 'parasitic gigantism', at least under some circumstances.

The effect of *S. mansoni* infection has also been examined on *B. glabrata* of varying age by Meier and Meier-Brook (1981). Although their experimental design was similar to Sturrock's, they obtained results strikingly different in some respects. The authors exposed groups of 90 snails to three miracidia each at five different ages (one, two, four, six, and eight weeks post-hatch) keeping a sixth group of 90 snails unexposed as a control. Snails exposed at age one week showed 81% mortality during the five-week prepatent period before the emergence of the first cercaria. In fact, five of the 17 snails surviving to age six weeks proved not to have been successfully infected. Snails exposed to schistosomes at age two weeks showed somewhat greater mortality than controls. But quite in contrast to Sturrock's findings, survivorship in snails infected at four, six and eight weeks was not strikingly different from control.

In another contrast with Sturrock, Meier and Meier-Brook observed no egg laying by any of their *B. glabrata* infected at age one week, nor by those infected at age two weeks. In snails infected at age four weeks and older, fecundity was reduced during the five-week prepatency, declining steeply in all treatments as cercariae began to emerge. All three older groups (exposed at 4, 6, and 8 weeks) were still laying eggs as the experiment ended at week 14, perhaps too early to observe total castration. In a final contrast with Sturrock, the authors found evidence that schistosome infection retards shell growth in *B. glabrata*; the effect was most pronounced in snails infected in their youth.

The maturity of the host snail at time of schistosome infection unquestionably influences the consequences of that infection (Fryer *et al.* 1990). But Meier and Meier-Brook's results strongly indicate that *S. mansoni* attack weighs most severely on small, juvenile *B. glabrata*, while Sturrock's results suggest just as clearly that schistosome infection is of more profound consequence to older, reproductively mature *B. pfeifferi*. Perhaps this is a function of host specificity; the relationships between schistosomes and their snail hosts are tuned at the population level. Meier and Meier-Brook used an *S. mansoni* strain derived from Liberian stock 50 years previous and *B. glabrata* from Puerto Rico. Sturrock's snails were Tanzanian, schistosome unspecified. The consequences of host specificity will be explored further in the section that follows.

Fasciola

The effect of *Fasciola* on English *Lymnaea truncatula* has been described by Hodasi (1972). He exposed three batches of 20 individual snails each to five newly hatched miracidia at ages two weeks (juvenile), four weeks ('adolescent'), and six weeks (adult), keeping a fourth batch of 20 uninfected as a control. He achieved successful infection in 18, 18, and 19 individuals, respectively. In a separate experiment, Hodasi exposed a second (apparently rather large) group at age two weeks to obtain a finer estimate of the effect of *Fasciola* on snail survivorship.

Several striking differences in the biology of the snail hosts of *Fasciola* and *Schistosoma* are evident from a comparison of Figure 6.5 with Figures 6.3 and 6.4. In the laboratory, *Bulinus* and *Biomphalaria* seem to reach reproductive age at about 6–7 weeks, while *L. truncatula* begins to lay eggs at week four. And while uninfected laboratory populations of *Bulinus* and *Biomphalaria* show very little mortality and uninterrupted reproduction through 20 weeks, control laboratory populations of *L. truncatula* seem to expire naturally.

The top half of Figure 6.5 shows a negligible effect of *Fasciola* on snail survivorship up to about week 12, at which time the decline ordinarily

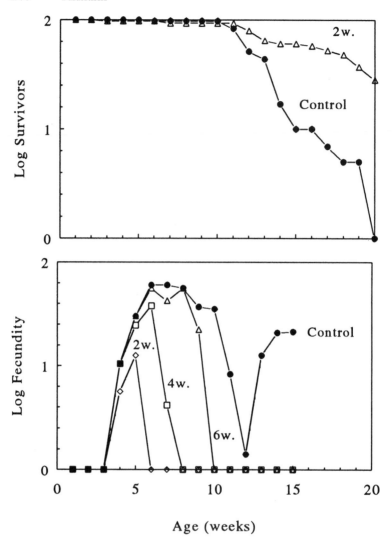

Figure 6.5. Survivorship and fecundity (eggs per survivor per week) in *Lymnaea truncatula* infected with *Fasciola hepatica* (data of Hodasi 1972). Survivorship data compare uninfected controls with snails infected at age two weeks. The fecundity of an uninfected control is compared with snails infected at ages 2, 4 and 6 weeks.

observed in *L. truncatula* is postponed. All uninfected snails were dead by week 20, while 38% of the snails infected at age two weeks remained. The last infected snail did not die until week 25 in Hodasi's survivorship experiment. The effect of parasitic infection on fecundity seems to be more severe than ordinarily observed from schistosomes. Snails infected at weeks two, four, and six were 100% castrated at weeks six, eight, and ten, respectively. But note that by week ten, individual *L. truncatula* will have already laid a large number of eggs. Given the four-week prepatent period typical of many trematodes, rapid maturation would seem to have considerable adaptive value as a defence against parasitism.

Hodasi showed that *Fasciola* infection may induce gigantism in *L. truncatula*, at least under some circumstances. His control animals reached a size plateau of about 7.5 mm at onset of egg laying, age 6–8 weeks. Snails infected with *Fasciola* at age two or four weeks evinced a growth spurt about 4–5 weeks post-infection, just as cercariae were beginning to emerge and reproduction was being suppressed. They reached a size plateau at about 9.5 mm, strikingly larger than the controls. But snails infected at age six weeks, after reaching maturity, remained at their fixed adult size and showed no evidence of gigantism.

Echinostomes

The echinostome *Ribeiroia marini* is reported to castrate *Biomphalaria glabrata* completely, as does *Echinostoma togoensis* its host *B. pfeifferi* (Combes 1982). Kuris (1980) introduced five miracidia of *Echinostoma liei* per snail into aquaria containing 50 *Biomphalaria glabrata* of three size classes: 1 mm, 2 mm, and 5 mm. He kept a second set of all three size classes unexposed as a control, and monitored survivorship over 30 days. While unexposed snails of all sizes suffered negligible mortality, Kuris did detect significantly increased mortality among 1 mm and 2 mm size classes exposed to *Echinostoma*. For unexposed snails initially 1 mm, mean shell diameter at the end of the experiment was 11.15 mm, significantly larger than the 9.15 mm averaged for similar snails successfully infected with *Echinostoma*.

Echinostomes have attracted considerable attention because of their especially aggressive interactions with other trematodes. In the event a single snail is doubly infected by an echinostome and, for example, a schistosome, the echinostome rediae often outcompete or even consume the larvae of the second species (Lim and Heyneman 1972). The effect of multiple trematode infections upon the molluscan host does not seem to have been well characterized, however. Interestingly, previous infection with *Echinostoma paraensei* seems to cause resistant strains of *B. glabrata* to lose much of their resistance to schistosome infection (Loker *et al.* 1986). Further discussion of the resistance of snails to trematode attack follows.

Resistance

Successful infection by a trematode parasite inevitably lowers the fitness of a snail host. Whether that fitness equals zero or some small fraction above zero depends on the age of the snail at time of attack, the number and species of the miracidia involved, and many other factors. In any case, it is clear that any defences a snail might mount to protect itself from invading trematodes would be most adaptive.

Our understanding of the mechanisms by which molluscs defend themselves from invasion by foreign bodies has progressed steadily for over 30 years, but details remain far from clear. Trematode miracidia are usually encapsulated and destroyed by mollusc haemocytes shortly after penetration. Only on those (exceptional) occasions when miracidia find a host for which they have evolved some special counter-measures can further development proceed (Coustau and Yoshino 1994). Mollusc plasma contains a protein (an 'opsonin') that makes foreign bodies more susceptible to attack by haemocytes (Bayne and Yoshino 1989, Yang and Yoshino 1990, Lodes *et al.*, 1991). Opsonins may belong to the general class of defence proteins found in invertebrate plasma, the lectins (Harris *et al.* 1993). It has been suggested that 'resistant' trematode sporocysts may have developed a mechanism to avoid or abort the process of opsonization. Lie and colleagues (1987) present evidence that resistant trematodes may have evolved the ability to modulate host haemocyte function by misdirecting haemocytes from the site of infection. But should they attract the attention of haemocytes, resistant miracidia and sporocysts seem to have evolved a variety of counter-measures (Connors *et al.* 1991). Given that primary sporocyst development proceeds normally, however, it is by no means certain that secondary sporocysts or rediae will find hospitable conditions in the host digestive gland. Less is known about the inhibition and even elimination of later larval stages, such as often seems to be observed in laboratory infections.

At this point we might profitably return to the three models of host/parasite coevolution outlined by Holmes (1983). Under the incipient mutualism model, where the host may accrue some benefit from parasitism, there would be selection to minimize host resistance to the parasite. This does not seem to be the case in trematodes and snails. Under the mutual aggression and prudent parasite models, selection acts to maximize host resistance, although selection is much stronger in the former than in the latter case. Should a parasite be able to evade the defences of a new host, with which it is not coevolved, the effect of a prudent parasite is predicted to be exceptionally pathological. But the effect of a parasite that has evolved by mutual aggression with its ordinary host may be lessened when first infecting a new host. We begin with data suggesting that *Fasciola* seems prudent from this viewpoint, but conclude with evidence for the aggression of *Schistosoma* and *Echinostoma*. This last situation is, again, usually observed in experiments with trematodes and snails.

Fasciola

In temperate regions of the old world, the usual intermediate host for *Fasciola hepatica* is the amphibious *Lymnaea truncatula*, while in the new world *Lymnaea bulimoides* is the most commonly reported host. In Australia, where the trematode has been introduced with livestock, its usual host is the slightly more aquatic *L. tormentosa*. *Fasciola gigantica* is more common in the tropics, hosted by *L. auricularia* in Asia and the related *L. natalensis* in Africa.

Boray (1966) exposed large numbers of lymnaeids collected worldwide to 25 miracidia from three *F. gigantica* strains (Kenya, Malaya and Java). Figure 6.6 shows selected results from experiments with the Kenyan strain. The author did not keep snails unexposed as controls, nor in fact are data available on snail survivorship. But as the normal host for African *Fasciola*, the $N = 30$ adult *L. natalensis* serve as a control for comparisons of infectivity.

Boray recovered viable metacercariae from 83% of the *L. natalensis* he exposed. It would also appear that his 100 adult *L. tormentosa* served as fairly efficient hosts for *F. gigantica*, even though this trematode is unknown in Australia. (The *L. tormentosa* strain used, from New South Wales, does, however, host *F. hepatica*.) Interestingly, such resistance as is shown by *L. tormentosa* seems to be post-redial. Of the 86 snails surviving exposure, 37 did develop rediae but never shed cercariae. The three other prospective hosts shown in Figure 6.6 are all European – *L. peregra* from Austria and *L. stagnalis* and *L. palustris* from Lower Saxony, Germany. All three of these were exposed as juveniles, aged less than 21 days in the first two cases and 21–70 days in the case of *L. palustris*. Only *L. palustris* proved entirely refractory. Eighteen *L. peregra* became infected

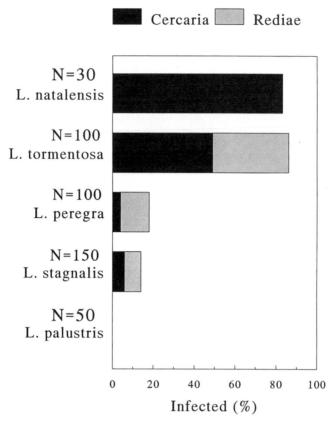

Cercaria ■ □ Rediae

Figure 6.6. Proportion of five *Lymnaea* populations infected with *Fasciola gigantica* in laboratory experiments of Boray (1966). Snails showing redial infection (only) are indicated by shaded bars, and blackened regions show those from which viable metacercariae were recovered.

with rediae, of which four produced viable metacercariae, and rediae were discovered in 21 *L. stagnalis*, nine of which carried their infections to fruit. This implies that although *F. gigantica* would meet substantial host resistance, it is not entirely excluded from Europe by incompatibility.

Once introduced, the rate at which *Fasciola* can adapt to new snail hosts may be extremely rapid. In a set of experiments mimicking the introduction of *F. hepatica* into Australia, Boray (1966) offered German *Fasciola* the opportunity to infect Australian *L. tormentosa* (NSW strain). The exposure of 200 snails to 20 German miracidia each initially resulted in a 93% mortality rate to the snails, which the author attributed to irregular larval development in foot, tentacles, and mantle. Normal redial

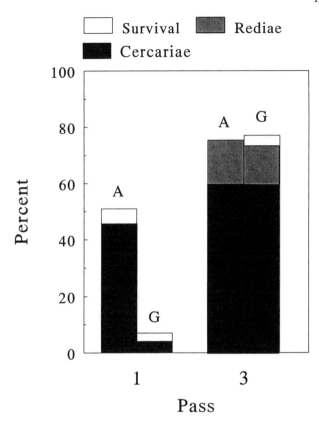

Figure 6.7. The infectivity of *Fasciola hepatica* from Australia (A) or Germany (G) on Australian *Lymnaea tormentosa* (data of Boray 1966). 'Pass 1' is at the start of the experiment ($N=200$ snails for both types of miracidia) and 'Pass 3' is at the third generation of culture in Australian snails ($N=39-47$). The height of each bar shows snail survivorship after exposure to miracidia, with shaded portion developing rediae (only) and metacercarial cysts recovered from the blackened portion.

development did proceed in 9 of the 14 survivors, leading to metacercarial production in all cases. By comparison, snail mortality was 49% when exposed to Australian *Fasciola*, and 92 of the 98 surviving snails developed rediae and ultimately produced metacercariae.

Boray harvested metacercariae from both experiments, fed them to rabbits, collected eggs from adult liver flukes, then used the miracidia produced to infect Australian *L. tormentosa* a second time. This procedure was repeated and the results of infectivity experiments at the third pass are shown in Figure 6.7. Survivorship of a somewhat smaller sample of snails infected by both strains of fluke was much improved, with 90% of the survivors developing German rediae and 100% redial development from Australian infections. Both *Fasciola* strains showed identical 61% success rates to metacercarial production at pass 3. This trend was somewhat confounded on additional passes (Boray 1969). But the speed at which a *Fasciola* population, even an effectively small laboratory population, can apparently adjust to a new host makes one consider non-genetic mechanisms of trematode adaptation.

Schistosomes

Much in contrast to the situation with *Fasciola*, *Schistosoma haematobium* is generally observed to display great specificity to individual populations of snail hosts. Sturrock (1967) exposed five groups of 50 *Bulinus nasutus productus* aged two, four, six, eight, and 14 weeks to five *S. haematobium* miracidia each. Although *B. nasutus* is in fact the normal host for *S. haematobium* in Tanzania, the author began his study with the knowledge that the snail strain he was using in this experiment was at least partially resistant. Ten snails aged two weeks at exposure did eventually shed cercariae, as well as a single individual exposed at four weeks. But to judge from cercarial shed, no infections developed in any of the remaining 239 snails.

Infections in the eleven snails progressed in a variety of fashions. Although Sturrock recorded no deaths among control snails or among snails exposed but not infected, seven of the infected snails died prior to week 28. One of these proved to have been entirely castrated – never producing an egg while shedding thousands of cercariae. Six infected snails did produce eggs, sometimes at the rate of hundreds per week, before their reproductive systems were finally overwhelmed and shut down. Two snails continued to produce viable eggs to week 28, although not at the rates of control snails, while shedding cercariae. Finally, two snails affected self-cures, shedding only a few cercariae early in their infections and ultimately regaining normal fecundity.

Sturrock's results illustrate three phenomena commonly observed in experiments involving trematodes and resistant snail hosts. First, as we also noted in Boray's *F. gigantica* data (Figure 6.6), it seems clear that snails mount multiple lines of defence. The 239 animals never shedding cercariae apparently defeated infection at some earlier stage than those shedding cercariae and subsequently affecting a cure. Second, the defences of resistant snails may not be 100% effective, but may be overwhelmed when the snails are young and/or exposed to large numbers of miracidia. I would speculate that all five miracidia successfully developed in the most severely affected snails among Sturrock's 11 hosts, and that snails able to recover were successfully penetrated by fewer miracidia. Finally, it is evident that the consequences of infection were not as severe as ordinarily observed in the 11 resistant snails ultimately hosting trematodes. This matches expectation from the mutual aggression model of host/parasite coevolution.

It is not entirely clear that in the arms race between parasite and host, the parasite will hold the upper hand. And in fact, it is not uncommonly observed that trematode populations show reduced infectivity to the snail population with which they are evolving. For example, the pomatiopsid *Tricula aperta* (or *Neotricula aperta*) is the (normally listed) host of *Schistosoma mekongi*. Three 'races' of *N. aperta* are recognized, called *alpha, beta,* and *gamma,* showing some morphological and chromosomal differences and quite possibly some reproductive isolation as well. The infectivity experiments performed by Kitikoon (1981b) were exceptional because they involved both wild-collected snails and wild-collected schistosomes, rather than laboratory-reared (and hence possibly inbred) lines.

Kitikoon collected his *gamma* snails at Khong Island near the Laos/Cambodia border. Here he also collected snails of a second pomatiopsid genus very similar to *Neotricula, Manningiella.* His miracidia were hatched from eggs collected from naturally infected Khong Island dogs. Kitikoon collected *alpha N. aperta* from the Mekong River near Khemmarat and *beta* snails from the Mune River near Phibun, both localities in Thailand about 200 km upstream from Khong Island. He held all snails in the laboratory for 60 days, checking for shed cercariae, and cracked a large sample of each to verify the absence of pre-existing trematode infection. He then exposed 200–250 snails of each of the four types individually to five *S. mekongi* miracidia, keeping about 100 unexposed controls, and monitored cercarial shed for another 60 days.

Pomatiopsids seem to be more difficult to culture than pulmonates. The survivorships to 60 days post-exposure shown in Figure 6.8, ranging from 79% to 95%, were comparable to those observed in unexposed controls. None of the controls shed cercariae, but all four exposed groups were successfully infected to some degree. About 13% of surviving *gamma T. aperta,* the nominal intermediate host for the Mekong schistosome at its Khong Island focus, did in fact shed cercariae. Remarkably, this figure was not significantly different from the 12% infection rate observed for co-occurring *Manningiella,* not ordinarily considered a host. The Khong island schistosome is better adapted to its local pomatiopsid

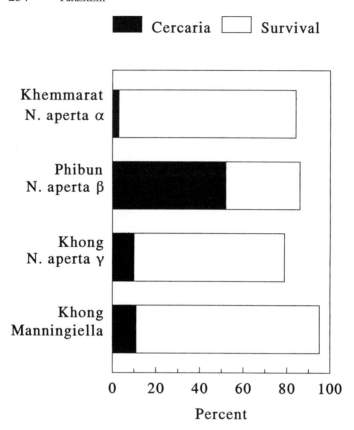

■ Cercaria ☐ Survival

Figure 6.8. The infectivity of Khong Island *Schistosoma mekongi* to four populations of Mekong River pomatiopsids: the *alpha, beta,* and *gamma* races of *Neotricula aperta* and *Manningiella conica.* (Data of Kitikoon 1981b.) The total length of each bar shows the survivorship of (*N*=200–250) snails to 40–60 days post-exposure, with blackened portions showing the survivors shedding cercariae.

populations than to the *alpha* snails of Khemmarat. But the 60% infectivity observed for *beta*-race snails collected 200 km upstream at Phibun is significantly greater than the Khong values. Clearly, the Khong Island schistosome could be doing worse, but it could also be doing better.

The relationship between schistosomes and their host populations may be as dynamic as that observed by Boray in *Fasciola* (Figure 6.7). Kagan and Geiger (1965) reported the results of three years of infectivity experiments involving laboratory lines of both *B. glabrata* and *S. mansoni* from both Puerto Rico and Brazil. The snails were mass-cultured. Every few months for 2–3 years, the authors exposed 100–200 snails 6–8 mm individually to 6-12 miracidia. Cercariae recovered 6 weeks post-exposure were used to infect mice and hamsters, which were necropsied 8 weeks subsequently to recover a new generation of eggs. It is not clear how many generations of schistosomes passed per year, but over a three-year period, the infectivity of Kagan and Geiger's Puerto Rican schistosome to its Puerto Rican host improved from 36% to 64%. The results from other combinations were mixed, and less dramatic. But at least occasionally, the mechanisms of trematode adaptation to a new host seem more efficient than one would expect from selection on available genetic variation. Again one might wonder about the relative heritabilities of trematode adaptation and snail defence.

The heritability of resistance to *S. mansoni* infection in *B. glabrata* certainly may be high (Webster and Woolhouse 1998). Richards and Merritt (1972) designed an elaborate breeding programme directed toward an understanding of the heritable basis for juvenile resistance to *S. mansoni* in *Biomphalaria glabrata*. Their schistosome was a strain that had been collected from Puerto Rico about ten years previously and maintained in mice. Several laboratory lines of snails were involved initially, carrying pigmentation markers to verify outcross. The authors had available Newton's (1955) 'M' *B. glabrata* line of mixed Puerto Rican/Brazilian origin, susceptible to infection and homozygous for the entirely recessive albinism allele. Richards and Merritt employed a subline of Newton's M developed from a spontaneous 'black-eye' mutant. (Black-eye is dominant over albinism but recessive to wild-type pigmentation.) The authors also used a pigmented *B. glabrata* strain from Puerto Rico resistant to schistosome infection, which they designated 'H'.

Richards and Merritt's first goal was to obtain lines as pure breeding for juvenile susceptibility or resistance as possible. Snails 1–4 mm in diameter were initially exposed to five miracidia each. Since the shells of the albinos and black-eyed individuals are translucent, individuals could be examined non-destructively after one week for signs of infection. Those not developing parasites were re-exposed to five miracidia each. Snails remaining uninfected after two exposures were isolated and allowed to self-fertilize, as founders of a new H-line ultimately proving to show 0% infectivity to the Puerto Rican strain of schistosome employed. The unexposed progeny of snails from which test exposures yielded high infection rates were allowed to self-fertilize to found a new M-line showing essentially 100% juvenile infectivity.

Results of hybridization studies are shown in Figure 6.9. Seven of 15 progeny produced by the M-line parent after outcross (47%) proved

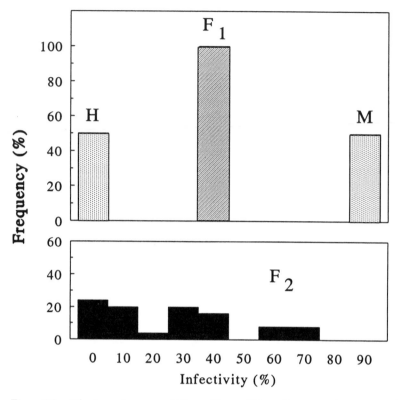

Figure 6.9. The juvenile susceptibility to Puerto Rican *S. mansoni* demonstrated by two parental stocks ('M' and 'H') and their F_1 and F_2 hybrid progeny (data of Richards and Merritt 1972). The abscissa shows the lower boundary for each category.

susceptible to schistosomes, as did 9/20 (= 45%) of the progeny produced by the H-line parent. The absence of any significant difference among the progeny indicates no maternal effects. A second (unexposed) sample of F_1 individuals, 25 of which could ultimately be verified (by the pigmentation of their progeny) as hybrids, were isolated and reared to self-fertilization. Infectivity among the F_2 progeny (typically tested in 10–20 snails) ranged from 0% (three snails) to 75%. In additional experiments, Richards and Merritt were unable to establish lines of snails showing intermediate infectivity. Selection for 50% susceptibility continued to produce wide ranges of susceptibility frequencies.

The authors concluded that juvenile susceptibility to schistosomes in their stocks of *Biomphalaria* was controlled by 'probably at least four' genes, and that both susceptibility and resistance might prove recessive at individual loci. No linkage could be detected between resistance genes and the pigmentation marker. More detailed analysis (Connors and Yoshino 1990, Connors *et al.* 1991) has subsequently shown that the hae-mocytes of Richard and Merritt's resistant snails demonstrate a greater capacity to generate destructive superoxide than susceptible snails, and that *S. mansoni* sporocysts ordinarily manufacture anti-oxidant molecules.

Employing some of the hybrid lines produced above, and using similar experimental procedures, Richards (1973) identified four patterns of life-time susceptibility in *B. glabrata*. His 'Type I' showed complete resistance, 'Type II' showed juvenile susceptibility but adult resistance, 'Type III' showed lifetime susceptibility, and 'Type IV' showed juvenile susceptibility but variable infectivity in adulthood. The adult resistance developed by Type II snails was shown to be conferred by a single dominant gene inherited in a Mendelian fashion.

Initially there was some effort to subdivide types II, III, and IV based on their variable responses to a second strain of *S. mansoni* from St Lucia (Richards 1975). But ultimately it became clear that lines of each of these four types were heterogeneous even in their responses to new schisto-somes from Puerto Rico. Many lines susceptible to infection with the original ('PR-1') schistosome strain were insusceptible to PR-2, not because PR-2 miracidia were encapsulated and destroyed, but because primary sporocytes failed to develop (Sullivan and Richards 1981).

Richards and Shade (1987) reported the results of exposures of lines from all four types of *B. glabrata*, plus Egyptian *B. alexandrina* and African *B. pfeifferi*, to 22 *S. mansoni* strains from Puerto Rico, St Lucia, Egypt, Kenya, and Zaire. The variety of responses was remarkable; juvenile and/or adult susceptibility or resistance was demonstrated by almost any of the lines tested. It is in fact true that Richards' laboratory lines seem to have retained some genetic variability, in spite of many generations of close inbreeding (Mulvey and Vrijenhoek 1981a). But most of the vari-ation in infectivity rate demonstrated in the cross-comparison table of Richards and Shade originates, I would speculate, from differences in the strains of *S. mansoni*.

The introduction of *S. mansoni* into the new world has been charac-terized as 'a gigantic though unplanned experiment' by Brown (1978). Basch (1976) tabulated the results of 38 studies involving experimental exposure of *Biomphalaria* to miracidia of allopatric *S. mansoni*, most involving multiple lines of snails and/or schistosomes, from both the old and new worlds. Techniques have been quite various. Exposure times and miracidium numbers have varied tremendously, as have the sizes, ages, and numbers of snails. Exposure has been singly or *en masse*, and even the criteria used by the authors to judge a snail 'infected' have varied. Basch's Table 1 included data sufficient to calculate infectivity on 86 exposures of old world snails to allopatric old world *S. mansoni* (old/old), 21 old world snails to new world *S. mansoni* (old/new), 24 of new world snails to old world *S. mansoni* (new/old), and 153 of new world snails to allopatric new world *S. mansoni* (new/new). One might hope that over such a large number of experiments, the consequences of the various methodological inconsistencies among the 38 studies might be mitigated.

The introduction of an old world strain of schistosome into an old world population of *Biomphalaria* to which it has no prior adaptation would seem to constitute something of a control for our 'gigantic though unplanned experiment' in the new world. The consequences of such an introduction appear to be unpredictable in the extreme. The old/old distribution shown in the top part of Figure 6.10 is the closest to uniform I can recall seeing in Biology. The largest number of observations are in the 0–9% infectivity range, with 16, and 90–100% range, with 14. The rarest observations were in the 70–80% and 80–90% classes, with four each.

The other three distributions were indistinguishable from each other by Kolmogorov–Smirnov tests, and have been combined in the lower part of Figure 6.10. All three (combined) show much higher prevalence in the 0–9% category than the old/old control (2×2 chi-square = 15.0). Apparently *S. mansoni* strains, regardless of whether they are fresh from Africa or have resided in the new world for 100–200 years, are significantly less likely to infect allopatric new world snail populations. And in fact, while authorities generally observe that all of the (about 10) African *Biomphalaria* species are potential hosts for *S. mansoni*, only about half of the (approximately 20) new world *Biomphalaria* are known to be susceptible.

There is no reason to think that new world *Biomphalaria* populations are more genetically diverse than those of the old world. So I suggest that in the mutually aggressive coevolution of fluke and snail, snail populations begin with a 'head start' such as shown in the lower part of Figure 6.10. Then not only do discrete populations of snails and flukes coevolve, flukes become better adapted to snail hosts generally through occasional dispersal events. Dybdahl and Lively (1996) offer some evidence that the dispersal capabilities of trematodes may exceed those of their snail hosts. Figure 6.10 also shows evidence that some generations of residence in the

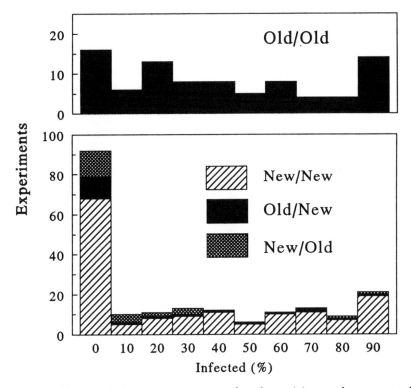

Figure 6.10. The infectivity (proportion of snails surviving to show trematode infection) observed in 284 separate exposures of *Biomphalaria* populations to allopatric *S. mansoni* (data tabulated by Basch 1976). 'Old' and 'new' refer to old world and new world snails and schistosomes, in the order given. The abscissa shows the lower boundary for each category.

new world may render schistosomes less infective back in the old world. I further suggest that new world *S. mansoni* populations may have suffered a bottleneck upon their introduction into the new world, so that they have lost whatever general adaptation they may have had for old world snails through reduced selection and genetic drift. Thus old/new data are indistinguishable from new/new or new/old.

Echinostomes

We began this section with a demonstration that a fluke with a Type I life cycle, *Fasciola*, seems to show broad adaptation to amphibious lymnaeid populations worldwide. We then found substantial host specificity in the Type II schistosomes, but noted that the host–parasite relationship seems dynamic, with both heritable and non-heritable components in both fluke and snail. What sort of relationship is be be expected in the Type III echinostomes?

Echinostoma liei was first described by Jeyarasasingam *et al.* (1972) parasitizing *Biomphalaria alexandrina* in the Cairo area. The authors considered that it showed an 'extreme degree of primary host specificity', apparently unable to infect Brazilian (Minas Gerais) *B. glabrata*, Brazilian *B. straminea*, or even a second 'intractable' Egyptian strain of *B. alexandrina*. Yet they did report successful infection of an 'NIH strain' of *B. glabrata*, more specific details on strain unavailable.

Lim and Heyneman (1972) summarized a series of single-trematode infectivity studies in *B. glabrata* (unspecified NIH strain) in preparation for echinostome double-infection experiments. They exposed snails of five size classes to five different numbers of miracidia and estimated rates of successful infection, to cercarial shed. No data are available on the survivorship or fecundity of the snails. Lim and Heyneman obtained fairly low rates of successful infection with single *S. mansoni* miracidia, perhaps around 20% for size classes 2–10 mm, dropping to 5% for their 20 mm snails. Their *E. liei* strain seems to have been slightly more successful than their schistosome in infecting smaller *B. glabrata*. Exposed to single *E. liei* miracidia, infection rates for the 2 mm snails were 26% and 5 mm snails 31%, with a drop-off to 6% in 20 mm snails comparable to the schistosome data. As might be expected, the Brazilian echinostome *Paryphostomum segregatum* was more infective to *B. glabrata* than either old world trematode. From single miracidia Lim and Heyneman achieved 40–64% infectivity in the size 2–15 mm, the rate dropping to 15% only, once again, in their largest 20 mm snails.

Christensen *et al.* (1980) exposed a wide variety of freshwater pulmonates (3–4 mm) to eight *E. liei* miracidia and evaluated infection rates 35 days post-exposure. Figure 6.11 shows results from five species: *B. alexandrina* (the normal Egyptian host), *B. camerunensis* (from quite distant Kinshasa), *B. glabrata* (an NIH strain from Puerto Rico), *Helisoma duryi* (Florida), and *Physa acuta* (Egypt). Although the authors did not keep unexposed controls, Figure 6.11 seems to reflect some evidence of decreased survivorship to day 35 among exposed snails. The effect is especially apparent in *B. camerunensis*; surely 11 of 43 snails 3–4 mm would not ordinarily die in five weeks of culture. Christensen and colleagues obviously achieved higher infection rates in *B. alexandrina* and *B. glabrata* with their eight miracidia than Lim and Heyneman achieved with their one. No individual snails from the remaining three species shed cercariae by day 35, although a single *B. camerunensis* was found to be har-

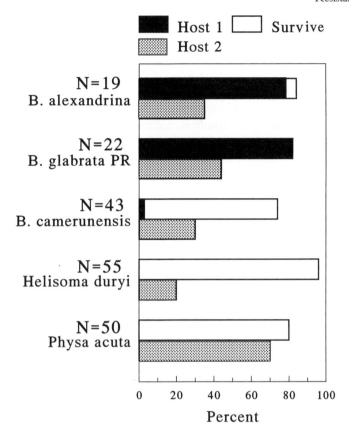

Figure 6.11. The infectivity of *Echinostoma liei* miracidia and cercariae to five species of pulmonate snails. (Data of Christensen *et al.* 1980.) The total length of each upper bar shows the survivorship of (*N*=variable) individual snails 35 days post-exposure to miracidia, with blackened portions showing the survivors shedding cercariae. From a second experiment, the shaded bars show the mean percentage of 20 or 25 cercariae exposed to each of six snails recovered as metacercarial cysts.

bouring rediae. The observation that new world *B. glabrata* is a better host than old world *B. camerunensis* to an old world echinostome is interesting, but not surprising in light of our experiences with schistosomes.

As has been observed in trematodes generally, the infectivity of *E. liei* is considerably less specific to its second intermediate host than to its first. Christensen and colleagues exposed six 3–4 mm snails of each species to 20 or 25 cercariae per snail and counted metacercarial cysts via dissection several days later (Figure 6.11). Data on snail survivorship are not available, and it is not clear whether complete resistance was demonstrated by any individual. But Christensen and colleagues considered *P. acuta*, harbouring a mean of 70% of the metacercaria to which it was exposed, a 'high susceptibility' second intermediate host. The remaining species were considered to show 'low susceptibility'.

The echinostome *Echinoparyphium recurvatum* successfully utilizes the entire mollusc community of Harting Pond, West Sussex, as second intermediate hosts: pulmonates, prosobranchs, and pisidiids (Evans *et al.* 1981). There seems to be little evidence of 'preference'. With the exception of two planorbids (*Planorbis planorbis* and *Gyraulus albus*), ranking by mean number of metacercaria per individual mollusc appears to follow their ranking by individual size: *Sphaerium corneum*, *Lymnaea peregra*, *Valvata piscinalis*, planorbids, *Pisidium subtruncatum* and *Potamopyrgus jenkinsi*. The authors note that *Pisidium* and *Valvata* might be considered the most important second intermediate hosts for *Echinoparyphium*, by virtue of their high population densities in the pond. Thus on the limited evidence available, it would appear that the relationship between echinostomes and their first molluscan host may be as close as that seen in schistosomes, but their relationship with their second intermediate host is non-specific.

Population regulation

To prospect for evidence of an effect of parasitism on snail density, one might begin where prevalence rates are unusually high. Rates ranging up to 60% have been reported by Fernandez and Esch (1991a,b) working with the *Helisoma anceps* population of a small North Carolina pond. The authors sampled snails at five localities along a 15 m shoreline, on ten occasions from April of one year to June of the next. Snails were hand picked, marked, assayed for parasitism, and returned to the water. Although no data were available on that portion of the population less than 3–4 mm, this approach gave an excellent picture of parasitism in the adult fraction.

The authors inferred that their population of *H. anceps* was primarily annual and semelparous, with a reproductive peak in April. Survival analysis was complicated by their observation that a small portion of the spring cohort grew sufficiently to reproduce in August. In any case, the prevalence of parasitism in adult snails (almost exclusively due to the hemiurid *Halipegus occidualis*) seemed positively correlated with the adult population density, offset by some months, as one might expect from predator–prey interaction. Adult *Helisoma* density was low in

April–June and peaked in October, while their proportion parasitized peaked in May.

But as is usually the case, *H. occidualis* infection generally leads to the castration of its host (Crews and Esch 1986), not its immediate demise. The rising densities of adult pulmonates generally observed in the summer, and their dramatic decline through winter and early spring, are conventionally ascribed to season and the life cycle of the snail. It is possible to view *Halipegus* prevalence as passively following *Helisoma* density, with no reference to an effect of parasite on host. Clearly more than a single generation of data will be required to detect an effect of a parasite on the density of its host.

Total trematode infection rates also occasionally top 50% in the *Lymnaea* populations of an English gravel-pit pond. Adam and Lewis (1993) monitored snail densities on 25 occasions over 2½ years, documenting along their way the sporadic occurrence of twelve trematode species. Their snail-collecting technique was more conventional (hand net) and more likely to include juvenile snails, although some size bias doubtless remained. The wonderfully complicating factor in Adam and Lewis' data is that during their study period, the *Lymnaea peregra* population more typical of their small body of water was completely replaced by the similar but large-lake adapted *L. auricularia* (see Chapter 5).

Since the *L. auricularia* population seems to have passed through a bottleneck upon first colonizing this gravel-pit pond, it seems unlikely that the snails could have carried their own trematode fauna with them. Thus one might expect a reduction in their parasite load. Yet almost all the local trematodes seem to have successfully switched hosts, and in fact the introduction of *L. auricularia* seems to have brought several new trematodes into the system. Figure 6.12 shows the combined prevalence of all twelve parasites together with the relative population densities of their two lymnaeid hosts. Prevalence rates appear to have increased since the replacement of *L. peregra* by *L. auricularia*. But it is difficult to detect a relationship between monthly snail densities and rates of parasitism. Snail populations peaked in August in all three years, corresponding directly to peak infection rates in years one and three. Parasitism lagged behind the snail population by a month or two in year two, when *L. auricularia* appeared. One is tempted to speculate that this delay may have represented a period of adaptation on the part of the trematode community. But again, the appearance is of a snail population responding to season, and of a parasite population responding to snail.

If seasonality could be factored out, perhaps an effect of parasite upon

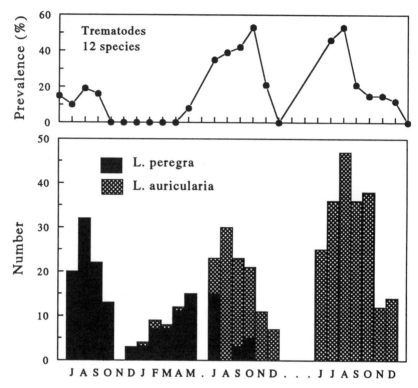

Figure 6.12. Abundance of two *Lymnaea* species in an English gravel-pit pond, together with their combined prevalence of trematode infection. In months where both snail species co-occurred, parasitism is the weighted average over both. (Data of Adam and Lewis 1993.)

host population could be more readily discerned. Dybdahl and Lively (1998) estimated the abundance of various *Potamopyrgus* clones in Lake Poerua, New Zealand, each February for five years (about 10–15 snail generations), noting the prevalence of trematode infections (8.5–16.9%). Their observation that clones common in one year tended to become overinfected by trematodes in the year following constitutes some evidence of time-lagged frequency-dependent selection.

Among the more detailed field studies of a trematode and its mollusc host are those of Pesigan *et al.* (1958) on *Schistosoma japonicum* and *Oncomelania quadrasi* in the Philippines. The authors selected ten sites within a few kilometres of Barrio Malirong, taking quantitative samples of *Oncomelania* from each on nine dates from May 1954 to November 1955. Their sampling device was a metal ring 13.5 cm in diameter, encircling an area of about 143 cm². Samples were located randomly on transects and collected until at least 300 individual snails had been encountered. (Over all 10 sites × 9 dates, the number of samples per site was usually well in excess of 30.) The authors corrected mean density at each site by discarding all samples not containing snails and not adjacent to samples with snails, on the assumption that a boundary to the suitable habitat had been passed. All snails collected were dissected and examined directly for evidence of parasitism.

Snails were picked by hand from the ring samplers, resulting in a systematic bias against smaller individuals. Nevertheless, the proportion of 'young' individuals (less than 2.5 mm shell length) seemed remarkably constant over the 18 months of the study, ranging only from a mean of 8.2% over all ten sites in July to 3.8% in August, both of 1955. Although perhaps stimulated to some extent by rainfall, reproduction apparently occurs year-round and generations (estimated to span about 12 weeks) seem to overlap entirely in this population. The constancy of the Philippine climate makes it possible to examine Pesigan's 18-month records of *Oncomelania* density for evidence of an effect due to parasitism with an improved sensitivity.

I analysed corrected mean population density (per sampling ring) from all ten sites and nine sample dates from Pesigan's Table XXIV, and corresponding mean infection rates (%) from his Table XLVII. Data analysis was through the 'Series' module of SYSTAT (Wilkinson 1988).

As the example data shown in Figure 6.13A and B illustrate, both mean snail density and mean infection rate varied slightly, although systematically, among sites. One might infer that the environment at site 3, for example, was more suitable for *Oncomelania* than that at site 9. The adjusted grand mean snail density over all 10 sites and 9 dates was 6.83 snails per 143 cm^2, with site means ranging from 4.0 to 10.8. The grand mean prevalence of detectable schistosome infection was 4.5%, with site means ranging from 1.9% at site 9 to 9.3%. These values are in line with trematode prevalence rates worldwide (Figure 6.1). Over all ten sites, the correlation between site mean snail density and site mean infection rate was 0.11, not significant. Pesigan hypothesized that schistosome infection rates might be higher at his five sites located near human habitation (including site 3) than at five remote sites (e.g. site 9) – a hypothesis he was unable to confirm. Whatever its nature, I removed the variation in habitat quality that apparently existed among the ten sites (for both snails and worms) by subtracting site means from both density and infection rates. Figures 6.13 C and D illustrate this procedure with example data.

As all ten sites were located quite near one another, it is reasonable to

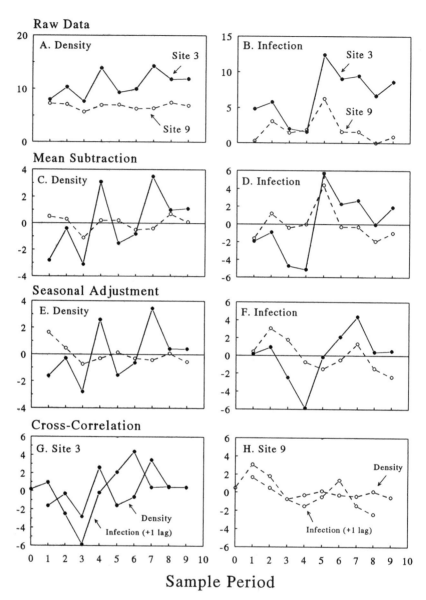

Raw Data

A. Density

Site 3

Site 9

B. Infection

Site 3

Site 9

Mean Subtraction

C. Density

D. Infection

Seasonal Adjustment

E. Density

F. Infection

Cross-Correlation

G. Site 3

Density

Infection (+1 lag)

H. Site 9

Density

Infection (+1 lag)

Sample Period

Figure 6.13. Mean density (individuals per 143 cm^2 ring) and mean prevalence of *Schistosoma japonicum* (%) in two (of ten) Philippine populations of *Oncomelania quadrasi*. (Data of Pesigan *et al.* 1958.) Abscissas give sample periods, approximately equally spaced over 18 months. Data from site 3 (Naliwatan upstream) is graphed with a solid line, and site 9 (Naliwatan downstream) graphed with a dashed line. Graphs (A) and (B) show the raw data, graphs (C) and (D) show data after subtraction of population means, graphs (E) and (F) show data arithmetically adjusted for seasonal trends over all ten sites, and graphs (G) and (H) illustrate the maximum cross-correlation discovered between density and infection within the ten sites, at a +1 sample-period lag for infection.

expect that the population dynamics of their snail and worm populations might be influenced by common environmental fluctuations. For example, Pesigan suggested that the heavy rains recorded in December 1954 and January 1955 (between sample periods 3 and 4) might both stimulate snail reproduction and wash contaminants into the rivers, increasing schistosome infection. And in fact, Figure 6.13D does show increased infection rates (in primarily adult snails) at both sites 3 and 9 for sample period 5 (late March). So I used the *adjseason* function of SYSTAT to arithmetically adjust the ten snail density data sets, and the ten infection rate data sets, for seasonality. The former did not appear to co-vary strongly by season. The largest seasonal index was only +1.18 (period 1), followed by −0.66 for period 9 and −0.58 for period 8. Thus Figure 6.13E does not differ markedly from Figure 6.13C, especially in middle sampling periods. The ten infection rate data sets did co-vary rather strikingly in most of the nine periods, however. Almost all showed large increases at period 5, engendering a −5.88 seasonal index. Seasonal indices for the infection data set were also substantial in periods 1, 2, 3, 7 and 9 (Figure 6.13F).

In order to explore the relationship (if any) between snail populations and the trematodes that parasitize them, I compared the adjusted snail density at each site with its adjusted infection rate using the CCF (cross-correlation) function of SYSTAT. I examined the contemporaneous Pearson correlation between density and infection ('lag = 0', $N=9$), as well as lags +1 and −1 ($N=8$) and +2 and −2 ($N=7$). Thus for each of ten comparisons, I calculated five values of Pearson's r. The most striking relationship was apparent at lag +1; density was positively correlated to infection, plus one time period, in nine of the ten comparisons. This relationship is illustrated in Figures 6.13G and H.

The binomial probability of an outcome as extreme as nine positives (i.e. the chance of 9 or 10 heads or tails) would be 0.021, putatively significant. However, there is a 10% chance of obtaining a result this extreme when examining five experiments (−2, −1, 0, +1, +2 lags), so its apparent 'significance' may be attributable to type I error. Summation of the values of p from the lag = +1 comparisons using the technique of Fisher (Sokal and Rohlf 1995:794) also leads one to a conclusion that the relationship between density and infection rate is suggestive but not quite significant. Setting aside the single negative correlation as spurious, and summing the nine remaining (one-tailed) ln P values, one obtains $\chi^2 = 27.2$, just barely non-significant with 18 degrees of freedom.

A positive correlation between the density of a snail population and its prevalence of infection about two months later is undeniably reminiscent of classical lynx-and-hare predator/prey oscillation. The generation time of *Oncomelania* is indeed slightly over two months, while cercarial shed begins five weeks post-infection, and peaks at about 10 weeks. But to the extent that one is entitled to draw a conclusion from this analysis, it would be that any effect parasitism may have had on the density of *Oncomelania* in the Philippines in 1954–55 was tenuous.

Aspidogastrids

Ogden Nash's observation that 'Big fleas have little fleas upon their backs to bite 'em' here finds some illustration. For unionacean mussels, which as larvae parasitize fish, are themselves the targets of a variety of parasites upon reaching adulthood. But the long-term culture of even healthy, unparasitized freshwater mussels presents a considerable technical challenge. We might not be surprised to learn that the consequences of parasitism to their survivorship and reproduction are not, as yet, well investigated.

Aspidogastrid trematodes are among the most common internal parasites of unionaceans, and may infect freshwater snails as well (Huehner and Etges 1981). The mixture of life cycles these worms display, involving both single hosts and multiple hosts, has prompted workers to recognize a small trematode subclass or order (the Aspidobothrea) to contain them (Cheng 1986). The eggs of *Aspidogaster conchiola* are shed into the water and carried by currents from a mussel host to the incurrent siphons of its neighbours, where they are directly infective. *Cotylaspis insignis* eggs, by contrast, develop into free-swimming larvae and enter the siphons of their hosts actively. Thus Huehner (1984) observed that *Aspidogaster* infections tend to be more common in populations of unionids inhabiting lotic waters, and *Cotylaspis* infections more common in the lentic.

Huehner examined 590 mussels from nine Missouri locations, representing 32 species. He found *Aspidogaster* infections in 213 individuals (23 species), *C. insignis* in 50 individuals (16 species), and *Cotylogaster occidentalis* in four individuals (three species). Duobinis–Gray *et al.* (1991) reported aspidogastrids in 30% of 219 unionids collected from Kentucky Lake, including 9 of the 10 mussel species present. Mussels usually contained only a few aspidogastrids each; the maximum intensity was 20 worms.

Adult worms (2–10 mm long) may be found throughout a bivalve host, especially in the intestines, digestive gland, and pericardial and nephridial cavities. They have been characterized as 'epithelial grazers' (Huehner *et al.* 1989). But even the short-term consequences of aspidogastrid infections to their hosts are poorly documented, and deserve additional attention. Unionids mount a defence by encapsulation (Huehner and Etges 1981).

Unionicolids

The life cycles of unionicolid mites are quite diverse. Although associated with a variety of freshwater invertebrates (sponges, chironomid midge larvae, gastropods), these mites are most commonly encountered in unionid mussels. *Unionicola aculeata* requires a mussel for oviposition but spends its adult life as a free-living planktivore. *Unionicola intermedia*, on the other hand, appears to be parasitic on mussels (primarily *Anodonta anatina*) at all life history stages (Baker 1976, 1977). Mites of this species (both nymphs and adults) seem to induce an inflammatory response in host tissues, then feed on haemocytes that congregate at the site of infestation. Mites locate mussel hosts by a combination of phototaxis and sensitivity to host-elaborated compounds, then enter through the siphons.

One or both sexes may display aggressive or territorial behaviours, a phenomenon that contributes to considerable variation in the intensity of mite infection. Edwards and Dimock (1988) reported that the entire adult population of *Anodonta imbecillis* inhabiting a small North Carolina pond appeared to be infected with *Unionicola formosa* year round. The mean density of female mites was 53.4/mussel, with but a single male generally in attendance. Approximately 90% of a large population of *Lampsilis siliquoidea* from a Michigan lake were infected with *Unionicola fossulata* (Mitchell 1965). The great majority of these mussels contained, however, only one or two females with their single males. At an average infection intensity of 1.96 mites/mussel, clearly this *Lampsilis* population is in no danger of succumbing to mite infection.

Unionicolids may show considerable host specificity (Edwards and Dimock 1995). Four species of mussels are common in the St. Marks River, Florida: *Elliptio icterina*, *Villosa villosa*, *Uniomerus declivus*, and *Anodonta imbecilis* (Downes 1986, 1989). *Unionicola abnormipes* is found most commonly in *Villosa*, *U. serrata* is almost entirely restricted to *Uniomerus*, and *U. formosa* is restricted to *Anodonta*. *Unionicola fossulata* is found in both *Villosa* and *Uniomerus*. (*U. formosa* aggressively excludes all other mite species from *Anodonta*.) Interestingly, Downes reported that no mites were ever found in *Elliptio*, and in fact mites would not enter her *Elliptio* in choice tests.

Vidrine (1990) examined 1000 individual mussels from a single site on

Village Creek in Hardin County, Texas, representing 20 species. In six visits during the course of a year he detected infection by one or more of 12 mite species in 499 mussels. Most of the unionicolids appeared to show striking preferences for individual unionid taxa. There has been some concern, however, that certain characters used for the identification of mite species may show host-induced phenotypic plasticity, confounding mite taxonomy (Downes 1990).

Some anecdotal evidence suggests that heavy mite infestation may weaken mussels to the point that they succumb to relatively minor environmental perturbation, such as handling (Majumder and Pal 1992). In any case, the occurrence of unionicolid mites would seem significantly widespread to warrant more detailed study of their effect on mussel hosts.

Summary

The same attributes that make freshwater molluscs attractive as organisms for ecological study, their predictable abundance and the ease by which they may be located also make them attractive hosts for parasites. Mollusc parasites are quite various; we concluded this chapter with brief reviews of the aspidogastrid trematodes and unionicolid mites that parasitize unionaceans. But the host/parasite relationship finds greatest development in the Digenea, an order of 40 000 fluke species entirely dependent on molluscs at some stage in their life cycle.

Some digenetic flukes produce eggs or miracidia that infect their mollusc host upon ingestion. In other cases, fluke eggs hatch into swimming miracidia that actively seek host molluscs, burrowing into body tissues. Through successive stages of development, trematode larvae convert their host's metabolic machinery to the production of cercariae, which typically begin emerging several weeks post-infection to seek second hosts (vertebrates, aquatic insects, or in some cases, other molluscs).

Host snails may be entirely castrated by digeneans of the genus *Schistosoma*, especially if infected as they approach adulthood and if multiple miracidia are involved. Castration may not be complete, however, depending on the age of the snail at infection. If infected as adults, snails sometimes demonstrate increased egg laying during the several weeks prior to the first emergence of cercariae, in an apparent effort to compensate for castration to come. Snails may also (perhaps rarely) outlive schistosome infection, regaining fertility after many weeks of shedding cercariae. But schistosome infection generally seems to depress the survivorship of host snails many times below that observed in uninfected controls.

Infection by *Fasciola* seems more likely to lead to castration than infection by *Schistosoma*, although the little lymnaeids hosting the former seem to mature more rapidly than most planorbid hosts of the latter, and thus still manage to reproduce. *Fasciola* infection may improve the survivorship of a host snail.

Trematode miracidia are usually encapsulated and destroyed shortly after penetration by the molluscs they attack. Generally, only in those circumstances where miracidia penetrate a host for which they have evolved special countermeasures can development proceed. Host specificity varies greatly; *S. haematobium* populations seem to form very close associations with particular *Bulinus* populations, while *Fasciola* miracidia generally seem infective to a broad range of lymnaeids. The infectivity of cercariae to molluscs as second intermediate hosts is considerably less specific than the infectivity of miracidia.

In the last century, *Fasciola hepatica* successfully invaded Australia, and *Schistosoma mansoni* the Americas. Once initially introduced to a new population of snails, trematode populations may show dramatic improvement in infectivity after just two or three worm generations. The defences mounted by snail hosts develop more slowly, but seem more heritable. Susceptible and resistant strains of *Biomphalaria* can be obtained by selective breeding, as well as strains showing different ages at onset of resistance. At least occasionally, the defences of coevolved hosts have reached such efficiency that their trematodes are more infective to closely related (but heretofore unchallenged) snail populations.

The introduction of *Schistosoma mansoni* into the new world has been characterized as a 'gigantic though unplanned experiment' in host/parasite coevolution. It appears that challenging a new world *Biomphalaria* population with an old world schistosome generally results in reduced infectivity. Results are approximately equivalent to challenging an old world *Biomphalaria* with a new world worm or even a new world *Biomphalaria* population with a new world worm adapted to a second snail population. Lessened resistance to infection is seen, however, when an old world *Biomphalaria* is challenged with an old world schistosome from a second population. This suggests that snails have a 'head start' in the race between host and parasite, but that this head start lessens when (as in the old/old comparison) trematodes become generally adapted to the range of potentially suitable hosts by interpopulation dispersal.

The relationship between trematodes and their mollusc hosts may be summarized as 'mutually aggressive', the parasites having evolved to maximize their total reproductive output, even at the expense of host viability. Trematodes usually show lesser effects upon hosts with which they have not coevolved.

Although trematode parasitism is widespread in freshwater mollusc populations generally, data on percentage prevalence are by far most commonly available for the hosts of human schistosomes. Considering dozens of published surveys from low latitudes worldwide, it appears that only about 1–5% of the individuals of an average snail population are infected with schistosomes. Occasionally, however, populations 20%, 50%, or even 90% infected may be encountered. Scattered data sets involving other taxa of snails and worms tend to confirm the schistosome figures.

Sampling a snail population for trematode parasites presents a considerable challenge in design. Multiple worm species may be involved, each of which will likely show a very patchy distribution in both temporal and spatial dimensions, influenced not only by the physical environment but by the biology of at least two hosts. The prevalence of trematode infection will generally be a function of the size and age of the host snails sampled.

The prevalence of parasitic infection may be related to snail population density. *Schistosoma japonicum* infection in ten Philippine *Oncomelania* populations ranged from 1.9% to 9.3%, averaged over 18 months. A time series analysis designed to prospect for a relationship between monthly percentage prevalence and snail population density returned some evidence that *Schistosoma* infection rate was positively correlated to *Oncomelania* density two months previous. The relationship was not quite significant, but reminiscent of predator–prey oscillation. Thus even apparently low parasitic infection rates may contribute to the regulation of freshwater mollusc populations.

7 · Predation

Molluscs are edible, and easily subdued by predators of even the most modest ferocity. Thus most aquatic vertebrates seem to eat molluscs, at least under some circumstances. I include under this sweeping generalization not only fish, but semi-aquatic vertebrates such as amphibians, certain reptiles, and a few mammals. In a charming review entitled 'Enemies of the land and freshwater Mollusca of the British Isles', Wild and Lawson (1937) listed 20 vertebrate predators of *Planorbis* alone, excluding fish but including such eye-catchers as the natterjack toad, pheasants, and bats. In addition, a large fraction of the macroinvertebrate predators of fresh waters recognize molluscs among their prey. Among the fauna of Oneida Lake, New York, F. C. Baker (1918) recognized six species of insect potentially dining on molluscs, one crayfish, eight leeches, and two of the molluscs themselves (large *Lymnaea*). To the Oneida rogue's gallery Baker added 46 species of fish, 8 amphibians, 7 reptiles, 6 birds, and 3 mammals. Michelson's (1957) review of possible biological control agents for pulmonates included single paragraphs on predatory flatworms, leeches, crustaceans, predatory molluscs, and mammals, plus two paragraphs on birds, three on reptiles and amphibians, and four on insects. Molloy and colleagues (1997) catalogued 176 predators of zebra mussels.

The present review will proceed from mammals through the vertebrates and into the invertebrates, as predator body sizes decrease and densities rise. We will focus on the evidence that predation may impact the size of freshwater mollusc populations, or their life histories, prospecting for community effects later in the chapter. We will find that the effects of large vertebrate predators (mammals, birds, reptiles and amphibians) are generally minor and patchy. Enclosure/exclosure experiments have, on the other hand, sometimes returned clear evidence of substantial predation effects due to fish and larger invertebrates such as crayfish and waterbugs. The impacts of the smaller but numerically more common invertebrates (sciomyzid flies, leeches, triclads, etc.) must also be important, at least on occasion, although more difficult to document. Immediate responses to predation threat may be characterized as 'resist or retreat': crawl out, burrowing, shell shaking, clamping, or enclosure. On a more evolutionary time scale, we will see that various modifications of shell shape and weight have led to great variability in the protection that shells offer their inhabitants. One might expect that all this variability in predation pressure and prey vulnerability might lead to a substantial effect on freshwater mollusc community structure. Such an effect, we shall see, is surprisingly difficult to establish.

The malefactors

Mammals

Mammalian predators of freshwater molluscs include mink, otter, raccoon, nutria, and various shrews, voles, and mice. All are dietary generalists, typically present in densities rather low to have substantial impact on mollusc populations. Exceptional circumstances arise, however. Wild Norwegian rats inhabiting the Po River valley, Italy, seem to prey heavily upon *Lymnaea stagnalis*, *Planorbis corneus*, *Viviparus ater*, *V. contectus*, *Anodonta cygnaea*, and *Unio pictorum* (Parisi and Gandolfi 1974). Nieder *et al.* (1982) observed the behaviour of rats in enclosures along the Po for a period of about four months. Although provided commercial pellet food, the rats seem to have hunted *Viviparus ater* rather avidly. The number of snails consumed rose along with the rat population, from about 5 snails/week for the founding pair of rats to 60 snails/week as the rat population increased to 16. The authors felt that the total of 502 broken shells ultimately recovered over the four months (mean size about 30 mm) would have been greater but for a 'dearth of *V. ater* available in the channel.'

Although primarily herbivorous, the North American muskrat (*Ondata zebithicus*) will under some circumstances prey heavily upon unionid mussels. Muskrats carry individual mussels to safe spots onshore where, after a number have been consumed, shells accumulate in identifiable 'middens'. They insert their incisors between the valves of their victims and pry upwards, typically breaking one of the valves of thin-shelled mussels but not of more heavily shelled species. As is true of predators generally, muskrats do not randomly sample all size categories of prey. Large mussels may be difficult to extract from the substrate, carry to the shore, and open, while small mussels may escape notice, or prove unworthy of the effort.

The effect of size-specific muskrat predation on mussel populations has been well documented by Hanson et al. (1989) and Jokela and Mutikainen (1995). Narrow Lake in central Alberta is home to large populations of both muskrats and *Anodonta grandis simpsoniana*. Hanson and his colleagues used an Ekman dredge to sample all depths of the north basin of Narrow Lake, in May through August of two successive years. They sieved samples through a coarse (6 mm) screen, forfeiting data on the young-of-the-year and 1+ cohorts which, in any case, were too small of size to be affected by muskrat predation. Hanson was able to estimate the age of the remaining mussels by reference to annual growth-stoppage lines on the shell. This technique seems to be especially reliable in Canadian populations of anodontines, where smooth and continuous shell growth during the summer is abruptly terminated by severe winter. Hanson standardized all data (both age and size) to the 1986 annulus, then combined to obtain the distribution shown in the lower half of Figure 7.1. I have converted the authors' original figures from percentage to absolute numbers, using their overall estimate of 1.23 million mussels in the study area.

Hanson and colleagues removed all old shells from the muskrat middens on the shores of Narrow Lake, then periodically resurveyed all middens during the remainder of the season. Surveys resumed when the ice cover opened the following spring. Shells collected during this 12-month period were measured and aged to their 1986 annulus. The authors considered that their total sample of 36 771 individual mussel remains (about 3% of the population aged 2+) underestimated the annual predatory impact of muskrat. Although there is some danger of double-counting should both valves remain intact but become separated, both valves may become lost or broken at greater frequency (Convey et al. 1989). The top half of Figure 7.1 shows the age distribution of the mussels consumed by muskrats in the north basin of Narrow Lake during the course of a single year.

The overall age distribution of the Narrow Lake *Anodonta* population was rather uneven; 1984 and 1981 seem to have been good years for recruitment, and 1982 and 1983 rather poor. Comparison of the two distributions in Figure 7.1 shows that muskrats made very little use of that large fraction of the *Anodonta* population younger than five years. Selection was doubtless more directly by size than by age, however. The

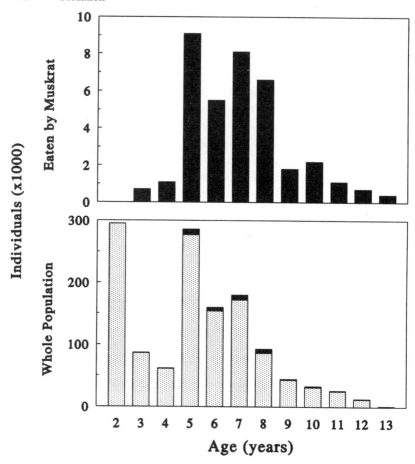

Figure 7.1. The 1986 age distribution of *Anodonta grandis* in Narrow Lake, Alberta. (Data of Hanson *et al.* 1989.) Mussels aged 0+ or 1+ were excluded. The fraction eaten by muskrats is graphed alone in the upper graph and cast onto the total population (as solid bars) in the lower graph.

authors noted that 86% of the mussels eaten were greater than 55 mm long, while only 37% of the natural population fell into this category. They felt that such predation pressures might result in selection for slow growth in this unionid population. One might argue, however, that a better life history 'strategy' would be to mature at a smaller size, and never reach 55 mm at all. (Mussels from the Narrow Lake population typically mature at age 4–5, or about 50 mm, and grow to about 70 mm.) We shall return to the effects of predation on life history later in this chapter.

Birds

Under some circumstances, ducks, geese, and swans (Anatidae) may consume substantial numbers of molluscs. For example, the blue-winged teal (*Anas discors*) is a surface- feeding, or 'dabbling' duck (subfamily Anatinae). Mabbott (1920) reported the contents of 319 blue-winged teal digestive tracts, collected in 29 states and four Canadian provinces, every month except June. All contained some vegetable matter, especially pondweeds and bullrushes. At most, 29 individuals contained any identifiable gastropod remains (*Valvata, Amnicola, Physa*, lymnaeids, planorbids) and at most seven contained identifiable bivalves (*Sphaerium, Pisidium*). However, there is evidence that during the nesting season, blue-winged teal may prefer invertebrate prey over vegetation, perhaps for its high protein content. Swanson and Meyer (1977) collected 44 blue-winged teal from nesting grounds in North Dakota, 24 of which were 'laying' (as judged by inspection of the reproductive tract) and the remainder of which were 'breeding'. Invertebrate remains constituted 99% of the gut volume of the former category, and 97% of the latter. The molluscan component of the diet was substantial, comprising 23% of the content of breeding blue-winged teal and 40% of the diet of laying females. Swanson and Meyer mention *Stagnicola caperata* and *S. palustris* as particularly important, with physids, planorbids, and sphaeriids also part of the diet.

Invertebrate prey also seem to be especially preferred over aquatic vegetation by juvenile ducklings at some developmental stages. Reinecke (1979) examined the diets of 41 juvenile black ducks collected from the Penobscot River valley in Maine. There was a pattern of decreased invertebrate food use with age, from 95% of the gut content dry weight for downy juveniles to 84% for partially feathered to 34% for fully feathered juveniles. Downy ducklings were surface feeders, and did not often capture molluscs. But molluscs (unspecified clams and snails) comprised 24% of the gut content dry weight of 12 partially feathered juvenile black ducks.

Diving ducks (subfamily Aythyinae) also may eat substantial numbers of molluscs as juveniles, and as egg laying females. In southwestern Manitoba, Bartonek and Hickey (1969) reported that snails were the second most common food item in the guts of female and juvenile canvasbacks (*Aythya valisineria*) and lesser scaups (*A. affinis*) collected during the breeding season. Males were much more vegetarian in diet. Cottam

(1939) summed data on gut contents from 1051 adult lesser scaup from 30 states of the United States and 5 Canadian provinces, over every month of the year. A great diversity of molluscs (both freshwater and marine) combined for 24.9% of the volume. Freshwater taxa mentioned specifically were *Amnicola* (1.7%), *Helisoma newberryi* (1.6%), and '*Planorbis*' (probably other *Helisoma*, 1.6%). The data of Rogers and Korschgen (1966) suggest that lesser scaup molluscivory may be a seasonal phenomenon. Few molluscs were found in the guts of their samples of lesser scaups breeding in Manitoba or overwintering in Louisiana. But the guts of 88 migrating individuals collected on the Keokuk Pool (Pool 19) of the Mississippi River between Iowa and Illinois contained, by volume, 85% mollusc remains.

Thompson (1973) identified nine taxa of gastropods and four of bivalves in the guts of 270 lesser scaup he collected from Pool 19. Of these, 76% had eaten *Sphaerium striatinum*, 40% had eaten *Musculium transversum*, and 10% had eaten unionids. Hydrobiids seem to have been the most important gastropod food, including *Somatogyrus* (33%), *Fontigens* (19%), and *Amnicola* (5%). *Campeloma* (21%), *Lioplax* (10%), and *Pleurocera* (8%) were also commonly eaten. Pulmonates were oddly rare in the diet of this sample of lesser scaup from the Mississippi, but may have fallen among the 'unidentified gastropods' found in 68% of the guts.

Large flocks of diving ducks visit Lake Erie in the vicinity of Point Pelee from late October until mid-December. Individuals are frequently observed surfacing with *Dreissena* in their beaks, which they manipulate and swallow. Hamilton et al. (1994) anchored exclosures of an unusual design (covered tops, open on all four sides) at ten sites in 5–7 m of water. Two separate rock beds were involved, 250–350 m offshore, surrounded by soft substrate largely uninhabited by *Dreissena*. Rocks were removed from each exclosure by s c u b a divers from September to November, and again in May. November samples did in fact reflect significantly greater mussel biomass (g/cm^2 rock surface area) under exclosures than in uncaged control areas. This seemed to be due to the selective removal of the larger size classes of mussels. But such losses as may have been inflicted on the largest mussels seem to have been compensated for by increased substrate availability for smaller individuals. Even in November, the differences in mussel density between exclosure and control were not significant, and overall population size distributions had readjusted by the spring. Hamilton concluded that 'ducks had little lasting impact on mussel populations, but mussel abundance may have determined duck concentration in the area.'

Among the birds there are several rather spectacular examples of diet specialization on freshwater molluscs. The limpkin (*Aramus guarauna*) is a wading bird of the Order Gruiformes (rails, coots, and cranes) with a narrow, hooked bill. It ranges from Florida to Argentina. Although some small quantities of insect larvae and vegetation (especially seeds) are occasionally found in stomach analyses, its normal diet seems to be almost entirely composed of large viviparid and ampullariid snails. According to Cottam (1942), it lifts its prey (typically *Pomacea paludosa*) from the substrate, turns it, and seats it back firmly in the mud aperture up. It then extracts the meat neatly from the shell, sometimes casting the operculum aside.

More famous than the limpkin is the snail kite (*Rostrhamus sociabilis*), a diurnal bird of prey (family Accipitridae) also ranging from Florida to Argentina. It has an oddly long and hooked beak (for a raptor) which it uses to great effect on *Pomacea*. Snyder and Snyder (1969) describe how it courses over the marshes until it spots a snail at the surface or on marsh vegetation. It then dives on the snail, siezes it in a talon, and carries it to a low tree. The bird perches on one foot, holds the snail aperture-up with its other foot, and extracts the meat cleanly with its unusual bill. For many years it was thought that the kite was a strict specialist on *Pomacea*. It is now clear that South American kites commonly attack *Marisa*, although their extraction efficiency is not as great (Snyder and Kale 1983). They may also infrequently attack small vertebrates, especially turtles.

Turtles

Returning to the subject of omnivory, four species of kinosternid turtles are common in Oklahoma: the yellow mud turtle (*Kinosternon flavescens*), the Mississippi mud turtle (*K. subrubrum*), the common musk turtle (*Sternothaerus odoratus*) and the Mississippi musk turtle (*S. carinatus*). Mahmoud (1968) examined the gut contents of several hundred individuals of all four species, collected statewide May to October. Mollusc remains were found in over 90% of the stomachs examined, ranging only from 24% of the gut volume in ($N = 121$) *K. flavescens* to 32% in ($N = 178$) *K. subrubrum*. The pie diagrams in Figure 7.2 show example data for the Mississippi musk and Mississippi mud turtles, combining crustaceans and insects as arthropods and carrion and aquatic vegetation as 'other'. Mahmoud considered all four turtle species 'euryphagic' bottom feeders, eating a wide variety of foodstuffs as available in their habitats.

Laboratory feeding trials suggested some differences in the predatory capabilities of mud and musk turtles, however. Mahmoud performed 47

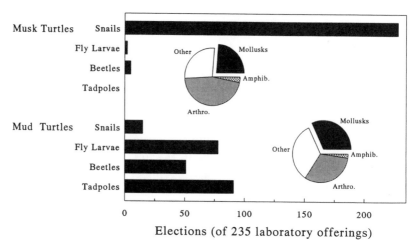

Figure 7.2. The bar graphs show the frequency at which musk turtles (*S. carinatus*) and mud turtles (*K. subrubrum*) fed on four types of prey in 235 feeding trials, and the pie diagrams compare wild-collected gut contents by volume. (Data of Mahmoud 1968.)

feeding trials, each involving 20 turtles (five of each species) and a bucket of mixed prey items: snakes, tadpoles, beetles, fly larvae and snails (*Physa*). The bar graphs of Figure 7.2 show the number of occasions (in $47 \times 5 = 235$ opportunities) upon which each of four prey items was eaten by two of Mahmoud's turtle species. Mud turtles fell first upon the tadpoles, beetles, larvae, and other squirming prey, while musk turtles seemed to prefer snails. So although their diets in the wild are not strikingly different, Mahmoud characterized mud turtles as 'more aggressive, fast-moving, and alert', suggesting that they may be less likely to pick up snails when other prey are available.

Given their omnivory, one might expect turtle populations to vary in diet as different prey becomes available in different environments. The yellow mud turtle is most common in Mexico and the southern United States; it reaches the northern limit of its range in Missouri. Kofron and Schreiber (1985) reported that the gut contents of 34 yellow mud turtles inhabiting a small pond in northeastern Missouri (51 captures) comprised primarily mollusc remains. Of 325 prey items recognized in faeces, 268 were *Physa*, *Helisoma*, pisidiids, or unidentified gastropods. The remainder were insects, fish, and crayfish. The predominance of molluscs in the faeces of this population does not seem to be an artifact of the authors' sampling method. Kofron and Schreiber found that the faeces of the co-occurring Blanding's turtle (*Emydoidea blandingii*) contained crayfish, insects, fish, and amphibian remains, but no mollusc shells at all.

In southern Mexico's Laguna Escondida, the mud turtle *Kinosternon leucostomum* eats very few molluscs (Vogt and Guzman 1988). By volume, freshwater snails comprised only 1.3% of the stomach flushings of 140 mud turtles trapped during the course of one year; by item count they comprised 10%. But unidentified molluscs were the majority prey item in the guts of 13 *Staurotypus triporcatus* from the Laguna, as judged by item count, second only to seeds and fruits in volume. In light of its 'large alveolar surfaces and massive jaw musculature', Vogt and Guzman considered *Staurotypus* to be a 'mollusc specialist'.

Salamanders

The term 'euryphagic' has found great use in reviews of salamander biology. Workers generally report that amphibians of the order Caudata will ingest any animal in their environment suitable to their mouth gape. For example, five species of salamanders inhabit the Hubbard Brook experimental forest in New Hampshire: the terrestrial *Plethodon cinereus*, terrestrial subadults of the newt *Notophthalmus viridescens*, and three species of more aquatic habit, the dusky salamander (*Desmognathus fuscus*), the brook salamander (*Eurycea bislineata*), and the spring salamander (*Gyrinophilus porphyriticus*). Burton (1976) reported at least occasional occurrence of gastropod and sphaeriid shells in the guts of adults of all three of these more aquatic species, and in the guts of larval *Desmognathus* as well.

Larval tiger salamanders (*Ambystoma tigrinum*) and larval newts (*N. viridescens*) are the only abundant aquatic vertebrates in McGuire's Pond, south of Carbondale, Illinois. Brophy (1980) reported that 57% of the larval newts and 21% of the larval *Ambystoma* he collected between March and September contained *Physa* in their guts. He recorded *Sphaerium* in 9% and 27%, respectively. By weight, freshwater molluscs were the second most important food category for both species, after ostracods, increasing in proportion through the season, as the larvae grew.

It would seem reasonable that the larger the salamander, and the more aquatic its habit, the greater danger it would pose to freshwater molluscs. A siren (*Siren lacertina*) measuring 369 mm from snout to vent, collected in a stagnant ditch near Norfolk, Virginia, contained 120 individual *Musculium*, 8 *Helisoma*, 5 *Gyraulus*, and 2 *Pseudosuccinea* (Burch and Wood 1955). The remainder of the ingesta were incidental. A 325 mm specimen collected in Orange Lake, Alachua County, Florida, defecated the shells of 323 *Planorbella scalare* (up to 9.1 mm diameter), 1 *Physa* and 1 *Musculium* (Moler 1994).

Fish

The vertebrates we have discussed thus far, mammals, birds, reptiles and amphibians, unquestionably do on occasion consume large numbers of freshwater molluscs. But these predators themselves are typically at low density and/or patchy in their impact. It is difficult to imagine that turtle or duck predation, for example, could measurably affect a freshwater mollusc population, simply because there are relatively few turtles or ducks per pond. Fish predation would seem to be a different matter.

Among the more thorough comparative surveys of fish diet is that of Maitland (1965), working in tributaries of the River Endrick, Scotland. The study streams were generally shallow, rocky, riffle-and-pool environments, somewhat acidic, with smallish mollusc faunas: *Ancylus*, *Lymnaea*, and *Pisidium* (species not identified). They comprise less than 1% of the benthos. The five most common fishes of this region are salmon (*Salmo salar*), trout (*Salmo trutta*), minnows (*Phoxinus phoxinus*), stone loach (*Nemacheilus barbatula*), and three-spined stickleback (*Gasterosteus aculeatus*). Maitland examined gut contents from several hundred individuals of most species, netted monthly at six sites. He reported some dietary specialization: trout ate most commonly at the surface, loach and stickleback rarely. But all five fishes fed heavily on the most common aquatic invertebrates (insects and crustaceans) and all except the loach ate molluscs, at least occasionally.

Hartley (1948) examined the stomach contents of all 11 fishes inhabiting the River Cam and one of its small tributaries in Barrington, England. He collected his fish by a variety of methods over several years, and his sample sizes ranged from 8 perch to 391 gudgeon. Molluscs seem to be much more common in the Cam and its tributaries (both quite alkaline) than in the rather acidic Endrick. And setting aside perch and loach for their small sample sizes, six of the nine fishes inhabiting the Cam can fairly be considered molluscivorous (brown trout, stickleback, eel, gudgeon, dace and roach). Indeed, *Potamopyrgus jenkinsi* was (by item count) the single most common food identified in the stomachs of the dace (*Leuciscus leuciscus*) and the roach (*Rutilus rutilus*). Numerical superiority for mollusc prey disappeared when Hartley summed the categories of insect larvae in his data set. But clearly the combination of fish abundance, their omnipresence, and their general willingness to consume

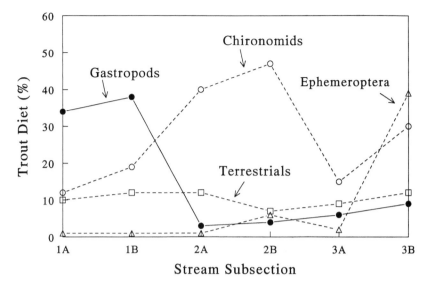

Figure 7.3. Frequencies of four common food items in the guts of cutthroat and rainbow trout captured April–October in sequential sections of Marshall Creek, British Columbia. (Data of Bryan and Larkin 1972.)

clams and snails, increases the likelihood that they may substantially impact mollusc populations.

The large literature on the diets of various trout species paints them as quite general predators. Brook, cutthroat, and rainbow trout are all common in Marshall Creek, British Columbia, downstream from a small trout hatchery. Bryan and Larkin (1972) divided a study section of the creek into six subsections of about 10–20 m each and sampled the trout population using electroshocking gear. Fish were anaesthetized and their stomach contents 'flushed out', then tagged and released. Figure 7.3 shows the proportions of four major food types observed in trout stomachs which Bryan and Larkin sampled down their transect between April and October, as item counts. Prey categories not shown included non-chironomid dipterans, fish and their eggs, tubificid worms, and non-tubificid oligochaetes. In any case, the spatial variation over this small study area is striking. 'Gastropoda' (unspecified) predominated in the two most upstream subsections, while insects of various sorts became more common in trout diet downstream. Trout population densities were greater in Bryan and Larkin's upstream sites, as were their fish capture rates. It is possible that the dietary shifts shown in Figure 7.3 are partially explained by severe cropping of the insect resource upstream.

Differing availability of prey types does not seem to be the entire story in Marshall Creek, however. For fish caught more than once, Bryan and Larkin calculated a correlation coefficient between first and second gut contents and converted this to its standard normal deviate 'within fish', (Z_w). The authors compared values of Z_w with similar statistics calculated between trout of the same size and species, caught at the same time (Z_b). In some circumstances they felt that their data indicated food specialization in individual trout, estimated as $Z_w - Z_b$. Individual specialization was most commonly on gastropod prey. The degree of food specialization in trout recaptured in the same subsection was not significantly greater than that shown by trout recaptured in different subsections. It would appear that at least part of the dietary variation displayed by fish populations may be due to differing individual foraging strategy or search image among individual fish.

Large, bottom-dwelling fish with powerful jaws seem especially dangerous to freshwater molluscs. At 16 cm standard length, roach begin to switch from diets of algae and macrophytes to zebra mussels in Poland's Lake Sniardwy (Prejs *et al.* 1990). By the time the fish reach 28 cm, they prey upon mussels almost exclusively. The massive jaws mounted by roach, freshwater drum, and round gobies have brought these species to the attention of researchers prospecting for controlling agents for zebra mussels in North America (French and Bur 1993, Nagelkerke and Sibbing 1996, French and Love 1995, Ghedotti *et al.* 1995). Jude (1973) examined the anterior guts of 344 gizzard shad netted from Pool 19 of the Mississippi River in Iowa. (This may be remembered as the site of several studies of duck diet.) Jude sampled stations many kilometres apart, over the course of 18 months, and collected fish 10–44 cm in standard length. Shad smaller than 15 cm seemed to eat little but algae. Older fish primarily ate pisidiids less than 5 mm long, which they crushed in their powerful gizzards. The hydrobiid *Fontigens* also appeared among shad gut contents occasionally, along with smaller numbers of microcrustaceans. *Cinetodus froggatti* is an ariid catfish inhabiting the Fly River of New Guinea. The gut contents of 16 specimens, ranging in length from 23 cm to 42 cm, were reported by Turner and Roberts (1978). Three guts contained no food, but from the remainder Turner and Roberts identified 14 species of molluscs, representing 13 genera and 8 families. The catfish had eaten, without crushing, heavily shelled thiarids, delicate planorbids, viviparids up to 22 mm, and brackish-water erodonid bivalves. No non-molluscan food items were observed.

Stein *et al.* (1975) found molluscs in the guts of 31.1% of 415 carp

(*Cyprinus carpio*) netted from Skadar Lake in the former Yugoslavia. On occasion some of these carp 'had their gut literally packed with shells and shell fragments'. The relative proportions of the molluscs in the diets of these carp did not match their proportions in Ekman grab samples of lake bottom. Stein and colleagues identified eight mollusc taxa, in order of abundance: *Pyrgula annulata* (from the minor rissoidean family Micromelanidae) 'form 1' (with carinae), *Valvata piscinalis, Dreissena, Pisidium* (perhaps several species), *P. annulata* form 2 (smooth), *Viviparus viviparus, Radix auricularia*, and the pleurocerid *Amphimelania holandri*. Carp seemed to prefer *Valvata* by a large margin. Though representing only a little over 14% of the mollusc fauna of Skadar Lake (by direct count), *Valvata* shells comprised about 82% of the mollusc remains identified in carp guts. Stein *et al.* attributed the high 'electivity' of *Valvata* to its convenient size, its thin shell, and the high proportion of living snails to dead *Valvata* shells on the lake bed. We will return to predator discrimination among mollusc species in a discussion of community effects later in this chapter.

The large and diverse family Cichlidae includes among its number quite a few fishes with specializations for molluscivory (McKaye *et al.* 1986, Witte *et al.* 1990). *Macropleurodus bicolor* is an 'oral sheller', seizing its prey and jerking violently to tear the mollusc's body from its shell. *Haplochromis ishmaeli* is a 'pharyngeal crusher', generally breaking its prey's shell entirely. Some interest has been directed toward such animals as possible biological controls against the pulmonate hosts of schistosomiasis (Slootweg *et al.* 1993).

Slootweg (1987) performed a series of laboratory feeding trials involving these species (and two others) wild-caught from Tanzanian Lake Victoria and cultured populations of *Biomphalaria glabrata*. Typical results from two individual fish are shown in Figure 7.4. Here fish isolated in tanks were offered 130 *Biomphalaria* per experiment; ten snails from each of 13 single-millimetre size classes. (Slootweg presented evidence elsewhere that food input at this level was more more than sufficient to satiate his study animals.) Experiments were performed four times over a study lasting several days.

Figure 7.4 shows that a larger (120 mm) *H. ishmaeli* preferred snails ranging from 7 mm to 9 mm in diameter, while an (88 mm) *M. bicolor* preferentially selected snails in the 4–5 mm range. These preferences fit Slootweg's predictions from a simple optimal foraging model based upon two functions of shell size: tissue dry weight of the snail and the handling time required by the two fish. Somewhat unexpectedly, the 'oral sheller'

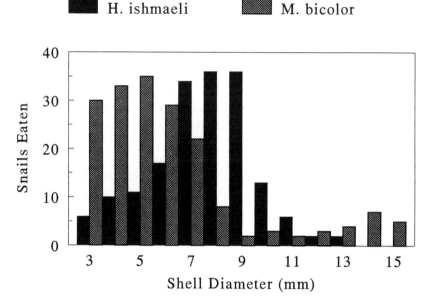

Figure 7.4. Diameters of the *Biomphalaria* consumed by individual molluscivorous cichlids, *Haplochromis ishmaeli* and *Macropleurodus bicolor*, in laboratory feeding experiments. (Data of Slootweg 1987.) A total of 40 snails of each size class was offered.

M. bicolor entirely crushed all prey smaller than 8 mm. Slootweg noted that previous observations had been of attacks on natural prey (usually *Melanoides* and *Bellamya*) with shells much stronger than *Biomphalaria*. Oral shelling may be specialization of the sort catalogued by Liem (1980): its inclusion in the repertoire does not involve trading off simpler, more generalized behaviours such as crushing. Indeed, by resort to oral shelling *M. bicolor* was able to eat a few of the largest *Biomphalaria* offered on occasion.

In the last twenty years a great deal of research has been directed toward the influences of predatory fishes upon benthic community structure (Gilinsky 1984, Power 1990, Diehl 1992). Among the fishes commonly involved in such studies have been sunfishes with demonstrated molluscivorous preference. Sadzikowski and Wallace (1976) compared the stomach contents of three size classes of the bluegill (*Lepomis macrochirus*), the green sunfish (*L. cyanellus*) and the pumpkinseed (*L. gibbosus*) wild-collected from two interconnected Michigan lakes over 40 summer days. Example data from ($N = 166$) bluegill and ($N = 122$) pumpkinseeds are shown in Figure 7.5.

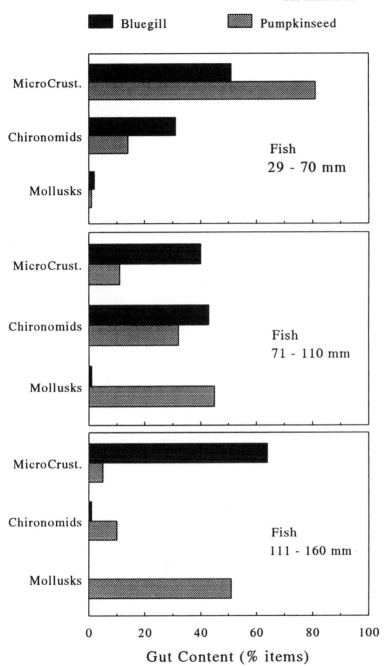

Figure 7.5. Frequencies of three common food items in the guts of bluegill and pumpkinseed sunfish, of three size classes. (Data of Sadzikowski and Wallace 1976.)

When young (29–70 mm fork length), both species of sunfishes primarily eat cladocerans and copepods, with a substantial chironomid component. Upon reaching the 'medium' size class, however, the pumpkinseed begins to switch to molluscs. Sadzikowski and Wallace reported 44.3% of the items in medium-sized pumpkinseeds were unspecified 'gastropods', and 0.7% unspecified bivalves. The distinction between the two sunfishes is even more striking in the 111–160 mm 'large' size class, where 50% of the items in pumpkinseed guts were molluscs, but only a 'trace' of such material was detectable in the bluegill diet.

It should be noted that adult bluegills have been found to eat *Physa*, *Helisoma*, and ancylids in other situations (Crowder and Cooper 1982). But pumpkinseeds have exceptionally wide pharyngeal arches and round, molar-like teeth or plates which are generally interpreted as specializations for molluscivory. They tend to inhabit weedier areas than other sunfishes. Molluscs seem to comprise a large portion of the pumpkinseed diet throughout its range, consistently (Laughlin and Werner 1980). There is some evidence of a positive relationship between pumpkinseed density and snail density (Mittelbach 1984).

Our attention now turns from studies of fish diet and gut content to experimental demonstrations of the impact fish may have upon mollusc populations. Since the 1970s we have been blessed with a wealth of enclosure/exclosure experimentation bearing on this question (Table 7.1). Some studies have been simple tests to see whether mollusc population densities may rise when predatory fish are excluded, while others have involved artificial enclosures of fish as well as exclosures. The enclosure of predators clearly increases the likelihood that a researcher will detect a predation effect, since even if overall natural predator densities are mimicked, the temporal and spatial patchiness of the predator population will be constrained, to the ultimate detriment of the prey. But enclosure of predators is nevertheless valid. Clearly if no effect is detected under such conditions, a genuine natural effect would seem quite unlikely.

Brown and DeVries (1985) investigated a phenomenon as well known in freshwater molluscs as elsewhere: high population density in a suboptimal environment. Although the author's field rearing experiments showed that *Lymnaea elodes* (of any origin) have better growth and fecundity in their larger, permanent pond ('F'), observed adult population densities were greater in their small, temporary pond ('A'). Their quantitative survey of mollusc predators suggested that predation by the mudminnow *Umbra lima* upon the egg and juvenile *Lymnaea* might be sufficient to

Table 7.1. *Exclosure tests of fish predation effects upon freshwater mollusc populations*

Workers	Population	Location	Effect on density
Exclosure (only)			
Robinson and Wellborn (1988)	*Corbicula*	Texas reservoir	Y
Osenberg (1989)	Snail community	Michigan lake	N
Bronmark (1988)	Snail community	Swedish lake	N
Walter and Kuiper (1978)	Pisidiid community	Swiss lake	Y
Thorp and Bergey (1981)	Mollusc community	S. Carolina reservoir	N
Keller and Ribi (1993)	*Viviparus*	Lake Zurich	Y
Enclosure/exclosure			
Brown and DeVries (1985)	*Lymnaea elodes*	2 Indiana ponds	Y
Bronmark *et al.* (1992)	Snail community	2 Wisconsin lakes	Y
Martin *et al.* (1992)	Snail community	Tennessee pond	Y
Gilliam *et al.* (1989)	Pisidiid community	New York stream	N

account for lower adult densities at Pond F. An artificial introduction of mudminnows into Pond A enclosures supported this hypothesis.

Pumpkinseed sunfish are the most conspicuous molluscivore in Lawrence Lake, Michigan. Here, as may be recalled from Chapter 5, Osenberg (1989) performed a cross-factored experiment testing both the importance of vertebrate predation and resource limitation to a community of eight gastropod species. Although the addition of phosphate fertilizers did seem to promote growth and reproduction of both the macrophytes and the snails that dwelled upon them, the effects of caging were not especially compelling. By snail species and size category, Osenberg found one cage effect on snail density nominally significant at the 0.05 level, which in nine tests may be attributable to Type I statistical error. The evidence for a caging effect on mean snail mass was more suggestive; four of nine tests were significant at the nominal 0.05 level. The *Valvata tricarinata* inhabiting caged sites appeared to have grown a little larger, and the *Amnicola limosa* may not have grown quite as much. Young-of-the-year *Helisoma* seemed to have grown larger inside the cages, although not *Helisoma* adults, a result that would depend on Osenberg's ability to distinguish between the two cohorts. Adult *Physa* seemed to have grown larger in the cages, although apparently not a cohort of *Physa* born during the experiment itself. Taken together, these results contain some evidence of a predator effect on snail growth rates, but little on density. The effect of predation on the growth and life history of freshwater molluscs is yet another subject to which we shall return.

The results of Bronmark and colleagues (1992) stand something in contrast with those of Osenberg. Bronmark set ten $3 \times 3 \times 2$ m cages in each of two Wisconsin lakes, one with a high natural density of pumpkinseeds (Mann Lake) and one characterized by a low natural density (Round Lake). *Amnicola limosa* and *Marstonia lustrica* were listed as the two 'dominant' snails in these lakes, and *Physa integra* mentioned in passing, but data by snail species were not collected. Cage meshes were very fine (0.5 mm) and would be expected to limit passage of even juvenile snails, as well as a wide variety of their predators. Into half their cages Bronmark and colleagues introduced three pumpkinseeds. Snails and periphyton in all cages were sampled through two successive summers, along with uncaged controls, and the contrasts in snail dry biomass between enclosure–exclosure–control blocks were examined for evidence of predation effects.

Bronmark and colleagues predicted that snail densities in the cages with pumpkinseeds enclosed should mimic Mann Lake controls, with fishless exclosures showing significantly higher densities. At Round Lake, the authors predicted that controls would match exclosures in snail density, and that cages with fish enclosed should show depressed snail densities. In fact all contrasts showed significant differences. Exclosures generally showed the highest snail biomass in both lakes throughout the study, enclosures the lowest biomass, and controls intermediate. This would seem to demonstrate both the potential of pumpkinseed sunfish to substantially impact snail populations, and the possibility that caging affords protection against other (perhaps invertebrate) predators as well.

Bronmark and colleagues reconciled their results with those of Osenberg by noting that Osenberg's study was of shorter duration, and the density of pumpkinseeds lower. But in general, the results of cage experiments designed to test the importance of predation to freshwater mollusc populations have in fact been quite mixed. Predation effects were weak in a smaller exclosure experiment Bronmark (1988) performed in a 1 ha pond in southern Sweden. His 1988 study site was inhabited by several species of waterfowl, roach, perch, tench and 12 gastropod species. Bronmark erected $1 \times 1 \times 1.5$ m exclosures on ten nearshore plots selected to vary in their macrophyte cover and caged each with 1.5 mm mesh. After three months he sampled each exclosure exhaustively, as well as an adjacent control plot of identical size. He recorded the abundance of each gastropod species and the dry weight of the three common macrophytes. Setting aside one pair of sites that had been lost from the study, Bronmark detected a significant correlation between overall snail density (per m^2) and abundance of macrophytes (per m^3) over $N = 18$ sites. As a consequence of the macrophyte correlation, the author felt 'the effects of predator exclosures on gastropod density could not be analysed straight forwardly'. But an indirect test suggested to Bronmark that the density of gastropods might be significantly higher in his exclosures than predicted from their macrophyte densities, implying evidence of a predator effect.

I re-examined Bronmark's (1988) data using analysis of covariance, with total snail density the independent variable, cage or no-cage the treatment, and macrophyte density the covariate. For consistency I expressed both snail and macrophyte density on a per m^2 basis. The multiple correlation was low (0.34) and none of the effects appeared significant ($P = 0.22$ for cage, $P = 0.79$ for macrophytes, and $P = 0.23$ for their interaction). Thus with a more direct analysis I was unable to confirm Bronmark's impression that total snail density in his small Swedish pond may be impacted by predation.

Martin et al. (1992) erected three blocks of four large (3 m × 4 m) cages in a 15 ha Tennessee pond. Cages received large redear sunfish, small redear sunfish, both, or neither, 'consistent with densities in the littoral zone' of the pond. Over 16 months, the negative impact of sunfish, at any size or density, upon enclosed snail populations was striking. While the combined 'dry mass' of Helisoma anceps, Physa heterostropha, and Gyraulus parvus rose to 3 g/m^2 each summer in the fishless exclosures, snail biomass in all other treatments remained quite small. Interestingly, periphyton coverage was much lower, and macrophyte biomass much higher, in the fishless, high-snail cages. It will be recalled from Chapter 3 that authors have often reported the effects of freshwater snails upon macrophytes in other situations to be markedly negative.

Par Pond, a South Carolina reservoir receiving heated water from a nuclear facility, is inhabited by high densities of Amnicola, Gyraulus, Helisoma anceps, H. trivolvis, Menetus, Physa, pisidiids and unionids. Thorp and Bergey (1981) located 36 exclosures of 4 m^2 in six locations on the shallow margins of the pond, covering each with 3 mm mesh. They compared the benthic faunas in their exclosures with a set of 36 adjacent control plots over three intervals: September to December, January to April, and May to August. Among all localities, temperatures and seasons,

Thorp and Bergey recorded striking depressions of mollusc density within their exclosures. They felt that many of the snails may have escaped sampling by migration to the mesh sides of their cages. But in any case, these data constitute no evidence of an effect of vertebrate predation on the molluscs of Par Pond.

The several studies that have specifically addressed vertebrate predation effects on freshwater bivalves have yielded equally mixed results. Walter and Kuiper (1978) were able to detect increased densities of *Pisidium* in exclosure cages they constructed in Lake Zurich. But Gilliam *et al.* (1989) found the densities of enclosed pisidiid populations unaffected by their manipulations of creek chub (*Semotilus*) densities.

Working with the *Corbicula* population of Fairfield Reservoir in Texas, Robinson and Wellborn (1988) reported some of the more striking effects on freshwater mollusc density thus far demonstrated by the use of predator exclosures. Their cages were unusually large (2×4 m) and constructed with unusually coarse mesh (12.7 mm). The authors also went to the unusual lengths of keeping aquatic vegetation constant across all sites, either by periodical removal or (later) by the uniform addition of 'artificial plants'. They set pairs of cages (one enclosed, one open control) at six sites in the main reservoir and six sites in a connecting cooling pond, taking Eckman grab samples from each six times over the course of a single year.

Over all sites and dates, Robinson and Wellborn counted 2605 *Corbicula* in samples from enclosed cages, but only 93 from similar-sized samples in open controls. As might be expected, however, clams were quite patchy in their distribution. The authors performed separate analyses of variance for each date, separately for the main reservoir and cooling pond. About half of these returned significant predation effects, while all showed significant variation among the six sites and significant predator $\times$ site interaction. Table 7.2 compares caged and control *Corbicula* densities in the main reservoir during the first three sample periods. It appears that several of these exclosures were, by chance, very lightly colonized during spring and summer reproduction. Exclosures number five and six, on the other hand, may have initially been colonized by densities of *Corbicula* too great to support. Thus the variance among treatment cages was of such magnitude as to obscure their difference with the open sites during May or July. Only in the September data were the predation effects demonstrably significant.

Clearly the density of the prey as well as the density of their predators will impact the ability of researchers to detect predation effects. If prey

Table 7.2. *The average number of* Corbicula *in 0.053 m² Ekman grab samples taken from six predator exclosures located in the main Fairfield Reservoir, Texas, compared with adjacent unmeshed controls*

Date	1	2	3	4	5	6
Open controls						
May	0.0	0.0	0.0	1.0	1.5	1.0
July	0.5	1.5	0.0	1.5	0.0	1.0
Sept	1.0	1.0	0.0	1.0	2.0	1.0
Exclosures						
May	3.5	1.0	19.0	1.0	71.0	246.5
July	0.5	4.5	12.5	0.0	87.5	10.0
Sept	6.5	5.0	9.0	8.5	56.0	50.5

Note:
Only the September difference is significant.
Sources: data from Robinson and Wellborn (1988).

densities and predator effects are temporally variable and spatially patchy, a researcher's difficulties are compounded. Where macrophytes add habitat structure to some sites on some dates but not to others, where mollusc populations reproduce, migrate and colonize, and where invertebrate predators and parasites complicate matters, it is a wonder that any significant effects of exclosure have been detected under any circumstances at all. The demonstration of such effects in the six studies cited in Table 7.1 stands as testimony to the potential for predation to regulate freshwater mollusc populations.

Crustaceans

Crayfish are omnivorous; their diet in the wild generally matches the availability of plants and animals (living and dead) in their immediate environment. If their immediate environment contains freshwater snails, crayfish seem to manifest a taste for them. Covich (1977) wild-collected ten *Procambarus acutus* from Massachusetts ponds and offered them *Physa gyrina*, leaves from *Elodea*, or both. In 60 one-hour trials, the crayfish never ate fewer than five (of ten) *Physa* offered, and often ate the entire lot. No individual crayfish preferred *Elodea* in mixed offerings, but several seem to have strongly preferred snails. Hanson *et al.* (1990) demonstrated a strong inverse relationship between the density of *Orconectes virilis* and that of snails (*Stagnicola elodes* and *Physa gyrina*) in small, artificially stocked ponds. Interestingly, the effect of crayfish density upon macrophyte growth was stimulatory in some circumstances, perhaps because the crayfish tended to reduce snail grazing pressure (Chambers *et al*. 1990).

Crayfish species may differ in their threat to snail populations. *Orconectes rusticus* consumed an average of about 150 *Amnicola lustrica* per day in the laboratory feeding trials of Olsen *et al*. (1991), significantly greater than the numbers consumed by similar-sized *O. virilis* (about 90 snails/day) or *O. propinquus* (about 60 snails/day). The large North American *Procambarus clarkii* has been implicated in the disappearance of *Biomphalaria* and *Bulinus* from regions of Kenya (Harper *et al*. 1990, Hofkin *et al*. 1991, 1992).

A number of laboratory feeding experiments have been directed toward understanding the potential impact of crayfish predation on populations of nuisance bivalves (Love and Savino 1993, Martin and Corkum 1994). *Orconectes limosus* are commonly observed eating *Dreissena* from steelon netting used to protect various works in Poland. Piesik (1974) found that crayfish in the 34–47 mm size range consumed 20–40 mussels per day, modally of the smallest (1–3 mm) size class. A 67 mm crayfish ate 48 mussels/day, modally of the 4–5 mm class, and a 90 mm crayfish ate 72 mussels/day, modally 6–7 mm. Most of Piesik's experiments reflected a much greater consumption of mussels by female crayfish than by males, a result he attributed to the transiently poor condition of males in mating season. MacIsaac (1994) performed similar experiments using larger mussels and *Orconetes propinquus* from the North American Great Lakes. Feeding rates were a more modest 4–5 mussels/day in his 3–5 mm size class, reduced to about a single mussel per day in size classes 8–14 mm. Crayfish size and sex were again important. MacIsaac found that predation rate upon (especially large) mussels could be reduced by the inclusion of alternative macrophyte foods, but that the crayfish continued to prefer mussels. Covich *et al*. (1981) reported that *Procambarus clarkii* consumed 11 *Corbicula* per day, in the size range 4–6 mm. Although *Procambarus* seemed unable to open *Corbicula* larger than 6 mm, *Cambarus bartonii* successfully attacked clams up to 9 mm.

Some of the increase in mollusc diversity observed in the exclosure experiments reviewed in the previous section (Table 7.1) may be attributable to reduced crayfish predation, particularly where the mesh size employed was fine. Enclosure/exclosure experiments involving the introduced *Orconectes rusticus* suggest that the crayfish may significantly reduce macrophyte and snail abundance in northern Wisconsin lakes

(Lodge and Lorman 1987), and snail and zebra mussel abundance in Michigan streams (Perry *et al.* 1997). The quantity of periphyton also seems to be negatively related to crayfish abundance, since periphyton is dependent on macrophyte surface area, but its quality (as chlorophyll a) increases as snail grazing pressure is reduced (Lodge *et al.* 1994).

Insects

Large predaceous hemipterans are fairly common in fresh waters, those hunting in the benthos constituting a natural threat to whatever molluscs they encounter. The North American water bug *Belostoma flumineum* (often exceeding 20 mm in length) seizes pulmonate snails between its front tibia and femur, plunges its beak into the exposed foot, and sucks all tissues cleanly from the shell. Crowl and Alexander (1989) reported laboratory predation rates of 3.5 *Physa*/day for male *Belostoma* encumbered with eggs, up to 5.5 *Physa*/day for unencumbered females. The laboratory observations of Kesler and Munns (1989) suggested 1.8 *Physa*/day for 4.5 mm *Belostoma*, up to 4.5 *Physa*/day for 18 mm bugs. Kesler and Munns cited other reports ranging from 4.1 small *Physa* and *Gyraulus*/day for 12–15 mm *Belostoma* to 0.3 *Physa*/day for the smaller belostomatid *Abedus*.

There is good reason to expect natural predation rates to be much lower than those observed in the laboratory. Snails will be more difficult to find in the wild, and other foodstuffs will be available to the bugs. Kesler and Munns implanted four pairs of plastic cans (40 cm diameter) in the grassy bottom of Peckham Pond, Kingston, Rhode Island. Into all eight they introduced four *Pseudosuccinea columella*, and into one of each pair they introduced a single adult *Belostoma*. (The enclosures also contained natural densities of *Pseudosuccinea* and *Physa vernalis*, as well as an occasional resident *Belostoma*.) After nine days the authors thoroughly censused the contents of all cans for bugs and freshly dead shells, concluding that a more natural estimate of *Belostoma* predation rate might be about 0.5 snails/bug/day. The authors could detect no predator preference for either snail species.

Kesler and Munns monitored the density of *Belostoma* in Peckham Pond, along with the densities of its two pulmonate prey *Pseudosuccinea* and *Physa*, biweekly over four years. Samples were taken principally by dip net along a 22 m transect. The authors felt that their method systematically underestimated *Belostoma* density, since the insects 'could dive below the volume swept by the net'. I should think the underestimate of snail density would be worse, as all snails inhabiting the bottom sediments

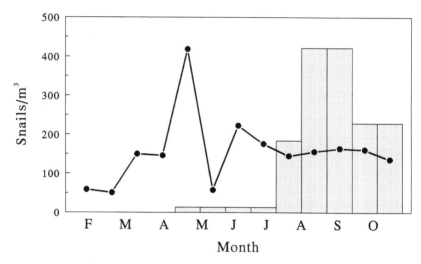

Month

Figure 7.6. The line shows the combined density of *Physa* and *Pseudosuccinea* (all age classes) in Peckham Pond, Rhode Island, averaged over 1982–84 and 1986 . Bars indicate the potential snail consumption rate (per m³) of the *Belostoma* population standing in 1983. (Data of Kesler and Munns 1989.)

will pass underneath a dip net as well. In any case, the *Physa* population appeared to be annual and semelparous, adults reproducing and dying in June. Two semelparous generations of *Pseudosuccinea* pass per year in Peckham Pond, one born in early May and reproducing in late July, the other born in July and reproducing in May.

Figure 7.6 shows the combined densities of *Physa* and *Pseudosuccinea*, over all age classes, averaged over four years of observations. The early May peak reflects the birth of the summer *Pseudosuccinea* cohort, added to overwintering *Physa*. The late May decline in snail density is due to the initially high *Pseudosuccinea* mortality, as the *Physa* population is laying eggs and expiring. Water bugs seemed to be rare during the early summer; Kesler and Munns recorded none at all in May and early July 1983, and only 1.3/m³ in June. But the *Belostoma* density increased dramatically in late July to 41 bugs/m³, and remained high in late August (about 15/m³), dropping back to zero in late October.

I derived the estimates of the '*Belostoma* demand' for snails shown in Figure 7.6 by integrating between the (only six) observations of *Belostoma* density provided by Kesler and Munns. The average bug density between each pair of observations was multiplied by the number of days in the interval and author's field predation estimate of 0.5 snails/bug/day. During the late summer and autumn, it appears that *Belostoma* could eat more snails than are present in Peckham Pond. Clearly snails are not the only prey item recognized by *Belostoma*. And it should immediately be noted that the snail densities are four-year averages, while the bug density is for 1983 only. Kesler and Munns state that the density of bugs in late July 1983 was exceptionally high. It is not clear why bug densities were not reported for other years, but one might expect a great deal of spatial and temporal variation in the density of such a mobile predator. (Discussion of habitat refuges in Peckham Pond will be postponed until later in the chapter.) In any case, the data of Kesler and Munns suggest that, at least occasionally, predaceous water bugs have the potential to exert substantial pressure on mollusc populations.

Diving beetles of the family Dytiscidae, numbering about 3000 species, are common inhabitants of ditches, lakes, and quiet streams worldwide. They are voracious predators, typically including on their menu all manner of aquatic insects and crustaceans, tadpoles and fish fry much larger than themselves. Their benthic larvae (often called 'water tigers') may pose an even greater threat to freshwater snails than the adult beetles, which primarily hunt in the water column. I am aware of several anecdotal reports of dytiscid predation upon freshwater snails (Bequaert 1925, Michelson 1957) but no formal studies.

There are about 1900 species in the coleopteran family Lampyridae (fireflies, lightning bugs), distributed worldwide. Larvae typically burrow in moist soils and are predatory upon soft-bodied invertebrates. Species adapted to especially muddy substrates along stream and pond banks may pose some threat to more amphibious freshwater snails. For example, lampyrids common in central New York include *Photinus consanguineus*, *P. scintillans*, *Photurus pennsylvanica*, and *Pyropyga fenestralis* (Hess 1920). *Photurus* larvae roam actively at night through moist, grassy fields, and the two *Photinus* species are generally subterranean, perhaps primarily hunters of earthworms. But according to Hess, *Pyropyga* larvae 'can be found wandering about on the ground at the edge of streams apparently in search of food.' Hess reported that *Pyropyga* will eat snails in laboratory feeding trials, although whether the victims were aquatic or terrestrial is not clear.

The larvae of the Asian lampyrid *Luciola* are semi-aquatic, and have been investigated for their potential as biological control agents for *Oncomelania* and the lymnaeid hosts of livestock flukes (Michelson 1957).

Kondo and Tanaka (1989) demonstrated that the larvae of *L. lateralis* may be reared on both *Semisulcospira libertina* and newborn *Pomacea canaliculata* (up to 11 mm) with good growth and survivorship. Second, third, and fourth instar larvae consumed an average of 0.7, 2.3 and 3.2 snails/day. *Luciola lateralis* was among four lampyrid species released in the Hawaiian islands during the 1950s, in unsuccessful efforts to control both *Lymnaea ollula* and the giant African land snail *Achatina*, an agricultural pest (Chock *et al.* 1961).

Dipteran flies of the family Sciomyzidae ('marsh flies') have also been the focus of considerable interest as possible biological control agents for pulmonates of medical and veterinary importance. Berg (1964, 1973) and his co-workers examined the life histories of a large fraction of the approximately 600 species of sciomyzids worldwide, concluding that the great majority are, as larvae, strictly predatory on freshwater snails. Adults typically lay eggs on emergent vegetation standing in very still, shallow water. The larvae (1 mm to 10 mm) have gut air-bubbles and hang from the surface film by posterior tufts of hydrophobic hairs, occasionally fastening onto emergent vegetation or crawling over moist soils at the water's edge. They sink their mouth hooks into the exposed foot of any snail they may encounter, including snails substantially larger than themselves. Prey typically draw the fly larvae into their shells by muscular contraction, where the larvae engorge themselves. In the laboratory, a single sciomyzid larva commonly ingests more than a dozen snails before forming a floating puparium.

An interesting exception is provided by the European and North American sciomyzid genus *Renocera*, whose newly hatched larvae swim to the bottom hunting pisidiid clams (Foote 1976). They pass their first instar entirely submerged, eating one or two clams. Second- and third-instar *Renocera* larvae hang inverted from the surface film, as usual, actively probing the bottoms of shallow pools for the 11–13 pisidiid clams they will need prior to pupation. Another class of exceptions are the sciomyzids, especially those relying on amphibious or terrestrial prey, that complete their larval development within a single snail. These are more accurately described as parasitoids, rather than predators.

Most sciomyzid larvae maintain contact with the atmosphere at all times. They do not typically dive to attack snails in the water below their grasp, and generally suspend their dying prey from the surface film while they feed. Eckblad (1973b) examined the effects of predator density, prey density, and water depth on the feeding rate of the North American sciomyzid *Sepedon fuscipennis*. Normal feeding trials were run at a water depth of 5 mm, under which conditions the larvae consumed averages of 14 to 24 small (2–4.5 mm) *Lymnaea palustris*, depending on densities. But both larval survival and kill rate were adversely affected when depth in the experimental chambers was increased to 50 mm. In field cages, larval survival halved when water depth was increased from 5 mm to 100 mm.

Sciomyzids occasionally manifest marked prey preferences (Manguin *et al.* 1988a,b). Neff (1964) exposed 13 species of medically important snails (*Biomphalaria*, *Bulinus*, and *Oncomelania*) to 10 species of sciomyzids. All larvae seemed to be able to attack almost all the pulmonates successfully. But the larger *Bulinus* seemed to repel or even suffocate smaller sciomyzids with copious mucus in some circumstances. Neff noted that in rapidly withdrawing from the initial bite, *Oncomelania* often pinched or injured its sciomyzid attacker with its operculum.

During the late 1950s and early 1960s, sciomyzids were widely introduced in efforts to control the lymnaeid intermediate hosts of the livestock fluke, *Fasciola*. The Central American *Sepedon macropus* did in fact become established on four of the Hawaiian Islands, and was subsequently observed eating *Lymnaea ollula* (Chock *et al.* 1961, Berg 1964). *Sepedon* was also introduced to Guam, although its fate is unknown. Researchers were unsuccessful in establishing a breeding population of the European *Pherbellia* (or *Sciomyza*) *dorsata* into Hawaii (Davis *et al.* 1961), nor (apparently) was the North American *Sepedon praemiosa* able to adapt to Australia (Berg 1964). It is unfortunate that no data seem to have been gathered regarding fly or snail population densities in any of these cases, either before or after introduction.

In addition to his laboratory studies, Eckblad (1973b) performed an enclosure/exclosure experiment similar to those more commonly attempted in studies of vertebrate predation. He erected 24 quadrats (0.6 m × 4.6 m) on the dry margins of Bool's Backwater in Ithaca, New York, enclosing 18 with very fine (110 μm) mesh. He then closed a small dam to flood the study area, and stocked his quadrats with natural densities of *Lymnaea palustris*, *Physa gyrina*, and *Gyraulus parvus*. The tops of six quadrats were left open to allow natural colonization by egg laying sciomyzids, six were enclosed entirely to prevent any sciomyzid predation, and six were enclosed and stocked with artificially reared *Sepedon* larvae above natural densities. Eckblad used an Ekman grab to sample snail densities in each quadrat on eight occasions from June through September.

Although very carefully designed and executed, Eckblad's experiments did not reveal especially convincing effects of sciomyzid predation of freshwater snail populations. No predation effect was detected on *Physa* or *Gyraulus* densities at any sampling date, possibly because these snails inhabited water too deep to be accessible to sciomyzid larvae. Analysis of variance did detect a predation effect on overall density of *Lymnaea* at one of the eight dates (14 July), as well as an effect on the density of small (less than 4.5 mm) *Lymnaea* on 29 July. These two results might at first appear spurious, seen together with the large number of non-significant results Eckblad also obtained. But only early in his experiments (6/28 and 7/13) was Eckblad especially successful in maintaining increased densities of sciomyzids in his high-predation treatment pens. There may also have been caging effects. Data are unavailable on the six unenclosed quadrats, but snail populations declined steeply through most of the summer in all three enclosed treatments. Eckblad concluded, in any case, that 'the overall effect of the predator on the *L. palustris* population by the end of the summer, either in reducing population density or changing its size–frequency distribution, appeared to have been minimal.'

Parashar and Rao (1988) used Ochterlony immunodiffusion tests in an effort to identify potential predators of the medically important *Indoplanorbis exustus*. They induced antibody production in mice by a series of injections with *Indoplanorbis* extract, then screened a variety of predatory insects co-occurring with the snail for immunological reaction with mouse antiserum. They obtained positive results only with the flesh fly *Sarcophaga misera*, a member of a family of dipterans not known previously as molluscan predators (although other sarcophagids have long been known as parasites or parasitoids of land snails; Bequaert 1925). Laboratory feeding experiments verified that larval *S. misera* killed *Indoplanorbis*, at an average rate of 6 snails/day. Parashar and Rao also reared *S. misera* on an artificial diet, repeated their Ochterlony test, and found no immunological reaction with mouse antisera, further confirming that *S. misera* is a natural predator of *Indoplanorbis*. One wonders how many more freshwater mollusc predators remain unrecognized in the vast diversity of the Class Insecta, if a fly from such a typically saprophagic and terrestrial family as the Sarcophagidae is a participant.

Molluscs

The strikingly negative impact that populations of large ampullariid snails may have upon pulmonates was reviewed in Chapter 5. They are prodigious consumers of macrophytic vegetation, and as such, their relationship with pulmonates may be described as competitive. There is evidence, however, that their impact as predators may not be inconsequential.

On many mornings, Demian and Lufty (1965a) arrived at their lab to find the bottom of their *Marisa* tank strewn with the shells of adult *Bulinus*. I quote the grisly details: 'The remains of the soft parts left in the shells showed no signs of decomposition, and when they were cut or injured by the scissors during dissection red blood frequently oozed out, indicating that death had occurred only a short time previously . . . Each day the aquaria were cleared of dead snails and empty shells, but a new crop of victims was found the following morning.' Suspecting foul play, Demian and Lufty posted a watch, and with patience it was discovered that *Marisa* will in fact prey upon *Bulinus* of all sizes nocturnally. *Marisa* do not seem to seek prey actively, but upon contacting a *Bulinus*, they sieze the shell of the smaller snail, invert it, and bite out such soft parts as they can reach. Very young snails are eaten whole. The rate of predation was highest in the absence of other food, but some predatory activity occurred regardless. Demian even observed his *Marisa* eating *Bulinus* eggs from *Elodea* leaves, leaving the *Elodea* untouched.

Lymnaea may also be included in the diet of *Marisa*, but *Biomphalaria* over about 5 mm are protected by the narrow aperture of their shells. Chi *et al.* (1971) extended the known diet of *Marisa* to include the mud-covered eggs of another medically important snail, *Oncomelania*, and Paulinyi and Paulini (1972) demonstrated that *Pomacea* will also eat *Biomphalaria* eggs. Indeed, the very large *Pomacea canaliculata* has been observed gnawing through the shell of adult (13 mm) *Biomphalaria* to consume the inhabitant (Cazzaniga 1990).

However, subsequent studies of *Marisa* by Demian and Lufty (1965b, 1966) showed that predatory behaviour seems to be learned; snails from the wild seem to be much less bloodthirsty. Cedeno-Leon and Thomas (1983) reported that such behaviour seems to be highly plastic, increasing with age, maturity, and (surprisingly) the presence of plant foods. Some individual *Marisa* do not seem to eat pulmonate eggs under any circumstances, even if previously conditioned by culture with *Biomphalaria*. Cedeno-Leon and Thomas concluded that *Marisa* might be effective as a predator of *Biomphalaria* only at high densities. That ampullarids have reliably eliminated *Biomphalaria* in Puerto Rican field trials suggests competition as the greater component of the interaction between these two groups.

Leeches

The leeches are a group of about 400 carnivorous annelid species completely oblivious to prey size. They attack the entire range of animal life inhabiting fresh waters. If the victim is large relative to the leech itself, observers generally describe the phenomenon as parasitism, otherwise it is predation; the leech itself seems to be unconcerned. Among the best-studied enemies of freshwater molluscs are 'snail leeches' of the family Glossiphoniidae, small animals with a muscular proboscis they can thrust into the tissues of their prey. Jawed leeches of the family Hirudinidae may also attack molluscs, as may erpobdellid leeches, which are characterized by a muscular, sucking pharynx (Klemm 1975). Even placobdellid leeches, generally considered ectoparasites of turtles, may feed upon molluscs (Moser and Willis 1994).

Among the earliest detailed descriptions of leech predation upon snails were derived from the laboratory observations of Chernin *et al.* (1956b) involving *Biomphalaria* and the glossiphoniid *Helobdella fusca* (a culture originating from New York). Adult leeches (about 10 mm) typically attach first to the shell of their victim with their posterior sucker, then probe exposed parts of the snail's body, ultimately inserting their proboscis. The leech then transfers its attachment from shell to substrate, and ingests the snail's tissues and body fluids while holding the shell suspended in the water column. Chernin was able to demonstrate predation of adult *H. fusca* on juvenile *Biomphalaria* up to about 5–6 mm shell diameter. Larger snails were by no means safe from attack, however. *Helobdella* broods its young (as do glossiphoniid leeches generally), and when their mother attacks a prey item, the juveniles rush forward on their mother's back to join the meal. Chernin reported that juvenile leeches may under some circumstances crawl entirely inside larger (12 mm) *Biomphalaria* and take up at least temporary existence as parasites.

As leeches typically ingest only liquefied meals, researchers have found serological techniques indispensable for the analysis of their gut content in the wild. Wrona *et al.* (1979) developed antisera against four putative prey groups for *Glossiphonia complanata* in Alberta, Canada: chironomids, oligochaetes, amphipods, and gastropods. While a locally common *Lymnaea* served as antigen for Wrona's approach, the resulting serum cross-reacted to *Physa* and *Helisoma* as well. Wrona tested for reactions in the stomachs of approximately 150 *G. complanata* collected through the course of a year at each of two sites near Calgary. Gastropods were in both cases the most common food item, at all times of the year except January and February. Across all months, mollusc reactions were detected in 42 of the 95 *G. complanata* guts testing positive for food of any sort at a temporary pond, and in 64 of 113 positive guts at a lake.

Glossiphonia complanata is among the more common macroinvertebrate predators (of any sort) inhabiting temperate fresh waters. Young (1981a) performed a serological study of *G. complanata* diet in Britain similar in many respects to that of Wrona in Canada. He tested each leech with a larger battery of antisera (ten in all) including an 'anti-mollusc' developed from a mixture of *Lymnaea peregra* and *Potamopyrgus jenkinsi*. His primary interest was in the difference between leech diets in unproductive lakes of northern Wales and productive lakes of Cheshire and Shropshire, England. Molluscs were once again the prevalent dietary item identified in the eight productive lakes, accounting for 31–58% of all positive reactions. Oligochaetes and chironomids were also commonly ingested, while the contributions of the seven other prey types were negligible. Molluscs did not, however, seem to figure heavily in the diets of *G. complanata* inhabiting the five unproductive lakes surveyed by Young, possibly because molluscan prey were unavailable.

Certainly among the most thorough studies focusing on the parasitic character of the snail–leech association was that of Klemm (1975). He took monthly samples of snails and pisidiids at ten sites in southern Michigan over the course of two years, carefully examining each individual snail or bivalve for leeches, both external and internal. Ultimately he discovered leech parasitism in 14 species of pulmonates, plus *Sphaerium simile*. Large *Helisoma* seemed especially vulnerable, small planorbids and amphibious lymnaeids somewhat less so, and seven prosobranch species appeared entirely invulnerable. Five glossiphoniid species were responsible: three *Helobdella*, *Glossiphonia heteroclita*, and *Marvinmeyeria*. *Glossiphonia complanata* was not among the leeches Klemm found parasitizing southern Michigan molluscs, but he verified its status as a mollusc predator with immunological tests. Klemm could not verify the molluscivory of *Hellobdella stagnalis* in southern Michigan, either as parasite or as predator, although Young (1980) subsequently demonstrated that *H. stagnalis* occasionally consumes molluscs in England.

Klemm's data set on the *Helisoma trivolvis* population of Bert Pond was especially large (Figure 7.7). Over all dates, 328 of 885 snails collected contained one or more leeches, for a summed prevalence of 37%. Quite a few snails were multiply parasitized, with 74 hosting two leeches, 23 hosting three, etc., up to one snail which contained 12 leeches. Klemm identified two species of leech among the 523 leeches recovered: 379 *Helobdella fusca* and 144 *H. lineata*. He found most individuals of both leech species between the snail's shell and mantle, although quite a few *H. fusca* occupied the kidney. Figure 7.7 indicates that the (combined) abundance of leeches in snails generally seemed lowest in the spring (especially May) when the new snail generation was born. Abundance

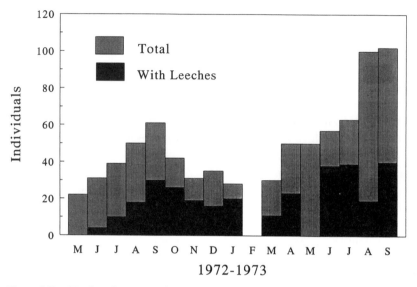

Figure 7.7. Total *Helisoma trivolvis* collected in 45-minute searches conducted at Bert Pond, southern Michigan, by Klemm (1975). The darkened portion of each bar shows those hosting internal leeches.

seemed rather constant throughout other seasons, however, even as snail population densities rose. I am unaware of any direct studies of the impact leech predation or parasitism may have upon molluscan populations, but it is difficult to discern evidence for a strong effect on the *Helisoma* population of Bert Pond.

Triclads

Young (1981b) extended his serological survey to include the gut contents of predatory triclads, as well as those of leeches. As in his earlier work (1980, 1981a), he sampled both productive English lakes and unproductive Welsh ones over the course of a year, testing the four triclad and three leech species he discovered with antisera to ten potential prey types. Summing over all dates and lakes, the triclad *Dugesia polychroa* proved to be a more specialized mollusc predator than *Glossiphonia complanata*. Young's samples of *D. polychroa* ranged from $N = 49$ to $N = 244$ over seven productive lakes, with from 70% to 91% testing positive for at least one food item. No less than 52% of all positive reactions were to Young's mollusc antiserum, as against 6–14% for his *Asellus* antiserum, the runner-up. The diet of *Polycoelis* (two species, *P. nigra* and *P. tenuis*) recalls that of *Helobdella stagnalis*, occasionally including molluscs but evincing a preference for oligochaetes. The two final species in Young's survey, the triclad *Dendrocoelum lacteum* and the macrophagous leech *Erpobdella octoculata*, did not seem to consume molluscs.

The studies of triclad gut content authored by Reynoldson and Piearce (1979a,b) are interesting from our perspective because they involved four separate mollusc antisera. The authors' anti-*Potamopyrgus* and anti-*Bithynia* sera were entirely specific, their anti-*Planorbis* serum more broadly reactive to four planorbids, and their anti-*Lymnaea* cross-reacting only slightly with *Physa*. (One wonders what fraction of positives might be missed with a general 'anti-mollusc' serum such as Young's, even if of complex origin.) Reynoldson and Piearce sampled four Shropshire meres and one Scottish loch over 6–24 months, collecting *Dugesia polychroa*, the very closely related *D. lugubris*, *Polycoelis tenuis*, and the (possibly exotic) *Planaria torva*. Three Welsh ponds were added for *Polycoelis nigra*. The diets of the two *Dugesia* were found to be indistinguishable with respect to their gastropod prey. Both triclads commonly seemed to ingest all four snail species in similar proportions, the overall percentages positive ranging from 38% to 71% across five lakes. At Cole Mere, however, *P. torva* appeared to prefer *Potamopyrgus* over the other three prey classes available. Again the two *Polycoelis* appeared less dangerous to molluscs than *Dugesia*, anti-snail serum of any sort accounting for only 2–14% of all positive reactions over eight sites.

Defences

Behaviour

Threatened with attack, the behavioural options available to most animals fall into the two categories, fight or flight. Snails differ from humans in this regard only by matters of scale; 'resist or retreat' may perhaps be more descriptive. In the 'resistance' category, Townsend and McCarthy (1980) have described a striking behaviour displayed by *Physa fontinalis* upon contact of its mantle with a small section of leech tissue (anterior sucker plus four segments). Typical responses recorded include vigorous shell shaking, followed by detachment of the foot from the substratum. Since the mantle cavity often happens to enclose an air bubble of some size at the time of leech contact, threatened snails commonly float to the surface. Alternatively, shell shaking occasionally spills the air bubble held by snails at the surface, causing them to sink. Such rising or falling through the water column are better counted as the consequences of 'resistance' than as the consequences of 'retreat', which in snails might be reserved to describe active, foot-to-the-substrate movement.

Townsend and McCarthy tested 30 *Physa* for their responses to each of 15 leech species (four families) and eight flatworm families (two families). Almost all 23 of these test predators, including large blood-sucking vertebrate parasites (such as *Hirudo medicinalis*) not considered especially dangerous to snails, at least occasionally elicited shell shaking from the snail contacted. But known snail predators, such as *Glossiphonia complanata*, elicited strong reactions from 100% of the snails tested, while the weakest response was evoked by *Erpobdella octoculata*, which is generally considered harmless to snails.

Bronmark and Malmqvist (1986) recorded similar resistance responses to *G. complanata* attack in *Lymnaea peregra* and *Planorbis planorbis*, although 'the reaction was not as vigorous as in *P. fontinalis*'. Two prosobranchs tested (*Bithynia tentaculata* and *Theodoxus fluviatilis*) simply withdrew into their shells, pulling their opercula behind them. It seemed to Bronmark and Malmqvist that the little limpet *Ancylus fluviatilis* had no behavioural responses available to it, and was consequently much more likely to be eaten by *G. complanata*. Wilken and Appleton (1991) compared four pulmonates (*Physa acuta*, *Aplexa marmorata*, *Bulinus tropicus* and *Lymnaea natalensis*) in their behavioural response to direct leech contact, using methods similar to those of Townsend and McCarthy. Neither *Bulinus* nor *Lymnaea* displayed any responses different from that evoked by contact with a blunt dissecting needle. But especially if the leech contact was made about the head or mantle fringe, Wilken and Appleton confirmed distinctive shell shaking behaviour for their two physids. Juvenile *Physa* may be less likely to fall to leech predation than juvenile *Lymnaea* or *Helisoma* for similar reasons (Bronmark 1992).

In passing, it might be noted that the reaction evoked when a snail is poked with an artificial instrument is also, in its own simple way, a resistance behaviour. Pulmonates in particular seem likely to withdraw their heads and pull their shells tightly to the substrate. This reaction has been called 'clamping' (Brown and Strouse 1988), and we will return to it before this chapter is concluded.

Over 24 hours, Bronmark and Malmqvist (1986) noticed an inverse relationship between the activity of *G. complanata* and the activity of *Bithynia tentaculata* in combined containers. They used time-lapse video to record movement in seven vessels containing ten *Bithynia* and an uncaged leech, seven containing ten snails and a caged leech, and six controls with no leech. Snail activity in the caged-leech treatment did not differ from control, but was much reduced in the unrestricted-leech treatments. Leeches seem to hunt, and snails seem to respond, via tactile cues almost exclusively.

Our review of the behavioural response to predation has thus far focused primarily on physical contacts between specific predators and their molluscan prey. The responses we have catalogued as 'resistance' include shell shaking, enclosure, and clamping, but not as yet active movement away from the threat. We now turn to 'retreat', the behavioural response most commonly observed when threat is received as a chemical cue.

Among the simpler manifestations of 'retreat' would be avoidance. Haynes and Taylor (1984) included several experiments relevant to the present discussion in their larger investigations of food finding in *Potamopyrgus*. They introduced over 100 snails into the centre of a circular 'choice chamber', recording the number of individuals ultimately entering five radially arranged rooms. Snails significantly avoided rooms containing filter-paper disks soaked with water from a leech tank (perhaps not entirely incompatible with the observations of Bronmark and Malmqvist), or with extracts of squashed *Potamopyrgus*. McCollum and colleagues (1998) reared *Physa* in aquaria with redear sunfish separated by permeable partitions. Although a few *Physa* were fed to the sunfish daily, the fish was not allowed into the main aquarium itself. Nevertheless, the chemical cues emitted by the sunfish seem to have prompted the snails to spend much more time under leaves and in dark corners, eating less and reproducing less than control snails reared without sunfish neighbours.

Snyder (1967) constructed a flowing-water apparatus that allowed fairly rapid and uniform introduction of varying concentrations of filtered crushed-snail extract to a chamber containing ten previously equilibrated test snails. Ultimately his experiments involved six concentrations of extract, 17 pulmonate species (3 families), and 10 freshwater prosobranchs (6 families), as well as a few marine species. Snyder characterized 16 snail species as 'reactive', 7 as 'questionably' or 'inconsistently' reactive, and 4 as 'nonreactive'. Prosobranchs typically dropped from the tank walls (which we have classified as 'resistance') and buried themselves, a retreat response that seemed to be extinguished by darkness. Pulmonates either dropped to the bottom (resistance) or crawled out of the water entirely (retreat).

Snyder also performed two sets of interspecific tests, one comparing representatives of six different families, the other four different species of

Helisoma. Although no reaction was seen in most of the former experiments, *Pseudosuccinea* did react to extracts from both *Physa* and *Viviparus*, and *Pomacea* reacted to *Pseudosuccinea* extract. At least occasional alarm reaction was elicited from any *Helisoma* by extract from any congener.

One reads something of a sense of frustration in Snyder's account of his experiments. Even among his 'reactive' species at high concentrations of extract, responses were often not displayed by substantial fractions of the animals tested. Snyder and Snyder (1971) showed that snail age should be figured among the many variables influencing the various behavioural responses to predation threat. Newborn *Pomacea paludosa* do not react to crushed intraspecific extracts; some sensitivity seems to develop after a week or two. Interestingly, reactions to turtle odour are present in newborn *Pomacea*, becoming extinguished at different rates for two different turtle species. Snyder and Snyder reported that *Pomacea* lose their sensitivity to the odour of the musk turtle (*Sternotherus minor*) when snails reach 3 g wet weight, while retaining sensitivity to the odour of snapping turtle (*Chelydra serpentina*) up to 20 g. Snapping turtles are indeed able to crush much larger snails than are musk turtles.

Triggers for the behavioural reactions displayed by freshwater gastropods in response to threat are probably much more complex than typical human observers recognize. Alexander and Covich (1991a) performed a series of experiments on the 'crawl out' retreat response manifested by *Physa virgata* when threatened by the crayfish *Procambarus simulans*. *Physa* did not seem to leave the water (at a frequency greater than control expectation) when crushed conspecifics were added to their vessel, nor when a caged crayfish was added, nor even when a caged crayfish was allowed to feed on crushed *Biomphalaria*. Only in response to the combination of a crayfish and crushed *Physa* did a significant number of snails (a mean of perhaps 25%) crawl out. Crawl-out duration seemed to extend for about two hours.

Helisoma trivolvis larger than about 8–9 mm in shell diameter are effectively protected from attack by *Procambarus*, while smaller *Helisoma* are vulnerable. Alexander and Covich (1991b) demonstrated a crawl-out response in 4–6 mm *Helisoma* quite similar to that displayed by *Physa*, lesser responses in 6–12 mm *Helisoma*, and no crawl out in 12–16 mm animals. The survey experiments of Covich et al. (1994) demonstrated significant crawl-out responses in three *Physa* species and *Lymnaea emarginata*, but not in the planorbids *Helisoma anceps* or *Gyraulus parvus* or the prosobranchs *Amnicola limosa* or *Campeloma decisa* (which burrowed).

In sum, resistance and retreat in their various forms may both be among the responses manifested by freshwater snails to threat of attack. Resistance seems to occur in response to physical contact, and retreat in response to chemical cue, although both may occur in tandem under some situations (e.g. 'drop and burrow'). The triggers for these behaviours are complex, very much dependent upon the nature of the cue and the snail receiving it.

Shell

The protection afforded by the molluscan shell is of two effects. Most obviously, especially large and strong shells may protect their wearers from crushing and/or engulfment by powerful, hard-bodied predators. Second, even the thinnest and lightest layer of calcium carbonate will be sufficient to insulate its wearer from at least some environmental fluctuations, pathogenic micro-organisms, trematode cercaria, calanoid copepods (Liebig and van der Ploeg 1995), and such soft-bodied predators as leeches, triclads, and other snails. These two effects will be examined in turn.

Regarding the likelihood of crushing and/or engulfment, some selection by predators on the size of their prey seems to be universal. We have already touched upon the size preferences displayed by muskrats foraging on mussels (Hanson et al. 1989, Figure 7.1), cichlids offered Biomphalaria (Slootweg 1987, Figure 7.4) and crayfish offered Dreissena (Piesik 1974). Other clear demonstrations of size-selective predation include Keller and Ribi's (1993) work on fish predation of Viviparus, Dudgeon and Cheung Pui Shan's (1990) laboratory experiments with a freshwater crab and five Hong Kong gastropods, and de Leeuw and van Eerden's (1992) data on diving duck predation of Dreissena.

It may be recalled that Covich et al. (1981) found crayfish able to eat only the smallest Corbicula offered them, typically ranging to no more than 6 mm standard length. Kennedy and Blundon (1983) crushed $N = 70$ Corbicula of varying size in an industrial instrument designed to measure the pressure applied to a surface. The regression of (log) shell length against (log) crushing force was quite significant ($r^2 = 0.73$) and is converted to a semi-log scale in Figure 7.8 (line 'C'). Kennedy and Blundon were quite impressed with the strength of Corbicula shells; their resistance to crushing seems to be significantly better than any of eight estuarine bivalves previously examined with similar methods.

The dashed line running obliquely across the lower corner of Figure 7.8 roughly traces the crushing strength of an average-sized adult crayfish (Procambarus clarkii). Force data, from Brown et al. (1979), reflect

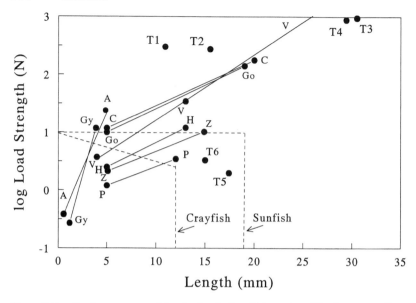

Figure 7.8. Resistance to crushing in the shells of 14 species of freshwater molluscs, as a function of maximum shell dimension ('length'). C, *Corbicula*, data of Kennedy and Blundon (1983). Go, *Goniobasis*, H, *Helisoma*, P, *Physa*, data of Stein *et al.* (1984). A, *Amnicola*, Gy, *Gyraulus*, V, *Viviparus*, data of Osenberg and Mittelbach (1989). Z, *Dreissena*, data of French and Morgan (1995). T1, T2, T3, T4, Lake Tanganyika thiarids; T5, T6, other thiarids, data of West *et al.* (1991). Dashed lines show approximate maximum size and strength crushable by the crayfish *Procambarus* (Brown *et al.* 1979) and the pumpkinseed sunfish (Osenberg and Mittelbach 1989).

decreasing strength but increasing gape from the base to the tip of the chiliped. The maximum gape shown, about 12 mm, is my own rough estimate. Figure 7.8 seems to indicate that even the smallest (5 mm) *Corbicula* should be invulnerable to crayfish attack. But in practice, Covich and colleagues report that crayfish typically chip at the edges of the clams with their mandibles, rather than their chilipeds, ultimately opening a hole large enough to consume the contents. The dashed lines shown in Figure 7.8 should be viewed as minima for refuge, not guarantees.

When they are offered snails with greatly different resistance to crushing, it is perhaps not surprising to observe that predatory sunfish consume the weakest. In laboratory choice tests, Stein *et al.* (1984) observed that redear sunfish slightly preferred *Physa integra* over *Helisoma* (two species) and that *Goniobasis* (='*Oxytrema*') *semicarinata* were either entirely ignored or rejected. Although no regression statistics were offered, approximations of the author's figures on crushing resistance as a function of shell size for these three snail genera are shown in Figure 7.8. *Goniobasis* (Go) seems to be indistinguishable from *Corbicula* in shell strength. *Helisoma* (H) is much weaker, and *Physa* (P) weaker still, recapitulating the preferences manifest by their fish predator.

Osenberg and Mittelbach (1989) performed tests similar to those of Stein *et al.*, involving 11 species of freshwater snails co-existing with pumpkinseed sunfish in three small Michigan lakes. The relationship between crushing resistance and shell length closely fit a semi-log linear model in all cases. I have selected data from *Gyraulus parvus* ($N = 20$, $r^2 = 0.93$), *Amnicola limosa* ($N = 43$, $r^2 = 0.92$) and *Viviparus georgianus* ($N = 20$, $r^2 = 0.82$) for inclusion on Figure 7.8 to reflect a range of shell morphology. (I have taken the liberty of extrapolating the *Viviparus* regression in Figure 7.8, as the authors tested only juveniles to 13 mm.) Although *Viviparus* seems comparable to most other species we have examined thus far, Osenberg and Mittelbach's data suggest that the smaller species demonstrate a much steeper increase in shell strength with increase in size than we have previously noted.

The figure also features approximate minimum boundaries for shell size and strength required for snails to obtain refuge from pumpkinseed sunfish. The 19 mm minimum size was calculated by substituting the length of the largest pumpkinseed examined into Osenberg and Mittelbach's Figure 7, solving for a 50% probability of escape, and converting snail mass to shell length using the regressions supplied. Minimum refuge in crush resistance (10 N) was estimated from the 50% survival rate in the author's Figure 8.

Clearly the prey populations will vary in their availability to the pumpkinseeds – larger snails escaping by virtue of the size and strength of their shells, smaller snails escaping notice. The authors predicted that pumpkinseeds ought to impose a 'hump-shaped' selection curve upon their prey population, with differing modes determined by the size of individual fish. But because of the rarity of large prey in the study lakes, the author's observations generally suggested that the likelihood that a snail was selected increased monotonically with size. The most important determinant of pumpkinseed diet in small Michigan lakes seems to have been the population dynamics of the snails upon which they feed.

Considerable attention has been directed toward the question of the vulnerability of *Dreissena* to predation by a variety of fishes. Nagelkerke and Sibbing (1996) reported that the maximum mussel size taken by common bream and white bream seems to be determined by their pharyngeal crushing power, but that the (somewhat greater) maximum size taken by roach is constrained by its oral gape. The common bream displayed a much lesser feeding efficiency on zebra mussels than either the white bream or roach. Redear sunfish can also feed on smaller *Dreissena* (French and Morgan 1995). When offered a choice among large *Helisoma* (12 mm), small *Helisoma* (8 mm) or zebra mussels (same sizes), adult sunfish always preferred the more energetically profitable *Helisoma*. French and Morgan's data on crushing resistance of *Helisoma* shells closely matched that previously published by Stein and colleagues (1984). Their data on the resistance of 5–15 mm *Dreissena* shells ($N = 50$, $r^2 = 0.59$) are added to Figure 7.8.

The thiarids of Lake Tanganyika are remarkable for more than their diversity of species and genera. The heavy calcification and armour of the fortresses carried by many of the denizens of that ancient lake remind all beholders of the seashells of the tropics (West and Cohen 1996). West *et al.* (1991) compared the crushing strength of six Tanganyika thiarids with that of five thiarids from elsewhere in Africa. All measurements were taken on about 25 adults fairly uniform in size, and tabulated as averages. The data from four Tanganyika species (T1 – *Spekia zonata*, T2 – *Lavigera* sp. A, T3 – *Paramelania imperialis*, T4 – *Paramelania crassigranulata*) and two more cosmopolitan species (T4 – *Potadoma ignobilis*, T6 – *Melanoides tuberculata*) are plotted in Figure 7.8. Although the latter two are unremarkable in shell strength, the four in the former category are indeed exceptional.

West and colleagues noted that predators, especially freshwater crabs of great size and strength, seem to have coevolved with the thiarids of Lake Tanganyika. In particular, the endemic crab *Platytelphusa armata* carries a crushing chiliped with broad, molariform dentition averaging about 6 cm long and opening perhaps 3 cm wide. No direct measures on the strength of this formidable armament are available, but the crabs do successfully crush *Lavigera* shells of 480 N average load strength, consuming the contents.

Entirely independent of their strength, shells provide protection from the generally softer and smaller predators that may invade. We have earlier mentioned the value of an operculum for the repulse of an invader, as witness Klemm's (1975) data on leech parasitism in Michigan prosobranchs and Neff's (1964) observations on *Oncomelania* under sciomyzid attack. But certain aspects of shell shape, particularly relative reduction of the shell aperture, seem to provide some protection from invasion even when an operculum is absent or ineffective.

Demian and Lufty (1966) performed a series of laboratory feeding trials involving four *Marisa* and 20 prey snails of varying size, either *Bulinus* or *Biomphalaria*, monitored over five days. Each day the authors inspected the tanks, counted surviving prey, and replaced any missing with snails of similar size. In a 24-hour period, four *Marisa* were generally able to eat most of the juvenile *Bulinus* offered, up to the largest size tested by Demian and Lufty (3–4 mm). Smaller prey were eaten whole, while larger snails were eaten from within their shells. But *Marisa* only seemed to prey effectively on *Biomphalaria* up to to 2–3 mm, the maximum size they could break or engulf whole. *Biomphalaria* as large as 5 mm in diameter seemed invulnerable to *Marisa* attack by virtue of both the strength of their shells and the narrowness of their apertures.

Snyder and Kale (1983) reported a similar phenomenon with an amusing turn-around. Here the predator is the snail kite and the prey, none other than *Marisa* itself. It will be recalled that the ordinary prey of the kite is *Pomacea*, possessing a globose shell with a wide aperture. But in Colombia, Snyder and Kale noted common attacks on *Marisa*, whose ability to pull its operculum deeply within a narrow aperture seems to afford considerable protection. Kites succeeded in about 92% of their extractions of *Pomacea*, but only 62% of *Marisa* extractions. As the birds often perch over water, dropped snails may be awarded a 'second chance'.

Freshwater molluscs are generally drab. Through most of its range, Israel's populations of *Theodoxus jordani* bear shells that are plain black. Yet on the shores of Lake Kenneret one will encounter populations of *Theodoxus* whose shells are decorated with genuinely festive stripes, checks, and zebra patterns. Heller (1979) reported a significant correlation between the relative frequency of dark-patterned shells in the lake and the distribution of black (basalt) background. His survey of fish gut contents identified two predators of *Theodoxus*: *Barbus longiceps* and (especially) the blenny *Blennius fluvialis*, neither of which occurs outside Lake Kenneret. Heller suggested that the polymorphic colour patterns may result from visually hunting predators within the lake, while outside the lake solid black is favoured by virtue of the superior insulation it affords against solar radiation.

Vermeij and Covich (1978) have pointed out that, setting aside the spectacular faunas endemic to ancient rivers and lakes, the shells of freshwater molluscs seem to show substantially fewer adaptations to thwart predators than do marine forms. Shells are generally thinner in fresh water, and more rarely show spines, knobs and other shell sculpturing. The weak, planispiral gastropod form is more frequently observed in fresh waters than in marine. Adaptations against invasive predators, such as the narrow, elongated apertures or aperture teeth shown by marine gastropods or the interlocking shell margins common in marine bivalves, also seem to be quite rare in freshwater molluscs. Vermeij and Covich felt that physiological limitations on shell deposition, such as lower temperatures and calcium availability, 'may be in part responsible'. They also noted that the predators of freshwater molluscs may be weaker and less diverse than those encountered by marine molluscs.

But I would add, by way of clarification, that predation pressure (the likelihood of being eaten) does not seem any less in freshwater than in the marine environment. If predation has had more impact on the evolution of shell form in marine environments, its impact on life history evolution may have been greater in freshwater. Rapid growth, large reproductive output and efficient dispersal may be seen as adaptations both to ephemeral environment and to predator escape. To this subject we now turn.

Life history

It may be recalled from previous discussions (Chapter 4) that selection seems to favour early maturation time if mortality in a critical window (early, but not newborn) is low, but a later maturation if such mortality is high (Figures 4.7 to 4.9). Crowl (1990, Crowl and Covich 1990) studied populations of *Physa* (*Physella*) *virgata* inhabiting four Oklahoma streams, designated PCF (permanent, with crayfish), PNCF (permanent, no crayfish), TCF (temporary, with crayfish) and TNCF (temporary, no crayfish). His temporary streams dried for three months in the late summer, inflicting heaviest mortality on the largest size classes of *Physa*. The crayfish common in his study streams was a known snail predator, *Orconectes*, preferring smaller (but not newborn) *Physa* and unable to handle the largest animals.

In the two streams from which crayfish were absent (PNCF, TNCF), *Physa* reached maturity at 5.4 mm (47.5 days) and 5.3 (42.5 days), respectively, and generations overlapped to the extent that Crowl could not distinguish cohorts. Time to maturity was similarly brief at the TCF site

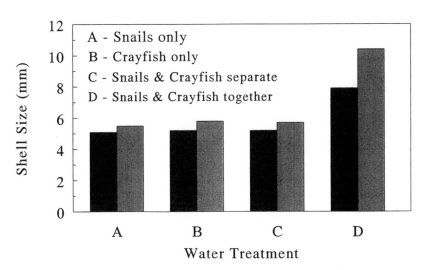

Figure 7.9. The effect of water from four different treatments on the mean size at first reproduction and death in *Physa virgata* (Crowl and Covich 1990). Each of the eight measures is a grand mean of 72 data, four populations × 18 snails per population.

(4.6 mm, 50.0 days), with the periodic nature of the environment imposing two distinct generations per year. But maturity was much delayed at the PCF site (9.3 mm, 117.9 days), so that a single generation passed annually. Crowl hypothesized that the larger size at reproduction observed in the PCF population might be a response to crayfish predation, a response precluded from the TCF site by abbreviated growing season.

Differences in age and size at maturity disappeared when snails from the four populations were reared in a common laboratory environment. Six groups of three snails were reared over their lifetimes for each of four populations × four treatments. All four populations matured at about 5 mm (45 days) when reared in water circulating from a reservoir that contained other snails, or that contained crayfish, or that contained snails and crayfish separated by meshwork. But most remarkably, all four populations delayed maturity to an average of about 8 mm (70 days) when reared in water circulating from a reservoir where crayfish were allowed to eat snails (Figure 7.9). Some chemical cue, apparently released by a crayfish attack on a conspecific, seems to delay maturity in Oklahoma *Physa virgata*. Clearly selection has not acted directly to fix adult size optima in these populations. But just as clearly, a trait that can delay maturity in response to an environmental cue, effectively promoting rapid growth through especially vulnerable size classes, has obvious adaptive value.

Consequences

Community composition

By this juncture the reader ought to be fairly well convinced that populations of molluscs commonly differ in their vulnerability to predation. The freshwater molluscan fauna is characterized by great diversity of size, shell strength, habitat and behaviour, and all of these variables may have demonstrable impacts on the likelihood of falling victim to a mob of predators that ranges from mink to flatworm. For example, it may be recalled that the carp inhabiting Skadar Lake, former Yugoslavia, seemed to select *Valvata piscinalis* over seven other molluscs available, including more common species of comparable size (Stein *et al.* 1975). Carp apparently selected *Valvata* for its shell, which was rather more fragile than those of other molluscs available.

The North Fork of the Holston River in southwestern Virginia is home to about 16 species of unionid mussels and a large population of muskrats. Neves and Odom (1989) surveyed the shell middens left behind by these rapacious rodents from 1979 to 1986, comparing their contents with quantitative samples taken from the stream bed in 1981. Although most mussel species seem to be taken according to their relative abundances, Neves and Odom recognized some preference for *Pleurobema oviforme* and the endangered *Fusconaia cor*, two species of 'medium' size (50–70 mm). Figure 7.10 compares the estimated abundance of the ten most common unionid populations in the North Holston Ford study area with their occurrence in muskrat middens 1981–86. (I have excluded midden data taken prior to the 1981 quantitative samples.) Overall, the six-year predation rate appears to have been about 3.8%.

My value of chi-square from the 2×10 contingency table corresponding to Figure 7.10 was 173.8, which with 9 degrees of freedom is significant at the 0.001 level. Combined by size class, Neves and Odom collected the shells of 408 individuals belonging to their two 'small' species, 573 individuals belonging to their five 'medium' species, and 135

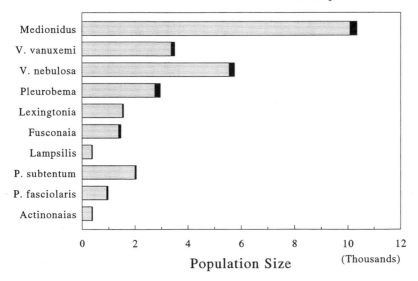

Figure 7.10. The ten most common unionid mussel populations inhabiting the upper North Fork Holston River, together with an estimate of their 1981 population sizes in the Neves and Odom (1989) study area. The top two species were categorized as 'small' by Neves and Odom, the middle five were 'medium sized', and the bottom three were 'large'. The fractions known to have fallen to muskrat predation 1981–86 are darkened.

individuals of the three 'large' species. My second (2×3) contingency analysis of these data returned a chi-square value of 28.5, which is again significant at the 0.001 level with 2 degrees of freedom. Muskrats took 3.9% of the three large species, 4.7% of the five medium-sized species, and only 2.9% of the two small species. These data are quite reminiscent of the Canadian *Anodonta* data collected by Hanson *et al.* (1989; Figure 7.1). They have been confirmed by Tyrrell and Hornbach (1998) working with a 37-species mussel community in eastern Minnesota (4 locations). We have previously hypothesized that small mussels may be more difficult for muskrats to find, or perhaps not worthy of return to the riverbank. In the diverse mussel faunas of the American interior, this phenomenon seems to have implications for unionid community composition.

The preferences manifest by the predators of freshwater molluscs may not originate entirely from the size and strength of the shells of their prey, but may also be influenced by their prey's behaviour. We have noted that leech attack tends to elicit shell-shaking behaviour from *Physa*, *Lymnaea*, and *Planorbis*, and enclosure from *Bithynia* and *Theodoxus*, but no apparent evasive behaviour at all from *Ancylus* (Bronmark and Malmqvist 1986). Brown and Strouse (1988) obtained similar results from a set of similar experiments involving six species of freshwater snails and the leech *Nephelopsis obscura*. The prosobranchs *Campeloma decisum* and *Amnicola limosa* withdrew behind their opercula, and were entirely invulnerable to leech attack. *Helisoma trivolvis* very rarely fell to the predator, at least partly because the snails were able to clamp so tightly to their substrate as to preclude entry by the leech. Brown and Strouse did not notice any of the shell-shaking behaviour reported by other observers of leech attack, but did show significantly higher frequency of clamping by pulmonates in tanks containing leeches than in those without. The three remaining species differed significantly in their vulnerability to leech predation: *Physa gyrina* most 'preferred', then *Helisoma anceps*, then *Lymnaea emarginata*. The authors concluded, however, with the reasonable suggestion that *Nephelopsis* 'does not really select prey, but merely feeds on what it can capture.'

Brown and Strouse's conclusion is quite likely generalizable to most freshwater predators. But given the great variation in the 'captureability' of freshwater molluscs of all sorts, by predators of all sorts, the potential effects of predation on community composition should be substantial. Thus it is surprising to note how rarely researchers seem to have demonstrated clear predation effects on mollusc community composition in the field. Among the ten predator exclosure experiments involving multiple mollusc species reviewed in the first section of this chapter (seven listed in Table 7.1, three involving invertebrate predators), only those of Eckblad (1973b) and Lodge and colleagues (1994) documented a community effect. Eckblad found some evidence that the preference of sciomyzid fly larvae for young *Lymnaea* might increase the (relative) abundance of *Physa* and *Gyraulus*, but the effect seemed weak and transient. Lodge reported that increased levels of crayfish predation in his enclosures resulted in fewer individual *Physa* and *Helisoma* relative to other snails (primarily *Amnicola*). Longer-term predation studies, more specifically directed to mollusc community composition, would seem in order.

Distribution

Although it certainly seems reasonable to expect some sort of relationship between the distribution of predators and their prey, the form of relationship is far from clear. Sih (1984) has outlined circumstances under which negative correlations, positive correlations, and no relationship at all may be expected between predator and prey densities. Negative correlations, to be expected either when prey outmanoeuvre their predators or when prey are effectively sessile, seem to me most likely to apply to freshwater molluscs. It is simply my impression that snails may indeed outmanoeuvre at least some of their invertebrate predators (ambushing leeches and flatworms, perhaps crayfish hunting from home burrows) but simultaneously are so slow relative to their vertebrate predators as to be effectively sessile. I should imagine that given the many other factors influencing distributions of both levels of the food chain, however, any sort of 'signal' would be extremely difficult to detect.

Most reports do, in fact, seem to suggest negative correlations. Such a relationship has been reported between the density of *Anodonta* and muskrat foraging range in a Finnish lake outlet (Jokela and Mutikainen 1995), and between the density of *Neritina latissima* and the molluscivorous fish *Sphoeroides annulatus* in a Costa Rican stream (Schneider and Lyons 1993). McKaye and his colleagues (1986) hypothesized that the absence of *Bulinus* from open, sandy areas of Lake Malawi might be attributable to predation by cichlids. Erection of fences to exclude the fish resulted in 40–60% increase in *Bulinus* density within the week.

Even though most species of snails inhabiting Trout Lake, Wisconsin, demonstrably preferred rocks with thick covers of periphyton, the quadrat samples of Weber and Lodge (1990) contained higher snail densities on more barren rocks. The authors suggested that this incongruous situation might arise from crayfish predation, since crayfish were more common among the densely packed, periphyton-covered rocks of the exposed shores than in the scattered, barren rocks found on protected shores. It is possible that crayfish may have been excluded from the richer snail-hunting grounds of Trout Lake by locally low dissolved oxygen. Covich (1981) reported increased density of *Physa gyrina* at the headwaters of an Oklahoma mineral spring, from which crayfish seemed excluded by water quality considerations.

At a larger scale, Crowl and Schnell (1990) reported a significant negative correlation betwen crayfish abundance and the density of *Physa virgata* over 11 streams in Oklahoma. Hofkin and colleagues (1991) surveyed 53 freshwater habitats in Kenya selected because they seemed suitable for *Biomphalaria*. They reported that 21 of these sites were inhabited by the large North American crayfish *Procambarus clarkii* but no pulmonate snails, 19 were inhabited by pulmonate snails but no crayfish, and only 4 sites contained both crayfish and snails, a highly significant negative correlation.

Moving to positive correlations, it may be recalled that Bronmark (1988) collected exhaustive samples from nine 1 m² predator exclosures and nine uncaged control plots of the same size at his small study pond in southern Sweden. Although his experiments were designed to estimate the importance of vertebrate predation on a diverse gastropod fauna, he also recorded data on the distribution of leeches (*Glossiphonia complanata*) in his benthic samples. He found leeches in ten of his 18 samples (5 exclosures and 5 controls). Over all sites, the median gastropod density (combining all 12 snail species) was between 445 and 402 individuals/m². Then Bronmark's data seem to reflect a significant positive association between snails and leeches; eight of the samples with high snail densities harboured leeches, while seven of the samples with low snail densities did not (Fisher's exact probability was 0.015). Both leeches and snails may here have been responding more directly to macrophyte densities, which were high in the snail-rich sites.

In addition to the enclosure/exclosure experiments previously discussed, Kesler and Munns (1989) took dip net samples from four habitat types in their small Rhode Island pond. Even given the apparently large impact that predatory bugs (*Belosoma*) seem to have on pulmonates at this locality, their single-date survey showed no correlation between bug abundance and the densities of either *Physa* or *Pseudosuccinea*. Their survey did seem to indicate an inverse correlation between bug abundance and the size of the potential gastropod prey, however.

At this stage it would be fun to introduce some data from Esrom Lake, Denmark, collected by Berg (1938). One of Berg's primary goals was to examine the relationship between depth and the bottom fauna of all sorts. He established a transect perpendicular to the shore and took seven samples of two Ekman grabs each on 12 dates from November 1932 to January 1934. He collected and identified a great variety of invertebrates, including *Dreissena*, one species of *Sphaerium,* several species of *Pisidium* (tallied together), and 12 species of snails (Table 7.3). I have graphed some of Berg's snail data in the upper half of Figure 7.11. The 'summer' plot is based on the samples of May 30, July 7, and August 2 combined, and shows the abundance of the four most common snail species as a function of depth. The 'winter' plot is identical except that the sample dates were February 15, December 5, and January 23. Although the snail species vary in the depths to which each extends, all reached maximum abundance in the shallowest (2 m) sample, regardless of season (as will be developed in Chapter 9).

In addition to molluscs, Berg tabulated the abundances of a great

Table 7.3. *Snails collected by Berg (1938) from Esrom Lake, Denmark, and 12-month totals, summing all depths*

Species	Total
Lymnaea ovata	85
Physa fontinalis	7
Planorbis planorbis	23
P. carinatus	4
P. albus	74
P. contortus	14
P. crista	47
Theodoxus fluviatilis	17
Bithynia tentaculata	641
B. leachii	66
Valvata piscinalis	740
V. cristata	51

variety of other benthic invertebrates, including four species of leeches (*Glossiphonia complanata, G. heteroclita, Helobdella stagnalis, Erpobdella octoculata*) and two species of triclads (*Planaria torva, Polycelis tenuis*) known to feed upon snails. No crayfish, sciomyzid flies, or any other benthic invertebrates preying upon snails were collected. The lower half of Figure 7.11 shows the summed abundances of these two classes of predators plotted on the same depth scale as their prey. The striking feature of these graphs is that the leeches show the same general distribution as most of their gastropod prey, reaching maximum abundances in the 2 m samples, while the two planarians are both most abundant at 8–14 m.

It is quite obvious that a few strokes of my calculator would return a positive correlation between leech abundance and the abundances of most snails, or a negative correlation with planarians. The opposite might be true for *Valvata* and its predators, whose range (especially in the summer) extends into deep water. Although consistent with a role for predation, clearly such distributional patterns are more easily interpreted as the result of snails, leeches, and planarians all responding independently to some environmental variable(s), such as depth or its correlates.

Then given the apparently high degree of spatial overlap between snails and their leech predators in Esrom Lake, one might be curious about temporal relationships. Do the leeches tend to be more common when snails are more common? Or have the life histories of the snails become

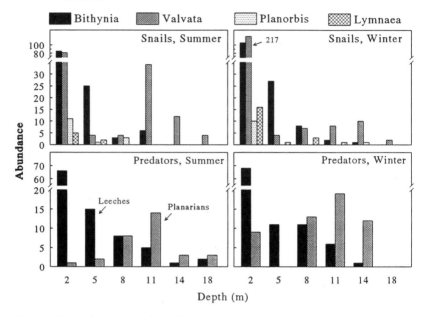

Figure 7.11. Abundance of the four most common snail species in Esrom Lake, Denmark, and two classes of their predators, as a function of depth (Berg 1938). Abundance is given as the total counts of individuals summed from two 225 cm² grab samples taken on each of three dates in the summer, or three dates in the winter.

modified to avoid leech predation, so that the correlation between leech and snail abundance is negative? To address this question, I totalled all snails and all snail-eating leeches in 2 m samples for each month. The Spearman rank correlation (R_s) was 0.43, non-significant for a two-tailed test with $N = 12$. I repeated this analysis for the 5 m samples, and obtained a second non-significant correlation ($R_s = -0.20$). Thus Berg's data seem to suggest no evidence of temporal relationship between the abundances of leeches and snails in Lake Esrom, but could be used to support any sort of spatial relationship one might wish to hypothesize.

Summary

The majority of the predators whose feeding habits have been touched upon here could safely be categorized as 'generalists'. And the feeding habits of the majority of the predators whose diets are touched upon here are difficult to observe directly. Thus most of the studies reviewed in the first section of this chapter were directed toward establishing the molluscan component of a wide sample of broad and changing diets. Indirect methods, especially gut content analysis and laboratory feeding trials involving suspected predators, have been favoured.

The molluscan fraction of the gut contents of such omnivores as turtles, amphibians, and many fishes seems quite sensitive to prey availability. Groups as diverse as ducks, sunfish, and crayfish seem to vary their mollusc consumption with age, gender, and reproductive condition. But by comparison with the gut contents of similar predators in similar habitats, it is possible to establish that some fish (e.g. large bottom-dwellers, some cichlids, pumpkinseed sunfish) as well as some invertebrates (particular species of leeches and triclads) specialize on freshwater mollusc prey.

While gut contents strongly reflect prey availability and weakly reflect preference, the data gathered from laboratory feeding trials leave just the opposite impression. Where both sorts of data may be compared, they may not agree. For example, under laboratory conditions crayfish may evince a preference for molluscs, as in fact may large ampullariid snails, even though both are typically considered herbivores. Predation rates are much more easily obtained in the laboratory, and are available for crayfish, sciomyzid and lampyrid larvae, and belostomatid bugs. Such figures are generally viewed as maxima, only weakly indicative of reality in the wild. Natural densities of *Belostoma* may peak at such levels as to consume all snails available, if one accepts experimentally estimated predation rates.

The importance of predation is much more difficult to establish than its widespread occurrence. Freshwater mollusc population densities seem positively affected by cages designed to exclude predators in about half the studies thus far published. Workers enclosing fish and crayfish predators with mollusc populations at natural densities have more reliably detected predation effects, although an enclosure of sciomyzid flies produced no result. Given their dependence on the coincidence of populations of two living creatures, predation effects would seem likely to be both temporally and spatially patchy. Artificial enclosure of a predator with its potential prey decreases such patchiness, perhaps increasing the apparent impact. For this reason, Robinson and Wellborn's (1988) demonstration of a (very patchy) predation effect in a Texas *Corbicula* population, using exclosures (but not enclosures) seems especially significant, as does Keller and Ribi's (1993) work with *Viviparus*.

Direct, especially long-term studies of freshwater mollusc densities, and those of their potential predators, are desperately needed. We know, for example, that as many as 52% of the *Dugesia polychroa* in productive English lakes may have eaten a freshwater mollusc recently, but we have no idea how many molluscs or *Dugesia* there may be in an English lake. I am aware of only one study, that of Hanson *et al.* (1989) on the muskrat predation of *Anodonta*, that begins to assess the impact of predation on freshwater mollusc populations directly. Programmes of this nature are not cheap, nor lightly entered into. Thus the artificial introductions of sciomyzids and lampyrids in biological control efforts, with no data taken on target snail population densities before or after, were opportunities lost.

The first line of defence for the typical freshwater mollusc is, and ever will be, its shell. Researchers report good correlations between shell size and (log) crushing resistance. Compared with the strength and gape of their typical crushing predators (fish, crayfish), many freshwater molluscs carry shells that afford excellent protection indeed. The shell strength of Lake Tanganyika thiarids seems to be greater than that displayed by other freshwater molluscs, and in fact, they are hunted by larger and more powerful crabs. It should be remembered, however, that crushing predators may be able to overcome shell defences by manipulation of the shell to attack its weak points, or by simple patience. There is some evidence that the shell markings carried by a few species (*Theodoxus*) may provide camouflage. There are also many freshwater molluscs, especially pulmonates and juveniles of all sorts, for which the shell constitutes no defence at all against crushing predators. But even the smallest and lightest shells may provide afford some protection against invading predators, such as insects, leeches, and other snails. The planispiral shell form so common in fresh water (planorbids, *Marisa*) presents an invading predator with a narrow opening, behind which the shell's inhabitant can withdraw deeply.

Withdrawing into the shell, or clamping tightly to the substrate, are two of the most obvious defensive behaviours displayed by freshwater molluscs. These we have categorized as 'resistance' responses, along with more spectacular shell shaking, dropping and rising reactions often displayed by freshwater snails. They seem to be elicited by direct contact with slow-moving predators, such as leeches, or sometimes by chemical cues, such as filtered extracts from crushed conspecifics. Snails may also display 'retreat' responses to chemical cues: avoidance, burrowing, or (for some pulmonates) crawling out of the water entirely. The triggers for all these defensive behaviours may be complex, depending certainly upon the species of snail and its age. The perception of chemical cues may be especially fine; responses in some circumstances seem to require active feeding on conspecifics by specific predators.

In at least one well-documented case, pulmonate snails seem to respond to chemical cues of predation by life history shifts. Growth is extended and reproduction delayed under long-term threat by crayfish to *Physa* at a young (but not newborn) age.

It is difficult to imagine the circumstances under which a predator population might take a random sample of a population of prey. For our purposes, it is immaterial whether predators actually choose among potential mollusc prey, or whether they attack everything but only ultimately prevail over some subset. Size-selective predation is a rule amply demonstrated by muskrats, cichlid fish, and leeches, among studies reviewed here. Other variables affecting 'handling time' besides overall size (including behaviour, shell shape, opercula, and defences such as mucus) combine to yield varying vulnerability to predation at the specific level. It is thus quite surprising to see how infrequently researchers have detected predation effects on freshwater mollusc communities. There is a little experimental evidence of a community effect due to sciomyzid flies and crayfish, and some analytical data on the varying impacts of muskrats to mussels. But review volumes on predation and aquatic communities (e.g. Kerfoot and Sih 1987) do not typically feature references to freshwater molluscs in this regard.

Turning to the potential for predators to affect freshwater mollusc distribution, the problem may be too much data, rather than too little. Most studies seem to suggest negative correlations between mollusc predators and their prey, although positive correlations, or no correlations at all, are often reported. Quadrat data from Esrom Lake, Denmark, can be viewed as supporting either positive or negative correlations, depending on one's selection of specific predator type (leech or flatworm) and prey type (*Bithynia* or *Valvata*). There seems to be no temporal correlation between predator and prey density in either direction. Distributional data are probably better interpreted as arising more directly from the environment. So although significant impacts of predation have been demonstrated upon specific molluscan populations, both directly and indirectly, by many methods on many occasions, the community effects of predation remain surprisingly elusive.

Although but briefly touched upon in the preceding chapter, the ability to disperse to new habitats may constitute some defence against predation, in an evolutionary sense. The many variables that impinge upon successful colonization of a new habitat, as well as the benefits that may accrue to successful colonists, will be discussed in Chapter 8.

8 · Biogeography

We begin this chapter with a review of the nominally abiotic factors that may contribute to the distribution of freshwater bivalves and gastropods on a regional scale. At least a dozen studies over the last 30 years have returned strong positive correlations between the abundance and/or diversity of freshwater molluscs and calcium concentration or related water quality variables. Laboratory experiments show both that calcium concentration in normal ranges can affect growth, survivorship, and fecundity, and that the various species often differ strikingly in their calcium optima. Some workers have suggested that calcium concentration (and correlated variables) may not act on abundance in the field, however, but rather serve as a 'filter for colonists' only. Others have suggested that the effects of calcium on natural abundances of molluscs are real but indirect, calcium concentration influencing the abundance of food.

We next review a relationship with analogues throughout biology, the number of mollusc species as a power function of habitat area. About a dozen studies worldwide have returned species/area regression coefficients somewhat less than or equal to the 'canonical' value of 0.25. Species vary in the mean areas of the habitats they occupy, just as they vary in calcium range. And an interaction is sometimes noticeable between area and calcium in their effects on the freshwater mollusc community, such that sites with larger areas tend to have more species than might be predicted from their calcium concentrations. This is again consistent with the hypothesis that the effects of calcium are indirect, through food abundance.

We review several other factors (isolation, current, substrate, etc.) for their regional effects on freshwater mollusc communities, generally finding them of lesser consequence than calcium and area. In the final section of this chapter, we examine evidence that variance in these two respects has its origin in life history adaptation. It will be recalled from Chapter 4 that, although most populations of freshwater molluscs seem to be undifferentiated with regard to reproductive effort, occasionally populations we have characterized as 'R-adapted' seem to allocate an order of magnitude greater energetic resource to reproduction and 'S-adapted' populations an order of magnitude less. It will also be recalled that, as a consequence of their resource allocation, both R-adapted and S-adapted populations were hypothesized to inhabit special environments. The present chapter concludes with a demonstration that such a model may in fact have some explanatory power in mollusc biogeography.

Environmental calcium

The influence of calcium availability on the distribution of freshwater molluscs has been recognized since well back into the nineteenth century. One should immediately note that the intercorrelations among total hardness, calcium concentration, pH, alkalinity, and conductivity are typically very high in natural fresh waters, and that workers have differed in their choice of variable(s) to emphasize. But regardless of the units in which this particular aspect of water chemistry is measured, whether it be acidity, buffering capacity, or the concentration of divalent cations, it has repeatedly been reported that freshwater mollusc species richness is a function of it.

The key seems to be substantial interspecific variance in tolerance for low calcium levels. For example, Boycott (1936) divided the British fauna into 30 'calciphile' species 'that ordinarily . . . need water with at least 20 mg calcium per litre and some of them more', and 26 species 'which we can find in soft water without surprise'. Boycott could not identify any 'certainly calcifuge' freshwater mollusc species among the British fauna; all 26 softwater species are also found in hard waters. (Extremely hard water seems uniformly detrimental.) Thus species richness would be expected to increase with calcium concentration, at least to a point.

Scattered through Boycott (1936) one can find 12 lists of species (bivalves and gastropods combined) from individual British lakes, accompanied by estimates of calcium concentration. In Figure 8.1 their number of species is graphed by calcium concentration on a double-log scale. If multiple estimates of calcium were provided by Boycott, I selected the lowest. Lakes Windermere and Leane are large and have exceptional mollusc faunas; Boycott chose these to make a point to which we shall return. Excluding them for the moment, over the other 10 lakes the

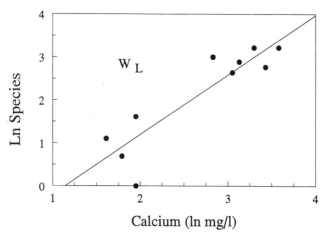

Figure 8.1. Number of mollusc species in British lakes as a function of calcium concentration (data of Boycott 1936). The larger lakes Leane (L) and Windermere (W) were excluded from the regression analysis.

correlation between (ln) species richness and (ln) calcium concentration is high: $r = 0.896$. (The natural logarithm seems best suited to small integers, such as species numbers.) The coefficient of regression is also highly significant (1.39 ± 0.24). Relationships such as that shown in Figure 8.1 are typical of freshwater mollusc faunas worldwide.

Laboratory studies

It appears that calcium concentrations within the normal ranges observed in fresh waters may have direct metabolic affects on molluscs. Greenaway (1971) held *Lymnaea stagnalis* in a medium loaded with ^{45}Ca tracer, coated their shells with wax, and measured the minimum equilibrium concentration of calcium – the concentration at which the net movement of tracer into and out of the animal balanced. Several snails failed to achieve calcium balance, continuing to show a net loss of calcium to the medium at low concentrations. Eight successful measurements of minimum equilibrium concentration ranged from 1.2 mg/l to 3.8 mg/l calcium, with a mean about 2.5 mg/l calcium.

Interestingly, *L. stagnalis* was among Boycott's 'calciphile' species, not normally found reproducing in waters less than 20 mg/l calcium. Greenaway notes that *L. stagnalis* are ordinarily not in calcium balance with their medium, but rather show net uptakes. Thus in the environmental concentrations of calcium where the snails are typically found, concentration gradients are absent or even slightly favourable for uptake.

This contrasts strikingly with the situation for sodium, for example, which *L. stagnalis* maintains in balance at considerable energetic cost (Greenaway 1970). It is also interesting to note that *L. stagnalis* seems to be able to use calcium as a cue for orientation, an ability more weakly expressed in the softwater *L. peregra* (Piggott and Dussart 1995).

Thomas and Lough (1974) placed individual *Biomphalaria glabrata* in variable volumes of water (down to 6.25 ml/snail) at calcium concentrations initially ranging from 0 mg/l to 80 mg/l. No attempt was made to control calcium input from shell dissolution or food, which may have been substantial (Young 1975). After 15 days, they found that final calcium concentrations in some of the smaller-volume experiments had decreased from 2.5 mg/l to about 0.5 mg/l calcium. They suggested that 0.5 mg/l might represent a minimum equilibrium calcium concentration for *Biomphalaria*, a figure substantially lower than Greenaway reported for *Lymnaea*. As the natural habitats of *B. glabrata* do often have much lower calcium concentrations than those of *L. stagnalis*, it seems likely that the calcium requirements of the two species may genuinely differ.

Thomas and colleagues (1974) reported some snail weight gain in almost all of these culture volumes and treatments, including '0' mg/l calcium, over their brief study. In larger volumes of water (400 ml/snail), growth seems to have been significantly lower at '0' and 2.5 mg/l calcium, but constant in 5–80 mg/l calcium. Egg production seemed to be more sensitive to external calcium concentration, on the other hand. No eggs were laid by any snail at '0' mg/l, very few were laid at 2.5 mg/l, and productivity increased steadily through the concentration series up to over 5 eggs/snail/day at 80 mg/l calcium.

The approach taken by Meier-Brook (1978) to measure the calcium requirements of the large ampullariid *Marisa cornuarietis* was more indirect. He prepared a 'standard fresh water' series based on mean world ratios of ions, ranging from 150 mg/l calcium to 10 mg/l. He then held 4–7 *Marisa* in individual beakers of water from this series for periods of several hours, measuring calcium concentrations before and after. Meier-Brook found that net calcium uptake increased with increasing external concentration up to about 50–75 mg/l. An especially striking peak uptake at about 75 mg/l was obtained from 'calcium-starved' snails – those which had been acclimated for 40 days at 25 mg/l calcium. Thus it appears that a calcium concentration as high as 25 mg/l may still be suboptimal for *Marisa*. Meier-Brook noted that the snails may be reared in a 25 mg/l medium, although their survivorship is initially much lower than observed at 100 mg/l, and shells are thinner.

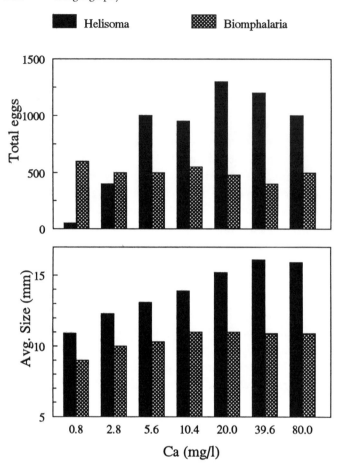

Figure 8.2. Average size (below) and net reproductive rate (above) for *Helisoma duryi* and *Biomphalaria camerunensis* reared 10 weeks in water of varying calcium concentration (Madsen 1987).

Such long-term studies are clearly desirable. Madsen (1987) performed 10-week experiments on growth and reproduction in four freshwater snail species at seven calcium concentrations. Figure 8.2 shows that the 10-week egg production of *Helisoma duryi* is much reduced at 0.8 mg/l and 2.8 mg/l calcium, while that of *Biomphalaria camerunensis* does not seem to be so affected. Shell growth in *B. camerunensis* seems to be adversely affected only by the lowest calcium concentrations, while shell growth in *H. duryi* seems sensitive over the entire range. Again one is impressed by the variability of calcium's effect on freshwater snail biology.

Madsen's culture media also varied in their pH (5.85–8.16) and concentrations of bicarbonate. It is not clear whether such effects as those illustrated in Figure 8.2 are directly due to shortages of calcium for shell building and normal physiological processes, or whether they may be indirect effects of the greater energetic cost for osmoregulation as a whole. Perhaps pH and the buffering capacity of the medium are the critical variables. The adverse effects of low pH, independent of calcium concentration, have been demonstrated by Hunter (1990).

The most thorough investigations into the relative importance of cation and bicarbonate concentrations to freshwater snails are those of Nduku and Harrison (1976, 1980a,b, and references therein). They reared *Biomphalaria pfeifferi* in a great variety of artificial media over periods of many weeks, combining age-specific survivorship and fecundity into estimates of the net reproductive rate (R). Populations of *B. pfeifferi* were viable in concentrations as low as 2 mg/l calcium in the laboratory, provided ample lettuce, from which they also extracted some calcium. Nduku and Harrison felt that 4 mg/l may constitute a more realistic lower limit in nature, however. Only at these very low levels of calcium did the concentration of bicarbonate seem to become significant – otherwise the buffering capacity of the water had little effect on R. The requirements of *B. pfeifferi* for all other ions were satisfied with Nduku and Harrison's typical laboratory diet; the snails grew and reproduced in deionized water to which $CaCO_3$ and food alone had been added. The addition of other cations, such as magnesium, sodium, and potassium, improved the performance of laboratory populations provided that the concentrations of these ions did not exceed that of calcium. The authors suggested that magnesium and possibly sodium may compete for calcium uptake sites.

Both Harrison *et al.* (1966) and Meier-Brook *et al.* (1987) have examined the effects of Ca/Mg ratio in more detail. Although Nduku and Harrison noted that low Ca/Mg ratios are rare in natural fresh waters, the concentration of sodium fairly often exceeds that of calcium as, for example, in coastal regions. As touched upon earlier, declines in freshwater mollusc abundance and diversity are usually noted when waters become excessively hard or approach 'saline'. In addition to the gross osmoregulatory problems one might expect in these situations, ionic ratios may play a role here.

Some information is available on calcium maxima, at least for *Biomphalaria* (Williams 1970b, Harrison *et al.* 1970). The intrinsic rate of natural increase in laboratory populations seems to be maximized at about 12 mg/l calcium, with a distinct negative response noted at about

40 mg/l. Williams suggested that from the standpoint of *Biomphalaria pfeifferi*, water less than 5 mg/l calcium ought to be considered 'soft' and water greater than 40 mg/l 'hard', with 'medium' waters considered optimal.

Turning to the bivalves, Hincks and Mackie (1997) held zebra mussels for 35 days in closed systems through which the waters of 16 different Ontario lakes were circulated. Calcium concentration, alkalinity, total hardness and pH explained most of the variance in their growth and survivorship. A calcium concentration of 32 mg/l seemed to give optimum growth; growth was actually negative at calcium concentrations less than 8.5 mg/l. Hornbach and Cox (1987) monitored growth and survivorship (very little reproduction was noticed) in populations of *Pisidium casertanum* maintained for 2½ years in natural waters and in artificial waters with four different concentrations of $CaCO_3$. Experimental animals were collected as juveniles from two Virginia ponds of very different hardnesses (calcium hardness 82 mg/l and 4 mg/l). Optimum growth for both populations occurred in the harder natural water. Both populations showed depressed survivorships in all artificial waters, but waters of medium hardness (about 160 mg/l as $CaCO_3$) were better than 10, 40, or 280 mg/l. The authors detected no 'clear-cut' influence of water hardness on growth rate, noting that differing survivorships tended to confound such results.

Field studies

Hornbach and Cox also took monthly quantitative samples of their two *P. casertanum* populations in the wild over a period of one year. The most striking difference was that individuals from the harder water population reached a much greater maximum size. But individuals of the softwater population seemed capable of reproduction at a smaller size, so both populations gave birth on the same schedule, summer and autumn. They counted a greater number of embryos per adult in the hardwater population, however, suggesting that fecundities in hard water may be greater. Probably because of their larger adult sizes, birth rates in caged individuals from the hardwater population were greater than those from the softwater population, even when transferred to soft water. The reverse experiment, softwater populations transferred to hard water, resulted in some minor improvement in birth rates. Thus it would appear that environmental calcium influences both the growth rates and the fecundity of *Pisidium* in the wild.

Kilgour and Mackie (1991) compared 10 demographic variables to 11 environmental variables in 17 Ontario populations of *P. casertanum*. Samples of clams, water, and sediment were collected on 12 occasions, April–December. It is difficult to assess the significance of the matrix of $10 \times 11 = 110$ correlation coefficients their analysis provided. (For 'table-wide' significance at the 0.05 level, Bonferroni correction would require an apparent *P* value of 0.00045.) Kilgour and Mackie performed two principal component analyses, reducing the environment to 4 variables and the demography to 3 variables, the interpretations of which were problematic. They then noted two 'significant' values in their new matrix of $3 \times 4 = 12$ correlations among principal components.

Inspecting the original matrix of 110 correlations directly, one notes that the highest correlations are positive ones between particle size or percentage coarse substrate lost on ignition (i.e. leaf fragments) and the mean adult length or weight of *Pisidium*. This would seem a local phenomenon, attributable to the food and habitat preferences we noted for *Pisidium* in Chapter 2. There are several large negative correlations between temperature and maximum adult shell size or weight or the adult length at which shelled larvae appear, as well as a positive correlation between temperature and abundance. This suggests that *P. casertanum* grows larger and reproduces less in cold waters, an effect that could be due both to local differences in the depths at which samples were taken and to larger climatic differences over 2° of latitude. Finally, maximum adult weight or proportion of the population with shelled larvae seem to be positively correlated with calcium hardness, alkalinity, or conductivity. This tends to substantiate the results of Hornbach and Cox.

Kilgour and Mackie reported no significant correlation between the abundance of *P. casertanum* and any water chemistry variable. However, the environments selected by them ranged in mean calcium hardness from 4 mg/l to 225 mg/l. We have noted that freshwater mollusc populations may be adversely affected by hardnesses that are too high as well as too low. Although there is considerable scatter, Figure 8.3 shows that mean *P. casertanum* abundances seemed to increase with hardness to a point, then decline. No monotonic relationship would seem expected over such a range, without data transformation.

If the optimum value for some water chemistry variable is W_{opt}, then the quantity

$$\mathbf{W} = \cos \left(W_{opt} - W_{obs} \right)$$

will equal 1.0 when the observed value (W_{obs}) matches W_{opt}. Values of $\mathbf{W}$ will be symmetrical around W_{opt}, decreasing to -1.0 as observed

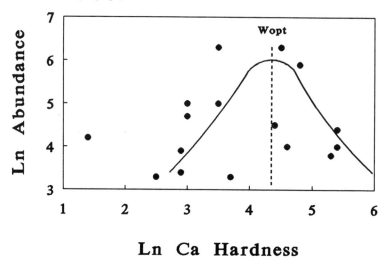

Ln Ca Hardness

Figure 8.3. Mean abundance of *Pisidium casertanum* (number of clams/m²) as a function of ln mean calcium hardness (mg $CaCO_3$/l) at 17 sites in Ontario (Kilgour and Mackie 1991). W_{opt} is the apparent optimum value of calcium hardness for this species, estimated as described in the text.

water chemistry values vary from optimum in either direction. So although one might not expect a significant monotonic regression between freshwater mollusc population density and W_{obs} when W_{obs} ranges over two orders of magnitude, such a relationship could be plausibly expected with **W**, calculated in this fashion. The quantity **W** will begin to increase again if the absolute value of $W_{opt} - W_{obs}$ exceeds 180 when coded in degrees or as π when radians. Thus log transformation, which is customary for water chemistry data in any case, would seem advisable before entering the data as radians into the calculation of **W**.

The correlation between **W** and *P. casertanum* abundance is maximized at $r = 0.49$ ($P < 0.05$) where W_{opt} is ln $4.4 = 81$ mg/l calcium hardness. This value of W_{opt} was obtained iteratively, and the curve in Figure 8.3 fitted by eye. Nevertheless, there is a remarkable correspondence with the 82 mg/l calcium hardness found by Hornbach and Cox to be most beneficial to *P. casertanum* in Virginia.

The range of *P. casertanum* extends into hardnesses lower than any sampled by Kilgour and Mackie. Papers by Rooke and Mackie (1984a,b) and Servos et al. (1985) focus on the biology of several *Pisidium* species in six Ontario lakes with mean calcium concentrations ranging from 3.15 mg/l down to 2.06 mg/l. Here the mean growth (in mm/day) of *P. caser-*

tanum and *P. ferrugineum* appeared to be correlated with hardness or alkalinity over one summer season, but not a second. The mean litter sizes of both species also seemed to be reduced in softer waters. Caged individual *P. equilaterale*, rare or absent from the softest of these lakes, showed reduced survivorship when transplanted into them, compared with caged controls.

Returning to snails, Rooke and Mackie also gathered several observations on the *Amnicola limosa* populations of these six lakes. Interestingly, during one season the *Amnicola* population of one of the softest lakes apparently failed to reproduce. The extreme rarity of *A. limosa* in their second very soft lake led Rooke and Mackie to suggest that its *Amnicola* population may have collapsed only recently. Caged populations of snails transplanted into these two soft lakes showed growth rates significantly lower than caged controls. Over all six lakes, *Amnicola* showed a significant correlation between hardness and both mean shell growth rates (mm/day) and population production (mg dry wt/m^2).

It may be recalled from Chapter 4 that *Viviparus ater* inhabits both the moderately hard Swiss Lake Zurich (8.3–14.5 mg CaCO$_3$/l) and the soft Lake Maggiore on the Swiss/Italian border (3.4–3.9 mg CaCO$_3$/l). We have previously reviewed the results of an extensive series of cage studies comparing the life histories of these two populations (Ribi and Gebhardt 1986, Ribi *et al.* 1986, Gebhardt and Ribi 1987). Not only are Zurich snails more fecund at each of their (five or more) reproductive seasons, the average size of the offspring they produce is larger. Ribi and Gebhardt found some evidence that the effects of low calcium on the life history of Maggiore *Viviparus* may be direct. Life expectancy is shorter in Lake Maggiore. The shells of older Maggiore snails are eroded, pitted, and sometimes perforated, but not those of Lake Zurich. While the fecundity of Zurich snails remains constant through their lifetimes, fecundity decreases steadily in Lake Maggiore. Kat (1982b) has also suggested that shell dissolution may have a direct impact on life histories of freshwater molluscs, his data coming from softwater populations of *Corbicula*.

Field studies have demonstrated fauna-wide positive correlations between the abundance or diversity of freshwater molluscs and calcium or related variables in Europe (Aho 1966, Dussart 1976, 1979a,b, Økland 1983), North America (McKillop and Harrison 1972, McKillop 1985, Pip 1987a), Asia (Palmieri *et al.* 1980, Ram and Radhakrishna 1984) and Africa (Williams 1970a). Workers have often measured a variety of water quality variables and reported high intercorrelation among them. Especially impressive have been studies, such as those of Dussart,

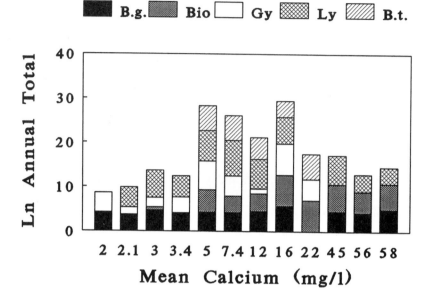

Figure 8.4. Total (usually over 12 months) individual snails counted for 5 species collected at 12 sites in Zimbabwe (Williams 1970a). Two pairs of sites (4 and 5, 6 and 7) were located in the same water bodies and hence had identical calcium values. These pairs are combined and their average snail densities graphed. B.g., *Bulinus globosus*; Bio, *Biomphalaria pfeifferi*; Gy, *Gyraulus spp.*; Ly, *Lymnaea natalensis*; B.t., *Bulinus tropicus*.

McKillop, and Williams, involving monthly samples of both water and mollusc populations over some lengths of time.

Williams (1970a), for example, established 14 stations in the vicinity of Salisbury, Zimbabwe, measuring pH, conductivity, and the concentrations of 10 ions monthly. He also took quantitative samples of the snail faunas using a drag scoop. Figure 8.4 shows the total number of individuals of each of five pulmonate species collected, summed over the year, as a function of the calcium concentration at 12 stations. *Bulinus globosus*, *Lymnaea natalensis*, and *Gyraulus* seem relatively unaffected by calcium concentrations at the lower end of this range, although *Gyraulus* disappears from the hardest stations. *Biomphalaria pfeifferi* and *Bulinus tropicus* seem to require about 5 mg/l calcium in the wild, but again, *B. tropicus* seems excluded from hard waters. Thus the overall impression from Figure 8.4 is that diversity is at a maximum in waters of about 5–40 mg calcium/l, in accord with other results from the laboratory and field.

Today the importance of environmental calcium for freshwater mollusc species richness is widely recognized. But before changing subjects, one should be careful not to leave the impression of unanimity regarding our understanding of the mechanism by which calcium acts. Some workers do not accept the notion that calcium concentration, in a density-independent fashion, can have any influence on population abundance. We note, for example, little evidence of an abundance effect in Figure 8.4. While recognizing that water chemistry may be manipulated to affect demographics in laboratory populations, one might argue that such observations are of limited relevance in the broader range of environments encountered naturally. Perhaps water chemistry 'acts as a filter for colonists' (Lodge *et al.* 1987) and nothing more.

The key may be a compromise offered by Boycott (1936:161). The third of his four reasons for 'the decided preference which most mollusca have for hard water' was: 'A good supply of soluble salts, of which calcium itself may not be the most important, is desirable for the growth of most water plants, and the vegetation of hard water is always richer than that of a corresponding habitat with soft water.' It seems most likely to me that in general, the effects of calcium, hardness, alkalinity, pH, and conductivity may be indirect. Higher levels of all (to a point) act to improve conditions for food organisms, which then enhance the growth of mollusc populations. Algal diversity and biomass are greater in hard waters (Blum 1956, Marker 1976). Detrital decomposition occurs more quickly in hard waters (Webster and Benfield 1986). Eriksson *et al.* (1983) reported that artificial liming of lakes and rivers improves the diversity and abundance of zooplankton, phytoplankton, and macrophytic vegetation after only a few years. Even host fishes for unionids seem more abundant in hard water (Strayer *et al.* 1981). Thus harder waters produce more food, and more food produces more molluscs subject to density-dependent control.

Evidence supporting the view that the effects of calcium are normally indirect comes from field studies involving both water chemistry and habitat size. The New River arises from the mountains of northwestern North Carolina, in a region underlain by gneiss and schist. After travelling for about 100 km it passes through a region of limestone and dolomite. The river and its tributaries upstream generally show 2–5 mg calcium/l, while calcium concentrations in some of the downstream tributaries reach over 40 mg/l. Dillon and Benfield (1982) sampled 87 sites in the New River catchment, recording five species of pulmonate snails (excluding limpets) at 26 sites. The abundance of each species was scored semi-quantitatively, 1–3, and summed for each site.

Calcium did not appear to effect the abundance of pulmonates in the New River directly, for snails were often rare in hard waters where the streams were small. Nor did calcium act as a simple filter for colonists, for snails were often found in very soft water where the rivers were large. In fact, a non-parametric analysis suggested that the interaction between alkalinity and stream drainage area (the product of the site ranks) accounted for more variance in pulmonate diversity and abundance than either alkalinity or drainage area separately. Dillon and Benfield thus suggested that both water chemistry and drainage area were affecting the abundance of food.

Area

Thus a second major factor influencing species richness, habitat area, comes to the fore. One of the primary goals Boycott (1936) set for himself was to establish the lower limits of calcium concentration tolerated by elements of the British mollusc fauna. As such, he concentrated on two softwater lakes with unusually diverse communities: Windermere ('exceptionally good . . . mature, sheltered with plenty of plants, silt and mud and varied subhabitats in different parts') and Leane ('particularly productive'). Figure 8.1 shows that these two lakes had many more species than one would predict from their water chemistry. The important factor seems to be that they are, with surface areas about 15 km² and 24 km² respectively, by far the largest lakes examined by Boycott.

At least four explanations have been advanced for the universal observation that larger habitats or islands have more species than smaller ones. They provide larger 'targets' for immigrants, their greater habitat complexity may facilitate coexistence once immigrants arrive, and the larger populations they support are less likely to go extinct stochastically. Trivially, a large sample from a fauna would be expected to contain more species than a small sample. Early work on the species–area relationship was reviewed by MacArthur and Wilson (1967), and more recently by Lomolino (1989). Workers have generally settled on a power function to model species-area relations, which together with its log transformation is given below:

$$S = cA^z$$
$$\ln(S) = \ln(c) + z \ln(A)$$

where S is the number of species on a habitat island, A is the area of the island, and c and z are fitted constants, typically obtained from a least-

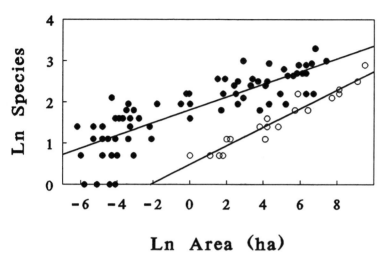

Ln Area (ha)

Figure 8.5. Number of snail species as a function of surface area in Danish lakes. (Data of Lassen 1975.)

squares linear regression fit to log transformed data. I have again elected to use ln (base-e logarithms) here rather than the more traditional base-10 logs as a matter of convenience; the scaling is better for small numbers like the species in the communities under consideration here. Much attention has been directed to the observation that z usually takes values between 0.20 and 0.35, with about 0.25 considered 'canonical' (Connor and McCoy 1979, Sugihara 1981, Martin 1981).

The species–area relationship in freshwater molluscs was first formally examined by Lassen (1975). Sidestepping the confounding effects of water chemistry, he compiled species lists separately for 68 eutrophic and 19 oligotrophic Danish lakes and ponds. He further divided his sample of eutrophic lakes into two groups, those smaller than 0.1 ha and those larger. His (double-log) regression analysis yielded slopes of 0.37 for small eutrophic lakes, 0.09 for large eutrophic lakes, and 0.25 for oligotrophic lakes.

Lassen did not report any regression statistics other than slope. I have reanalysed the data from reproductions of his figures 1 and 2. In keeping with more conventional procedures, I have not divided his eutrophic lakes into area classes. Eutrophic and oligotrophic lakes are replotted together in Figure 8.5, and regression statistics are given in Table 8.1. The

Table 8.1. *Statistics from regression analyses of mollusc species number on habitat area. Other independent variables as noted*

	N	r	C	z (SE)	F
Denmark (data of Lassen 1975)					
Oligotrophic	19	0.96	1.6	0.23 (0.015)	218★★★
Eutrophic	68	0.83	6.1	0.16 (0.013)	149★★★
Finland (Aho 1978a)					
	43	–	1.1	0.15	–
New York snails and unionids (Browne 1981)					
	37	0.93	1.2	0.23	–
Sweden (Bronmark 1985a)					
	115	0.40	2.8	0.11	–
Southern New England (Jokinen 1987)					
Low calcium	23	0.85	4.0	0.19 (0.026)	–
High calcium	41	0.65	5.6	0.12 (0.022)	–
Northeastern New York (Jokinen 1987)					
Low calcium	14	0.11	4.6	−0.02 (0.043)	–
High calcium	35	0.47	4.1	0.11 (0.034)	–
Connecticut (data of Jokinen 1983)					
Area only	90	0.48	3.8	0.14 (0.027)	26.6★★
Calcium only	90	0.26	3.5	0.21 (0.085)	6.2★★
Multiple	90	0.56	0.9	0.14 (0.025) a	20.0★★★
				0.24 (0.073) c	
English Lake District (data of Macan 1950)					
Area only	53	0.63	2.5	0.22 (0.042)	26.1★★★
Calcium only	53	0.20	0.26	0.34 (0.26)	1.65
Multiple	53	0.69	−0.03	0.23 (0.040) a	17.33★★
				0.47 (0.20) c	
Distance only	41	0.31	0.79	0.011 (0.005)	4.15★
Africa (data of Brown 1994)					
Area only	18	0.79	1.1	0.27 (0.052)	27.0★★★
Conductivity only	18	0.73	−0.47	3.32 (0.78)	18.1★★★
Multiple	18	0.86	−0.65	0.19 (0.055) a	21.2★★★
				1.85 (0.74) c	
Unionacea, rivers of Eastern US (data of Sepkoski and Rex 1974)					
Area only	41	0.67	0.0	0.32 (0.058)	31.9★★★
pH only	41	0.19	2.44	−0.10 (0.085)	1.51
Distance only	41	0.27	2.73	−0.14 (0.082)	2.99
Multiple	41	0.79	−0.19	0.37 (0.049) a	32.4★★★
				−0.24 (0.054) d	

Table 8.1 (*cont.*)

	N	r	C	z (SE)	F
Unionacea, Ohio and Maumee Rivers (Watters 1992)					
Area	47	0.92	174	0.34	–
Fish species richness	40	0.96	0.046	0.45	–

Notes:
N, number of habitat 'islands'; r, correlation coefficient; C, y intercept, expressed in hectares for simple area regressions (back-transformed from the log scale); z, regression coefficient (and standard error); a, area; c, calcium or conductivity depending on context; d, distance; F, value of F ratio from ANOVA ($P<0.05$*, $P<0.01$**, $P<0.001$***). All data are for snail faunas inhabiting lakes and ponds, unless otherwise noted.

correlations between (log) number of snail species and (log) lake area are extremely high in Denmark. The slope of the regression over all eutrophic lakes combined seems to be about 0.16, a fair compromise between the 0.09 and 0.37 reported by Lassen. The difference between the 0.25 reported by Lassen for oligotrophic lakes and the 0.23 calculated here is doubtless due to small errors in data transcription.

Figure 8.5 shows striking effects due to general trophic level, probably equivalent to hardness or pH here. Oligotrophic lakes generally have fewer species than eutrophic lakes, but the slopes of the two species–area relationships are different. The y-intercepts reported in Table 8.1 (anti-logs) show that 1 hectare oligotrophic ponds have fewer than a third of the species expected in eutrophic lakes of similar size. Yet the species richness of large oligotrophic lakes approaches that of large eutrophic lakes. Here clearly displayed in Scandinavian lakes is the same interaction reported by Dillon and Benfield in a southern Appalachian river. Again the suggestion is that both water chemistry and habitat area affect the food supply for freshwater molluscs, and it is to the food supply that diversity responds.

The best fit to a mollusc species–area regression ($r = 0.93$, Table 8.1) of which I am aware was obtained by Browne (1981). His 37 sites in upstate New York were distributed fairly evenly from springs less than 0.001 km² up to Lake Erie, at over 25 000 km², and included from 1 to 78 species of snails and mussels. In contrast, the 115 Swedish ponds sampled by Bronmark (1985a) ranged only from 0.003 ha to a little over 2.3 ha, with from 2 to 14 snail species. Table 8.1 shows that Bronmark's species–area correlation was much lower, although still highly significant ($r = 0.40$, $P < 0.001$).

Bronmark did not measure any water chemistry variables, noting that his ponds were all eutrophic and underlain by a 'homogeneous geological background'. He did, however, have access to previously published lists of macrophyte species at 45 of his ponds. The number of macrophyte species ranged from 4 to 30, and Bronmark showed a significant species–area relationship for the plants as well, fitting the equation 0.129 log $A + 0.629$ with $r = 0.43$ ($P < 0.001$). It would seem that both habitat complexity and food resources, at least for snails able to exploit macrophytes, can be significantly correlated with lake size. Bronmark obtained a very nice multiple regression ($r = 0.902$) using log number of snail species as the dependent variable and pond area, number of macrophyte species, and a measure of isolation to be discussed presently, all log-transformed, as independent variables.

Clearly a more realistic representation of the effect of area on species will be obtained if the effects of water quality can be taken into consideration. Analytically, one might perform a multiple regression including both area and calcium content, or some other water chemistry variable correlated with calcium, as independent variables. Aho (1978b) surveyed the gastropods in four 'subareas' of the lake district of southern Finland. Although diversity in his subareas I and II was very low (5 species), his regions III and IV had enough variance in snail species number to warrent examination (12 and 15 species, respectively). The 20 lakes of subarea III were all small, ranging from 0.003 km^2 to 0.79 km^2, with calcium ranging from about 3 mg/l to 10 mg/l (converting from German degrees of hardness). Here Aho reported a high correlation between number of species and hardness ($r = 0.740$, $P < 0.001$) and a lesser one with area ($r = 0.482$, $P < 0.05$). A multiple regression (not on a log scale, unfortunately) found hardness and area together accounting for 73% of the variance in species number. Lakes were larger in subregion IV, ranging from 8.4 km^2 to 119 km^2, with calcium ranging from about 5 mg/l to 16 mg/l. Aho considered 11 of these lakes polluted and 11 'in more or less natural condition'. There was a highly significant correlation between hardness and species number among the 11 unpolluted lakes ($r = 0.848$, $P < 0.001$), although none with area.

Aho (1978a) reported the results of a simple log–log regression analysis of the species–area relationship over 43 lakes; essentially his combined samples from subareas III and IV. Water chemistry was not included in this model, although the fit would clearly have benefited, and a value of *r* was not given. Nevertheless, such statistics as were supplied by Aho are added to Table 8.1.

Lodge *et al.* (1987) examined previously published data on snail species richness and physico- chemical variation in 64 northern Wisconsin lakes. Two independent variables were included in their stepwise regression of species number, area ($F=16.2$, $P<0.001$) and alkalinity ($F=14.7$, $P<0.001$), the result being a very good fit ($r=0.65$). These results compare very well to those of Aho. All variables were untransformed, however, and thus it is difficult to compare Lodge's findings to previous species–area regressions as collected in Table 8.1. Jokinen (1987) also performed stepwise multiple regressions of species number on area and seven other independent variables. Her analysis was of two data sets: 46 lakes in southern New England and 29 lakes in northeastern New York, all greater than 10 ha. Lake area ($F=7.25$, $P<0.01$) and Na/Ca ratio ($F=5.88$, $P<0.05$) were selected in the former analysis, and lake area ($F=38.4$, $P<0.001$) and altitude ($F=8.44$, $P<0.01$) in the latter. The stepwise procedure may have been misleading, however, because Na/Ca ratio was highly correlated with calcium concentration in New England ($r=-0.42$, $P<0.01$), as might be expected, as well as with altitude in New York ($r=-0.59$, $P<0.01$). Again neither species number nor lake area was transformed before analysis.

However, Jokinen also performed more conventional double-log species–area analyses of a larger data set of ponds and lakes, further dividing each of her two regions into high-calcium (>5 mg/l) and low-calcium sets. Statistics for these four regressions are shown in Table 8.1. The situation in New England is reminiscent of that reported by Lassen in Denmark 10 years previously (Figure 8.5). Jokinen found excellent fits to the (log)linear model for both hard (eutrophic) and soft (oligotrophic) New England lakes, the former showing a higher intercept but lower slope than the latter. Thus one expects more species in small hardwater lakes than in small softwater lakes, but the difference disappears with increasing size. Jokinen's high-calcium regression for New York matches her high-calcium New England regression rather well. The low-calcium New York regression is non-significant, perhaps reflecting 'the isolation of the Adirondack lakes, with their depauperate fauna'.

Jokinen's southern New England data deserve a more detailed examination. Data on a large fraction of the 64 lakes she analysed in 1987 are available in her (1983) monograph on the freshwater snails of Connecticut. The appendices of that earlier work list 230 sites, 92 of which are lakes or ponds with a least one snail species present, with measurements of area (in hectares) and calcium (mg/l) available. Number of species ranged from 1 to 17, area from 0.1 ha to 370 ha, and calcium from 0.8 mg/l to 35 mg/l. (There is clearly little danger that calcium concentrations below 35 mg/l will affect freshwater snails adversely.) Setting aside two outliers, Table 8.1 shows that the simple regressions of log area on log species and log calcium on log species were both significant. The area-only simple regression seems a fair compromise between the two New England regressions of Jokinen (1987) previously tabulated, showing an intermediate slope of 0.14 and a substantially lower r. It is interesting to note that this slope of 0.14 is preserved in the multiple regression, while r increases substantially, to 0.56 (probably as significant with $n = 90$ as the overall New England values).

So far in discussing the effects of lake area, we have proceeded as though all species were equally likely to inhabit lakes of any size. Indeed to the extent that the species–area relationship is due to non-deterministic factors such as sampling bias or 'target size', this would be true. But judging from actual data, it appears that species vary in their ability to colonize small, possibly unstable habitat islands, just as they appear to vary in their tolerance for low calcium concentrations. One tends to find the same subset of species in the smallest ponds, and as pond size increases, the addition of species does not appear to be entirely random. The data of Macan serve as an illustration.

Over 10 years, Macan (1950) made a series of qualitative and quantitative samples of molluscs from the English Lake District. His studies centred about the large, soft Lake Windermere, (1500 ha, about 5 mg/l calcium). Macan catalogued 11 snail species in Windermere, the same number Boycott obtained from the literature. (Note that Figure 8.1 includes bivalves as well.) Interestingly, Macan found four species not reported by Boycott: two planorbids, the lymnaeid *Myxas glutinosa*, and *Potamopyrgus jenkinsi*. He failed to collect two *Ancylus* species (he felt his equipment was inappropriate for limpets): the deep water *Valvata cristata*, and *Planorbis crista*, whose occurrence in Windermere he questioned. Macan also collected snails in 12 other nearby lakes, in 36 of 41 bodies of water he considered 'tarns' (14 ha to 0.08 ha), and in 4 of 15 'pools' (less than 0.08 ha). Taken altogether Macan's data set included 53 bodies of water harbouring 16 species.

Macan measured calcium concentration (mg/l) on three occasions at most tarns and pools, discounting the spotty data from August 1947. Data were available for only two sampling periods in some cases, and only single estimates were presented for the 12 lakes. Such data as were available I have averaged. The maximum average calcium concentration (22.4 mg/l) seems to be well below any reasonable estimate of W_{opt}, and thus I judge cosine transformation to be unnecessary. Macan provided estimates of surface area for all tarns, but not for lakes. I estimated lake areas by clipping and weighing an enlargement of his Figure 1.

Table 8.1 shows statistics for the simple regressions of area and calcium on number of snail species in 53 bodies of water from the English Lake District. The species–area regression was highly significant and rather canonical ($r = 0.63$, $F = 26.1$, $P < 0.001$). Although the simple regression of species number on log mean calcium concentration was not significant, calcium did contribute significantly to a multiple regression when entered as a second independent variable along with log area, as we have seen in previous analyses. Table 8.1 shows that the multiple r improved from 0.63 to 0.69, while the species–area regression coefficient remained unaffected.

Of Macan's 16 species, *Myxas*, *Potamopyrgus*, and *P. carinatus* were found only in Windermere, at least as of 1950. *Valvata cristata* and *Lymnaea stagnalis* were found in but two sites. Then setting these aside, Figure 8.6 shows the mean, SEM, and range for the log surface area of the bodies of water inhabited by the remaining 11 species. The error bars are rather large for many of these, especially those only rarely collected by Macan. However, it is clear from Figure 8.6 that significant differences do exist in the mean lake sizes inhabited by the various taxa. The mean size of the 6 tarns inhabited by *Lymnaea glabra* was aln $(-1.53) = 0.22$ ha, while the maximum size was only 1.6 ha. Boycott states that it 'has been generally recognized that *L. glabra* characteristically occurs in mean places where few if any other Mollusca care to live – shallow grassy ditches and ponds which usually dry in the summer.' It is interesting to note that the first eight species listed in Figure 8.6 are lymnaeids and planorbids, and that generally only in the larger (and more diverse) lakes does one add other families. At the other end of the spectrum we see *Physa fontinalis*, found in 13 lakes averaging about aln $(4.34) = 77$ ha, although ranging down to 1.3 ha. Boycott says it is 'not particular about volumes' but is 'never found in closed ponds or static water.' It is also significant that ranges, with just a couple of exceptions, are restricted at the low end, not at the high. The snails all generally seem to be able to inhabit large bodies, but vary in their capacities to colonize smaller, perhaps more unstable habitats successfully.

The number of species on a habitat island is usually, today, thought of as a function of immigration and extinction. But from a community

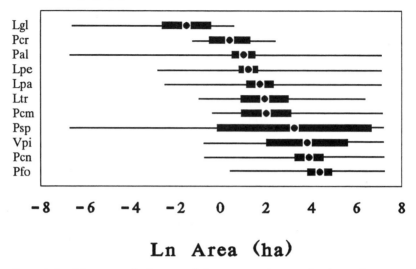

Ln Area (ha)

Figure 8.6. Means, standard errors of the mean, and ranges for the surface area of the water bodies inhabited by the 11 most widespread gastropods in the English Lake district (data of Macan 1950). Lgl, *Lymnaea glabra* ($N = 6$); Pcr, *Planorbis crista* (4); Pal, *P. albus* (35); Lpe, *Lymnaea pereger* (45); Lpa, *L. palustris* (22); Ltr, *L. truncatula* (8); Pcm, *Planorbis complanatus* (8); Psp, *P. spirorbis* (4); Vpi, *Valvata piscinalis* (14); Pcn, *Planorbis contortus* (15); Pfo, *Physa fontinalis* (13).

standpoint, the *in situ* evolution of a new species would be equivalent to immigration. Brown (1994, table 12.5) has compiled the number of snail species inhabiting 21 natural African lakes. Three of these contain no gastropods, while species range from 1 to 57 in the remaining 18. To biologists more familiar with the rest of the world, the striking feature of these data is the degree of endemism. Brown states, for example, that 'Endemism is not strongly marked amongst the living gastropods of Lake Albert', referring to a table showing only 8 (of 15) species endemic. Brown's table 12.5 also includes data on lake area (km²) and 'salinity', reported as conductivity in μmhos. Although one would expect diversity to increase with conductivity to a point, it is quite clear that at conductivities ranging up to 3300 in some lakes, the gastropod fauna may be adversely affected.

Figure 8.7 (upper) shows the species–area relationship for Brown's 18 lakes. As shown in Table 8.1, the double-log fit to the regression was very good ($r = 0.79$), and the slope very close to 'canonical'. Judging from the figure, lakes Rudolf, Nabugabo, Kivu and Bangweulu do not seem to have as many species as expected for their sizes. The median conductivity of the water is quite low in Nabugabo and Bangweulu (25 and

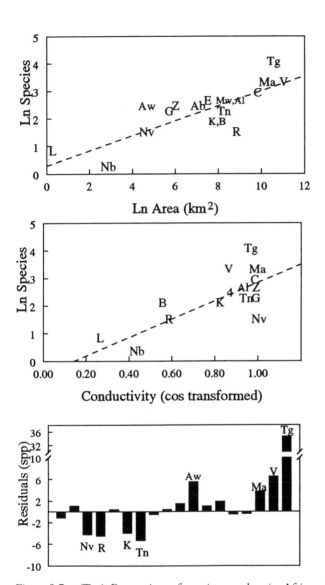

Figure 8.7. (Top) Regression of species number in African lakes on lake area. (Middle) Regression of species number in African lakes on $\log_{10}$ conductivity (cosine transformed). (Bottom) Residual variation in the number of snail species in African lakes not explained by a multiple regression on area and conductivity. The lakes are arranged by the number of species they contain, left to right. The abbreviations for the individual lakes are: Ab, Abaya (Margherita); Al, Albert; Aw, Awasa; B, Bangweulu; C, Chad; E, Edward; G, George; K, Kivu; L, Lungwe; Ma, Malawi (Nyasa); Mw, Mweru; Nb, Nabugabo; Nv, Naivasha; R, Rudolf (Turkana); Tg, Tanganyika; Tn, Tana; V, Victoria; Z, Zwai. The number '4' marks lakes E, Mw, Ab, and Aw. (Data of Brown 1994.)

33 μmhos, respectively), while conductivity is very high in Kivu and Rudolf (1240 and 2750 μmhos).

The results of a simple regression of species number on transformed median conductivity are also given in Table 8.1 and shown in Figure 8.7 (middle). Values of conductivity were $\log_{10}$ transformed (rather than $\log_e$, given their range) and cosine transformed, as detailed in the previous section. Again the value of W_{opt} was obtained iteratively. The best fit to a linear regression ($r = 0.73$) was obtained when $W_{opt} = 2.5$, that is, when optimum conductivity is taken as about 316 μmhos. In the middle section of Figure 8.7 one especially notices that high- conductivity lakes K and R fit the regression much more closely.

Finally, Table 8.1 shows the results of a multiple regression including both area and transformed conductivity as independent variables. Both variables contributed significantly to the $r = 0.86$ fit, with the log–area slope reduced from 'canonical' to 0.19, a value more consistent with the other values in Table 8.1. The residual variance from this multiple regression, shown in the lower section of Figure 8.7, is in many ways the most interesting component of the data. Here the ordinate has been converted to read in number of species directly by subtracting the anti-$\log_e$ of the regression estimate from the value observed. One might be surprised, judging from the scales of the upper graphs in Figure 8.7, that Lake Tanganyika has 34 species more than the regression estimate. Also showing an unusual 'excess' of species, above what one might expect from their sizes and conductivities, were Lakes Victoria ($+6.5$), Awasa ($+5.5$), and Malawi ($+3.7$). Showing unexpected deficits of species were lakes Naivasha, Tana, and the two high-conductivity lakes Kivu and Rudolf.

These 18 lakes range in median latitude from about 13° N for L. Chad to 12° S for L. Malawi. Given the generality of the observation that diversity increases toward the equator, one might speculate that latitude could account for some of the residual variation shown in Figure 8.7. But simple regressions of species number on latitude, both log-transformed and untransformed, were not significant, nor did latitude contribute significantly to a multiple regression with area and conductivity. In fact the mean latitude of the four negative-residual lakes is lower than the mean latitude of the four positive-residual lakes.

Noting that the deeper lakes are generally the oldest and most permanent, Brown speculated that snail diversity might be a function of depth. Lake Tanganyika is in fact the deepest of the African lakes, at 1435 m maximum. The simple regression of log depth on log species number in the 18 African lakes was slightly more significant than the untransformed regression ($F=5.75$, $P<0.05$, $r=0.53$). But because log depth is highly correlated with log area in this data set ($r=0.63$, $P<0.01$), its contribution to a multiple regression was not significant. As a third independent variable, log depth had a regression coefficient of 0.078 ± 0.070, improving the multiple r only from 0.86 to 0.90.

D. Brown (pers. comm.) has further suggested a role for altitude. Lakes Tanganyika, Victoria, Awasa, and Malawi are low enough to enjoy truly tropical climates, while Lake Tana, for example, is quite high. But it seems unlikely to me that any single variable acting on a biogeographical scale can be found to contribute much beyond the $0.86^2 = 74\%$ of the variance explained by area and water quality here. For an explanation of the variance shown in Figure 8.7, one might better look within-lakes than among-lakes. Lake Naivasha, for example, dried out completely in fairly recent times.

It is interesting to note that Brown counted 30 endemic species among the Lake Tanganyika snail fauna, almost exactly the residual number in Figure 8.7. The fit is not as good elsewhere in Africa. Lake Malawi has 3.7 extra species but 17 of its 27 species are endemic, while Victoria has 6.5 extra species with 13 of its 28 endemic. Lake Awasa, with its residual of +5.5 species, has no endemics at all. Gorthner (1992) has suggested that some fraction of the speciation seen in ancient lakes may be non-adaptive. Regardless of the explanation one may favour for the data of Figure 8.7, it is fair to speculate that if competition is to be found, it is in lakes such as Awasa, Victoria, Malawi, and Tanganyika.

Isolation

Among the contributions made by MacArthur and Wilson (1967) to the theory of island biogeography was their analysis of the effects that distance from a mainland source area ought to have on the equilibrium number of species. In addition to his observations on the effects of lake area and water chemistry, Aho (1978a,b) examined his data on the gastropods of southern Finland for evidence of distance effects. He chose to concentrate on the 20 small lakes of his 'subarea III' in the vicinity of large, diverse, Lake Pyhäjärvi. Aho detected a significant negative correlation between species number in these lakes and their distances to Pyhärjärvi, just as would be expected if the larger lake were serving as a source of colonists. But he also found a highly significant positive correlation between distance from Pyhäjärvi and water hardness. Since no distance-diversity effect was apparent when the effects of water chemistry were partialled out, Aho did not pursue this line of inquiry further.

Aho did report, however, a significant positive correlation between species number in his low-diversity 'subarea I' and the percentage of the total drainage basin occupied by lakes, which might be taken as an indirect measure of the importance of isolation. Bronmark (1985a) also included several variables designed to estimate indirectly the significance of isolation in his analysis of Swedish gastropod faunas discussed previously. For about 100 ponds he measured distance to the closest pond (nearest-neighbour distance), mean distance to the five closest ponds, and number of ponds in a 50 ha square centred about the pond under consideration. Two additional variables, number of species in the nearest pond and in the five nearest ponds, could be taken either as measures of isolation or as measures of the general suitability of the local environment. Of these five variables only the last, mean number of species in the five nearest ponds, was entered (along with pond area and macrophyte diversity) into Bronmark's stepwise multiple regression of gastropod species number.

The most influential examination of the effects of isolation on the number of freshwater mollusc species to date may have been that of Sepkoski and Rex (1974). They analysed lists of unionid species for 49 'coastal rivers as biogeographic islands', ranging from the Penobscot River of Maine to the Escambia River on the Gulf Coast of Florida. Estimating area as land drained (rather than land inundated, as customary for lakes), they obtained a double-log fit ($r = 0.60$★★★) to the equation $S = 0.78A^{0.316}$.

Sepkoski and Rex obtained values of 10 environmental variables for their 49 rivers: temperature (both mean and variance), dissolved solids, and seven water chemistry measures. They also estimated degree of isolation as the number of stepping-stones to each of four presumed source rivers: the St Lawrence (north of the study area), the Alabama-Coosa (south of the study area), and two internal sources, the Appalachicola and the Savannah, chosen by their diversity. The number of steps to nearest source area was included as well. Thus Sepkoski and Rex had in their quiver 16 independent variables (area, 5 stepping-stone measures, 10 environmental measures) with which to predict the dependent variable, number of unionid species. At the end of 12 multiple regressions using various combinations of these, both log-transformed and untransformed, they obtained a multiple $r = 0.89$ with an 'interactive model' involving hydrogen ion concentration (antilog$_{10}$ of $-$pH), dissolved solids, nitrate concentration, and A/AC and A/SV, drainage area divided by stepping stone distances to the Alabama-Coosa and the Savannah, all untransformed.

I would prefer a simpler approach. From the author's appendix 1 I selected area and hydrogen ion concentration (data were unavailable for calcium or alkalinity), two variables which we have good reason to expect will influence freshwater mollusc diversity. In addition, over the distances encompassed by this survey (15 degrees of latitude) one might expect climate or coastal distance from the tropics to influence diversity. Thus I selected stepping-stone distance from the Alabama-Coosa River (ranging from 1 in the south to 33 in the north) not as a measure of isolation but as a measure of latitude. These three variables were log-transformed (effectively reconverting hydrogen ion concentration to pH) and used as independent variables in an analysis of unionid diversity.

Three outliers were identified in the initial regressions: the Blackstone River with an extremely small drainage area and the Satilla and Waccasassa Rivers with very low numbers of unionid species (two each). The results of my regression analyses based on 41 rivers are shown in Table 8.1. My simple species–area regression was not substantially different from that obtained by Sepkoski and Rex, with an identical estimate of 0.32 for the slope and an improved r. (Note that all values of C in Table 8.1 have been converted to hectares, here from the mi^2 used by Sepkoski and Rex.) The simple regression of log species number on pH was not significant, nor did pH add significantly to subsequent multiple regressions. The simple regression on log stepping-stone distance from the Alabama-Coosa River approached significance, and when included in a multiple regression with log area improved the multiple r from 0.67 to 0.79. Thus a regression with two independent variables, area and an estimate of climate or latitude, is only slightly less predictive than that of Sepkoski and Rex, with five independent variables. One must judge the evidence weak that isolation and dispersal have played a role in the diversity of eastern United States unionid faunas.

We now return to the data of Macan (1950) on the gastropods of the English Lake District. In the previous section it was demonstrated that the number of species in 53 bodies of water in this region seemed to be a function of habitat area and, to some extent, calcium concentration. Five of the 12 lakes surveyed by Macan were located within 10 km of large, diverse Lake Windermere, as were all 36 of the tarns. Among this subset of 41 water bodies, Macan found five gastropod species absent from his Lake Windermere list, as might have been guessed from Figure

8.6. Subtracting these species from the several sites where they occurred, it is reasonable to hypothesize that Windermere has served as the source for the remaining 11 elements of the Lake District gastropod fauna.

The relationship between species numbers and distance-from-source is customarily modelled logarithmically (e.g. Lomolino 1989):

$$S = e^{-kI**2}$$

where I is the measure of isolation (generally distance), e is the base of natural logs, and k is a fitted constant. Taken together with the effects of habitat size and water chemistry, species richness can now be modelled:

$$S = cA^z \mathbf{W}^m e^{-kI**2}$$
$$\ln(S) = \ln(c) + z \ln(A) + m \ln(\mathbf{W}) - kI^2$$

where area is A, the measure of water chemistry is $\mathbf{W}$, and c, z, m, and k fitted constants. In its log-transformed version, a multiple regression may be used to fit the constants by least squares.

I estimated dispersal distance as the minimum linear distance from Windermere to each of the 41 smaller lakes and tarns, taken from an enlargement of Macan's figure 1. Before including these data in a multiple regression with area and calcium as above, I performed a simple regression analysis. I found a significant *positive* regression of log species on squared distance from Lake Windermere (Table 8.1)! It happens that by chance of geology, regions outlying Windermere tend to have harder water than those nearby. The correlation between calcium concentration and distance (squared) to Windermere is 0.55 ($P < 0.001$). Thus diversity tends to increase away from the putative 'source', a reverse of Aho's findings.

Clearly these data show no more evidence of isolation effects than any other we have examined. We must conclude this section by noting that it has generally proved difficult to identify source areas from which to test the effects of isolation on the diversity of freshwater mollusc communities. The small number of tests that have been made to date have yielded unspectacular results.

Other environmental factors

Continuing in the tradition of Boycott, quite a few authors have catalogued the various environmental variables that may affect freshwater mollusc populations (Macan 1950, Malek 1958, Appleton 1978, Lodge *et al.* 1987). Depth, substrate, flow, and vegetation are always mentioned. Økland (1983, 1990) surveyed the gastropod faunas of almost 1500 lakes, ponds, and rivers in Norway. In addition to significant positive correlations between abundance or species richness and hardness or pH, he reported relationships with macrovegetation (3–4 categories) and substrate (4–5 categories). Bronmark (1985a) and Pip (1987a) also found relationships between snail diversity and macrophytes. While such factors may certainly influence freshwater mollusc diversity and abundance, the largest fraction of their variance is measured among habitats, rather than among populations. Detailed discussion of these variables will be deferred to Chapter 9.

Under some circumstances, however, substrate and current may vary regionally. The distribution of unionacean mussels seems particularly sensitive to large-scale hydrological variation (Bauer *et al.* 1991, DiMaio and Corkum 1995, Morris and Corkum 1996). For example, Strayer (1983) analysed the unionid faunas of three rivers in southeastern Michigan, dividing them into 75 segments of 8 km each. The rivers originate in glacial outwash plains and then pass through an end moraine/till plain region and a lake plain region before reaching their mouths in the Great Lakes. Mussel species richness is positively correlated with river drainage area, as expected. But Strayer also showed that surface geology (acting through current and turbidity) influenced the unionid species present on a regional scale. Strayer (1993) performed a similar analysis on the mussel fauna of the Susquehanna, Delaware, and Hudson River drainages of Pennsylvania and New York. Species richness was low over his 141 sites (a mean of about 3 species per site, 10 species maximum). In addition to a significant (semi-logarithmic) relationship with stream drainage area, Strayer reported that species richness seemed to be related to the presence or absence of a tide.

In a similar fashion, Haynes (1985) surveyed the gastropod faunas of about 10 rivers around the island of Viti Levu, Fiji. The rivers are generally swift and stony at their headwaters in the central highlands of the island, slowing and becoming more muddy as they approach the sea. Haynes attributed the increasing species richness she recorded in downstream sites to the regional effects of current and substrate, although again drainage area might also enter into the explanation.

Watters (1992) reported a strong correlation between number of unionacean species present and drainage area in the Ohio and Maumee River systems ($r = 0.92$, Table 8.1). The slope of the species–area relationship for fish ($z = 0.32$) was indistinguishable from that for mussels ($z = 0.34$), and the linear correlation between mussel species richness and fish species richness was high ($r = 0.96$, Table 8.1). Watters felt that mussel diversity might be dependent on fish diversity, rather than both being independent functions of the same environment.

The effects of climate on the regional distribution of freshwater molluscs have been touched upon briefly in our reviews of works on Ontario *Pisidium* (2° of latitude), the unionids of the eastern United States (15° of latitude) and the snail faunas of African lakes (25° of latitude). Researchers who have monographed freshwater mollusc faunas on a sufficiently large scale (e.g. Økland 1990, Brown 1994) have often attributed specific ranges to summer maximum temperatures, or winter minima.

Community composition

R. H. Green's (1971) 'multivariate approach to the Hutchinsonian niche' must be numbered among the more influential of all studies involving freshwater molluscs. In this first use of multiple discriminant analysis in community studies, Green analysed the 10 most widespread bivalves of central Canada: 2 *Sphaerium* species, 6 *Pisidium*, *Lampsilis radiata* and *Anodonta grandis*. His environmental variables were about half of the between-lake and half of the within- lake variety.

For each of 32 lakes surveyed, Green took one water sample above the thermocline and one below, or only a single such sample if stratification was absent. From these samples he measured pH, alkalinity, calcium, total hardness less calcium, and sodium chloride. These last four were log transformed. He then took a large number of qualitative bottom samples from each lake. For each sample containing bivalves he measured depth (log transformed), mean sediment particle size (also log transformed to 'phi units'), the standard deviation of the particle size (in phi units – the 'sorting coefficient'), and percentage loss on ignition (arcsine transformed). Knowing depth, Green could accord each sample estimated values for the five water chemistry variables.

The most widespread species, *Anodonta*, was found in 62 samples (18 lakes), next was *Pisidium casertanum* in 52 samples (12 lakes), and so on, down to *Sphaerium lacustre* in 11 samples (5 lakes). Green's aim was to use discriminant analysis to distinguish samples containing the 10 species as far as possible: to separate the 62 *Anodonta* samples from the 52 *P. casertanum* samples, and so forth. Some overlap would seem inevitable, however, since many of Green's samples doubtless contained more than one species. Further, his 62 *Anodonta* samples, for example, did not contain 62 different values of the five water chemistry variables, but at most 2(18) = 36. Since many of the 18 lakes containing *Anodonta* were surely also among the 12 lakes containing *P. casertanum*, the potential for overlap on the water chemistry 'axes' of niche dimension would seem great.

Five discriminant functions (DF) were required to account for 95% of the among-species variance. Interestingly, only two variables showed high coefficients in the first DF (45%): total alkalinity (−0.77) and calcium (0.61). The *Pisidium* species scored low on this DF, and the unionids high. Regarding the opposite symbols of these two variables, Green noted that in small, closed basins with high hardness and little water turnover, calcium carbonate precipitates before other salts. A number of such ponds were included among Green's sample sites, and they apparently contained *Pisidium* but no unionids. Thus Green interpreted his first DF as a function of lake size or water turnover. It is interesting that lake area, neglected in the introduction and methods sections of Green's paper, came to figure so prominently in the discussion.

Most of the remainder of the discrimination was due to within-lake factors. Depth showed a very high coefficient (0.822) in Green's second DF (24% of the variance), and (decreasing) particle size a rather smaller one. Here the two unionids seem to inhabit shallower water with coarser substrate, with the *Pisidium* species (especially *P. conventus*) deeper. Green's DF 3 was also primarily interpretable in terms of sediment characteristics. Such within-lake variables were the subject of our discussion in Chapter 2.

In studies of this sort it is customary to furnish statistics regarding the power of the discriminant functions obtained. One often sees classification matrices giving percentage of correct assignments or scatter plots with species means and confidence limits. The absence of such information in Green's work leads me to speculate that he uncovered a phenomenon well known by freshwater mollusc ecologists since the nineteenth century. Across any typical sample as environmentally diverse as Green's 32 lakes seem to have been, the general trend may have been for bivalves to occur together, not apart.

Boycott felt it so unlikely that British species of freshwater molluscs typically interact in any way that he declined to use the term 'association', preferring 'concurrence' instead. Nevertheless he was quite interested in 'concurrences'. Boycott tabulated data on the distribution of *Pisidium* from 466 sites in Britain provided by C. Oldham. For each of 15 species he calculated the 'average number of companions': the average number of other species of *Pisidium* in habitats occupied by the species

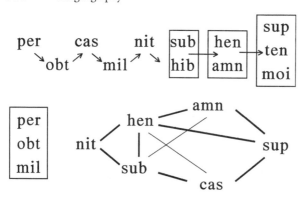

Figure 8.8. (Above) Boycott's 'hierarchy of gentility' showing British *Pisidium* species ranked by their 'average number of companions'. (Below) the positive associations among many of the same species sampled from southern England by Bishop and Hewitt (1976). Species connected by dark lines are associated at the 99% confidence level (by chi-square), and those with narrower lines at the 95% level. amn, *P. amnicum*; cas, *P. casertanum*; hen, *P. henslowanum*; hib, *P. hibernicum*; mil, *P. milium*; moi, *P. moitessierianum*; nit, *P. nitidum*; obt, *P. obtusale*; per, *P. personatum*; sub, *P. subtruncatum*; sup, *P. supinum*; ten, *P. tenuilineatum*.

in question. These figures varied from 1.1 and 1.3 for *P. obtusale* and *P. personatum* up to 6.1 and 6.3 for *P. moitessierianum* ('*torquatum*') and *P. tenuilineatum*. He then proposed the 'hierarchy of gentility' shown at the top of Figure 8.8, from *P. personatum* ('which lives in the slums') up to '*henslowanum* and *amnicum* which insist on the best places with 4 or 5 other species and *supinum, tenuilineatum,* and '*torquatum*' which are even more particular.'

This is a different (but not less valid) approach from the one I have pursued in this chapter thus far. Rather than inferring the suitability of a water body by its size and water quality, Boycott effectively let the clams themselves tell him which habitats were superior. In so doing, he reduced our multiple dimensions (habitat size, water quality, and perhaps other variables as well) down to a single one: 'average number of companions'.

Some 40 years later, Bishop and Hewitt (1976) confirmed Boycott's judgments. They took quantitative samples of *Pisidium* from 182 localites in eastern England. Three rare species, *P. hibernicum, P. moitessierianum,* and *P. pulchellum* were found less than five times. They then performed simple chi-sqare tests for association between all pairs of the remaining nine species. The result (bottom of Figure 8.8) is quite strikingly similar to Boycott's 'hierarchy'. Three species, *P. milium, P. personatum,* and *P. obtusale* were not significantly associated with any other. A putatively significant association was detected between *P. nitidum, P. subtruncatum,* and

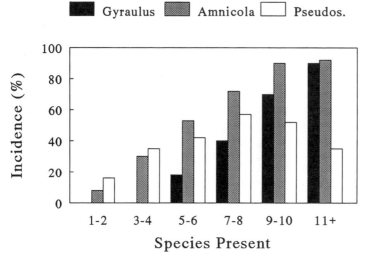

■ Gyraulus ▦ Amnicola ☐ Pseudos.

Figure 8.9. Three example incidence functions compiled by Jokinen (1987) from the freshwater snail faunas of New York and New England. *Gyraulus deflectus* is a 'high-S' species, *Amnicola limosa* a 'C–D tramp', and *Pseudosuccinea columella* a 'super-tramp'.

P. henslowanum, but not between *nitidum* and *supinum*, although *henslow-anum* and *supinum* were associated, and so on. It should be noted that Figure 8.8 is based on a matrix of $(N^2 - N)/2 = 36$ values of chi-square, and the levels of 'significance' should be viewed accordingly.

According to Bishop and Hewitt, 'Calcium availability is not a limiting factor in the study area, which is highly calcareous'. They did, however, recognize eight habitat categories among their sample sites (e.g. river, stream, canal, ditch, pond, etc.). They interpreted the lower half of Figure 8.8 primarily as the result of nine diverse species responding to these habitat variables independently. They did not view their results as evidence of competition or community structure.

We have previously reviewed the work of Jokinen (1987) on species–area relationships in the freshwater snail faunas of northeastern New York and southern New England. Dividing her data into two regions and two calcium levels, she reported rather high correlations between the surface area of lakes and ponds and the number of snail species they contained (Table 8.1). Jokinen subsequently went on to combine all her data and sort her 23 most common species into 'incidence categories' according to Diamond (1975), graphing the percentage occurrence of each species in water bodies of different diversities. For example, Figure 8.9 shows that *Amnicola limosa* occurred in about 5% of all habitats found to contain but 1 or 2 species, in about 30% of the habitats with 3–4 species total, and onward up to a 90% incidence in habitats containing 11 or more species.

Jokinen divided her species into four categories. *Amnicola limosa* was a 'C–D tramp', present in sites of all diversities at increasing incidence. There were 12 species fitting this description. An additional 4 species had similar 'incidence functions' but were absent from the lowest diversity (1–2 species) sites. These were categorized as 'A–B tramps'. The 5 species absent from both 1–2 species habitats and 3–4 species habitats (for example *Gyraulus deflectus*, Figure 8.9) were called 'high-S species'. All 21 high-S species showed peak incidence in water bodies of the highest diversity (11+). Jokinen did find, however, 2 species with highest incidence in the 7–8 or 9–10 species habitats. These 2 species, *Ferrissia fragilis* and *Pseudosuccinea columella* (Figure 8.9), she termed 'supertramps'.

The principle behind Jokinen's analysis was identical to that of Boycott's. She judged the suitability of habitats by the number of species they contained, not by independent environmental measurement. Jokinen's main conclusion was also identical to that of Boycott: species vary in their apparent ability to colonize low-species sites, i.e. sites that are inhospitable for some reason. In examining the entire shape of the 'incidence function', Jokinen obtained some information lost by Boycott when he calculated his 'average number of companions'. She identified two 'supertramp' species that she suggested might be crowded out of the best sites. This could be a statistical artifact, but it is consistent with the hypothesis that competition may play at least some role in species distribution. Yet one might also note that *P. columella* and *F. fragilis* are among the most widespread (and hence, in a sense, successful) of all Connecticut species. Perhaps 'adapted to unusual environments' would do more justice than 'crowded out'.

Over the period 1972–85, Pip (1978, 1985, 1986a,b, 1987a, 1988) surveyed the gastropods of 430 sites in Manitoba and elsewhere in central Canada – essentially the same area surveyed by Green in his study of bivalves. Pip's sites were quite various, located on small ponds, large lakes, creeks, and rivers, as were her collecting methods: wading, canoeing, dredging and s c u b a. Large lakes were assigned multiple sites. Most sites were visited only once, with a search effort limited to one hour, but many were resampled in different years. A variety of environmental measurements were taken at each site, including pH, alkalinity, chloride, nitrate, phosphate, sulphate, dissolved organics, total dissolved solids, and macrophytes. The inclusion of both lakes and rivers in a single study, however, precluded the measurement of habitat size or surface area.

Early in these studies (e.g. 1978, 1985) it became clear from joint-occurrence data (generally pairwise chi-square tests) that the (over 40) gastropod species inhabiting the waters of central Canada were not assorting independently. From tabulations of t-statistics comparing sites where each species was present with those where it was absent, it also seemed likely that the physico-chemical environment was responsible. Pip (1988) demonstrated extremely high intercorrelations ($P = 0.001$ or less) among alkalinity, chloride, sulphate, phosphate, and total dissolved solids, suggesting that these water chemistry variables could be 'regarded as a block with many similar trends'.

Pip used two different clustering techniques to construct dendrograms of the 'mean niche position' and 'niche congruency' of her 36 most widespread species. In the former analysis, she used Euclidean distances over the mean values for each of six water chemistry variables, standardized to Z-scores. In the latter analysis, her metric was the overlap between each pair of species for each of the variables, divided by their combined range, summed to a maximum value of six. Some rather distinct clusters emerged from these analyses, although their significance is difficult to judge. At a minimum these results confirm that one may expect subsets of snail species to appear positively associated when sampling from geographically large areas, and that water chemistry may have substantial influence on the list of taxa one might compile from any lake or pond.

Økland's (1990) monographic treatment of the freshwater snails of Norway was similar in some respects to that of Pip. From 1953 to 1973 he performed a semi-quantitative survey of 1498 localities in six categories (lakes, ponds, ditches, mires, puddles, and rivers), identifying 27 species. Although a formal treatment of the effects of habitat area was not attempted, Økland recognized one 'running water species' (*Ancylus fluviatilis*), one 'pond species' (*Armiger crista*) and two species of ditches, mires and puddles (*Lymnaea glabra* and *L. truncatula*). His stepwise multiple regression analysis of lake data from southeastern Norway suggested that water hardness (and, to a lesser extent, macrovegetation) were the primary factors in determining both number of species and time-catch abundance. He performed cluster analyses to group his sites by their hardness, pH, aquatic macrovegetation (3 states), geology (3 states), and terrestrial vegetation (5 states). Certain details of the mapping of snail abundance and diversity on the results of this analysis suggested to Økland that macrophyte abundance might have effects independent of total hardness and pH.

I am aware of only one previous study in which the effects of both water chemistry and habitat area on the mollusc fauna have been explored simultaneously, that of Aho *et al.* (1981). Earlier we reviewed Aho's (1978a,b) research on the distribution of freshwater snails in four 'sub-areas' of Finland, noting that water hardness, water colour, and habitat area seemed to account for a large portion of the variance in species number. Aho *et al.* (1981) combined these four subareas into one data set (72 lakes, about 400 sample plots) and examined the distribution of some 20 species on these three environmental axes. The overall impression was similar to that to be gained from the present Figure 8.6: considerable variation in the mean, due primarily to variation at the lower ends of the distributions, not at the upper. Aho also performed an incidence analysis similar to that of Diamond (1975), although his incidence functions were plotted with absolute frequencies on the ordinate, not percentage, and are difficult to interpret. He finished with a cluster analysis based on species co-occurrence data that showed several distinct clusters, which he attributed to shared environmental preferences.

Such ordinations as performed by Pip and Aho have considerable heuristic value in understanding community composition. The techniques employed, however, need not be so fancy. In this chapter we have repeatedly found that just two variables, habitat area and water quality (measured as calcium concentration or something related), exert significant influence on the number and identity of the freshwater mollusc species present. I would rather not collapse these two variables to a single axis by calculating 'incidence functions'. Nor do I feel that the use of matching coefficients, distance metrics, co-occurrence data or 'concurrence' data is necessary or even desirable. Rather, I suggest that maximum information may be transmitted by a simple plot of each species in a fauna by the average area and hardness of the habitats it occupies.

As an example I return to the data of Jokinen (1983) on the freshwater snails of Connecticut. It will be recalled that this data set includes the lists of species collected from 90 lakes and ponds, where both surface area and calcium concentration show a significant effect on the local diversity (Table 8.1). Jokinen reported that the entire freshwater snail fauna of Connecticut included 35 species, but 11 of these were found in at most three of the sites under study, and I have excluded them from my analysis. Two of the 90 sites were inhabited only by excluded species, and so themselves were eliminated. Then the 24 species remaining are listed in Table 8.2, along with statistics on the surface area and calcium concentrations across those of the 88 lakes and ponds in which they were found. The species are ranked in Table 8.2 by the mean surface area of their lakes and ponds. These figures range widely. It will be noticed, for

Table 8.2. *Summary statistics for log$_e$ surface area and log$_e$ calcium concentration at the lakes and ponds inhabited by the 24 most widespread snail species in Connecticut*

Species are ranked by the average area of their habitats

Species	N	ln area (ha)			ln calcium (mg/l)		
		Min.	Mean (s.e.)	Max.	Min	Mean (s.e.)	Max.
Stagnicola elodes	6	−2.53	0.62 (1.07)	4.40	1.16	2.64 (0.31)	3.14
Helisoma trivolvis	12	−2.12	1.28 (0.66)	4.39	0.87	2.28 (0.19)	2.98
Planorbula armigera	13	−3.91	1.34 (0.73)	4.76	0.79	1.67 (0.16)	2.77
Physa vernalis	5	−3.00	1.38 (1.67)	5.33	1.22	1.78 (0.24)	2.49
Pseudosuccinea columella	41	−3.91	2.15 (0.34)	5.33	0.09	1.66 (0.11)	3.08
Lyogyrus granum	15	−3.91	2.23 (0.57)	4.70	0.26	1.54 (0.16)	2.48
Ferrissia fragilis	37	−4.61	2.33 (0.38)	5.33	0.10	1.53 (0.12)	3.14
Physa heterostropha	36	−3.22	2.38 (0.40)	5.91	0.79	2.47 (0.10)	3.56
Micromenetus dilatatus	45	−3.91	2.59 (0.35)	5.62	0.09	1.46 (0.11)	3.08
Gyraulus parvus	30	−2.53	2.70 (0.36)	5.92	1.10	2.27 (0.13)	3.56
Helisoma anceps	41	−3.91	2.78 (0.32)	5.41	0.47	1.85 (0.12)	3.56
Lyrogyrus pupoidea	7	1.81	2.82 (0.29)	3.91	0.47	1.57 (0.32)	2.76
Physa ancillaria	24	−3.91	2.88 (0.41)	5.43	0.47	1.72 (0.17)	3.56
Physa gyrina	13	−1.72	3.01 (0.52)	4.79	0.69	1.16 (0.11)	2.10
Promenetus exacuous	20	−0.49	3.05 (0.36)	5.91	0.99	2.21 (0.15)	3.56
Fossaria modicella	14	0.53	3.10 (0.34)	5.43	0.99	1.98 (0.18)	3.56
Laevapex fuscus	23	1.17	3.24 (0.17)	4.40	0.47	1.74 (0.16)	3.08
Campeloma decisum	22	−0.92	3.47 (0.25)	5.62	0.09	1.73 (0.14)	3.56
Cipangopaludina chinensis	8	0.60	3.52 (0.53)	5.43	1.96	2.65 (0.20)	3.56
Amnicola limosa	45	1.31	3.57 (0.16)	5.62	0.09	1.74 (0.14)	3.56
Ferrissia parallela	5	1.55	3.65 (0.64)	5.43	0.69	2.54 (0.49)	3.56
Gyraulus deflectus	23	1.81	3.74 (0.21)	5.43	0.47	2.00 (0.17)	3.56
Helisoma campanulata	16	1.41	3.74 (0.30)	5.92	0.26	2.10 (0.26)	3.56
Valvata tricarinata	6	2.24	4.16 (0.49)	5.43	2.48	3.03 (0.15)	3.56

Note:
N, lakes and ponds inhabited by the 24 most widespread snail species.
Source: data from Jokinen (1983).

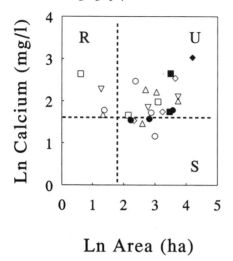

Ln Area (ha)

Figure 8.10. Twenty-four Connecticut snail species, plotted by the mean area and calcium concentration of their habitats (Table 8.2; data of Jokinen 1983). Closed symbols are prosobranchs: diamond *Valvata*, square vivipariids, circle hydrobiids. Open symbols are pulmonates: squares lymnaeids, circles physids, diamonds ancylids, erect triangles small planorbids, and inverted triangles *Helisoma*. Dashed lines divide the species into three groups referred to in the text as **U**, **S**, and **R**.

example, that the 3 smallest mean habitat areas, plus 2 of their standard errors, do not overlap the 3 largest mean habitat areas minus 2 standard errors. This phenomenon does not seem attributable to variation in the maximum lake size inhabited; almost all species seem to have been found in large lakes. Rather, the species vary strikingly in the minimum size of the lake or pond in which they were found. Although not as obvious in Table 8.2, mean calcium concentration varies as widely as mean habitat area, again the result of differing species minima. The 3 smallest mean calcium concentrations (those of *Physa gyrina*, *Micromenetus dilatatus*, and *Ferrissia fragilis*), plus 2 standard errors, do not overlap the 3 largest (*Valvata tricarinata*, *Cipangopaludina chinensis*, and *Stagnicola elodes*) minus 2 standard errors.

The 24 species are plotted in Figure 8.10 by their mean habitat areas and mean calcium concentrations. I have drawn a dashed line through an apparent gap in the distribution on the abscissa, while the ordinate is divided by a dashed line at 5 mg/l calcium. These lines are not meant to imply absolute boundaries. Rather, they are drawn to divide the graph into four regions for the sake of further discussion.

The region I have labelled **R** is occupied by the first four species listed in Table 8.2. These species seem to be adapted to habitats of small surface area, probably subject to freezing and drying and possibly temporary. The mean calcium concentration in these habitats is, on the other hand, at least fairly high, suggesting that they are productive. Notice for example that *Stagnicola elodes* inhabits ponds of the smallest mean surface area, but with the third highest mean calcium concentration. I therefore suggest that, in the main, Connecticut populations of *S. elodes*, *Helisoma trivolvis*, *Planorbula armigera*, and *Physa vernalis* are **R**-adapted, in the sense of Chapter 4. We have previously reviewed a considerable body of evidence suggesting that at least some Indiana populations of *Lymnaea (Stagnicola) elodes* show **R**-adaptation (population 'F' of Brown 1982). A New York population of *Helisoma trivolvis* was, in fact, one of two populations known to have demonstrated an order of magnitude greater reproductive effort than predicted for body size (see Figure 4.3). So for the four species in region **R** of Figure 8.10 I predict a high reproductive effort, a short lifespan, and semelparity (see Table 4.2).

The species occupying the region I have labelled **S** in Figure 8.10 inhabit regions low in calcium, perhaps relatively unproductive. However, these sites are larger, and in no case temporary. **S**-adapted populations seem to tolerate low food and perhaps osmoregulatory stress in exchange for environmental predictability, and as a result their overall reproductive capacity may be reduced. Notice that the species with the lowest mean calcium concentration, *Physa gyrina*, has a mean habitat area above average for the 24 species. It may be recalled that Indiana populations of *P. gyrina* are unusually broad in their diet and habitat use (Brown 1982) and that laboratory populations show unusual temperature tolerance (van der Schalie and Berry 1973, figure 4.1). Thus I suggest that Connecticut populations of *Physa gyrina* and to a lesser extent *Micromenetus dilatatus*, *Ferrissia fragilis*, and the two *Lyrogyrus* species tend to be **S**-adapted in the sense of Chapter 4. I hypothesize that their reproductive effort, relative to their body size, may be reduced.

Region **U** includes species typically found in habitats good in all physical respects. **U**- populations are undifferentiated with regard to life history traits, and as such can mature slowly or quickly and reproduce semelparously or iteroparously. But because gastropod diversity will be expected to reach a maximum in large, hard water habitats, **U**-populations would seem most likely to encounter interspecific competition. This theme will be pursued in the next chapter.

We have noted that in an 'incidence function' researchers effectively collapse the two axes graphed in Figure 8.10 to a single axis. Scoring lowest on such axes one finds 'supertramps', followed by 'A–B tramps', 'C–D tramps', and finally 'high-S species'. One might imagine a diagonal line running through Figure 8.10 and expect correspondence between projections of the 24 species on it and Jokinen's (1987) classification of them into incidence categories. To offer some anecdotal evidence, the square symbol very close to the intersection of the two dashed lines plots the average environment of *Pseudosuccinea columella*. One might argue that *Pseudosuccinea* occupies the worst average environment occupied by any of the 24 species; both rather soft and rather small. Of all the incidence functions plotted by Jokinen, that of *P. columella* (Figure 8.9) best fits the description of 'supertramp'. Similarly, the species with the most strikingly 'high-S' incidence function identified by Jokinen, *Valvata tricarinata*, would appear to score highest on a diagonal axis drawn in Figure 8.10.

Summary

The number of mollusc species present in a lake or river tends to increase as a function of a set of variables normally intercorrelated in natural fresh waters: calcium concentration, hardness, alkalinity, pH, and conductivity. Laboratory studies have confirmed years of field observations showing that calcium concentration in particular may directly affect both the growth and the reproduction of freshwater molluscs. Importantly, it appears that populations vary in their adaptation to calcium stress. Thus as the field biologist samples lakes and rivers of increasing hardness, the species count generally increases, up to a point. But laboratory studies have also shown that water of very high hardness (for example, greater than 40 mg/l calcium) may be detrimental to freshwater molluscs. Cosine transformation of the dependent variable can be employed to analyse the diversity or abundance of molluscs as a function of water quality ranging into very high hardnesses.

There is evidence, however, that the effects of calcium and related water chemistry variables on mollusc populations may be largely indirect. Harder waters are generally richer, supporting more primary productivity, greater biomass, and higher levels of decomposition. Perhaps it is to the greater availability of food, rather than to the improved osmotic environment, that natural populations of freshwater molluscs are responding.

Habitat area, measured as surface area for lakes and ponds or drainage area for rivers and streams, is a second variable well known for its influence on freshwater mollusc diversity. Table 8.1 shows the results of 14 double-log species–area regressions, including four data sets from Scandinavia, one from England, one from Africa, and the remainder from the eastern United States. The correlation coefficients of 13 are highly significant. Values of the slope (z) approach a 'canonical' value of 0.25 in four of these cases, seem substantially less in seven cases and greater for the two unionacean data sets. Just as the case with calcium, individual mollusc species seem to vary in their adaptation for smaller rivers or ponds, so that species are not added randomly as area increases.

Table 8.1 also shows that trophic level or water quality often significantly affect the relationship between habitat area and freshwater mollusc diversity. Harder, more eutrophic ponds have more species, but the slope of their species–area regressions are generally shallower. Hence the diversity of a large oligotrophic lake may approach that of a large eutrophic lake. The inclusion of a measure of water quality along with habitat area in a multiple regression analysis of mollusc species number may improve the fit significantly.

These two independent variables, water chemistry and habitat area, are as useful for explaining the diversity of the ancient African lakes as they are for the recently colonized lakes of Scandinavia. In general, other important variables are difficult to identify. Isolation, measured as distance from presumed sources of colonists, would seem a likely candidate for freshwater molluscs. Little evidence of an isolation effect was obtained in an analysis of the snail faunas in small lakes and ponds near England's Lake Windermere or in the unionid faunas of eastern United States coastal rivers. Some latitude effect may be present in the latter data set. Most of the data sets reviewed here were collected at scales too small to reflect substantial influence from climatic variation, and large enough to remove the influence of depth, substrate, and current.

Workers have often found that freshwater mollusc species do not appear to assort independently on a geographic scale, but rather seem to form 'clusters', 'concurrences' or 'associations'. There seems to be little evidence that such associations reflect any sort of interaction among the species, however. Rather, it appears that the distribution of each species is an independent consequence of the environment. I suggest that not only may the number of species inhabiting a particular lake or river be modelled as a function of water chemistry and habitat area, the composition of the molluscan community may be as well.

To illustrate this approach, 24 species from the freshwater snail fauna of Connecticut were plotted by the mean size and mean calcium concentration of their habitats (Figure 8.10). Three categories were recognizable: **R**-populations adapted to small (i.e. ephemeral) but productive habitats, **S**-populations adapted to soft (i.e. unproductive) but permanent habitats, and **U**-populations adapted to permanent, productive habitats. In the final chapter of this work, the potential for interactions among such mollusc populations as may co-occur in permanent, productive habitats will be examined in additional detail.

9 · *Communities*

In this chapter we will examine the relationships among populations within freshwater mollusc communities. We define the term 'community' broadly, noting as we do that its component mollusc populations may be of such diverse ecological character that the likelihood of interactions among them may be less than the likelihood of interactions between them and other elements of the benthos. The artificiality of the concept of the freshwater mollusc community does not, however, erase its utility.

Developing a theme first opened in Chapters 2 and 3, here we review a large body of literature approaching diet and habitat from a comparative standpoint. Differences in the gut content of co-occurring gastropod species can be substantial, but this seems primarily due to differing habitat choice, rather than selective grazing. Variation in substrate preference seems central in gastropod communities, with other aspects of the local environment (especially depth) important as they impact substrate. The overlap in diet and habitat observed within bivalve communities seems great.

We next review the several studies that have applied ordination techniques to variation in species distribution within freshwater mollusc communities, and offer several original analyses. The elements of most communities seem to associate into subsets according to features of the habitat. We also find one situation where species seem to aggregate apart, and cases where little structure of any sort is apparent. Sampling scale seems to be a key to detecting interspecific pattern in distribution. Both taxonomic relatedness and overall morphological similarity have been used as indirect measures of ecological similarity. In at least one well-documented community where species do not seem to be assorting independently, co-occurring freshwater gastropod species tend to be less similar to each other than one would expect. The association of morphologically diverse species seems to reflect some evidence that competition is structuring this community.

Our review of the 'distribution of commonness and rarity' in freshwater mollusc communities finds that relative species abundances often differ significantly from random. Most distributions fit the 'canonical lognormal' model, suggesting some minimal competition for resources. And there is some evidence of unexpectedly high taxonomic diversity among communities, suggesting that species distributions may be influenced by ecological overlap.

Terms and conditions

I date the current rigour by which studies of community ecology are judged to about 1981, with the symposium organized by Strong and his colleagues (Strong et al. 1983). They subscribed to the loose definition of ecological communities fairly common now, 'groups of species living closely enough together for the potential of interaction'. This is the definition I adopt here, noting as I do that no reference is made to organization or structure of any sort, that 'interaction' could include any sort of symbiosis, competition, predation and/or parasitism, and that indeed any interaction at all is only 'potential'. Fauth and colleagues (1996) have suggested 'assemblage' to describe a community of phylogenetically related organisms, a term that summarizes the subject matter of this chapter rather well. 'Mollusc assemblage' offers little advantage over 'mollusc community', however, and is more unfamiliar.

Strong and his colleagues invited 29 primarily analytical and experimental contributions to their 1981 symposium, testing for organization in communities with considerable rigour. No study involving freshwater molluscs was invited, and this was no oversight. Research on mollusc communities lags behind that on communities of most other sorts. In any case, a twofold mission for community ecologists emerged. First we must establish that the collection of organisms before us has some organization, then secondly we may try to establish the cause of that organization. The great majority of this chapter is devoted to mission one. But on the rare occasions when mission two comes to the fore, the reader may perceive a bias toward competition as a community organizer. This is an artifact of the data at hand. It is possible to prospect for evidence that competition structures freshwater mollusc communities in data on the molluscs alone, and as such more data have accumulated regarding this answer to the question. Tests for the potential community effects of predation and disturbance require data on predators and the environment, and such data are more difficult to obtain, as Chapters 6, 7 and 8 have testified.

We have already reviewed a body of evidence regarding ecological overlap (Chapters 2 and 3) and competition (Chapter 5) among freshwater molluscs. Introduced populations of bivalves may compete for space and/or food with native populations, as the North American experience with *Corbicula* and *Dreissena* has demonstrated (Gardner *et al.* 1976, Boozer and Mirkes 1979, Schloesser *et al.* 1996). A number of studies (e.g. Lassen and Madsen 1986) have demonstrated food competition and physical interference between *Helisoma* and *Bulinus*, at least in the laboratory. We also saw evidence of competition among wild (although in some cases introduced) populations of ampullariids or thiarids and *Biomphalaria* (Radke *et al.* 1961, Pointier *et al.* 1988, 1989). We might return, for a moment, to the experiments of Brown (1982) regarding northern Indiana populations of *Helisoma trivolvis*, *Aplexa hypnorum*, *Physa gyrina*, and *Lymnaea elodes*.

As may be recalled from Chapter 5, Brown estimated habitat overlap among his four species as the similarity of their distributions over eight Indiana ponds. He estimated dietary overlap using an arena with four compartments (carrion, pond vegetation, decaying leaves, control) to determine food preference. Brown then calculated niche overlap among the four species over both habitat and dietary dimensions, finding the greatest value between *Lymnaea elodes* and *Physa gyrina*. This prompted him to design enclosure experiments to estimate the potential for competition between these two in greater detail (see Figure 5.12). But were I to select a pair of Brown's populations for detailed study, I would have tended to set aside 'habitat overlap' data (co-occurrence across eight ponds in this case). It is entirely possible that low habitat overlap may be a consequence of competition, rather than evidence that it does not occur. I would have concentrated on dietary overlap instead. *Lymnaea* strongly avoided detritus and preferred carrion (an unusual foodstuff, but a legitimate measure of preference), while *Aplexa hypnorum* avoided carrion to prefer detritus. Since neither *Physa* nor *Helisoma trivolvis* showed any significant preference among the four compartments, they would have been my nominees for 'most likely to compete'.

Brown considered his entire set of species a 'guild of pulmonate snails'.

In providing evidence of interspecific competition between pairs of populations not even especially likely (in my view) to compete, Brown's studies persuade one that competition may, at least on some occasions, structure freshwater mollusc communities. The concept of the 'guild', a group of organisms sharing a resource, is considerably narrower than the subject matter of this chapter. It is clear that among the freshwater molluscs may be found several guilds, none of which is exclusive.

The ecological similarities between snails and grazing mayfly and caddisfly larvae, for example, are striking. The caddisfly *Helicopsyche* even carries a coiled 'shell' of sand grains good enough to fool students on their invertebrate zoology exams. In Chapter 3 we mentioned studies by Cuker (1983b), Cattaneo and Kalff (1986) and Hawkins and Furnish (1987) where freshwater snails were excluded or removed from a habitat, and in each case, an increase was noted in the population densities of such grazers as oligochaetes, microcrustaceans, and insect larvae (most notably the larvae of the dipteran family Chironomidae). Hill (1992) reported a great deal of dietary overlap between *Goniobasis* and caddisfly larvae of the genus *Neophylax*. Periphyton biomass, *Neophylax* biomass and lipid content were all greater in Tennessee streams lacking *Goniobasis* than in those where *Goniobasis* was present. A review of 89 experimental studies involving stream grazers and their periphyton resources has been offered by Feminella and Hawkins (1995).

High levels of ecological similarity certainly also exist between bivalves and insect filter feeders, such as *Hydropsyche* and other net-spinning caddisflies. In fact, the effects of zebra mussel invasion are far-reaching, extending much beyond their adverse impacts on unionids and other filter feeders (MacIsaac 1996, Karatayev *et al.* 1997). Zebra mussels clarify the water (Lowe and Pillsbury 1995) and alter the physical character of the substrate (Botts *et al.* 1996, Ricciardi *et al.* 1997), with consequence to entire lake ecosystems.

Aneuran tadpoles compete with snails, while snails may in fact facilitate tadpole growth (Bronmark *et al.* 1991). In a carp pond in India, *Bellamya bengalensis* reportedly almost outcompeted the stocked fish for their food (Raut 1986). Freshwater molluscs do not live in isolation from the resources upon which they feed, the other consumers with which they compete, or the predators, parasites, and pathogens which feed upon them (Thomas 1990). But as long as it is recognized that the 'freshwater mollusc community' under study in this chapter is an artificial construct, erected for our convenience, it would seem safe to proceed.

Interspecific ecological overlap

Gut content

Dudgeon and Yipp (1985) collected samples of nine Hong Kong gastropods: the prosobranchs *Melanoides tuberculata*, *Bithynia* spp., and *Sinotaia quadrata*, and the pulmonates *Physa acuta*, *Lymnaea ollula*, *Radix plicatulus*, *Biomphalaria straminea*, *Gyraulus* sp., and *Hippeutis cantonensis* (a planorbid). Ten sites were visited, with most species collected at more than one site. They dissected stomachs from approximately 20 individuals for each species per site, and estimated percentage volume from microscopic examination. Intersite differences were striking. Dudgeon and Yipp compiled a series of tables comparing *B. straminea* gut contents over all pairs of the five sites it was collected, *L. olivula* over all pairs of the seven sites it was collected, and so forth, noting significant intraspecific differences in almost all comparisons. The only gut content data published were pooled over all sites, however, rendering interspecific comparison difficult.

By volume, detritus comprised the majority of the average gut content in all nine Hong Kong species (Figure 9.1, upper). The most striking contrast is the high proportion of green algae in the stomachs of *P. acuta*, *B. straminea*, and *L. ollula*. This may certainly be due at least partly to food availablity at different sites. These three species were found together in at least four of five to seven sites, so the difference between their gut contents and those of the other six species may be an artifact. Fortunately, three species (*R. plicatulus*, *Gyraulus*, and *Bithynia*) were found only at a single site, so their comparison is a fairer one. While *Bithynia* and *Gyraulus* contained 95–100% detritus, *Radix* guts contained a strikingly large fraction of macrophyte tissue. So at least some of the variation seen in Figure 9.1 seems to reflect dietary differentiation in the snails.

Reavell (1980) surveyed the gut contents of 20 British species of freshwater gastropods collected from over 20 sites. He estimated percentage total squashed area for each component of the diet in a fashion similar to that of Dudgeon and Yipp, but presented data on the gut contents of snails collected at individual sites. The Shropshire Union Canal is reported to have both high nutrient levels and high hardness, while the upland Tannyrallt stream has low levels of both. Nevertheless the main organic fraction of gastropod gut contents at both sites was detritus, as in Hong Kong (Figure 9.1, lower). Shropshire species as a whole seemed to

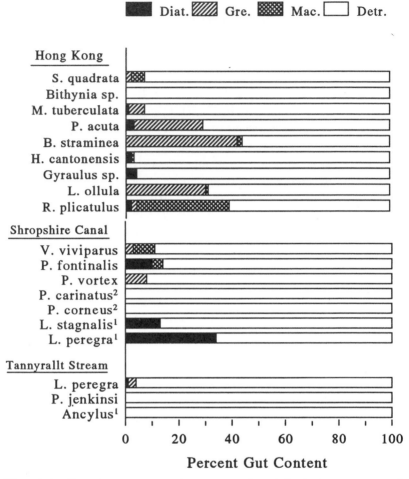

Figure 9.1. Organic gut contents (percentage volume) for freshwater gastropods from three areas. Most data from Hong Kong are pooled for several sites (Dudgeon and Yipp 1985). Data from the hardwater canal and softwater upland stream in Britain are from Reavell (1980). Data with known month of collection are superscripted: [1]April, [2]August. Results are given for diatoms (Diat.), green algae (both unicellular and filaments combined, Gre.), macrophytes (Mac.), and Detritus (Detr.).

ingest more diatoms, green algae, and macrophytes, as might be expected from the greater abundance of such resources in richer environments. It is instructive that the guts of the only species present at both sites, *L. peregra*, included on average 34% diatoms in Shropshire but only 1% diatoms at Tannyrallt.

Several finer-scale studies lead one to believe that both the broad overlap and the specific differences apparent in Figure 9.1 may be more than an artifact. Kesler and colleagues (1986) have compared in some detail the diets of *Physa vernalis* and the lymnaeid *Pseudosuccinea columella* in a small Rhode Island pond. They quantified gut contents as the proportion of 40 microscope fields containing each of five dietary categories (plus sand), and again found detritus to be more common than any other foodstuff in both species, over nine samples taken through one year. Diatoms were especially common in the guts of both species during the spring. Kesler and colleagues introduced equal numbers of the two species into a 5 litre enclosure which had been exposed to natural conditions for 37 days previously. After two days, the snails were removed and gut contents compared. They found a significantly higher proportion of detritus in the guts of *Physa* and a similarly high proportion of filamentous algae in *Pseudosuccinea*. They suggested that since the latter has stronger jaws than *Physa*, larger teeth with thicker cusps, and a gizzard equipped with sand, *Pseudosuccinea* is better adapted to ingest tough algal filaments.

Lodge (1986) obtained similar results working in a rather different situation. His initial observation was that *Lymnaea peregra* was generally found on submerged macrophytes (especially *Elodea*) in Radley Pond, a very small (1 ha) eutrophic English pond, while *Planorbis vortex* was generally found on emergent macrophytes (especially *Glyceria*). Discussion of Lodge's entire set of observations on the substrate preferences of the snail fauna of this particular pond will be saved for later in this chapter. Here it is important to note that neither *Lymnaea* nor *Planorbis* seem to be eating the macrophytes themselves, but rather grazing upon the attached epiphyton.

Lodge found that organic detritus was significantly more abundant upon *Glyceria* than upon *Elodea* per unit area, but that filamentous green algae commonly attached only to *Elodea*. Detritus comprised about 30% by volume of the epiphyton collected on *Elodea*, with diatoms and filaments amounting to about 20% each. But although detritus also amounted to about 30% of the gut volume of *L. peregra* grazing on *Elodea*, filaments comprised over 50%, a significantly greater proportion. Once again we see evidence of a lymnaeid selecting for filaments, here both by habitat choice and differential grazing. That no evidence of this sort of preference appears in either *L. peregra* population shown in Figure 9.1 serves as a lesson. A second lesson can be gathered from Lodge's observation that, although detritus comprises about 60% of the gut volume of *P.vortex*, detritus amounts to 80% of the epiphyton volume on *Glyceria*, so that the snail actually seems to be selecting against detritus and for diatoms.

Underwood and Thomas (1990) subsequently completed a thorough laboratory study that nicely complemented the field observations of Lodge. Rather than allowing English *L. peregra* and *Planorbis planorbis* to select different substrates, in effect they 'forced sympatry'. (*P. planorbis* is very similar to Lodge's *P. vortex*, although with a broader shell.) Underwood and Thomas placed 16 individual *Lymnaea* or *Planorbis* in beakers with previously ungrazed shoots of the macrophyte *Ceratophyllum demersum* for two days. (The unpalatability of *Ceratophyllum* may be recalled from Chapter 3.) Gut contents were not strikingly different except in that *Lymnaea* had (perhaps previously) ingested sand while *Planorbis* had not. Expressed as a percentageage of the total volume, the guts of both species contained primarily algae, followed by amorphous detritus and dead macrophytic tissue, with no evidence that any living macrophytic tissue had been ingested.

They divided the algal component of the ingesta into 15 categories, and calculated 'electivity indices' based on relative volumes. Comparison with ungrazed *Ceratophyllum* showed that both snails tended to leave the same algal taxa behind, especially small and adnate diatoms. Although the contents of *Lymnaea* and *Planorbis* guts were not compared directly, the tabulations show some striking differences. It appears, for example, that *Lymnaea* more efficiently grazed the adnate diatom *Epithemia turgida*, with an electivity index of $+0.47$, than did *Planorbis* ($E = -0.04$). So although these results may not be directly relevant to the situation in the wild, they demonstrate the possibility that different taxa of freshwater snails may graze in different fashions.

In general, however, field evidence is weak for selective ingestion and stronger for selective microhabitat choice. Barnese and colleagues (1990) placed nutrient-releasing clay pots in Douglas Lake, Michigan, for 25 days, keeping them enclosed for protection from macrograzers. They then placed individual snails on small fragments of these pots and allowed them to graze a single trail over the surface. Setting aside detritus and other potential foodstuffs, they compared the percentage algal cell bio-volume on grazed and ungrazed surfaces in six categories: overstorey diatoms, understorey diatoms, the blue-green *Schizothrix*, and the greens *Characium* and *Coleochaete*. Very little difference was apparent in the grazing effects of five local pulmonate species (2 lymnaeids, 2 physids, 1 planorbid). A sixth species, the pleurocerid *Goniobasis livescens*, differed significantly from the five pulmonates. Most of the difference seemed to be due to a generally lower grazing efficiency for *Goniobasis* on all categories of algae, rather than selective ingestion.

Barnese and colleagues made the reasonable suggestion that the difference in grazing efficiency between the pulmonates and *Goniobasis* was due to their striking radular differences. As may be recalled from Chapter 3, pulmonate radulae have many more teeth per row than *Goniobasis*, and each cusp is much smaller. But it is also possible that the difference Barnese and colleagues report may be an artifact of unnaturally high food availability and unnaturally frightened *Goniobasis*. Under normal circumstances, *Goniobasis* may have the ability to compensate for its coarser radula behaviourally, as has been suggested for marine prosobranchs (Hawkins *et al.* 1989). Some observations of my own suggest that undisturbed freshwater snails of many diverse types generally tend to wipe the substrate clean.

A problem common to most of the gut content studies discussed thus far has been quantification of such ingesta as algal filaments, detritus, and macrophyte tissue. Visual estimates of 'percentage gut volume' occupied by these food categories must be very approximate – perhaps so approximate as to miss real differences between the available periphyton and the contents of various snail guts. 'Percentage algal biovolume' as calculated by Barnese *et al.* is an improvement, but count data would be superior. So for reasons of quantification, Dillon and Davis (1991) performed a detailed analysis on the numbers of diatom cells found in the guts of snails collected from single square metres of homogeneous substrate, comparing them with diatoms collected artificially.

The diatom flora of the Mitchell River in Surry County, North Carolina, is typical of rapidly flowing, nutrient-poor waters, containing about 20 species ranging in size from about 10 μm to about 100 μm. Cell shape is also quite various, from the slender, threadlike *Nitzschia* to *Cymbella*, about the shape of an American football. Also inhabiting the Mitchell River are three gastropod species: *Physa* sp. and two pleurocerids, *Leptoxis carinata* and *Goniobasis proxima*, the latter in high density. The cusps on *Goniobasis* radulae seemed perfectly scaled to collect 100 m diatoms such as *Synedra* and *Cymbella*, while the much finer radula of *Physa* seemed better scaled to collect 10 μm diatoms, such as *Achnanthes*. Might this be reflected in gut contents?

Dillon and Davis selected a 1 m^2 sample site at mid-river, with a uniformly shallow, rapid current and rocky bottom. Snail density was high (certainly hundreds per square metre) and the rocks appeared to have very little periphyton of any sort. We collected 5 adult *Goniobasis*, 10 juvenile

Table 9.1. *Abundance of the most common diatoms found on substrate and in guts of adult* Goniobasis proxima, *adult* Physa, *and juvenile* Leptoxis *in two samples of the Mitchell River, NC*

	Riffle				Pool	
	Substrate	Goniobasis	Leptoxis	Physa	Substrate	Goniobasis
Achnanthes deflexa	1090	1117	1090	1114	853	900
A. microcephala	42	41	62	43	0	0
A. minutissima	0	1	0	1	183	141
Cocconeis placentula	35	34	31	37	8	9
Cymbella tumida	21	1	6	7	9	11
Gomphonema parvulum	13	13	16	13	10	11
Navicula cryptocephala	17	24	25	26	78	53
N. decussis	31	30	51	33	53	41
Nitzschia acicularis	16	19	24	15	4	13
Synedra fasciculata	15	21	31	25	66	67
S. rumpens	156	135	95	113	126	90
S. ulna	61	54	52	62	75	106

Source: data of Dillon and Davis (1991).

Goniobasis, 10 juvenile *Leptoxis*, and 2 adult *Physa*, preserving each immediately in 70% ethanol. We also scraped several rocks and washed the loose material into vials. We then selected a second 1 m² site in a slow, deep, silty pool near the riverbank, overshadowed by trees. We again collected and preserved 5 adult *Goniobasis*, and sucked a sample of the surface sediment. A total of 1500 diatom cells were identified from each sample.

Example results are shown in Table 9.1. In spite of striking radular differences, no significant difference is apparent between the gut contents of riffle-dwelling adult *Goniobasis*, *Leptoxis*, or *Physa* in the abundance of any diatom species. The only difference between these three samples and the sample artificially collected with a scalpel was that two larger diatom species, *Synedra* and *Cymbella*, tended to be underrepresented in snail guts. The relative frequency of the diatom taxa identified from the guts of juvenile *Goniobasis* was rather distinctive, but an analysis of frustule fragments led us to conclude that this phenomenon was due to rougher processing by juvenile *Goniobasis*, not a different diet.

Table 9.1 also shows that the diatom flora from the calm, muddy pool was quite different from that of the rocky riffle, but once again, adult *Goniobasis* seemed to sample the diatoms randomly, about as one might with a pipette. The diatoms ingested by *Goniobasis* also vary quite widely by season in the Mitchell River. Although *Achnanthes* seemed to be the most common diatom in snail guts year around, we found that second place went to *Cymbella* in the fall (11%) and *Synedra* in the winter (23%).

It should be remembered that we did not examine any other component of the diet of these gastropods, and thus cannot address the question of dietary overlap directly. But we were surprised to find that radular morphology seemed to have such a small effect upon the material ingested. The careful studies of Smith (1989a,b) also failed to turn up evidence of feeding selectivity across five New York populations of *Helisoma trivolvis*, in spite of striking diversity in radular tooth morphology. It may well be that the evidence for selective grazing gathered by Lodge, Kesler, Calow and others applies to rich, diverse pastures, not the poorer, possibly overgrazed Mitchell River. In any case, it might be best to assume as a starting hypothesis that snails are randomly sampling whatever they happen to be crawling upon, but that diverse snail populations may select different surfaces upon which to crawl. In a sense, diet and habitat are synonymous.

Feeding strategy

Daldorph and Thomas (1988) compared six British freshwater snails in their responses to a variety of short-chain carboxylic acids and maltose, known to be released from decaying plant tissues. All the chemicals attracted and/or arrested *Lymnaea peregra*, all but the smallest carboxylic acid (C_2) attracted *Planorbis contortus* and *P. planorbis*, and all but the larger carboxylic acids (C_5–C_8) attracted *Physa fontinalis*. Very little response was noticed in *Planorbis vortex* or *Bithynia tentaculata*.

I am intrigued by the available data on diurnal variation in feeding rates or activity rates (the distinction is a fine one). A heroic study was made of a *Biomphalaria glabrata* population inhabiting a St Lucia marsh by McKillop and Harrison (1982). They followed individual snails using a square grid made of thread and a flashlight, and observed nine times more feeding activity at night than during daylight. This same phenomenon was noted in a laboratory population of *B. glabrata* by Rotenberg *et al.* (1989). A related phenomenon, also demonstrated in both laboratory and field populations, is a significant tendency for *Biomphalaria* to float to the surface during dark hours (Pimentel and White 1959b, Bao 1985). The phenomenon may be seasonally reversed, however (Pimentel-Souza *et al.* 1984).

Williams and Gilbertson (1983) verified nocturnal increases in the heartbeat, locomotion, and feeding rates of healthy *Biomphalaria* but noted that these differences were extinguished in parasitized snails. Infected snails ate continually, presumably in an effort to keep up with the increased metabolic demands of a parasite load. In a similar fashion, McDonald (1973) found two to three times more feeding at night in a laboratory population of *Lymnaea stagnalis* at a comfortable 20 °C. Feeding periodicity was extinguished at 30 °C, however, as the snails endeavoured to balance their energy budgets. Health, food, or temperature stress may explain Veldhuijzen's (1974) failure to note any feeding rhythm in his population of *L. stagnalis*. We have previously noted what a fragile thing is behaviour.

In direct contrast, laboratory populations of *Bulinus* are much more active and feed at increased rates during daylight hours. This has been shown for *B. africanus* by Morgan and Last (1982a) and for *B. tropicus* by Chaudhry and Morgan (1983). To my knowledge, the only other freshwater snail that has been well studied in this regard is *Melanoides tuberculata* (Beeston and Morgan 1979a,b; Morgan and Last 1982b). This snail generally spends daylight hours buried and feeds at night, but has two peaks of activity, one at dawn and the other at dusk, the latter perhaps more associated with return to hiding. It would be fascinating to put *Biomphalaria*, *Bulinus*, and *Melanoides* together and observe if they continue to display three such different cycles of activity when under identical conditions.

Depth, temperature, and oxygen

It seems quite evident that all three of these factors are related and that all, usually in combination, must affect freshwater mollusc distribution. Substrate and current are also generally related to depth in natural environments, but I will attempt to discuss their effects separately, in sections to follow.

It may be recalled from Chapter 2 that habitat overlap is often extensive in populations of freshwater bivalves. For example, Strayer (1981) surveyed the microhabitat choice of 22 species of mussels in Michigan streams. His techniques were qualitative but thorough. He estimated depth, substrate type, vegetation, distance to shore, and current speed for 2161 individual mussels at 37 sites. He concluded: '*Villosa iris* was usually found nearer shore and in shallower water than were other species . . . Other than this, I could discern no consistent differences among the microhabitats of the various species.' Strayer and Ralley (1993) were largely unable to distinguish the microhabitats of a much smaller assemblage of mussels in New York (6 species), with an even larger number of sites (270) and variables (9). This led the authors to 'question the adequacy of a traditional microhabitat approach to unionacean ecology.'

Salmon and Green (1983) took 240 quadrat samples of 0.5 m² each from Ontario's Middle Thames River, recording water velocity, depth, percentage vegetative cover, distance to shore, and substrate. They performed a preliminary principal component analysis on the substrate data to convert percentageages by weight of five substrate categories into scores on three principal components (PC), and did a second principal component analysis on these three substrate PCs plus four other variables to extract five new PCs, upon which they performed discriminant function analysis. There is no way to know if any of the (rather few) interpretable results of this analysis are significant or not. But broadly, it was concluded that mussels 'are most often found in vegetated shallow areas with low current, but their occurrence is also biased toward substrates with a substantial component coarser than gravel'. Salmon and Green also performed a discriminant function analysis to distinguish the habitats of the seven mussel species of the Middle Thames, and derived a gradient of unknown significance from the mid-channel, coarse *Villosa iris* ($N = 108$) to the nearshore, sandy *L. radiata* ($N = 2$). The discrepancy between these results and the meagre results of Strayer should sound a cautionary note to future workers in the community ecology of mussels.

It may in fact be possible to detect some evidence of habitat differentiation in the simpler unionoid communities of lakes. Stone and colleagues (1982) established five 60 m transects toward the centre of Budworth Mere, Cheshire, and sampled the three populations of unionids it contained with SCUBA. Mussel densities exceeded 100/m² in the shallow samples, decreasing to negligible below 3 m. Stone *et al.* attributed this decrease to many factors, including soft substrate, high levels of inorganic particles suspended with the food, and too little light, oxygen or current. They further noted that mussels in the very shallowest margins of the lake run greater risks of predation and exposure, as we have noted in Chapter 2. *Unio tumidus* was most common at depths of 0.5–1.0 m in Budworth Mere, where coarse sediments seemed to suit its heavy shell. *Anodonta cygnea* seemed to be relegated to depths in the 1.0–2.0 m range, where it survives on fine sediments by virtue of its lighter shell. *Anodonta anatina*, with an intermediate shell, may (I might speculate) be on its way to exclusion in Budworth Mere. One might wish for an update.

Britannia Bay is located on the Ottawa River about 10 km upstream from the site studied by Hamill and colleagues (1979) featured in our

Table 9.2. *Number of individual pisidiids in samples at selected depths along a transect in Britannia Bay, Ottawa River*

	Depths (m)				
	0.25	0.36[a]	3.0	6.0	Total
Mean particle (mm)	0.17	0.18	0.15	0.29	(all 8 depths)
Musculium securis	2	0	9	4	39
M. transversum	50	24	0	0	80
Pisidium casertanum	106	165	16	0	437
P. dubium	2	1	0	0	8
P. ferrugineum	0	2	1	0	8
P. henslowanum	0	17	0	0	37
P. lilljeborgi	14	54	4	4	230
P. nitidum	26	85	2	3	291
P. punctatum	15	38	1	1	75
P. variabile	12	9	0	0	44
P. walkeri	0	1	0	0	1
Sphaerium striatinum	0	0	26	2	64

Notes:
[a] Only four grab samples. Multiply N by 1.25 to compare with other depths.
Source: Kilgour and Mackie (1988).

Chapter 2 discussion of corbiculoid habitat. Here Kilgour and Mackie (1988) took four or five Ekman grab samples at each of eight depths down a transect. Their results (Table 9.2) are strongly reminiscent of the distribution of unionids in lakes, on a somewhat smaller scale. All nine *Pisidium* species show reduced abundance in the 0.25 m sample, maximum abundances at 0.36 m or 0.5 m, and a gradual reduction as depth increases beyond 0.5m. This could be due to any of several environmental variables. Kilgour and Mackie found significant positive correlations between depth, percentage organic component in the substrate, and algal biomass.

The clear calm waters of Lake Zurich, Switzerland, stand in contrast to the environment of Kilgour and Mackie's Ottawa River bay. Walter and Kuiper (1978) took quantitative samples of the pisidiid fauna at depths of 3 m, 6 m, 9 m and 15 m from Lake Zurich, noting that the border between littoral and profundal zones is about 9 m. The contrast between their results (Table 9.3) and those of Kilgour and Mackie is striking. Over all pisidiid species, the modal abundance occurred at 6 m, just where the Britannia Bay fauna is disappearing. And in Lake Zurich,

Table 9.3. *The 14 pisidiid species collected in quantitative samples from Lake Zurich, categorized by depth of occurrence*

	Modal abundance (m)	Depth range (m)	Total individuals
Littoral			
Musculium lacustre	3	3–12	127
Sphaerium corneum	6	6	4
Pisidium milium	6	3–12	48
Transitional			
P. amnicum	9	9–12	8
P. pulchellum	9	3–9	16
Profundal			
P. conventus	12	12	1
P. personatum	15	6–15	1200
Eurytypic			
P. casertanum	6	3–15	1075
P. henslowanum	6	3–15	457
P. hibernicum	6	3–15	1249
P. moitessierianum	6	3–15	9631
P. nitidum	6	3–15	2008
P. subtruncatum	6	3–15	2920
P. tenuilineatum	12	3–15	1036

Source: Walter and Kuiper (1978).

several *Pisidium* species (most strikingly *P. personatum*) do not occur in shallow water at all.

The depth distributions of freshwater gastropods are generally much more similar to those of the pisidiids of Britannia Bay than Lake Zurich. Berg's (1938) data on snail distribution in Danish Esrom Lake (Figure 7.11) are typical. Regardless of season, all nine Esrom species reached maximum abundance in Berg's shallowest sample, 2 m. These data are complemented nicely by Økland's (1964) study on the large (2.08 km²) eutrophic Lake Borrevann, Norway. Økland's samples were not as temporally comprehensive as Berg's, but included a much greater range of depths. He took 14 samples with a 0.1 m² metal frame from a depth of 0.2 m, and 170 Ekman grab samples of 200 m² each over depths of 1.5 m, 2.0 m, 3.0 m, 5.0 m, 6.0 m, 7.0 m, 10.0 m, and 15.0 m. Økland's 0.2 m samples were taken on a rocky, wave-exposed environment; only *Anisus contortus* had its modal abundance in this region. Table 9.4 shows

Table 9.4. *Gastropods collected in quantitative samples from five depths in Lake Borrevann, Norway with total collected*

	Average density (per m²)					
	0.2 m	1.5 m	2.0 m	3.0 m	5.0 m	Total
Lymnaea peregra	6	7	4	0	0	20
L. auricularia	8	423	0	0	0	180
Physa fontinalis	3	100	58	0	0	75
Anisus contortus	20	0	0	0	0	105
A. crista	32	703	0	0	0	390
A. complanatus	31	17	98	0	0	55
Acroloxus lacustris	3	7	0	0	0	8
Valvata cristata	0	0	504	33	0	300
V. piscinalis	16	707	236	27	6	495

Note:
A few *Valvata* were found at depths below 5.0 m.
Source: Økland (1964).

that all the remaining eight species reached maximum abundances at either 1.5 m or 2.0 m, just as was the case in Esrom Lake. Only the two *Valvata* species were found deeper than 2 m, and only very few *Valvata* were collected below the 3 m littoral boundary. The similarity between this distribution pattern and that observed for lake-dwelling bivalves (e.g. Figure 2.2, Table 9.2) is striking.

The distribution of snails in lakes such as Esrom and Borrevann has often been interpreted as a function of dissolved oxygen concentration. Berg himself published one of the more influential works on comparative gastropod respiration, using the Esrom Lake species (Berg and Ockelmann 1959). The subject has been well reviewed subsequently (Ghiretti and Ghiretti-Magaldi 1975, Russell-Hunter 1978, Aldridge 1983, McMahon 1983a). It has now been demonstrated that a snail's oxygen requirement is a function of everything: its acclimation, size, sex, and health, the temperature and osmolarity of the medium, the surrounding oxygen tension (in some cases), and even illumination. It has also been established that freshwater snails may respire in all fashions imaginable: over their general body surfaces, through lungs filled with air (even some ampullariids can breath at the surface), through lungs filled with water (many pulmonates never surface), through aquatic gills (not to exclude quite a few gill-like structures in pulmonates) and even terrestrial gills (certain pomatiopsids). The literature seems to suggest that the respiratory rates of freshwater snails are generally adjustable within the range of dissolved oxygen concentrations they normally encounter. This does not seem to be as important a habitat variable as others we will discuss.

In spite of the great respiratory variability that can be induced by the environment, Russell-Hunter (1964:96) has written 'As a group, freshwater snails seem to have a fairly uniform respiratory rate'. In fact, depth distributions of snails in Scandinavian Lakes do not appear especially variable among species. Such results are not unique to Scandinavia. Dudgeon (1983a) found that *Thiara, Melanoides, Sinotaia*, and *Radix* were all most abundant in the shallows of Hong Kong's Plover Cove Reservoir, and followed the populations through an unusually severe water level fluctuation.

Horst and Costa (1971) took Ekman grab samples at 20 sites across two transects of McCargo Lake, New York, much smaller (32000 m²) and shallower (maximum depth 5.57 m) than Lakes Esrom or Borrevann. The result was similar, however. All five common species, *Amnicola limosa, Valvata tricarinata, Physa sayii, Gyraulus parvus* and *Promenetus exacuous*, reached maximum abundance at the edges of the lake, in water 0.75–1.25 m deep. Snails were completely absent from the 12 stations deeper than 2.5 m, the top of the McCargo Lake profundal zone during the summer these collections were made. Horst and Costa observed much lower levels of dissolved oxygen in the profundal samples (usually 0.0 ppm), and high levels of hydrogen sulphide.

There is no question that low levels of dissolved oxygen effectively exclude gastropods from the deeper waters of these three lakes. The hypolimnion apparently formed a barrier at about 3 m in Lake Borrevann and 2.5 m in McCargo Lake during the sampling periods involved, below which snails rarely, if ever, penetrated. It is not immediately clear why snails should be as aggregated in shallow water in Esrom Lake, however, since the top of the profundal zone here is not reached until about 12 m (Jonasson 1972). Perhaps, as we suggested for bivalves in Chapter 2, snails of nearly all types prefer to be as close to the surface as possible, without exposing themselves to disturbance by wave, wind, and predator.

Two qualifications should be noted immediately. First, on some occasions gastropods have been found at great depths, including *Lymnaea ovata, Valvata piscinalis*, and several others (e.g. Lake Tanganyika species). Mouthon (1986) listed several citations to the occurrence of gastropods at depths of 100 m. I am not certain whether there are any truly profundal freshwater gastropods, however, as we saw in some *Pisidium*. Additional research on gastropod distribution in the ancient lakes would be welcome.

Second, there is some evidence of seasonal depth migration in freshwater gastropods. In some situations (e.g. *Bulinus* in Zimbabwean ponds) snails may rise in the winter to find warmer temperatures in the shallows (Shiff 1966). But most pulmonates, especially from temperate areas, seem to respond to lowered temperature in the laboratory by moving to greater depths, at least on a small scale (Boag and Bentz 1980, Boag 1981). Interestingly, *Lymnaea elodes* seems to rise in the aquarium when temperatures are lowered, in keeping with its tendency to aestivate dry. Among the earliest to note seasonal pulmonate migrations was Cheatum (1934), working at Douglas Lake, Michigan. His casual observations were confirmed, at least for *Physa integra*, by Clampitt (1972, 1974). Clampitt observed a springtime migration from sandy shoals 10–20 m offshore (1 m deep) to a cobble bottom adjacent to the water's edge, where egg laying took place. In the autumn, after death of the parental generation, the juveniles returned offshore.

But Clampitt did not confirm migration for *Helisoma antrosa*, which seemed to remain offshore in Douglas Lake, depositing most eggs on the shells of other snails. Wall (1977) detected only random movements in a population of *Lymnaea catascopium* from Michigan's Lake Ann, over three years of study. I am unaware of studies showing evidence of seasonal migration in any prosobranch population.

Before we propose that snails, as a natural law, generally live in the shallowest regions of lakes where they may remain undisturbed, the distribution of snails in Douglas Lake ought to be examined more closely. In the summer of 1982, Laman and colleagues (1984) used SCUBA to collect four transects parallel to the Douglas Lake shore, at depths of 0.5 m, 3.0 m, 6.0 m, and 9.0 m. Their data show that populations of five pulmonate species seemed to prefer 6.0 or 9.0 m depth (Figure 9.2). It is amusing to note that both populations of prosobranch, which are occasionally considered better adapted to the aquatic environment by virtue of their gills, seem to prefer shallower water than the pulmonates. Pace and colleagues (1979) used SCUBA to collect two transects in New Mission Bay, Lake Michigan, both ranging from shore to about 9 m. The maximum abundances of the three most common species, *Gyraulus parvus*, *Valvata tricarinata*, and the hydrobiid *Marstonia decepta* were not at 1.5 m, but at 3–4 m.

The works of both Pace and Laman confirm the generalization that snails do not prefer, or even often inhabit, the profundal zone. Laman and

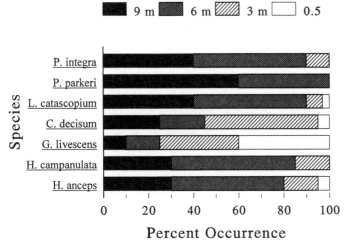

Figure 9.2. Relative proportions of 7 gastropod populations inhabiting 4 depths in Douglas Lake, Michigan. (Data of Laman *et al.* 1984.)

his colleagues reported that temperature is highly stratified in Douglas Lake in the summer, with a thermocline at about 10 m, and that no collections were taken at lower depths because virtually no snails were seen. The densities of all three of Pace's snail populations had dropped to very low levels at their deepest sites, which were still well up in Lake Michigan's epilimnion. These snail populations do not, however, reach maxima especially close to the surface. Where snail populations reach maximum density at depths of 3–9 m, attention might profitably turn to current and substrate.

Apparently the bottom of Douglas Lake is quite sandy in the region sampled by Laman *et al.* to a depth of several metres. Their explanation for the dominance of *Goniobasis* and *Campeloma* in the 0.5 m and 3.0 m samples was that 'pulmonate snails, with thinner shells and no operculum could probably not withstand such wave action'. They observed that submerged macrophytes were most common at 2–5 m, and that a silt and detritus bottom predominates thereafter. Current and substrate conditions in the New Mission Bay of Lake Michigan may be similar. In fact, substrate (broadly speaking) may explain why the snails of Esrom Lake seem to be crowded into the upper 2–5 m of a 12 m littoral zone. Jonasson (1972) indicates that the aquatic vegetation generally extends only to about 5 m.

The effects of depth and substrate are quite commonly confounded. Turning briefly away from the subject of lakes, Caquet (1990) sampled down the 400 m length of a ditch southwest of Paris, monitoring the densities of four pulmonate populations over the course of 16 months. The gradual grade of the ditch was reflected by a gradation of macrophytic vegetation, from riparian herbs and grasses in regions subject to summer drying to submerged hydrophytes *Ceratophyllum* and *Callitriche*. Caquet reported that *Lymnaea palustris* was uniformly distibuted down the length of the ditch, but documented a clear preference of *Anisus rotundatus* for the shallow end, and the preferences of *A. albus* and *Physa fontinalis* for the deeper. Whether such relationships may be due to depth *per se*, to differing macrophytic vegetation, degrees of exposure, or possibly even competitive exclusion, cannot of course be discerned.

Substrate

Our discussion of gastropod habitat has been facilitated in no small measure by the happy chance that the University of Michigan Biological Station is situated on Douglas Lake, inhabited by healthy populations of a number of snail species. Clampitt (1973) quantitatively sampled four transects from shore to 30–150 m offshore, some transects uniformly shallow throughout, others falling to depths of 10 m, carefully noting substrate. Example results from transects 'a' and 'b' are shown in Figures 9.3 and 9.4, for two different sampling dates. Figure 9.3 shows that in the summer of 1969, *Physa integra* seemed to show a slight preference for algae-covered cobbles near shore, while *Helisoma antrosa* strongly preferred the detritus–covered sand found 10–20 m offshore. All of these samples were taken at a depth of about 1 m. The preference of *P. integra* for cobbles was much more pronounced upon resampling in May 1970. That the distribution of *H. antrosa* in transect 'a' was not a function of distance to shore can be seen in transect 'b', where in 1969 the *H. antrosa* population clearly preferred sandy substrate close to shore.

The instability of freshwater snail distributions is an important second message from these data. Although it is not clear why *H. antrosa* seems to have disappeared in 1970, both sets of observations at transect 'a' are in accord with Clampitt's (1972) results on seasonal migration in these two species previously discussed. But the contrast between the July 1969 and July 1970 data collected at transect 'b' is striking. Further, the collections of Laman *et al.* (1984) shown in Figure 9.2 were made very near Clampitt's transect 'b', in June through August 12 years later. Laman *et al.* noted many discrepancies between his data and Clampitt's, but could offer no explanation other than 'populations of snails in Douglas Lake are not stable over long periods'. The lesson to be learned from this is that surveys of freshwater snail distribution should never be repeated.

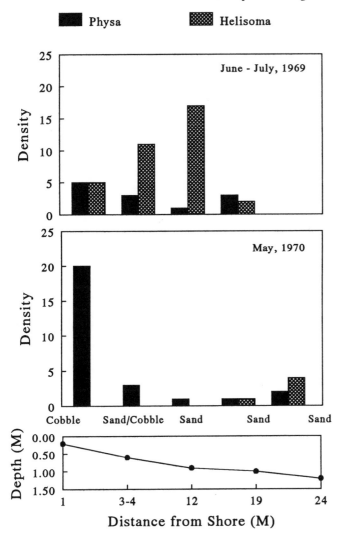

Figure 9.3. Clampitt's (1973) transect 'a'. Density of *Physa integra* and *Helisoma antrosa* (N/m²) as a function of distance to shore, depth, and substrate on two dates in South Fishtail Bay, Douglas Lake, Michigan.

Returning to the thread of our discussion, Clampitt followed his field observations by performing substrate-choice experiments in an oval chamber half lined with algae-covered stones and half with algal-incrusted sand. On the border between these two regions he placed a dish containing 20–50 snails from one of four Douglas Lake species. He allowed 30 minutes for the snails to leave the central dish and then

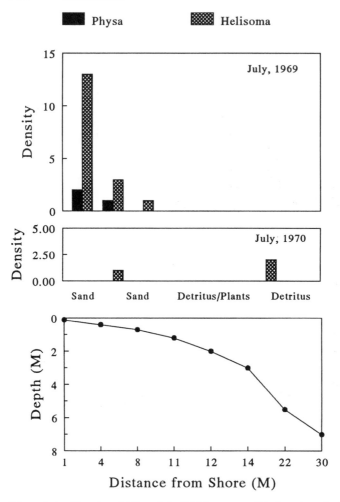

Figure 9.4. Results of Clampitt's (1973) transect 'b'. Density of *Physa integra* and *Helisoma antrosa* (N/m²) as a function of distance to shore, depth, and substrate on two dates in North Fishtail Bay, Douglas Lake, Michigan.

recorded their positions at 15 minute intervals. Clampitt found highly significant preferences for cobble in both *Physa integra* and *Physa parkeri*, no preference in *Helisoma campanulata*, and significant preference for sand in *H. antrosa*.

Substrate preference, often varying among species, is a generally reported phenomenon. Substrate apparently influences the distribution of both pulmonates and prosobranchs in the marshes of Carribean islands (McKillop *et al.* 1981), the rivers of Hungary (Toth and Baba 1981), and the lakes of north Africa (Leveque 1972). Vaidya (1979) used choice experiments to show that Indian populations of both *Viviparus bengalensis* and *Melania scabra* strongly preferred sand over stone, even if the sand was acid washed and the stone was covered with food. The 'architecture' of the substrate, its shape and placement, may have effects at least as important as its composition (Kershner and Lodge 1990).

Among the most interesting of these investigations are those that compare snail abundance on bottom with abundance on macrophytic vegetation, a task properly requiring two sampling gears. Macan (1950) used a pond-net to collect 8 'submerged vegetation' stations, and trawls and grabs to collect 18 'exposed shore' stations, 12 'reed bed on sandy bottom' stations, and 19 'reed bed on muddy bottom' stations in Lake Windermere, England. He reported apparently strong preferences in *Valvata piscinalis* and *Physa fontinalis* for submerged vegetation over any bottom sediment. *Planorbis contortus* seemed to show some preference for vegetation, but no substrate preference of any kind was detected in *Lymnaea pereger*, *Planorbis carinatus*, or *P. albus*. Soszka's (1975) results from Mikolajskie Lake, Poland, were more striking. All of the gastropods for which an adequate sample could be obtained seemed to show a significant preference: the pulmonates *Radix ovata*, *Armiger crista*, *Physa acuta* and *Acroloxus lacustris* 'associated with macrophytes', the prosobranchs *Bithynia tentaculata* and (alas!) *Valvata piscinalis* were more commonly in the benthos. A complicating factor in such studies is that the surface area of macrophyte samples often greatly exceeds the bottom area sampled, biasing comparisons between macrophytes and sand, mud, or cobble (Brown and Lodge 1993).

As may be recalled from earlier in this chapter, Lodge's (1985, 1986) observations at Radley Pond suggested considerable habitat specialization in the gastropod fauna. The pond apparently does not have any cobble, sand, or even exposed mud bottom, but rather is lined with dense stands of macrophytes and allochthonous leaf litter. Lodge (1985) took random samples within four habitat types: submersed macrophytes, emergent macrophytes, water lilies, and allochthonous detritus. Almost all of the 12 Radley Pond gastropod populations showed some significant preference. Among the more striking associations were *Lymnaea peregra* and *Valvata piscinalis* on submersed macrophytes, and *Planorbis vortex*, *P. contortus* and (alas!) *Bithynia tentaculata* on emergent macrophytes. *Acroloxus lacustris* was the only species preferring water lilies (by a non-significant margin over emergents), and no gastropod population preferred a detrital substrate.

Lodge went on to perform some rigorous substrate-choice experiments exploring the association between *P. vortex* and one particular emergent macrophyte, *Glyceria maxima*. He found that the strong attraction of *P. vortex* to this plant was neither influenced by the presence of other snail populations nor by prior conditioning of the plant by snails, but apparently both by *Glyceria*'s emergent nature (as a route to the surface for breathing) and by its peculiar collection of epiphytic detritus and algae.

Relationships between particular macrophyte species and particular populations of snails have been reported on several other occasions (e.g. De Coster and Persoone 1970). In some cases, however, it is more a matter of diet than habitat. Pip and Stewart (1976) found *Physa gyrina* seasonally associated with *Potamogeton pectinatus* and *Lymnaea stagnalis* associated with *P. richardsonii* in a marsh on the shore of Lake Manitoba, Canada. The snails seem to be most common on their respective host plants at times of carbohydrate maxima, corresponding to periods of active plant growth. Pip and Stewart note that the snails inflict heavy grazing damage upon their hosts, and so suggest that their periodic association may result from the snails' attraction to particular metabolites given off by plants at their most vulnerable. The complexity of the interaction between freshwater snails and macrophytes has been reviewed by Thomas (1987) and Bronmark (1989). Such associations as may be shown are often transitory, weak in their effect when compared with those of site and year (Vincent *et al.* 1991).

Gregg and Rose (1985) transplanted aquatic macrophytes into substrate trays and placed the trays into several regions of an Idaho stream. They found strong positive associations between the macrophytes, *Gyraulus* and *Physa*. Evidence suggested that the snails were not attracted to the macrophytes as food or substrate, but rather as effective buffers of current velocity.

Current

Most of the studies discussed in the previous two sections have taken place in ponds and lakes. But in lotic environments, although it is difficult to separate the effects of depth, substrate, and current, I would suggest that current velocity may be of primary importance. In my experience, few pulmonate species (in particular) can be said to inhabit lotic environments at all; their occurrence in rivers being restricted to calm backwaters and eddys. I am unaware of any studies on the effect of current as rigorous as many of the studies on substrate we have discussed: for example, combining field observation with laboratory experiments. But the phenomenon may be so striking and so obvious as to discourage study. One rarely sees study on the importance of water to fish.

There have been many good laboratory investigations showing that most pulmonates can be dislodged by fairly low current speeds, typically 0.6–0.9 m/sec, depending on snail size and substrate (Jobin and Ippen 1964, Moore 1964, Dussart 1987). But the measurement of actual micro-current speeds *in situ* is difficult, especially on the irregular surfaces where snails reside. Thus most authors who have considered current speed critical in the distribution of freshwater snail populations have relied on rather indirect evidence (e.g. Appleton 1975, 1976, Thomas and Tait 1984). Pimentel and White (1959a) measured current velocity at two-thirds depth in midstream for over 1000 square yard (0.91 m²) samples from a small tributary of the Quebrada Sabana Llana, Puerto Rico. They also measured shade, turbidity, depth, and substrate, relating each of these factors to the density of *Biomphalaria glabrata*. Pimentel and White suggested that stream velocity and depth were among the more important determinants of *Biomphalaria* distribution, and that bottom material was secondary.

As we have seen in Chapter 5, striking differences are often noted between the life histories of snail populations in the field and laboratory. Loreau and Baluku (1987b) found a much lower intrinsic rate of increase in the population of *Biomphalaria pfeifferi* of the Virunga Stream at Lwiro, Zaire, than had been observed in laboratory populations previously. They offered good evidence that neither temperature, water hardness, nor food limitation could be responsible for the lower survivorship and fecundity seen in the stream, and considered current velocity (as affected by rainfall) likely to be the critical determinant of population dynamics.

To be complete, it should be noted that preference for current may vary seasonally, just as depth preference. Rowan (1966) noticed autumn migration in a population of *Helisoma trivolvis* inhabiting a small pond in Montana. The snails did not move to the deeper regions of the pond, but rather to the pond inlet, where increased current would provide additional protection from freezing.

Models of species distribution

In the first chapter of the present work we noted that from a human perspective, molluscs are half plant. They move, but not much. True, it is easier to sample the plants in an old field than the clams in a river. But with greater certainty than is possible for plants, an ecologist may assume that the molluscs he or she has collected in a small sample have elected to be there, because they would leave if they did not. So within a lake or a stream, the distribution of snails or clams seems a potentially valuable source of ecological information.

In one sense, however, the distributions of freshwater molluscs may contain too much information. A clam or snail may appear in a sample because it is physically adapted to that particular microhabitat and/or attracted to some resource and/or avoiding predators and/or competitors elsewhere. In another sense, the distribution of animals in a population may contain no information at all. The 'null hypothesis' must not be forgotten. Snail and clam populations may also be distributed randomly around a river or lake, without reference to any biological processes whatsoever.

Freshwater environments certainly offer mosaics of habitats, with population 'sources' and population 'sinks'. Pulliam (1988) has prominently pointed out that in some situations, population densities may be highest in 'sink' habitats. For example, it may be recalled from Chapter 3 that *Physa integra* seems most common on mats of *Ceratophyllum* in Iowa ponds (Clampitt 1970), even though this plant constitutes very low quality food (Sheldon 1987). Floating mats of *Ceratophyllum* may be population sinks. Similarly, Horvath and colleagues (1996) have gathered substantial evidence that small streams may function as sinks for zebra mussels. But in general, few studies of freshwater mollusc ecology have been sufficiently detailed to address Pulliam's concern. Controlled experiments are ultimately required, just as they have been necessary in all other areas of scientific inquiry.

In any case, keeping all caveats in mind, I am surprised that mollusc communities have seldom been analysed by ordination techniques. A great variety of such methods have been developed by ecologists to classify plant communities (e.g. Gauch 1982, Digby and Kempton 1987), and the ordination literature is vast. For organisms suited to this sort of analysis, ordination offers a beginning to the study of the interactions among them and their environment. Ordination studies of variation between mollusc communities (e.g. Green 1971, Sepkoski and Rex 1974) have been discussed in Chapter 8. I am aware of only five analyses within single mollusc communities, however, two involving sites in England: Malham Tarn (Calow 1973c) and the River Hull (Storey 1986); and three works in French: Lake Chad (Leveque and Gaborit 1972, Leveque 1972), the St Lawrence River (Lamarche *et al.* 1982), and France's Aube River (Mouthon 1979).

The two English studies were similar methodologically. Both Calow and Storey established a series of sample sites and recorded presence or absence of a variety of freshwater snails. Calow found 13 species in Malham Tarn and Storey found 15 species in the River Hull (including 12 of the Malham species). Both authors calculated an unusual matching

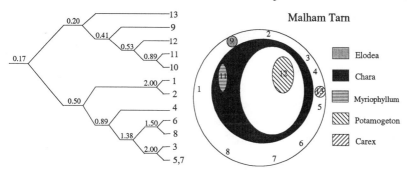

Figure 9.5. Schematic drawing of Malham Tarn, England, and an ordination of 13 collection sites based on its gastropod fauna (13 species) as performed by Calow (1973c).

coefficient due to Mountford (1962) between all pairs of sites, and then used the symmetric matrix of site similarites in a cluster analysis.

Both Storey and Calow found that the presence of gastropod species can be used to classify freshwater sites into distinct clusters well correlated with major environmental features. Storey could distinguish four zones in the River Hull drainage. Zone 'A' sites were unstable and subject to drying, and were characterized by *Lymnaea peregra*, while 'B' sites were small but permanent, being characterized by *Potamopyrgus jenkinsi* and *Ancylus fluviatilis*. High-diversity zone 'C' incorporated sites in the middle reaches of the River Hull, inhabited by these three species plus as many as 11 others. No snails were found at zone 'D' sites, admittedly heterogeneous environmentally. Calow's analysis of Malham Tarn resulted in two very distinct clusters – one of weed beds and the other of rocky shores, as may be seen in Figure 9.5.

The studies in French were also similar to each other methodologically. Leveque and Gaborit (1972) found seven mollusc taxa in their quantitative samples of Lake Chad: the viviparid *Bellamya*, the thiarids *Melania* and *Cleopatra*, the unionid *Caelatura*, and corbiculoids *Byssanodonta*, *Corbicula*, and *Pisidium*. Rather than cluster analysis, the authors used factor analysis to characterize their approximately 100 sites by the abundances of these seven taxa. They noted striking regional differences in the mollusc fauna, and recognized clusters of sites that appeared related to substrate. Mouthon (1979) identified 28 species of gastropods and 12 bivalve species at his 17 sites on the River Aube. He used the abundances of these species as variables in a factor analysis, plotted his sites on the first two factors, and recognized five major groups of sites.

Such studies as those of Storey, Calow, Leveque and Gaborit, and Mouthon are certainly informative regarding environmental conditions. But from the standpoint of freshwater mollusc ecology, there may be more interest in associations of species than associations of sample sites. Since distinct regions of rivers or lakes seem to be inhabited by fairly distinct mollusc faunas in these studies, I would expect the results of a reanalysis of molluscan associations to be similar (but certainly not identical) to the original site analyses of the authors. So I have reanalysed Mouthon's data, using sites as the variables and species as the objects to be clustered. Further, I would prefer not to perform principal component analysis, factor analysis, or indeed any sort of technique that involves the calculation of a variance/covariance matrix from data sets where most observations take the value 0.0. So I set aside the five species Mouthon found only at his site 13 and converted the abundance data on his remaining 35 species to the presence/absence matrix shown in Figure 9.6.

Choice of ordination technique has always been a matter of taste. Jackson and colleagues (1989) have reviewed the properties of eight similarity coefficients that may be applied to presence/absence data, the coefficient of Mountford used by Calow and Storey not among them. (Mountford's similarity coefficient can range to infinity, whereas all the coefficients conventionally used for ordination range from 0.0 to 1.0 or − 1.0 to 1.0). I prefer the simplest coefficient in the list of Jackson and his colleagues, the Jaccard coefficient, calculated as the number of species shared by two sites divided by the total species in the two sites considered together. 'Negative matches', sites where neither species is present, are not considered in the calculation of the Jaccard coefficient. This might seem undesirable at first. We are assuming no barriers to dispersal in these communities, so the observation that two species are both absent at site A would seem to contain as much information as their mutual presence at site B. But while presence can be reported with certainty, absence has a large error term. Negative matches seem much more likely to reflect sampling error than joint occurrences. So from the River Aube data I calculated a 35×35 symmetric matrix of Jaccard matching coefficients using the CORR module of SYSTAT (Wilkinson 1988).

Although cluster analysis as applied by Calow and Storey is certainly a useful way to examine data for groups, the conversion of a complex matrix of similarities to a linear diagram necessarily involves considerable distortion. A representation in more dimensions is obviously preferred. Again, many techniques have been employed for this purpose over many years, but the review of Kenkel and Orloci (1986) persuades me that

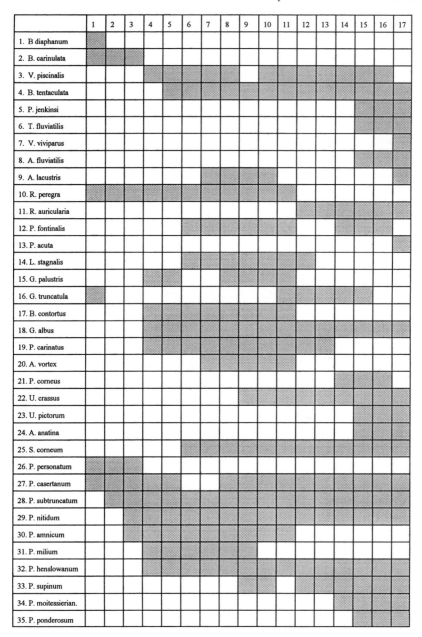

Figure 9.6. The distribution of 35 mollusc species at 17 sites (numbered from upstream down) on the River Aube, France. (From Mouthon 1979.)

non-metric multidimensional scaling is a suitable method of summarizing overall interspecific relationships. So I used the MDS module of SYSTAT (Wilkinson 1988) to scale monotonically the 35×35 symmetric matrix of species similarities to the two-dimensional graph shown in the upper half of Figure 9.7. After 50 iterations, Kruskal's stress was 8.6%, a low value indicating a good fit.

Figure 9.7 shows six groups with Jaccard coefficients of 0.8 or more. Starting in the lower right of the figure, we see (A) the two species restricted to stations 1–3, (B) the three species characteristic of the upriver sites 4–11 or 13, (C) the cluster that includes the six species distributed over the entire Aube, (D) the two species restricted to the downstream sites 9–17, (E) the five species restricted to sites 15–17, and finally, closest to the lower left corner, (F) the two species found only at site 17.

Cluster (C) is not a distinct community itself, but is in a sense an artifact of the analysis of similarity coefficients (Jackson et al. 1989). These six widely dispersed species would be expected to have high Jaccard coefficients with each other and at least some similarity to all other species. Although they are expected to cluster together in the middle of the ordination, they are members of all the other communities, not a separate community unto themselves. Similarly, clusters D, E, and F are not distinct, but nested. The species of cluster (E) are found with the species of cluster (F), and those of (D) are found with both (E) and (F). So I would recognise three distinct mollusc communities in the Aube River: (A), (B), and (D–E–F). The five species found only at site 13 (not figured) would constitute a fourth.

Thus we find ordination studies of freshwater mollusc communities generally in accord. Mollusc species seem to assort themselves into discrete communities by the environment, on substrates or along stream gradients, for example. In lakes and rivers generally, is there some particular environment better for all populations? Mouthon's upstream cluster (A) has two species, midstream cluster (B) has three species, and downstream cluster (D–E–F) has nine species. Storey's (1986) results were comparable. One is reminded here of the 'river continuum concept' (Vannote et al. 1980, Minshall et al. 1985) that species richness should increase from headwaters to midorder streams. In lentic environments, the studies we have reviewed both in this chapter and in Chapters 2 and 3 suggested that depths of about 1 metre seem preferred by a wide variety of bivalves and gastropods. On the other hand, we have also reviewed several cases of apparent habitat segregation in freshwater molluscs, such as Clampitt's

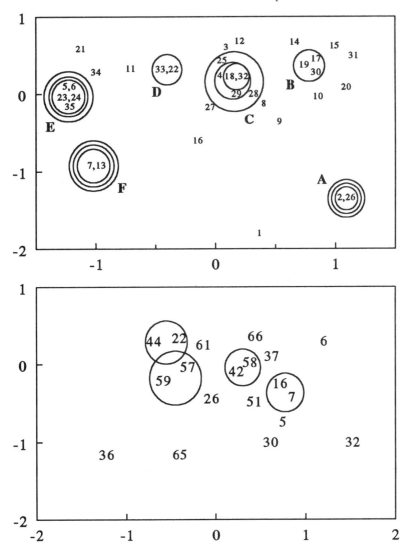

Figure 9.7. (Top) Non-metric two-dimensional scaling of mollusc associations in the River Aube, France (data of Mouthon 1979). Contours show Jaccard similarities 1.0 to 0.8 at 0.1 intervals. Refer to Figure 9.6 for key to species numbers. (Bottom) Non-metric two-dimensional scaling of unionid associations in the upper Mississippi River (Ellis 1931a,b). Contours group species within 10–20 standard deviations of abundance over 14 sites. Refer to Table 9.5 for key to species numbers.

(1970) work with *Physa* and the observations of Ross and Ultsch (1980) on *Goniobasis*.

Here I introduce a data set collected by Ellis (1931a,b) and republished by van der Schalie and van der Schalie (1950) showing little tendency for association in the unionid mussel fauna of the upper Mississippi River. More recent surveys have been reported by Holland-Bartels (1990) and Hornbach and colleagues (1992). Ellis sampled 254 stations down some 1100 river km from Point Au Sable, Minnesota, to Cairo, Illinois, primarily using a 0.56 m^2 self-closing dredge. 'The sand, mud, (and) gravel of the bottom with all the contained mussels . . . were then carefully examined by trained observers, these data are considered quite reliable. Thousands of individual dredgings were made.' Only 86 of the 254 stations yielded mussels, and these were grouped into 14 regional zones stretching some 900 km by van der Schalie and van der Schalie. The 39 species recognized are listed in the 'Ellis' column of Table 9.5, with overall distribution and abundance. Even combining stations into zones, it seems fairly clear that these mussel populations show aggregated distributions, as is most commonly observed in biotic surveys. All $N=279$ *Lampsilis ovata*, for example, were collected in just 7 of the 14 zones. But do mussel populations generally tend to occur in the same subset of sites?

The author's sampling technique was probably sufficiently unbiased that his data can be analysed quantitatively. So rather than a matching coefficient, I calculated a standardized Euclidean distance between pairs of species, again using the CORR module of SYSTAT. This introduces a new twist to an old problem, however. Even if they never co-occur, all rare species will appear to clump together near the origin when distance is Euclidean. Thus I eliminated the 18 species in Table 9.5 with abundances less than 100. I also combined the sympatric subspecies '*Lampsilis anodontoides*' and '*L. anodontoides fallaciosa*' (now generally considered *L. teres*) and eliminated this taxon as a severe outlier.

The lower half of Figure 9.7 shows a monotonic two-dimensional scaling of the remaining 19 species on Euclidean distances over the 14 zones. Kruskal's stress reached a minimum of 8.4% after only 15 iterations – a good fit. There is some evidence of a few two-species 'clusters'. *Quadrula nodulata* (58) and *Obliquaria reflexa* (42) are both uniformly distributed over all 14 sites, and thus their apparent association is an artifact. But the other three associations are genuine – two of these, *Fusconaia ebenus/Obovaria olivaria* (22 and 44) and *Elliptio dilatata/Anodonta corpulenta* (7 and 16) are graphed in the upper portions of Figure 9.8.

However, in contrast to the findings of Calow, Storey, and Mouthon

Table 9.5. *Unionacean mussel populations inhabiting several rivers of the North American interior*

Species	Ellis — Mississippi		Ahlstedt — Clinch		Ahlstedt — Elk	Havlik — Mississippi	v.d.S. — Grand
	Ni	Ns	Ni	Ns	Ns	Ni	Ni
1) *Actinonaias carinata*	49	7	326	104	1	149	499
2) *A. ellipsiformis*	7	1	0	0	0	0	5
3) *A. pectorosa*	0	0	161	101	40	0	0
4) *Alasmidonta marginata*	4	1	abs	37	0	1	123
5) *Amblema plicata (peruviana)*	477	14	27	67	14	1457	309
6) *Anodonta imbecillis*	181	8	0	0	0	27	0
7) *A. grandis (corpulenta)*	215	9	0	0	1	3	2
8) *Arcidens confragosus*	7	4	0	0	0	9	0
9) *Carunculina parva*	6	1	0	0	0	3	0
10) *Conradilla caelata*	0	0	1	11	2	0	0
11) *Cumberlandia monodonta*	0	0	8	21	0	0	0
12) *Cyclonaias tuberculata*	2	2	59	75	57	0	69
13) *Cyprogenia irrorata*	0	0	3	30	0	0	0
14) *Dromus dromus*	0	0	1	16	0	0	0
15) *Elliptio crassidens*	1	1	abs	11	6	44	0
16) *E. dilatata*	137	4	94	77	16	13	363
17) *Epioblasma brevidens*	0	0	3	31	0	0	0
18) *E. capsaeformis*	0	0	27	43	0	0	0
19) *E. triquetra*	0	0	4	28	1	1	0
20) *Fusconaia barnsiana*	0	0	37	61	10	0	0
21) *F. cuneolus*	0	0	20	55	4	0	0
22) *F. ebena*	246	9	0	0	0	1424	0
23) *F. edgariana*	0	0	8	30	10	0	0

Table 9.5 (cont.)

Species	Ellis		Ahlstedt			Havlik	v.d.S.
	Mississippi		Clinch		Elk	Mississippi	Grand
	Ni	Ns	Ni	Ns	Ns	Ni	Ni
24) F. flava	0	0	0	0	0	431	396
25) F. subrotunda	0	0	68	64	21	0	0
26) F. undata	346	14	0	0	0	0	0
27) Lampsilis fasciola	0	0	9	72	13	0	0
28) L. higginsii	3	2	0	0	0	33	0
29) L. suborbiculata	0	0	abs	1	0	0	0
30) L. ovata (ventricosa)	279	7	20	60	36	88	177
31) L. teres (anodontoides)	953	14	0	0	1	17	0
32) L. radiata (siliquoidea)	333	4	0	0	0	4	4
33) Lasmigona complanata	61	7	0	0	2	5	0
34) L. costata	0	0	64	103	42	3	36
35) Lastena lata	0	0	5	20	2	0	0
36) Leptodea fragilis	697	13	1	25	8	43	0
37) L. laevissima	193	10	0	0	0	6	0
38) Lexingtonia dollabelloides	0	0	abs	1	27	0	0
39) Ligumia recta	46	10	4	28	0	54	22
40) Medionidus conradicus	0	0	46	57	7	0	0
41) Megalonaias gigantea (nervosa)	81	11	0	0	17	21	0
42) Obliquaria reflexa	222	14	0	0	12	251	0
43) Obovaria subrotunda	0	0	0	0	4	0	0
44) O. olivaria	387	12	0	0	0	291	0
45) Plagiola lineolata	28	7	0	0	4	7	0
46) Plethobasus cyphyus	8	4	1	28	0	5	0

Species							
47) Pleurobema cordatum (coccineum, sintoxia)	10	2	1	8	1	105	98
48) P. oviforme	0	0	4	27	2	0	0
49) P. plenum	0	0	abs	3	0	0	0
50) P. rubrum	0	0	abs	3	0	0	0
51) Potamilus alatus	259	11	5	67	32	17	0
52) P. capax	47	9	0	0	0	0	0
53) Ptychobranchus fasciolaris	0	0	19	72	1	0	0
54) P. subtentum	0	0	76	78	0	0	0
55) Quadrula cylindrica	0	0	15	32	13	0	0
56) Q. intermedia	0	0	0	0	8	0	0
57) Q. metanevra	180	8	0	0	7	68	0
58) Q. nodulata	104	12	0	0	0	344	0
59) Q. pustulosa	344	13	3	30	26	680	442
60) Q. sparsa	0	0	abs	1	0	0	0
61) Q. quadrula (fragosa)	280	11	0	0	17	54	40
62) Simpsoniconcha ambigua	1	1	0	0	0	0	0
63) Strophitus undulatus (rugosus)	67	8	?	7	0	27	36
64) Toxolasma lividus	0	0	0	0	17	0	0
65) Tritogonia verrucosa	344	11	0	0	6	24	0
66) Truncilla donaciformis	219	14	0	0	0	430	0
67) T. truncata	78	9	4	31	0	210	0
68) Villosa iris	0	0	9	46	18	0	1
69) V. perpurpurea	0	0	1	4	0	0	0
70) V. vanuxemi	0	0	abs	22	0	0	0

Notes:
N_i, number of individuals collected; N_s, number of sites where each population occurred; abs, a species found in qualitative samples only. Synonyms in parentheses.
Source: data of Ellis (1931a,b), Ahlstedt (1986), Havlik and Marking (1980) and van der Schalie (1948).

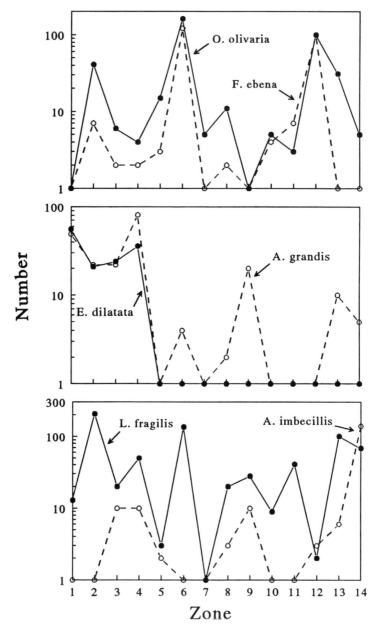

Figure 9.8. Abundance of selected unionid mussels in regional zones of the upper Mississippi River (Ellis 1931a,b; van der Schalie and van der Schalie 1950).

and others, it is clear from Figure 9.7 that the majority of the mussel species of the upper Mississippi River tend to vary independently, scattered in clumps about the river in what appears to us to be a random fashion. To illustrate this point I have also plotted in Figure 9.8 the abundances of two outliers from Figure 9.7 – *Anodonata imbecillis* (6) and *Leptodea fragilis* (36). The former fluctuates narrowly, from 0 to 10, through the first 13 sites, then increases tenfold at site 14. The latter fluctuates widely through all 14 sites (dramatic even on a log scale), with no trend apparent.

Freshwater mollusc populations may aggregate together, they may aggregate independently, and they may also aggregate apart. We now turn to a remarkable data set collected by F. C. Baker (1918) from the Lower South Bay of Oneida Lake, New York (Dillon 1981). Baker made 162 quantitative benthic samples using a dredge, a long-handled dipper, or direct inspection of measured stones, recording depth, substrate, and distance to the shore and counting all macroinvertebrates and macroalgae. A total of 21 of his samples either contained no snails or were omitted from the report. Baker recognized 37 species and subspecies in the remaining 141 samples, although most workers today would recognize somewhat fewer. In their reanalysis of this fauna 50 years later, Harman and Fourney (1970) revised the number of snail species to the 29 I have listed in Table 9.6. (Omitted are several species Baker recorded from the lake but did not collect in quantitative samples.) The table shows clear evidence of the aggregation that is so common in benthic samples (Caquet 1990). One might hypothesize that each Oneida Lake snail population may be aggregating in a preferred habitat type of some sort, just as Calow found in Malham Tarn.

Baker's 141 samples included 23 with just one snail species present, 24 with two species present, and so on, up to a single site with ten different species of snails. This 'S_o' distribution (the number of species observed) is shown in Figure 9.9. The question is whether the many separate snail populations of Oneida Lake generally tend to find the same habitat favourable, or whether they tend to aggregate in different habitats. Is the S_o distribution approximately what one would expect from randomly casting the 29 species into 141 samples?

I designed a Monte Carlo simulation of Baker's sampling process, creating a species pool in which each species was represented according to its abundance in the lake overall. For example, the probability of selecting *Campeloma decisum* from this pool was 17/5582, since 17 individual *C. decisum* were collected among the total of 5582 snails (Table 9.6).

Table 9.6. *Snail populations collected by Baker (1918) from Oneida Lake, New York*

'Sites expected' column derived from a Monte Carlo simulation, as described in the text

Species	Individuals	Sites observed	Sites expected, mean (s.d.)
1) *Campeloma decisum*	17	10	13.3 (3.2)
2) *Viviparus georgianus*[a]	1	1	
3) *Valvata bicarinata*[a]	4	3	
4) *V. sincera*	49	6	31.9 (3.5)
5) *V. tricarinata*	545	36	97.5 (5.0)
6) *Amnicola limosa*	39	2	24.5 (3.6)
7) *A. lustrica*	2841	78	133.6 (1.9)
8) *Cincinnatia binneyana*[a]	13	5	
9) *Gillia altilis*[a]	11	4	
10) *Somatogyrus subglobosus*[a]	3	3	
11) *Bithynia tentaculata*	220	15	70.4 (5.3)
12) *Goniobasis livescens*	82	11	42.8 (4.6)
13) *Physa integra*	210	22	69.5 (5.0)
14) *P. sayii*	160	38	62.6 (4.1)
15) *Lymnaea catascopium*	455	72	95.0 (4.4)
16) *L. columella*	17	2	13.4 (3.1)
17) *L. haldemani*	6	2	5.1 (2.0)
18) *L. humilis*	3	1	2.8 (1.4)
19) *L. stagnalis*	20	1	16.3 (4.3)
20) *Gyraulus deflectus*	1	1	1.2 (1.0)
21) *G. hirsutus*	47	18	30.7 (4.3)
22) *G. parvus*	581	60	99.1 (4.6)
23) *Helisoma anceps*	44	23	28.8 (4.0)
24) *H. campanulata*	67	24	37.9 (5.0)
25) *H. trivolvis*	50	12	29.6 (6.9)
26) *Planorbula jenksii*	1	1	1.1 (1.0)
27) *Promenetes exacuous*	82	34	41.8 (4.3)
28) *Ferrissia parallela*	45	17	29.1 (6.9)
29) *Laevapex fuscus*[a]	2	1	

Notes:
[a] Omitted from morphological analysis.

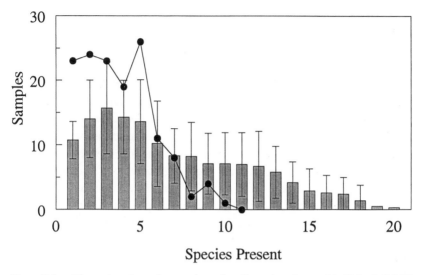

Figure 9.9.   The points show the number of snail species observed in Baker's (1918) samples from Oneida Lake (the S_o distribution), and the histrogram shows the number expected (the S_e distribution) with 95% confidence limits. (Analysis of Dillon 1981.)

From this pool I randomly drew 141 samples identical to Baker's in number of individual snails sampled, with replacement. This entire 141-sample procedure was performed 30 times. I totalled the number of sites where each species appeared, and the total number of sites with one species, two species, etc., to derive the S_e (species expected) distribution shown in Figure 9.9.

Our original intuition regarding the distribution of the individual species may be confirmed from the third column of Table 9.6. Almost all the snail populations actually occurred in far fewer sites than expected from the random simulation of Baker's sampling process. *Valvata sincera,* for example, would be expected to occur in about 32 of Baker's 141 sites, not 6. Aggregation seems to be a general phenomenon. Moreover, different snail populations tend to aggregate at different sites, so that Baker's samples generally included far fewer species than expected. Figure 9.9 shows that simulations routinely included samples with 12–14 species present, and occasionally up to 20 different species. So there is evidence that in the freshwater gastropod community of Oneida Lake, populations tend to specialize on different habitats, so that interspecific overlap is minimized. This would seem to constitute some evidence of interaction between them, historical if not ongoing.

Among the most prominent themes of Chapter 2, Chapter 3, and the present chapter up to this point has been the high level of ecological overlap generally displayed by freshwater mollusc populations. Now here we have reviewed two data sets, those of Ellis and Baker, that seem to suggest very diverse environmental preferences. The apparent contradiction is probably due to simple differences in the scale and sampling. Mollusc populations will appear aggregated together if sampled over radically heterogeneous environments, but will aggregate randomly with respect to one another, or possibly even aggregate apart, in homogeneous environments. Baker took only 22 samples of floating vegetation from Oneida Lake, and his 18 'deep water' samples were generally in the range of 3–6 m. Had he taken 70 samples from weeds and 70 samples from rocks, his results might have looked like Calow's (Figure 9.5). If Baker had replicated transects into deeper parts of Oneida Lake, he almost certainly would have noticed associations like Berg's (Figure 7.11) or Laman's (Figure 9.2). And similarly, if Berg had taken 100 samples at 2 m in Esrom Lake, he might have seen dissociations like Baker's (Figure 9.9). Storey and Mouthon (Figure 9.7) studied the upper parts of river drainages, where a few kilometres may make a great deal of difference in the environment, and zones of faunal association appeared. But the 1100 km stretch of Mississippi studied by Ellis was, again speaking grossly, homogeneous, and mussel distribution was a function of other things.

The work of Whittaker (1956) is now generally recognized as a turning point in community ecology. Prior to 1956, ecological communities were seen as discrete, interacting units, but subsequently they have generally been viewed as rather arbitrary samples on a continuum. Whittaker found that the plants growing at any spot in the Smoky Mountains are each independently reacting to environmental gradients, so that no discrete, reproducible associations are to be found. However, the freshwater molluscs studied by Calow, Storey, and Mouthon do seem to form associations. The snails of Oneida Lake (and possibly the mussels of the Mississippi) seem to form *dis*sociations. I have reviewed several other raw data sets on the distributions of freshwater molluscs in individual rivers and lakes (Baba 1967, Bishop and deGaris 1976, Toth and Baba 1981, Ham and Bass 1982, Laman *et al.* 1984, Chatfield 1986), and these data combined with my own experience lead me to conclude that freshwater mollusc species rarely appear to vary independently of each other on environmental gradients.

I am not suggesting a reversion to community ecology before Whittaker. I believe that mollusc species appear to form discrete groups because their freshwater habitats generally vary in a discrete fashion, not in a continuous one. Figure 9.7, for example, shows discrete mollusc faunas for different regions of the Aube River. This is likely because streams do not grow larger gradually, but rather empty their contents into much larger rivers quite suddenly. And a snail in a lake is either on a rock, or mud, or vegetation. There is no rock – mud gradient. So the existence of discrete associations (or dissociations) of freshwater molluscs cannot be considered evidence of any interaction between them or of any 'community organization'. Each mollusc species may well be reacting independently to environments they perceive as discrete.

Models of species similarity

Even when interspecific competition is strong, one is not likely to discover that species distributions differ significantly from random (Hastings 1987). So to probe further for evidence of community structure, one might examine more than simple species aggregation and co-occurrence. I am attracted to an idea first proposed by Elton (1946) that the taxonomic similarity of co-occurring animals can provide an 'entering wedge'. Elton noted that species in the same genus tend to be morphologically similar to each other, and thus have similar ecological requirements. He suggested therefore that congeneric pairs of species should be more likely to compete – a red-tailed hawk is more likely to compete with a red-shouldered hawk, and a myrtle warbler is more likely to compete with a blackburnian warbler. So Elton counted congeneric pairs of species in a variety of bird faunas, noted fewer than he expected, and considered this to be evidence of competition.

Elton's original analysis has since been shown to have been flawed, but the underlying principle retains its validity. We stand at the edge of a lake suspecting strongly that the molluscs it contains are clumped into different habitats. There are three possible outcomes from the taxonomic analysis of such aggregations: unexpected dissimilarity, unexpected similarity, and the expected – no taxonomic relationship whatsoever. The first result may be interpreted as evidence of competition, although other reasonable interpretations exist. The second result could also be interpreted as the consequence of similar ecological requirements for similar species, but such requirements are not limiting, so no competition occurs. At first glance, the upper half of Figure 9.7 might suggest unexpected dissimilarity – no congeneric pairs at all are to be found in local clusters (setting aside cluster C and nesting). In fact, only two confamilial pairs are present – the planorbids *Planorbis carinatus* and *Bathyomphalus* (= *Anisus*) *contortus* in cluster (B) and the unionids *Unio pictorum* and *Anodonta anatina* in cluster (E). But here we run into the major difficulty with this sort of analysis – few congeneric species occur in the Aube River as a whole, so few pairs would be expected if species were cast into sites randomly.

The analysis of congeneric pairs in faunal samples is statistically difficult. How does one handle congeneric triplets, for example? For this reason, authors more recent than Elton have turned to the species-to-genus ratio (Simberloff 1970). But S:G ratios convert data that are fundamentally discrete in nature into a variable that appears to vary continuously from zero to one, but does not. So to avoid this pitfall, I have proposed a Monte Carlo method to analyse taxonomic similarity in faunal samples (Dillon 1987). The technique also has weaknesses, which we shall address presently. But to illustrate the technique I introduce a new, larger set of data on the distribution of riverine molluscs: the unionids of the Clinch and Elk Rivers of the southeastern United States.

In the late 1970s the Tennessee Valley Authority initiated its Cumberlandian Mollusc Conservation Program to survey the unionid mussels of nine tributaries of the Tennessee River, along with their habitats and fish hosts. The data, reported by Ahlstedt (1986), include a fair number of quantitative samples, taken with a metal frame of 0.25 m^2. The majority of the data published by Ahlstedt were, however, qualitative. For each of the nine rivers, a crew of four collectors searched a series of sampling stations along some prescribed stream length using SCUBA, snorkle, and hand rake, and even searching muskrat middens for fresh-dead shells. This technique certainly introduces bias against small, inconspicuous species, and probably biases for individuals of the endangered target taxa. But if relative abundances are set aside, the list of species collected at each sample site is probably fairly accurate.

Here I analyse the two largest data sets collected by Ahlstedt. The Clinch River is one of the major tributaries of the upper Tennessee River in mountainous east Tennessee. Ahlstedt took 141 samples, ranging from Clinch River mile 150.8 to mile 323.8, a distance of 279 km (his Table 2). The Elk River of central Tennessee and northern Alabama enters the Tennessee River over 600 km downstream from the last Clinch sample. Ahlstedt took 110 samples from the Elk, ranging from mile 28.0 to mile

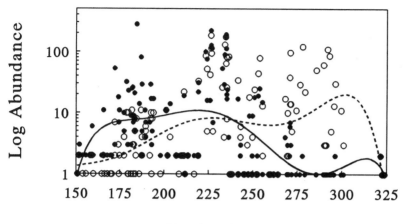

Figure 9.10. The abundance of *Actinonaias carianta* and *Actinonaias pectorosa* at 141 sites on the Clinch River, Virginia and Tennessee. (Data of Ahlstedt 1986.) Curves (solid for *A. carinata*, dashed for *A. pectorosa*) fit using a polynomial.

200.0 (277 km total – his Table 14). The Elk River data have been published separately by Ahlstedt (1983).

Table 9.5 summarizes the distribution and abundances of unionid mussels in these two rivers. Both mussel faunas showed at least some zonation along the lengths of stream sampled. For example, the two species of 'muckets' (genus *Actinonaias*) appeared to be almost equally common in the Clinch River. These are equally large, heavy, conspicuous mussels, distinguished only by minor details of shell shape and pigmentation. There should be no bias to collect one over the other. Figure 9.10 shows, however, that *A. pectorosa* is much more common upstream and *A. carinata* is more common downstream. How general is this phenomenon? It is clear from the raw data that at least some zonation occurs throughout the unionid communities of the Clinch and Elk, just as we saw in the data of Storey and Mouthon. Do the most similar mussel species tend to assort along the stream gradient to minimize competition present or potential?

I have categorized 128 Clinch River samples simultaneously by the number of species and the number of genera each contained (6 samples had no mussels and 7 samples had but one species) in Table 9.7. The number of samples containing two species, three species, and so forth, is

Table 9.7. Ahlstedt's (1986) samples of the Clinch River mussel fauna, categorized by the number of species and genera they contained

Number of species	Number of genera																							S_o
	1	2	3	4	5	6	7	8	9	10	11	12	13	14	15	16	17	18	19	20	21	22	23	
2		7																							7
3		1	1																						2
4				3																					3
5				1																					1
6					2	2																			4
7						4	3																		7
8							2	2																	4
9							2	7	2																11
10						1	2	5	4																12
11							1		6	1	3														11
12								4	1	2															7
13								1		1		1	1												4
14										4	1	3	1												9
15										2		2		1											5
16										1		3	1	1											6
17											1	3		2											6
18												2		1	1										4
19														1	4										5
20													1												1
21															1	2									3
22															1	1	1								3

	0	8	1	4	2	7	10	19	13	11	5	13	7	5	3	4	1	2	1	2	1	2
23																		2				2
24																		1				2
25						1											1					2
26																						0
27															1							1
28																1						1
29													1									1
30											1	1										1
31															1							1
32																						0
33												1								1		1
34																				1		1
T_o	0	8	1	4	2	7	10	19	13	11	5	13	7	5	3	4	1	2	1	2	1	2

totalled along the right margin of the table – this is the S_0 distribution we encountered previously. In a similar fashion, summing the columns of this table to the bottom margin results in the T_0 distribution, the higher taxa (genera in this case) observed. Is this T_0 distribution approximately what one would expect given 128 samples of this species number from this fauna?

If there is no tendency for co-occurring mussels to be more or less similar to one another taxonomically, a random sample of species from the Clinch River fauna using the S_0 distribution should give a distribution of genera (T_e, higher taxa expected) indistinguishable from T_0. But if co-occurring mussels tend to be taxonomically dissimilar, for example, the T_0 distribution will tend to be higher than the randomly generated T_e distribution.

I used a Monte Carlo simulation to generate T_e distributions over 12 800 samples. Two simulations were performed, under different assumptions. In the 'weighted' simulation, the probability of selecting each species was proportional to the number of sites at which it was collected. For example, Table 9.5 shows that in the Clinch River, *Actinonaias carinata* was found at 104 sites, *A. pectorosa* at 101 sites, *Alasmidonta marginata* at 37 sites, and so on. The total occurrences of all mussels at all sites were $104 + 101 + 37 + \ldots = 1694$. So the probability of selecting *A. carinata* in the weighted simulation was $104/1694 = 0.061$. (Contrast this with the simulation used to generate the S_e distribution for Oneida Lake (Figure 9.9), which was weighted by the number of individuals collected, not the number of sites.) But the degree to which each species is widespread may be a function of the community processes that this analysis is trying to uncover. So a second 'unweighted' simulation was performed, where the probability of selecting each of the 43 species was equal. These two null hypotheses are analogous to those commonly derived from simulations used to test island biotas (e.g. Simberloff 1978).

I repeated this entire analysis using Ahlstedt's data from the Elk River. Table 9.5 shows many contrasts between the Clinch and the Elk faunas. Collections from the Elk included 38 species, only 26 of which appear among the 43 species listed from the Clinch. Abundances were lower in the Elk. The largest number of different species at any site on the Elk was 19, compared with the 34-species samples of the Clinch (Table 9.7).

Using one-sample Kolmogorov–Smirnov tests, I found no differences between the observed distributions of genera in the Clinch River and either weighted or unweighted simulation. From the weighted simulation, for example, one would expect 0.3 samples with one genus, 6.9 samples with two genera, 2.4 samples with three genera, etc. The fit to

T_o (Table 9.7) was very good. Nor was any difference apparent between the Elk T_o and either weighted or unweighted T_e. So the mussel species do not generally seem to sort by genera along stream gradients. It should be noted that this mussel fauna includes a large proportion of monospecific genera, rendering taxonomic tests such as mine somewhat insensitive. But as far as can be told, Ahlstedt's mussel samples seem to be taxonomically random.

It is certainly possible, however, that any competition between mussel species may be for fish hosts for their glochidia. Although most mussels seem to be able to use a variety of fish hosts, as we saw in Chapter 2, some species seem to show considerable host specificity. But since the fish hosts are known for only about half of the 43 Clinch River species, any prospecting for host competition might more profitably be directed toward other communities of unionids. Barr and colleagues (1986) reported another activity of the Cumberlandian Mollusc Conservation Program, a survey of macrofauna at 15 sites in nine rivers potentially harbouring endangered mussel species. They used a variety of techniques (snorkelling, seining, electrofishing) to collect both qualitative and quantitative samples of the fish fauna at each site during the three-month period glochidia are believed released. They also took quantitative samples of the mussel faunas to supplement those of Ahlstedt. By good fortune, the fish hosts have been described for 6 of the 8 mussel species (74% of the individuals) collected by Barr and colleagues in quantitative samples from a site (river mile 27.8) on the Nolichucky River, Tennessee.

What sort of host overlap is apparent, given the relative abundances of both mussels and fish available? Barr obtained 15 fish species in quantitative samples at the Nolichucky River site of interest to us, and recorded an additional 8 rarer species in qualitative samples only. Not a single one of the 15 most common fish species has ever been implicated as a host for any of the 6 mussel species. Watters (1994) lists 13 known hosts for *Actinonaias ligamintina* ($=carinata$), 12 fish not collected by Barr and the smallmouth bass, found in qualitative samples only. None of the 13 known fish hosts for *Amblema plicata* ($=costata$) seems to be found in this section of the Nolichucky. The problem is doubtless due in no small part to the diversity of fish. Simply for reasons of availability, researchers have generally experimented with larger, more common fish such as catfish, bass, sunfish, and crappie. Until some brave soul selects a place like the Nolichucky and quantitatively tests all 8×23 combinations, the question of host competition in unionid communities will remain open.

There is a another method of measuring the similarity of co-occurring species, simpler and more practical than trying to estimate resource use and more direct than taxonomic relatedness. One could take direct morphological measurements. This practice is as old as the method of congeneric pairs, originating with Huxley (1942) and Lack (1947), and being brought to the fore in a seminal paper by Hutchinson (1959). Hutchinson noted that congeneric species of mammals and birds tended to be of unexpectedly diverse sizes when occurring sympatrically, and went so far as to suggest that the ratio of the larger to the smaller species tends to be about 1.3. He considered this effect an adaptation to minimize competition. Although some attention has continued to centre on body size ratios, many workers (e.g. Ricklefs *et al.* 1981, Travis and Ricklefs 1983) have measured a greater variety of morphological attributes, especially including limb sizes and aspects of the trophic apparatus. The assumptions are the same as those that are made for taxonomic analysis, and the question remains the same. Do co-occurring animals tend to be morphologically similar, morphologically different, or random in this regard?

Let us return to Baker's (1918) data on the snails of Oneida Lake to address this question. Of the dozens of morphological attributes one could measure in an effort to characterize the ecological requirements of a snail, aspects of the radula would seem to be among the most important. (Some caveats may be recalled from Chapter 3, however.) It also seems reasonable to expect that shell morphology might give clues about a snail's ecological requirements. Diet studies lead us to expect that the ten-fold size range seen in the Oneida Lake gastropod fauna may have ecological importance. I was able to obtain living or preserved specimens of the 23 species not marked with a footnote superscript in Table 9.6. (The 6 species excluded from this analysis accounted for only 35 individuals in total.) For each species I measured shell length, body whorl length, shell width, aperture length, aperture width, spire angle, radula length, and radula width, and counted radula rows, teeth per row, and cusps per 30 teeth. Further details regarding measurement techniques, as well as means for all 11 morphological variables measured over all 23 species, are available in Dillon (1981).

The data set included variables that were skewed, measured in all manner of units with all manner of variances, and highly intercorrelated. So I performed a principal component analysis on the between-species correlation matrix of log-transformed character means. I accepted the six principal components that together accounted for the first 95% of the variance. I then calculated two statistics to measure overall morphological dissimilarity within each of Baker's 111 samples containing two to seven species present. (Samples of eight, nine, and ten species were

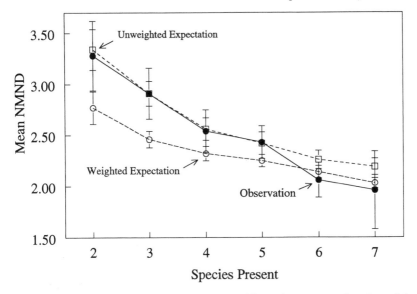

Figure 9.11. Mean nearest morphological neighbour distance as a function of the number of snail species present in samples from Oneida Lake (analysis of Dillon 1981). Baker's (1918) samples are graphed as closed circles, the results of an abundance-weighted simulation as open circles, and an abundance-unweighted simulation as open squares. Bars are 95% confidence limits for the means.

excluded due to small sample size.) First, if competition plays a role in determining the species present in each sample, it seemed possible that the most similar animals would be most severely affected. So a measure of the dissimilarity of the most similar species, mean nearest morphological neighbour distance (NMND) was calculated for each of the 111 sites (Ricklefs and Travis 1980). Where N species were present in a sample, NMND was calculated as the average of the N Euclidean distances to nearest neighbours in six-dimensional space. Another possibility would seem to be that competition may be most severe among individuals in communities with the smallest overall morphological dimensions. So a measure of the dissimilarity of the most dissimilar species, mean furthest neighbour distance (FMND) was calculated for Baker's 111 samples in a fashion analogous to NMND.

The NMND in Baker's samples is shown in Figure 9.11 as a function of species present. Also shown in Figure 9.11 are the results of two Monte Carlo simulations of Baker's sampling process. The lower data set comes from an abundance-weighted simulation analogous to the one shown in Figure 9.9. Here for $N = 2 - 7$, N different snail species were randomly drawn from an abundance-weighted species pool, with replacement.

Both NMND and FMND were calculated for each sample, to a total of 100 iterations for each N. The uppermost data set is the result of an unweighted Monte Carlo simulation. As I argued in the taxonomic analysis of mussel distribution, relative abundances may be a function of species interactions such as competition, and weighting by abundance may introduce bias. In all other details of computation, however, the unweighted simulation was performed in a manner identical to the weighted simulation described above.

Baker's samples with two, three, four, and five species present are all significantly more morphologically diverse than abundance-weighted expectation (Dillon 1981). This is true regardless of whether the statistic is NMND (Figure 9.11) or FMND (not shown). 'Morphological space' seems to fill up rapidly, so that the phenomenon disappears in samples of six or seven species. (Might this be the reason that Baker's samples contain fewer species than expected, as witnessed in Figure 9.9?) So by comparison with the weighted null hypothesis, the snail species co- occurring in small bottom samples of Oneida Lake apparently tend to be morphologically dissimilar.

The fit in both NMND and FMND between Baker's data and unweighted expectation was very good, however. But using the tabular method we have just applied to the Clinch River mussel data, Dillon (1987) was able to reject the unweighted null hypothesis taxonomically. The T_o distribution based on genera co-occurring in Baker's samples did indeed seem to be shifted to the right of unweighted T_e, with 0.107 the value of D from a Kolmogorov–Smirnov one-sample test. Thus the snails that co-occur in bottom samples from Oneida Lake seem to be significantly dissimilar, regardless of the method by which similarity is measured.

The sizes and shapes of the freshwater snails of Oneida Lake are so various that there must be hundreds of combinations resulting in very little morphological overlap. Perhaps each benthic sample is unique. Or perhaps the snails are assorted into just a few stable assemblages of co-adapted species with dissimilar morphology. To address this question, I performed a two-dimensional scaling of the 16 most common Oneida Lake species, using the same techniques described for Mouthon's data earlier in this chapter. The 13 species that occurred in five sites or less were excluded. For this analysis the Jaccard similarity coefficient was calculated over all Baker's 117 sites with two or more species present. After 45 iterations, the final stress for the two-dimensional scaling shown in Figure 9.12 was 0.151.

The contrast between Figure 9.12 and Figure 9.7 is striking. A single

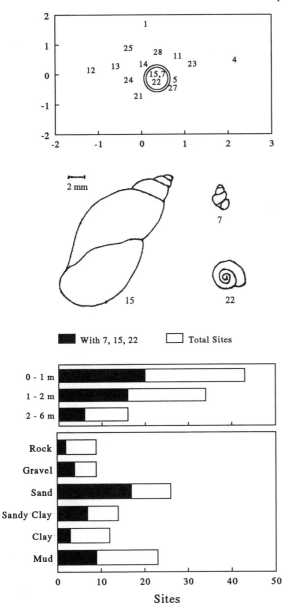

Figure 9.12. (Top) Non-metric two-dimensional scaling of the associations between the 16 most common snail species inhabiting Oneida Lake, NY (data of Baker 1918). Contours show Jaccard similarities at 0.5 and 0.4. (Middle) Scale outlines of *Lymnaea catascopium* (15), *Amnicola limosa* (7) and *Gyraulus parvus* (22). (Bottom) The samples from Oneida Lake containing more than two snail species categorized by depth and substrate, with the proportion containing the association of *Amnicola lustrica*, *Gyraulus parvus*, and *Lymnaea catascopium* darkened.

cluster is apparent in the Oneida Lake data, perhaps not surprisingly involving the three most common species. Note in Table 9.6 that *Amnicola lustrica, Gyraulus parvus,* and *Lymnaea catascopium* are by far the most widespread species in Oneida Lake, found in 60–70 sites each, while the next most common species (e.g. *Valvata tricarinata, Physa sayii,* and *Promenetes exacuosus*) were found in only 30–40 sites. So if *A. lustrica, G. parvus,* and *L. catascopium* were cast randomly into 117 sites, the Jaccard coefficients between them would be expected to be high. They would be expected to form a somewhat artificial cluster in the middle of the MDS ordination, in the same fashion as the central cluster in Mouthon's data (Figure 9.7). So at first glance, Figure 9.12 would appear to result from a random distribution of species in samples, with no evidence of community structure.

But upon closer inspection, Figure 9.12 does show structure – a structure that helps explain the significant morphological dissimilarity of co-occurring snails in Oneida Lake. Consider, as an example, Baker's 17 samples with four species present. Weighted by abundance, the probability of *A. lustrica* appearing in a sample is $p_1 = 2841/5582 = 0.509$ (Table 9.6), the probability of *G. parvus* is $p_2 = 0.0104$, the probability of *L. catascopium* is $p_3 = 0.082$, and the probability of all other species is $q = 0.304$. The chance of *A. lustrica* – *G. parvus* – other – *L. catascopium*, in that order, would be:

$$p_1[p_2/(1-p_1)][q/(1-p_1-p_2)][p_3/(1-p_1-p_2)]$$

Note here that *A. lustrica, G. parvus* and *L. catascopium* are sampled without replacement, but the selection of one of the other 13 species with combined probability q would not markedly affect q, so that this probability stays constant.

Now considering order, there are $4! = 24$ ways to select four species. Calculating individual probabilities of each outcome as above, and summing the probabilities of all 24 such outcomes, the chance that three of the species in a sample of four different species would be *A. lustrica, G. parvus,* and *L. catascopium* is 0.205. In 17 four-species samples, one would expect $0.205(17) = 3.5$ to have those species. But in fact, 10 of the 17 four-species samples have *A. lustrica, G. parvus,* and *L. catascopium*. One would expect 1.8 of the 23 three-species samples to have this association, but Baker found 3. This phenomenon appears to extend throughout Baker's data. So although one would expect these three species to cluster in the centre of Figure 9.12 as a simple consequence of their abundances, a Jaccard coefficient as high as 0.50 suggests some biological interaction.

Perhaps *Amnicola*, *Gyraulus*, and *Lymnaea* are among the least likely snail species to compete. They are very different morphologically and taxonomically. *Amnicola lustrica* is a small, tall- spired prosobranch (shell about 3.3mm×2.2mm), *G. parvus* is a small planispiral pulmonate (1.0 mm×3.1 mm) and *L. catascopium* is a large, high-spired pulmonate (14.5 mm×8.5 mm). A large fraction of the morphological dissimilarity within Oneida Lake samples (Figure 9.11) may result from their association.

Dillon (1981) did not find any relationship between morphological similarity and depth, substrate, or any other environmental variable. And in fact, the *A. lustrica* – *G. parvus* – *L. catascopium* association does not seem to be restricted to any particular habitat type. Figure 9.12 also shows the 93 sites with three or more species categorized by substrate type and depth. The small preference of the three species for sand and their rarity at rock and clay sites probably results from the tendency for rock and clay sites to have fewer species overall. So unlike the associations found in rivers, *A. lustrica*, *G. parvus*, and *L. catascopium* seem to associate independently of environmental variables.

I am not sure of the generality of these results. Oneida Lake in 1918 was home to one of the largest snail faunas of which I am aware worldwide, setting aside the ancient lakes with endemic radiations, and the data set is large. For comparison, I have analysed the taxonomic similarity within Økland's (1964) samples of Lake Borrevann, discussed earlier in this chapter (Table 9.4).

Økland took 85 quantitative samples at depths of 0.2 to 3.0 m, collecting nine species of snails in five families. But only 47 samples contained more than one species. I categorized these 47 samples simultaneously by the number of species and families they contained, and compared the resulting T_o to T_e distributions for both abundance-weighted and unweighted simulations. Quite in contrast to the Oneida Lake results, values of D were -0.12 for the comparison with the weighted simulation and -0.13 for the unweighted comparison. The negative values indicate that snails co-occurring in Lake Borrevann tend to be more similar taxonomically than random expectation, although not significantly. (The critical value with $N=47$ would be -0.20.) My impression is that the little tendency there seems to be is due to the co-occurrence of the two common *Valvata* species in the deeper samples, where other gastropods are rare.

I have treated the Oneida Lake data as though every sample were collected simultaneously. In fact Baker states that his field work 'was con-

Table 9.8. *Above the diagonal — observed frequency of co-occurrence between five species of* Planorbis *in 2 m samples from Esrom Lake, over 12 months. Below the diagonal — frequency expected from independent assortment*

	P. planorbis	P. carinatus	P. albus	P. contortus	P. cristata
P. planorbis	–	0	5	1	5
P. carinatus	1.0	–	1	1	1
P. albus	4.5	1.5	–	2	6
P. contortus	1.0	0.3	1.5	–	2
P. cristata	3.0	1.0	4.5	1.0	–

Source: data of Berg (1938).

ducted during the month of July', so this assumption is probably a fair one. But separation on a temporal scale may be as important as separation on a spatial one. For this reason I would like to return to the other data set we have from a Scandinavian lake, the work on Esrom Lake by Berg (1938).

Is there any tendency in Esrom Lake for morphologically dissimilar snails to co-occur, or similar snails to assort separately, as we found in Oneida Lake? Or is Berg's data set more like Økland's? It would be very difficult to detect any assortment on a spatial scale in Berg's data, since only seven samples were taken each period. And a glance at Figure 7.11 shows that all species would show positive associations, simply because few snails of any species occur in the three deepest samples. But considering only Berg's 2 m sample, a weak test is possible for temporal separation.

Berg's samples included five species of *Planorbis* (or *Anisus*). *Planorbis albus* occurred in 2 m samples in 9 months, *P. planorbis* and *P. crista* in 6 months, and *P. carinatus* and *P. contortus* in 2 months each. Then the probability that *P. albus* and *P. planorbis* co-occur in a given month in the 2 m sample would be $9/12 \times 6/12 = 0.375$, and one would expect them to co-occur in 0.375 (12) = 4.5 monthly samples if they were assorting randomly. The expected number of co-occurrences between all five *Planorbis* species over the 12 months are shown below the diagonal in Table 9.8.

Above the diagonal in Table 9.8 are given the actual number of months that these pairs of *Planorbis* species were observed co-occurring in Berg's 2 m samples. Although the significance of these results is difficult to assess with $N=12$, in general the match between observation and random expectation is very good. The greatest deviations are positive. *Planorbis planorbis* and *P. crista* show two more co-occurrences than expected, and *P. albus* and *P. crista* show 1.5 more co-occurrences. These results would seem to be due to the fact that all planorbids are rarer in the winter, and more common in the summer. Thus there is no evidence of temporal separation.

Models of species abundance

In the rest of this chapter I take, as a point of departure, a 'distribution of commonness and rarity' for each of the mollusc communities examined. This is another matter dating back to the 1940s (Preston 1948) that subsequently attracted a great deal of attention (reviews by May 1975, Pielou 1975). It has generally seemed clear to ecologists that distributions of species abundance are not arbitrary, but seem to contain some information about community organization. Some distributions (e.g. all species about equally abundant) are almost never encountered. But a rigorous test of the suspicions of generations of community ecologists was over forty years in coming – the bull's eye method of Hopf and Brown (1986).

Hopf and Brown used as their null hypothesis MacArthur's notion that abundances can be modelled as a 'stick' of unit length which is broken at random into S pieces, where S is the number of species in a community. The sum of the lengths of the S pieces ($f_1 + f_2 + f_3 + \ldots + f_S$) would equal 1.0. They proposed a test statistic:

$$R = \left[\left(\sum_{i=1}^{S} f_i^2 - 1/S \right) / (1 - 1/S) \right]^{1/2}$$

and provided a 'lookup table' relating any value of R from $S=1$ to 50 to a 'score'. With some minor reservations, a score of 20 is equivalent to a significance level of $1/20 = 0.05$.

We have discussed thus far three large, quantitative data sets on snail abundance, Berg's on Lake Esrom (Chapter 7), Økland's on Lake Borrevann (Table 9.4), and Baker's on Oneida Lake (Table 9.6). We have also examined a fair number of bivalve data sets in this chapter: those of Kilgour and Mackie (1988) on the abundance of pisidiid species in the Ottawa River (Table 9.2), Walter and Kuiper (1978) on pisidiids in Lake Zurich (Table 9.3), Ellis on the unionids of the upper Mississippi, and

Table 9.9. *Unionid mussels collected in quantitative samples at seven sites on the Altamaha River, Georgia*

Species	Individuals
Alasmidonta arcula	13
Anodonta couperiana	4
A. gibbosa	10
Canthyria spinosa	88
Elliptio dariensis	6
E. hopetonensis	237
E. shepardianus	8
Lampsilis dolabraeformis	495
L. splendida	15

Source: Sickel (1980).

Ahlstedt on unionids in two tributaries of the Tennessee River (Table 9.5). Although we have concentrated on the number of sites where mussels occurred in Ahlstedt's qualitative samples, here we will examine relative abundances from 385 quantitative samples of 0.25 m^2 each taken from the Clinch River (his Table 3).

In addition, Table 9.5 shows two data sets on mussels of the North American interior not previously mentioned. Van der Schalie (1948) hired a commercial clammer to sample 45 stations on the Grand River, a tributary of Lake Michigan. These samples were taken with a crow-foot dredge, and must be biased at least somewhat against small individuals. The sample of Havlik and Marking (1980), from the Mississippi River at Prairie du Chien (about zone 4 of Ellis), is remarkable for its lack of bias in this regard. Havlik and Marking sieved 10 cubic metres of material from a spoil bank dredged to improve navigation. Finally, as an interesting contrast to these four interior unionid faunas, Table 9.9 shows the relative abundances of a smaller (but certainly no younger or less 'stable') fauna from the Altamaha River, an Atlantic drainage of Georgia (Sickel 1980).

The relative abundances are plotted on a $\log_2$ scale for the five mussel communities in Figure 9.13, the two pisidiid communities in Figure 9.14, and the three snail communities in Figure 9.15. The use of base 2 logarithms is largely historical. The incidence of any species whose abundance was equal to an even power of two was split between two categories. For example, van der Schalie collected $N=1$ *Villosa iris* and $N=2$

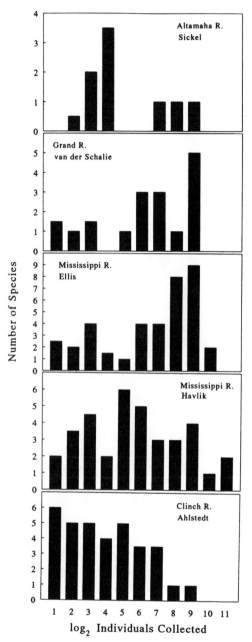

Figure 9.13. Number of unionid mussel species in log$_2$ categories (octaves) of species abundance. Category '1' includes species with from 0 to $2^1 = 2$ individuals; category '2' are species with from 2 to $2^2 = 4$ individuals, etc. (Data are from Sickel 1980, van der Schalie 1948, Ellis 1931a,b, Havlik and Marking 1980, and Ahlstedt 1986.)

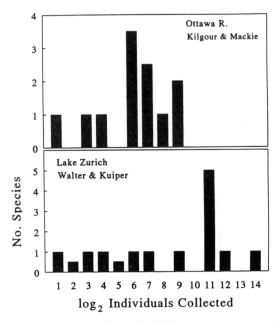

Figure 9.14. Number of pisidiid species in octaves of species abundance (data of Walter and Kuiper 1978 and Kilgour and Mackie 1988.) Abscissa scaled as in Figure 9.13.

Anodonta grandis (Table 9.5). Then Figure 9.13 shows 1.5 species in incidence category 1: the *Villosa* and half the *Anodonta*.

Table 9.10 shows values of R and 'score' from bull's eye tests of Hopf and Brown. (The tiny discrepancy between the value of R I have calculated from Baker's data here and that calculated by Hopf and Brown from the same data comes from my use of the entire $S_o = 29$ data set, not $S_o = 23$.) The data of Ellis, van der Schalie, Kilgour and Mackie, and Økland all had scores less than 20, suggesting that the relative abundances of molluscs in these four communities might be modelled as random processes. The figures show that in spite of the log transformation, all four of these distributions have long tails to the left and high modal abundances. We now turn our attention to the six data sets that do not seem as likely to be random: Baker, Berg, Walter, Sickel, Havlik and Ahlstedt.

Two mathematical distributions developed to describe species abundances have attained prominence because some ecological meaning has been accorded them — the lognormal and the log-series. Although there is some controversy, the lognormal is generally viewed as fitting abundances in large, stable, equilibrium communities, while species abun-

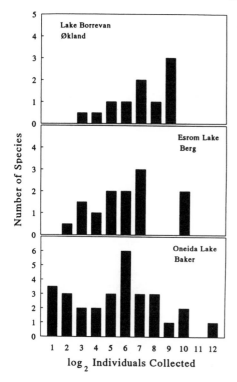

Figure 9.15. Number of snail species in octaves of species abundance (data of Baker 1918) Berg 1938 and Økland 1964.) Abscissa scaled as in Figure 9.13.

dances in small, stressed, or pioneer communities are predicted to fit the log-series. The only other model with a biological rationale is Hughes' (1986) 'dynamics' model, where the abundances of i species at time t are based on simulations of recruitment rate, survival (including catastrophe), and 'gregariousness', parameterized arbitrarily.

If one ranks the species in a community from most common to most rare and plots the log of their abundance (a 'dominance-diversity curve'), a data set fitting a log-series is easily recognized by its linearity. But abundances fitting the dynamics model will appear 'deeply concave, especially in the upper part' (Hughes 1986). A data set fitting the lognormal will skew to the right in the lower part, as rare species are added and the mode of the distribution is approached. But the lognormal distribution is difficult to fit to a data set, as a large portion of its left side (species that are very rare) often does not appear in collections. (Preston said that these species are hidden behind a 'veil line'.) Sometimes, in fact, ecologists

Table 9.10. *Species–abundance statistics for the data sets shown in Figures 9.13 to 9.15*

	S_o	R	Score	S_e	σ_o	σ_e
Oneida snails (Baker)	29	0.512	20	29	2.95	3.19
Esrom snails (Berg)	12	0.501	20	12	2.19	2.85
Zurich pisidiids (Walter)	14	0.477	20	14	3.91	2.91
Altamaha mussels (Sickel)	9	0.573	20	10	2.82	2.73
Mississippi mussels (Havlik)	36	0.329	20	42	3.50	3.26
Mississippi mussels (Ellis)	38	0.184	19	–	–	–
Ottawa pisidiids (Kilgour)	12	0.361	19	–	–	–
Grand mussels (van der Schalie)	17	0.275	17	–	–	–
Borrevann snails (Økland)	9	0.323	13	–	–	–
Clinch mussels (Ahlstedt)	34	0.339	20	–	–	–

Notes:

S_o, number of species observed; R and 'score', bull's eye statistics of Hopf and Brown (1986); S_e, number of species expected given a log normal model; σ_o, observed variance; σ_e, variance expected for canonical log normal model fit to observed.

consider that more than half of the species in a community are missing from a sample hypothesized to fit the distribution, such that the abundance of no species can be considered modal. Hughes (1986) argued convincingly, however, that it is unwise to even attempt to fit a lognormal to any distribution that does not reveal a mode.

Hughes collected 222 data sets on species abundances (no freshwater molluscs) and compared them for their fit with each of these three models. He plotted number of species against $\log_2$ number of individuals and examined for a mode. Hughes found modes in 16 of his 31 data sets for vertebrates, but only 24% of his 182 invertebrate data sets. Over all 222 sets, only 28% could thus be considered lognormal. To distinguish between data fitting log-series and dynamics models, Hughes examined the upper part of their dominance-diversity curves to see if they were deeply concave. He found that the log-series distribution proved a better prediction than the dynamics model in only 4% of the samples. Thus his dynamics model, based from the start on a simulation of the behaviour of natural populations, emerged as the clear victor over the two older models based on theory.

In apparent contrast to the majority of invertebrate communities, five of the six non-random sets of abundance data gathered in Figures 9.13 to 9.15 are strongly modal. The lognormal model would seem to fit these freshwater mollusc abundances rather well. The exception is Ahlstedt's Clinch River mussel community, which seems to have relative abundances fitting a log-series. (The dynamics model would predict more very abundant species, and thus a longer tail on the right of the Clinch River distribution.) A traditional interpretation is that such a log-series may in fact be a lognormal with a mode at the veil line, and that the mode would be revealed if more individuals were collected. This may be the case – Ahlstedt's samples come from a much larger fauna than Sickel's but were much less extensive than those of Havlik and Marking.

I hesitate to fit a lognormal distribution to Ahlstedt's data or to the abundance data that appear to be random. But Table 9.10 shows variance fits to the other five data sets. Notice in Figure 9.15 that more snail species occur to the left of the modes than to the right. The same observation could be made for Walter and Kuiper's pisidiid data (Figure 9.14). If we assume that these distributions are in fact lognormal, this implies that all species have been collected. So the observed variances given in Table 9.10 for the Baker, Berg, and Walter data were calculated directly from the $\log_2$ abundances. In contrast, some of the species seem to be obscured behind a 'veil line' in the collections of Havlik and Sickel. I assumed the true mean of these two distributions to be the average of the (transformed) modal abundances, and thus inferred that the true number of mussel species in the Altamaha would be 10 and in Havlik and Marking's Mississippi sample 42. I then plotted the known portions of these two distributions on probability paper and estimated the variances graphically (Sokal and Rohlf 1995:116).

Preston (1962) pointed out that the variance of lognormal abundance distributions seems to be 'canonically' related to the number of species in communities. The standard deviation of a canonical lognormal distribution, σ_e, is related to S by the following relationship (Sugihara 1980):

$$S = \sigma_e \sqrt{\pi/2} \exp\left[\frac{(\sigma_e \ln 2)^2}{2}\right]$$

May (1975) thought that this canonical relationship might be a trivial mathematical property of all lognormal distributions for large S, but it has since been shown fairly convincingly that natural communities seem to fit the canonical relationship more closely than can be explained by mathematical generalities (Sugihara 1980, May 1986).

The last column of Table 9.10 shows the standard deviation expected from a canonical lognormal distribution for S the same as observed by Baker, Berg, Walter and Kuiper, Sickel, and Havlik. In general, the fit between observed and canonically expected standard deviation is very good. Judging by the rather broad error bars associated with such estimates of variance (Sugihara 1980, fig. 7), I am sure that four of the five data sets do not differ significantly from canonical. The pisidiid data set of Walter and Kuiper seems to show rather more variance than would be predicted from a canonical lognormal.

To explain the canonical lognormal distribution of species abundances, Sugihara proposed the 'sequential breakage model'. This differs from the 'broken stick' null hypothesis that we have rejected for these data in an important respect. Rather than breaking up the community resources (the 'stick') randomly, Sugihara suggests that the first species takes a random portion, and then the second species takes a random portion *of the remainder*, and then the third species takes a random portion of the resources not taken by the first two, and so on.

I am prepared to accept this model as a generally applicable to freshwater mollusc communities. Two of the three snail communities seem to fit the sequential breakage model, and two quite different mussel communites as well. I consider it likely that Ahlstedt's abundance data would have been found to be canonically lognormal if more samples were taken. The pisidiid community of Lake Zurich is anomalous. Abundances appear to be more evenly distributed than expectation, much as MacArthur's original 'broken stick'. This requires further study, but I have no idea what sort.

It is certainly possible that species abundances in the other two mussel communities, Økland's snails, and Kilgour and Mackie's Ottawa pisidiids are genuinely random. But in light of the fact that Havlik's clear lognormal distribution was obtained from a sample taken at one of the 14 zones Ellis collected, it also seems possible that the Ellis data, at least, is an artificial admixture of several distinct communities. Disturbance may have played a role. The mussels of the Grand River were, like those of the Mississippi, exploited commercially, and fishermen remove certain heavy, pearly species at disproportionate freqencies. The northerly populations of snails and pisidiids may have been founded only recently, and the communities not in equilibrium. Thus we are brought back to the subject of variation in species composition *between* communities, first discussed in Chapter 8.

The assembly of communities

We noted in Chapter 8 that sets of faunal lists are commonly analysed by pairwise tests for association between species. At least two problems arise with this approach, however. First, even a fairly small number of species in the 'pool' will call for a large number of pairwise tests, rendering type I statistical error very likely. Second, we have emphasized that putatively significant associations are an expected consequence of the independent responses of different species to a common environment, and hence do not imply any interaction among them. What is needed is a test designed more specifically to detect species interaction, yielding a single estimate of p.

In this chapter we have introduced the notion of using the taxonomic similarity of a pair of species as a measure of their ecological similarity, and hence of the likelihood that they may compete. Notice, for example, that the four **R** species found among Jokinen's (1983) Connecticut snails (Table 8.2, Figure 8.10) include a large planorbid, a small planorbid, a physid and a lymnaeid. Could such taxonomic diversity be a coincidence? The 'large planorbid' in region **R** is *Helisoma trivolvis*, which Jokinen noted almost never co-occurs with *Helisoma campanulatum* over her entire 230-site Connecticut data set. Boerger (1975) reported the same phenomenon in Ontario. Pip (1987b) has performed a great many chi-square tests on the co-occurrence and environmental preferences of the five *Helisoma* species found in central Canada, noting many putatively significant differences. Might the sort of phenomenon we found evidence of within Oneida Lake be expressed among lakes as well?

The 24 Connecticut snail species analysed to construct Figure 8.10 included 1 valvatid, 2 viviparids, 3 hydrobiids, 3 lymnaeids, 4 physids, 3 ancylids, and 8 planorbids, inhabiting 88 sites. Since an assumption of this analysis is that taxonomic similarity estimates ecological similarity, I subdivided the planorbids into two groups: the large *Helisoma* (3 species) and the small *Gyraulus*, *Planorbula*, *Micromenetus* and *Promenetus* group (5 species). Seven of the 88 sites under study here were inhabited by only 1 species. Then the remaining 81 sites are cast into Table 9.11 by the number of species and the number of 'families' (dividing the Planorbidae) they contained.

In Figure 9.16 the T_o distribution from the bottom margin of Table 9.11 is compared with T_e taken from a weighted Monte Carlo simulation performed as outlined earlier in this chapter. Jokinen's lakes and ponds do seem to contain more higher taxa than one would expect from random. This is most apparent in the 6– and 7-family categories. The value of D from a Kolmogorov-Smirnov one-sample test is 0.12, however – probably not quite significant. (For $\alpha = 0.05$, the critical value of D is 0.151). I think it possible that random processes (colonization since the Pleistocene, subsequent extinction due to weather or human

Table 9.11. *Lakes and ponds (total 81) from Connecticut, categorized simultaneously by the number of snail species and number of snail 'families' they contained (planorbids divided — see text)*

Number of species	Number of Families							S
	2	3	4	5	6	7	8	
2	10							10
3	2	3						5
4		1	8					9
5		1	3	6				10
6		1	4	4	3			12
7			1	2	6			9
8				2	1	3		6
9				1	5	3		9
10					4	1		5
11					1	1		2
12						1		1
13						1	1	2
16						1		1
T_o	12	6	16	15	20	11	1	81

Source: data of Jokinen (1983).

disturbance, etc.) may have combined to obscure a slight tendency for ecologically similar gastropod species not to co-occur in Connecticut lakes. After all, human disturbance has by now almost certainly obscured the significant relationship between distribution and taxonomy observable in Oneida Lake in 1918.

Interestingly, the result of a second, unweighted run of the Monte Carlo simulation did not differ significantly from those of the weighted run shown in Figure 9.16. In this chapter we have noted that within lakes it is normal to see tremendous ranges in the relative abundance of species. Within Oneida Lake, for example, over three orders of magnitude separated the abundance of the most common species from the abundance of the rarest (Table 9.6). But Table 8.2 shows much less variation in the number of lakes and ponds inhabited by the 24 Connecticut species. The five species of small planorbids, for example, occurred in 13, 20, 23, 30, and 45 sites, and hence the influence of weighting by number of lakes occupied approaches negligible. In any case, we must conclude that there is no significant tendency for the snails co-occurring in Connecticut lakes and ponds to be more dissimilar taxonomically than random expectation

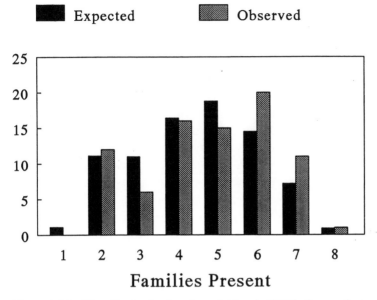

Figures Present

Figure 9.16. The T_o distribution from Jokinen's (1983) Connecticut snail data (Table 9.11) compared with T_e from Monte Carlo simulation.

regardless of the null hypothesis tested. The influence of competition on the distribution of freshwater molluscs at the biogeographic scale generally remains undemonstrated.

Summary

The distinction between diet and habitat is a fine one for snails (and to some extent, for the bivalves too). Researchers generally report broad similarities in the gut contents of diverse gastropod species, detritus almost always the primary component by volume. Yet specific differences, especially regarding the proportions of macrophytes and filamentous algae in the diet, are occasionally striking. This seems to arise not from selective retention of foodstuffs, but from differing habitat choice and differing size and strength of various gastropod jaws. Some snails are large and have strong mouth parts, some are not, and some macrophytes are tough while others are tender. But if a snail crawling over a macrophyte can rip up some plant tissue, the snail will not stop to consider whether this constitutes 'high quality food'. The snail will eat it. Thus little snails, such as the ancylids and smaller planorbids, do not eat aquatic vegetation, and big snails, such as the ampullariids, do.

Most diatoms fit in typically sized snail mouths, bigger diatoms often probably not as well as small diatoms. But some diatoms are stuck so tightly to rocks that they cannot be prised loose, and typical snails find that entire rocks do not fit in their mouths at all, and pass on. As much as I would like to, I cannot find any evidence that the actual morphology of the radula matters. The overall size of the trophic apparatus does, certainly. But if two snails have radular ribbons that are in the same order of magnitude, whether they have 7 teeth or 100 teeth per row seems to be immaterial. Dietary overlap is a likely consequence.

Substrate preference has been demonstrated in a wide variety of snails, both in the field and in the laboratory. Populations have been found preferring vegetation, cobble bottom, or sand. Some snails manifest preferences for macrophytes of particular species or life forms, in response to the varying architectures of the leaves, the palatabilities of the plants themselves, or the palatability of their epiphyton. Such relationships are very dynamic in the wild. Other snail/macrophyte associations may result as much from the protection such cover affords. The distribution of snails in lotic environments seems to be strongly affected by current. Again, snail populations differ at least somewhat in the current speed they can tolerate. In lakes, however, almost all mollusc populations reach maximum densities in the shallows. There are just a few systematic exceptions; *Valvata* among the gastropods and *Pisidium conventus* among the clams. Evidence suggests that it is the substrate/current variable that seems to render the shallows more attractive to most species, rather than dissolved oxygen, distance to shore, or depth *per se*. Seasonal depth migration has occasionally been documented, and snail distributions in general seem quite volatile. But again, the overall impression is one of general habitat overlap, with particular exceptions.

Just like most other organisms, freshwater molluscs have aggregated distributions. If an ecologist siezes a sampler and rushes down to the nearby river or lake with intent to sample a heterogeneous environment or an environmental gradient, mollusc populations will appear clumped together. I think this is because, with some exceptions of course, freshwater molluscs have broadly similar ecological requirements, at least on the scale a human might notice, and they perceive their environment more discretely than we do. But if the ecologist declares, instead, an intention to sample in an environment that seems homogeneous, he or she will find the species of molluscs clumped without much regard to each other, or (possibly even significantly) clumped apart from each other. Again, the molluscs perceive their environment in what appears to us to be a discrete fashion, so what a human would judge to be grossly uniform will constitute a patchwork to the mollusc community.

In at least one unusually large and diverse snail community of which we have record, the snail species that occur together tend to be unexpectedly diverse morphologically, no matter how this is measured. The phenomenon seems to be due to the dominance of three very different taxa, a fairly large lymnaeid, a small planorbid, and a small prosobranch, and to their tendency to occur together. I do feel that this is fairly convincing evidence of community structure, most likely to minimize competition. But no such evidence of competition is apparent in the smaller snail faunas of two Scandinavian lakes, nor in lists of snail species when compared among Connecticut lakes, nor in the large unionid faunas of the North American interior.

A survey of 10 freshwater mollusc communities (three of snails, two of pisidiids, and five of unionids) suggests that the relative abundances of the species in most are not random. They tend to be distributed in a 'canonical lognormal' fashion, which Sugihara has attributed to 'minimal community structure'. With the caveat that random processes may often combine to leave communities far from equilibrium, 'minimal community structure' has great appeal as a model for freshwater molluscs generally.

Literature cited

Abbott, T. (1979) Asiatic clam (*Corbicula fluminea*) vertical distributions in Dale Hollow Reservoir, Tennessee. In *Proceedings, First Int. Corbicula Symposium,* ed. J. Britton, pp. 111–8. Fort Worth, Texas: Texas Christian University.

Abdel-Malek, E. (1952) The preputial organ of snails in the genus *Helisoma* (Gastropoda: Pulmonata). *Am. Midl. Nat.,* **48**, 94–102.

Adam, M. and Lewis, J. (1992) The lack of coexistence between *Lymnaea peregra* and *Lymnaea auricularia* (Gastropoda: Pulmonata). *J. Molluscan Stud.,* **58**, 227–8.

Adam, M. and Lewis, J. (1993) The role of *Lymnaea auricularia* (L) and *Lymnaea peregra* (Muller) (Gastropoda: Pulmonata) in the transmission of larval digeneans in the Lower Thames valley. *J. Molluscan Stud.,* **59**, 1–6.

Ahlstedt, S. (1983) The molluscan fauna of the Elk River in Tennessee and Alabama. *Am. Malacol. Bull.,* **1**, 43–50.

Ahlstedt, S. (1986) *Cumberlandian Mollusk Conservation Program Activity 1: Mussel Distribution Surveys,* TVA/ONRED/AWR-86/15. Knoxville, TN: Tennessee Valley Authority. 125 pp.

Ahmed, M., Upatham, E., Brockelman, W. and Viyanant, V. (1986) Population responses of the snail *Bulinus (Physopsis) abyssinicus* to differing initial social and crowding conditions. *Malacol. Rev.,* **19**, 83–9.

Aho, J. (1966) Ecological basis of the distribution of littoral freshwater molluscs in the vicinity of Tampere, South Finland. *Ann. Zool. Fenn.,* **3**, 287–322.

Aho, J. (1978a) Freshwater snail populations and the equilibrium theory of island biogeography. I. A case study in southern Finland. *Ann. Zool. Fenn.,* **15**, 146–54.

Aho, J. (1978b) Freshwater snail populations and the equilibrium theory of island biogeography. II. Relative importance of chemical and spatial variables. *Ann. Zool. Fenn.,* **15**, 155–64.

Aho, J., Rania, E. and Vuorinen, J. (1981) Species composition of freshwater snail communities in lakes of southern and western Finland. *Ann. Zool. Fenn.,* **18**, 233–42.

Albrecht, E., Carreño, N. and Castro-Vasquez, A. (1996) A quantitative study of copulation and spawning in the Southern American apple-snail *Pomacea canaliculata* (Prosobranchia: Ampullaridae. *Veliger,* **39**, 142–7.

Aldridge, D. (1982) Reproductive tactics in relation to life-cycle bioenergetics in three natural populations of the freshwater snail, *Leptoxis carinata. Ecology,* **63**, 196–208.

Aldridge, D. (1983) Physiological ecology of freshwater prosobranchs. In *The Mollusca, Vol. 6*, ed. W. Russell-Hunter, pp. 329–58. Orlando, FL: Academic Press.

Aldridge, D., Russell-Hunter, W. and Buckley, D. (1986) Age-related differential catabolism in the snail *Viviparus georgianus*, and its significance in the bioenergetics of sexual dimorphism. *Can. J. Zool.*, **64**, 340–6.

Alexander, J. and Covich, A. (1991a) Predator avoidance by the freshwater snail *Physella virgata* in response to the crayfish *Procambarus simulans*. *Oecologia*, **87**, 435–42.

Alexander, J. and Covich, A. (1991b) Predation risk and avoidance behavior in two freshwater snails. *Biol. Bull.*, **180**, 387–93.

Allen, E. (1924) The existence of a short reproductive cycle in *Anodonta imbecilis*. *Biol. Bull.*, **46**, 88–94.

Aloi, J. and Bronmark, C. (1991) Effects of snail density on snail growth and periphyton. *Verh. Internat. Verein. Limnol.*, **24**, 2936–9.

Anderson, R. and Crombie, J. (1984) Experimental studies of age-prevalence curves for *Schistosome mansoni* infections in populations of *Biomphalaria glabrata*. *Parasitology*, **89**, 79–104.

Anderson, R. and May, R. (1979) Prevalence of schistosome infections within molluscan populations: Observed patterns and theoretical predictions. *Parasitology*, **79**, 63–94.

Andrews, E. (1964) The functional anatomy and histology of the reproductive system of some pilid gastropod molluscs. *Proc. Malacol. Soc. Lond.*, **36**, 121–40.

Andrews, E. (1965) The functional anatomy of the gut of the prosobranch gastropod *Pomacea canaliculata* and of some other pilids. *Proc. Zool. Soc. Lond.*, **145**, 19–36.

Antheunisse, L. (1963) Neurosecretory phenomena in the zebra mussel *Dreissena polymorpha* Pallas. *Arch. Neerl. Zool.*, **15**, 237–314.

Appleton, C. (1975) The influence of stream geology on the distribution of the bilharzia host snails *Biomphalaria pfeifferi* and *Bulinus (Physopsis)* sp. *Ann. Trop. Med. Parasitol.*, **69**, 241–55.

Appleton, C. (1976) Observations on the thermal regime of a stream in the Eastern Transvaal with reference to certain aquatic pulmonata. *S. Afr. J. Sci.*, **72**, 20–3.

Appleton, C. (1978) Review of literature on abiotic factors influencing the distribution and life cycles of bilharziasis intermediate host snails. *Malacol. Rev.*, **11**, 1–25.

Araujo, R. and Ramos, M. (1997) Evidence of intrafollicular fertilization in *Pisidium amnicum* (Muller, 1774) (Mollusca: Bivalvia). *Invertebr. Reprod. Devel.*, **32**, 267–72.

Arey, L. (1923) Observations on an acquired immunity to a metozoan parasite. *J. Exp. Zool.*, **38**, 377–81.

Arey, L. (1932) A microscopical study of glochidial immunity. *J. Morphol.*, **53**, 367–79.

Avelar, W. (1993) Functional anatomy of *Fossula fossiculifera* (D'Orbigny,1843) (Bivalvia: Mycetopodidae). *Am. Malacol. Bull.*, **10**, 129–38.

Avelar, W. & de Mendonca, S. (1998) Aspects of gametogenesis of *Diplodon rotundus gratus* (Wagner 1827) (Bivalvia: Hyriidae) in Brazil. *Am Malacol. Bull.*, **14**, 157–63.

Avelar, W., daSilva, A., Colusso, A. and Dal Bo, C. M. R. (1991) Sexual dimorphism in *Castalia undosa undosa* Martens, 1827 (Bivalvia: Hyriidae). *Veliger*, **34**, 229–31.

Avise, J. C. (1994) *Molecular Markers, Natural History and Evolution*. New York: Chapman and Hall.

Avolizi, R. (1971) Biomass turnover in natural populations of viviparous Sphaeriid clams: Interspecific and intraspecific comparisons of growth, fecundity, mortality, and biomass production. Ph. D. Diss., 169 pp. Syracuse, NY: Syracuse University.

Avolizi, R. (1976) Biomass turnover in populations of viviparous sphaeriid clams: Comparisons of growth, fecundity, and biomass production. *Hydrobiologia*, **51**, 163–80.

Baba, K. (1967) Malakozonologische zonenuntersuchungen im toten tiszaarm bei szikra. *Tiscia (Szeged)*, **3**, 41–55.

Bagge, P. and Jumppanen, K. (1968) Bottom fauna of Lake Keitele, Central Finland, in relation to hydrography and eutrophization. *Ann. Zool. Fenn.*, **5**, 327–37.

Bailey, R. (1988) Correlations between species richness and exposure: Freshwater molluscs and macrophytes. *Hydrobiologia*, **162**, 183–91.

Bailey, R. (1989) Habitat selection by a freshwater mussel: An experimental test. *Malacologia*, **31**, 205–10.

Baily, J., Jr. (1931) Some data on growth, longevity, and fecundity in *Lymnaea columella* Say. *Biologia Generalis*, **7**, 407–28.

Baker, F. (1918) *The Productivity of Invertebrate Fish Food on the Bottom of Oneida Lake, NY, with Special Reference to Mollusks*, Technical Publication 9. Syracuse: NY State College of Forestry. 264 pp.

Baker, R. (1976) Tissue damage and leukocytic infiltration following attachment of the mite *Unionicola intermedia* to the gills of the bivalve mollusc *Anodonta anatina*. *J. Invertebr. Pathol.*, **27**, 371–6.

Baker, R. (1977) Nutrition of the mite *Unionicola intermedia*, Koenike and its relationship to the inflammatory response induced in its molluscan host *Anodonta anatina* L. *Parasitology*, **75**, 301–8.

Baker, S. and Hornbach, D. (1997) Acute physiological effects of zebra mussel (*Dreissena polymorpha*) infestation on two unionid mussels. *Can. J. Fish. Aquat. Sci.*, **54**, 512–19.

Baluku, B. and Loreau, M. (1989) Comparative study of the population dynamics of *Biomphalaria pfeifferi* (Gastropoda: Planorbidae) in two streams in eastern Zaire. *Revue de Zoologie Africaine*, **103**, 311–25.

Baluku, B., Josens, G. and Loreau, M. (1989) Preliminary study of the density and distributions of molluscs in two streams in eastern Zaire. *Revue de Zoologie Africaine*, **103**, 291–302.

Bao, C.-Y. (1985) Diel changes in the floating behavior of *Biomphalaria straminea* (Pulmonata: Planorbidae) in the laboratory. In *Proceedings of Second International Workshop on the Mollusca of Hong Kong and Southern China*, ed. B. Morton and D. Dudgeon, pp. 585–90. Hong Kong: Hong Kong University Press.

Barbosa, F. (1973) Possible competitive displacement and evidence of hybridization between Brazilian species of planorbid snails. *Malacologia*, **14**, 401–8.

Barbosa, F., Pereira da Costa, D. and Arrunda, F. (1984) Competitive interactions between species of freshwater snails. 1. Laboratory studies: 1b. comparative studies of the dispersal and the vagility capabilities of *Biomphalaria glabrata* and *Biomphalaria straminea*. *Mem. Inst. Oswaldo Cruz*, **79**, 163–7.

Barnes, G. (1955) The behavior of *Anodonta cygnea* L. and its neurophysiological basis. *J. Exp. Biol.*, **32**, 158–74.

Barnes, G. (1962) The behavior of unrestrained *Anodonta*. *Anim. Behav.*, **10**, 174–6.

Barnese, L., Lowe, R., & Hunter, R. D. (1990) Comparative grazing efficiency of pulmonate and prosobranch snails. *J. N. Am. Benthol. Soc.*, **9**, 35–44.

Barnhart, M. (1999) Potential hosts and reproductive characteristics of some unusual unionoids. *Proc. Freshwat. Molluscan Conserv. Soc.*, **1**.

Barnhart, M. and Roberts, A. (1997) Reproduction and fish hosts of unionids from Ozark Uplifts. In *Conservation and Management of Freshwater Mussels. II: Initiatives for the Future*, ed. K. Cummings, A. Buchanan, C. Mayer and T. Naimo, pp. 16–20. Rock Island, IL.: UMRCC.

Barr, W., Ahlstedt, S., Hickman, G. and Hill, D. (1986) *CMCP Activity 8: Analysis of Macrofaunal Factors*, Report TVA/ONRED/AWR-86–22. Knoxville, TN: Tennessee Valley Authority. 96 pp.

Barrow, J., Jr. (1961) Observations of a haplosporidian, *Haplosporidium pickfordi* sp. nov. in fresh water snails. *Trans. Am. Microsc. Soc.*, **80**, 319–29.

Bartonek, J. and Hickey, J. (1969) Food habits of canvasbacks, redheads, and lesser scaup in Manitoba. *Condor*, **71**, 280–90.

Basch, P. (1970) Relationships of some larval strigeids and echinostomes (Trematoda): Hyperparasitism, antagonism, and 'immunity' in the snail host. *Exp. Parasitol.*, **27**, 193–216.

Basch, P. (1976) Intermediate host specificity in *Schistosoma mansoni*. *Exp. Parasitol.*, **39**, 150–69.

Bauer, G. (1987) Reproductive strategy of the freshwater pearl mussel *Margaritifera margaritifera*. *J. Anim. Ecol.*, **56**, 691–704.

Bauer, G. (1994) The adaptive value of offspring size among freshwater mussels (Bivalvia: Unionoidea). *J. Anim. Ecol.*, **63**, 933–44.

Bauer, G., Hochwald, S. and Silkenat, W. (1991) Spatial distribution of freshwater mussels: The role of host fish and metabolic rate. *Freshwat. Biol.*, **26**, 377–86.

Baur, B. (1998) Sperm competition in molluscs. In *Sperm Competition and Sexual Selection*, ed. T. Birkhead and A. Moller, pp. 255–305. New York: Academic Press.

Bayne, C. and Yoshino, T. (1989) Determinants of compatibility in mollusc-trematode parasitism. *Am. Zool.*, **29**, 399–407.

Bayomy, M. and Joosse, J. (1987) The effects of isolation, grouping and population density of fecundity of *Bulinus truncatus*. *Int. J. Invertebr. Reprod. Dev.*, **12**, 319–30.

Beeston, D. and Morgan, E. (1979a) A crepuscular rhythm of locomotor activity in the freshwater prosobranch *Melanoides tuberculata* (Müller). *Anim. Behav.*, **27**, 284–91.

Beeston, D. and Morgan, E. (1979b) The effect of varying the light:dark ratio within 24-h photoperiod on the phase relationship of the crepuscular activity peaks in *Melanoides tuberculata*. *Anim. Behav.*, **27**, 292–9.

Belanger, S., Farris, J., Cherry, D. and Cairns, J. (1985) Sediment preference of the freshwater Asiatic clam, *Corbicula fluminea*. *Nautilus*, **99**, 66–72.

Bequaert, J. (1925) The arthropod enemies of mollusks, with description of a new dipterous parasite from Brazil. *J. Parasitol.*, **11**, 210–12.

Berg, C. (1964) Snail control in trematode diseases: The possible value of sciomyzid larvae snail-killing diptera. *Adv. Parasitol.*, **2**, 259–309.

Berg, C. (1973) Biological control of snail-borne diseases: A review. *Exp. Parasitol.*, **33**, 318–30.

Berg, K. (1938) *Studies on the Bottom Animals of Esrom Lake.* Naturvidenskabelig Og Mathematisk Afdeling 9 Raekke, vol. 8. Copenhagen, Denmark: Det Kongelige Danske Videnskabernes Selskabs Skrifter.

Berg, K. (1952) On the oxygen consumption of Ancylidae (Gastropoda) from an ecological point of view. *Hydrobiologia,* **4,** 225–67.

Berg, K. (1953) The problem of respiritory acclimatization illustrated by experiments with *Ancylus fluviatilis* (Gastropoda). *Hydrobiologia,* **5,** 331–50.

Berg, K. and Ockelmann, K. (1959) The respiration of freshwater snails. *J. Exp. Biol.,* **36,** 690–708.

Berrie, A. (1968) Prolonged inhibition of growth in a natural population of the freshwater snail *Biomphalaria sudanica tanganyicensis* (Smith) in Uganda. *Ann. Trop. Med. Parasitol.,* **62,** 45–51.

Berrie, A. and Visser, S. (1963) Investigations of growth-inhibiting substances affecting a natural population of freshwater snails. *Physiol. Zool.,* **36,** 167–73.

Berry, A. (1974) Freshwater bivalves of Peninsular Malaysia with special reference to sex and breeding. *Malay. Nat. J.,* **27,** 99–110.

Berry, A. and Kadri, A. (1974) Reproduction in the Malayan freshwater cerithiacean gastropod *Melanoides tuberculata. J. Zool.,* **172,** 369–81.

Bickel, D. (1965) The role of aquatic plants and submerged structures in the ecology of a freshwater pulmonate snail, *Physa integra* Hald. *Sterkiana,* **18,** 17–20.

Bilgin, F. (1973) Studies on the functional anatomy of *Melanopsis praemosa* (L) and *Zemelanopsis trifasciata* (Gray). *Proc. Malacol. Soc. Lond.,* **41,** 379–93.

Binder, E. (1959) Anatomie et systematique des Melaniens d'Afrique occidentale. *Rev. Suisse Zool.,* **66,** 1–68.

Bishop, M. and DeGaris, H. (1976) A note on population densities of mollusca in the River Great Ouse at Ely, Cambridgeshire. *Hydrobiologia,* **48,** 195–7.

Bishop, M. and Hewitt, S. (1976) Assemblages of *Pisidium* species (Bivalvia: Sphaeriidae) from localities in eastern England. *Freshwat. Biol.,* **6,** 177–82.

Blinn, D., Truitt, R. and Pickart, A. (1989) Feeding ecology and radular morphology of the freshwater limpet *Ferrissia fragilis. J. N. Am. Benthol. Soc.,* **8,** 237–42.

Blondeaux, A. (1974) Recherches sur les populations de paludines (*V. viviparus*) et sur la croissance des individus des deux sexes. *Bull. Biol.,* **108,** 81–95.

Bloomer, H. (1939) A note on the sex of *Pseudoanodonta* Bourguiquat and *Anodonta* Lamarck. *Proc. Malacol. Soc. Lond.,* **23,** 285–97.

Blum, J. (1956) The ecology of river algae. *Bot. Rev.,* **22,** 291–341.

Boag, D. (1981) Differential depth distribution among freshwater pulmonate snails subjected to cold temperatures. *Can. J. Zool.,* **59,** 733–7.

Boag, D. and Bentz, J. (1980) The relationship between simulated seasonal temperatures and depth distributions in the freshwater pulmonate, *Lymnaea stagnalis. Can. J. Zool.,* **58,** 198–201.

Boerger, H. (1975) Movement and burrowing of *Helisoma trivolvis* (Say) (Gastropoda, Planorbidae) in a small pond. *Can. J. Zool.,* **53,** 456–64.

Bohlken, S., Joosse, J., van Elk, R. and Geraerts, W. (1986) Interaction photoperiod and nutritive state in female reproduction of *Lymnaea stagnalis. Int. J. Invertebr. Reprod. Dev.,* **10,** 151–7.

Bohlken, S., Joosse, J. and Geraerts, W. (1987) Interaction of photoperiod grouping and isolation in female reproduction of *Lymnaea stagnalis. Int. J. Invertebr. Reprod. Dev.,* **11,** 45–58.

Boltovskoy, D., Izaguirre, L. and Corres, N. (1995) Feeding selectivity of *Corbicula fluminea* (Bivalvia) on natural phytoplankton. *Hydrobiologia*, **312**, 171–82.

Bondesen, P. and Kaiser, E. (1949) *Hydrobia (Potamopyrgus) jenkinsi* Smith in Denmark illustrated by its ecology. *Oikos*, **1**, 252–81.

Boom-Ort, B. (1992) Uptake of sand-grains into the stomach of *Biomphalaria glabrata*. In *Proceedings of Tenth International Malacological Congress*, ed. C. Meier-Brook, pp. 109–14. Tubingen: Unitas Malacologia.

Boozer, A. and Mirkes, P. (1979) Observations on the fingernail clam *Musculium partumeium* (Pisidiidae), and its association with the introduced Asiatic clam, *Corbicula fluminea*. *Nautilus*, **93**, 73–83.

Boray, J. (1964a) Studies on the ecology of *Lymnaea tormentosa*, the intermediate host of *Fasciola hepatica*. I. History, geographical distribution, and environment. *Aust. J. Zool.*, **12**, 217–30.

Boray, J. (1964b) Studies on the ecology of *Lymnaea tomentosa*, the intermediate host of *Fasciola hepatica*. II. The sexual behavior of *Lymnaea tomentosa*. *Aust. J. Zool.*, **12**, 231–7.

Boray, J. (1966) Studies on the relative susceptibility of some Lymnaeids to infection with *Fasciola hepatica* and *F. gigantica* and on the adaption of *Fasciola* spp. *Ann. Trop. Med. Parasitol.*, **60**, 114–24.

Boray, J. (1969) Experimental fascioliasis in Australia. *Adv. Parasitol.*, **7**, 95–210.

Borcherding, J. (1991) The annual reproductive cycle of the freshwater mussel *Dreissena polymorpha* (Pallas) in lakes. *Oecologia*, **87**, 208–18.

Botts, P., Patterson, B. and Schloessen, D. (1996) Zebra mussel effects on benthic invertebrates: Physical or biotic? *J. N. Am. Benthol. Soc.*, **15**, 179–84.

Bousfield, J. (1979) Plant extracts and chemically triggered positive rheotaxis in *Biomphalaria glabrata* snail intermediate host of *Schistosoma mansoni*. *J. Appl. Ecol.*, **16**, 681–90.

Bovbjerg, R. (1965) Feeding and dispersal in the snail *Stagnicola reflexa* (Basommatophora: Lymnaeidae). *Malacologia*, **2**, 199–207.

Bovbjerg, R. (1968) Responses to food in lymnaeid snails. *Phys. Zool.*, **41**, 412–23.

Bovbjerg, R. (1975) Dispersal and dispersion of pond snails in an experimental environment varying to three factors, singly and in combination. *Phys. Zool.*, **48**, 203–15.

Boycott, A. (1919) Parthenogenesis in *Paludestrina jenkinsi*. *J. Conchol.*, **16**, 54.

Boycott, A. (1936) The habitats of freshwater mollusca in Britain. *J. Anim. Ecol.*, **5**, 116–86.

Brackenbury, T. and Appleton, C. (1991a) Morphology of the mature spermatozoan of *Physa acuta* (Draparnaud 1801) (Gastropoda: Physidae). *J. Molluscan Stud.*, **57**, 211–18.

Brackenbury, T. and Appleton, C. (1991b) Effect of controlled temperatures on gametogenesis in the gastropods *Physa acuta* (Physidae) and *Bulinus tropicus* (Planorbidae). *J. Molluscan Stud.*, **57**, 461–70.

Brackenbury, T. and Appleton, C. (1991c) Morphology of the mature spermatozoon of *Bulinus tropicus* (Krauss, 1848) (Gastropoda: Planorbidae). *Malacologia*, **33**, 273–80.

Brendelberger, H. (1995) Growth of juvenile *Bithynia tentaculata* (Prosobranchia, Bithyniidae) under different food regimes: A long-term laboratory study. *J. Molluscan Stud.*, **61**, 89–95.

Brendelberger, H. (1997) Coprophagy: A supplementary food source for two freshwater gastropods. *Freshwat. Biol.*, **38**, 145–57.

Brendelberger, H. and Jurgens, S. (1993) Suspension feeding in *Bithynia tentaculata* (Prosobranchia, Bithyniidae), as affected by body size, food, and temperature. *Oecologia*, **94**, 36–42.

Bronmark, C. (1985a) Freshwater snail diversity: Effects of pond area, habitat heterogeneity and isolation. *Oecologia*, **67**, 127–31.

Bronmark, C. (1985b) Interactions between macrophytes, epiphytes, and herbivores: An experimental approach. *Oikos*, **45**, 26–30.

Bronmark, C. (1988) Effects of vertebrate predation on freshwater gastropods: An exclosure experiment. *Hydrobiologia*, **169**, 363–70.

Bronmark, C. (1989) Interactions between epiphytes, macrophytes, and freshwater snails: A review. *J. Molluscan Stud.*, **55**, 299–311.

Bronmark, C. (1992) Leech predation on juvenile freshwater snails: Effects of size, species and substrate. *Oecologia*, **91**, 526–9.

Bronmark, C. and Malmqvist, B. (1982) Resource partitioning between unionid mussels in a Swedish lake outlet. *Hol. Ecol.*, **5**, 389–95.

Bronmark, C. and Malmqvist, B. (1986) Interactions between the leech *Glossiphonia complanata* and its gastropod prey. *Oecologia*, **69**, 268–76.

Bronmark, C., Rundle, S. and Erlandsson, A. (1991) Interactions between freshwater snails and tadpoles – competition and facilitation. *Oecologia*, **87**, 8–18.

Bronmark, C., Klosiewski, S. and Stein, R. (1992) Indirect effects of predation in a freshwater, benthic food chain. *Ecology*, **73**, 1662–74.

Brophy, T. (1980) Food habits of sympatric larval *Ambystoma tigrinum* and *Notophthalmus viridescens*. *J. Herpetol.*, **14**, 1–6.

Brown, D. (1978) Freshwater molluscs. In *Biogeography and Ecology of Southern Africa, Part 2*, ed. M. Werger, pp. 1153–80. The Hague: W. Junk.

Brown, D. (1979) Biogeographical aspects of African freshwater gastropods. *Malacologia*, **18**, 79–102.

Brown, D. (1988) *Sierraia*: Rheophilus West African river snails (Prosobranchia: Bithyniidae). *Zool. J. Linn. Soc.*, **93**, 313–55.

Brown, D. (1994) *Freshwater Snails of Africa and Their Medical Importance*, Revised Second Edition. London: Taylor and Francis.

Brown, D. and Wright, C. (1980) Molluscs of Saudi Arabia. Freshwater molluscs. In *Fauna of Saudi Arabia, Vol. 2*, ed. W. Wittmer and W. Battiker, pp. 341–58. Basle: Pro Entomologia c/o Natural History Museum.

Brown, J. and Davidson, D. (1977) Competition between seed-eating rodents and ants in desert systems. *Science*, **196**, 880–2.

Brown, K. (1979) The adaptive demography of four freshwater pulmonate snails. *Evolution*, **33**, 417–32.

Brown, K. (1982) Resource overlap and competition in pond snails: An experimental analysis. *Ecology*, **63**, 412–22.

Brown, K. (1985a) Intraspecific life history variation in a pond snail: The roles of popluation divergence and phenotypic plasticity. *Evolution*, **39**, 387–95.

Brown, K. (1985b) Mechansims of life history adaption in the temporary pond snail *Lymnaea elodes* (Say). *Am. Malacol. Bull.*, **3**, 143–50.

Brown, K. and Lodge, D. (1993) Gastropod abundance in vegetated habitats: The importance of specifying null models. *Limnol. Oceanogr.*, **38**, 217–25.

Brown, K. and DeVries, D. (1985) Predation and the distribution and abundance of a pond snail. *Oecologia*, **66**, 93–9.

Brown, K. and Strouse, B. (1988) Relative vulnerability of six freshwater gastropods to the leech *Nephelopsis obscura* (Verrill). *Freshwat. Biol.*, **19**, 157–65.

Brown, K., Leathers, B. and Minchella, D. (1988) Trematode prevalence and the popluation dynamics of freshwater pond snails. *Am. Midl. Nat.*, **120**, 289–301.

Brown, K., Varza, D. and Richardson, T. (1989) Life histories and population dynamics of two subtropical snails (Prosobranchia: Viviparidae). *J. N. Am. Benthol. Soc.*, **8**, 222–28.

Brown, K., Carman, K. and Inchausty, V. (1994) Density-dependent influences on feeding and metabolism in a freshwater snail. *Oecologia*, **99**, 158–65.

Brown, K., Alexander, J. & Thorp, J. (1998) Differences in the ecology and distribution of lotic pulmonate and prosobranch gastropods. *Am. Malacol. Bull.*, **14**, 91–101.

Brown, S., Cassuto, S. and Loos, R. (1979) Biomechanics of chilopods in some decapod crustaceans. *J. Zool. (Lond.)*, **188**, 143–59.

Browne, R. A. (1978) Growth, mortality, fecundity, biomass and productivity of four lake populations of the prosobranch snail, *Viviparus georgianus*. *Ecology*, **59**, 742–50.

Browne, R. (1981) Lakes as islands: Biogeographic distribution, turnover rates, and species composition in the lakes of central New York. *J. Biogeogr.*, **8**, 75–83.

Browne, R. and Russell-Hunter, W. (1978) Reproductive effort in molluscs. *Oecologia*, **37**, 23–7.

Bruenderman, S. and Neves, R. (1993) Life history of the endangered fine-rayed pigtoe *Fusconia cuneolus* (Bivalvia: Unionidae) in the Clinch River, Virginia. *Am. Malacol. Bull.*, **10**, 83–91.

Bryan, J. and Larkin, P. (1972) Food specialization by individual trout. *J. Fish. Res. Board Can.*, **29**, 1615–24.

Buckley, D. (1986) Bioenergetics of age-related versus size-related reproductive tactics in female *Viviparus georgianus*. *Biol. J. Linn. Soc.*, **27**, 293–309.

Bull, J. (1983) *Evolution of Sex-Determining Mechanisms*. Menlo Park, CA: Benjamin/Cummings.

Burch, J. (1960) Chromosomes of *Pomatiopsis* and *Oncomelania*. *Ann. Reports Am. Malacol. Union*, **26**, 15.

Burch, J. (1975) *Freshwater Sphaeriacean Clams of North America*. Hamburg, Michigan: Malacological Publications.

Burch, J. and Tottenham, J. (1980) North American freshwater snails: Species list, ranges, and illustrations. *Walkerana*, **3**, 1–215.

Burch, P. and Wood, J. (1955) The salamander *Siren lacertina* feeding on clams and snails. *Copeia*, **3**, 255–6.

Burky, A. (1971) Biomass turnover, respiration, and interpopulation variation in the stream limpet *Ferrissia rivularis* (Say). *Ecol. Monog.*, **41**, 235–51.

Burky, A. (1974) Growth and biomass production of an amphibious snail, *Pomacea urceus* (Müller) from the Venezuelan savannah. *Proc. Malacol. Soc. London*, **41**, 127–43.

Burky, A. (1983) Physiological ecology of freshwater bivalves. In *The Mollusca, Volume 6, Ecology*, ed. W. Russell-Hunter, pp. 281–327. Orlando, FL: Academic Press.

Burky, A., Benjamin, R., Conover, D. and Detrick, J. (1985a) Seasonal responses of filtration rates to temperature, oxygen availability, and particle concentration of the freshwater clam *Musculium partumeium* (Say). *Am. Malacol. Bull.*, **3**, 201–12.

Burky, A., Hornbach, D. and Way, C. (1985b) Comparative bioenergetics of permanent and temporary pond populations of the freshwater clam *Musculium partumeium* (Say). *Hydrobiologia*, **126**, 35–48.

Burla, H. (1972) Die abundanz von *Anodonta*, *Unio pictorum*, *Viviparus ater*, *Lymnaea auricularia* and *Lymnaea ovata* im Zurichsee, in abhangigkeit von den wassertiefe und zu verschiedenen jahreszeiten. *Vjschr. Naturforsch. Ges. Zurich*, **117**, 129–51.

Burton, R. (1983) Ionic regulation and water balance. In *The Mollusca, Volume 5. Physiology, Pt. 2*, ed. A. Saleuddin and K. Wilbur, pp. 292–352. Orlando, FL: Academic Press.

Burton, T. (1976) An analysis of the feeding ecology of the salamanders (Amphibia: Urodela) of the Hubbard Brook Experimental Forest, New Hampshire. *J. Herpetol.*, **10**, 187–204.

Buse, A. (1971) Population dynamics of *Chaetogaster limnaei vaghini* Gruffydd (Ologachaeta) in a field population of *Lymnaea stagnalis* L. *Oikos*, **22**, 50–5.

Byrne, R. and Dietz, T. (1997) Ion transport and acid-base balance in freshwater bivalves. *J. Exp. Biol.*, **200**, 457–65.

Byrne, R., McMahon, R. and Dietz, T. (1989) The effects of aerial exposure and subsequent reimmersion on hemolymph osmoloity, ion composition, and ion flux in the freshwater bivalve *Corbicula fluminea*. *Physiol. Zool.*, **62**, 1187–202.

Cain, G. (1956) Studies on cross-fertilization and self-fertilization in *Lymnaea stagnalis appressa* Say. *Biol. Bull.*, **111**, 45–52.

Calow, P. (1970) Studies on the natural diet of *Lymnaea pereger obtusa* and its possible ecological implications. *Proc. Malacol. Soc. Lond.*, **39**, 203–15.

Calow, P. (1973a) The food of *Ancylus fluviatilis* (Müller), a littoral, stone-dwelling, herbivore. *Oecologia*, **13**, 113–33.

Calow, P. (1973b) Field observations and laboratory experiments on the general food requirements of two species of freshwater snail, *Planorbis contortus* and *Ancylus fluviatilis*. *Proc. Malacol. Soc. Lond.*, **40**, 483–9.

Calow, P. (1973c) Gastropod associations within Malham Tarn, Yorkshire. *Freshwat. Biol.*, **3**, 521–34.

Calow, P. (1974) Evidence for bacterial feeding in *Planorbis contortus* L. (Gastropoda: Pulmonata). *Proc. Malacol. Soc. Lond.*, **41**, 145–56.

Calow, P. (1975) The feeding strategies of two freshwater gastropods, *Ancylus fluviatilis* and *Planorbis contortus* L. in terms of ingestion rates and absorption efficiencies. *Oecologia*, **20**, 33–49.

Calow, P. (1978) The evolution of life-cycle strategies in freshwater gastropods. *Malacologia*, **17**, 351–64.

Calow, P. (1979) The cost of reproduction – a physiological approach. *Biol. Rev. Cambridge Philos. Soc.*, **54**, 23–40.

Calow, P. (1981) Adaptational aspects of growth and reproduction in *Lymnaea peregra* (Gastropoda: Pulmonata) from exposed and sheltered aquatic habitats. *Malacologia*, **21**, 5–13.

Calow, P. and Calow, L. (1975) Cellulase activity and niche separation in freshwater gastropods. *Nature*, **255**, 478–80.

Calow, P. and Calow, L. (1983) Ecology, ontology, and reproductive biology of

Sinotaia quadrata, a viviparid prosobranch, with special reference to the costs of brooding. In *Proceedings of Second International Workshop on the Malacofauna of Hong Kong and Southern China*, ed. B. Morton and D. Dudgeon, pp. 479–89. Hong Kong: Hong Kong University Press.

Calow, P. and Fletcher, C. (1972) A new radiotracer technique involving ^{14}C and ^{51}Cr, for estimating the assimilation efficiencies of aquatic, primary consumers. *Oecologia*, **9**, 155–70.

Caquet, T. (1990) Spatial distribution of four freshwater gastropod species in a ditch near Orsay, France. *Hydrobiologia*, **203**, 83–92.

Caraco, N., Cole, J., Raymond, P., Strayer, D., Pace, M., Findlay, S. and Fischer, D. (1997) Zebra mussel invasion in a large turbid river: Phytoplankton response to increased grazing. *Ecology*, **78**, 588–602.

Carman, K. and Guckert, J. (1994) Radiotracer determination of ingestion and assimilation of periphytic algae, bacteria, and absorbed amino acids by snails. *J. N. Am. Benthol. Soc.*, **13**, 80–8.

Carriker, M. (1946) Observations on the functioning of the alimentary system of the snail *Lymnaea stagnalis appressa* Say. *Biol. Bull.*, **91**, 88–111.

Cattaneo, A. and Kalff, J. (1986) The effect of grazer size manipulation on periphyton communities. *Oecologia*, **69**, 612–7.

Cazzaniga, N. (1990) Predation of *Pomacea canaliculata* (Ampullaridae) on adult *Biomphalaria peregrina* (Plenorbidae). *Ann. Trop. Med. Parasitol.*, **84**, 97–100.

Cedeno-Leon, A. and Thomas, J. (1982) Competition between *Biomphalaria glabrata* (Say) and *Marisa cornuarietis* (L): Feeding niches. *J. Appl. Ecol.*, **19**, 707–21.

Cedeno-Leon, A. and Thomas, J. (1983) The predatory behaviour of *Marisa cornuarietis* on eggs and neonates of *Biomphalaria glabrata*, the snail host of *Schistosoma mansoni*. *Malacologia*, **24**, 289–97.

Chamberlain, T. (1934) The glochidial conglutinates of the Arkansas fanshell, *Cyprogenia alberti* (Conrad). *Biol. Bull.*, **66**, 55–61.

Chambers, P., Hanson, J., Burke, J. and Prepas, E. (1990) The impact of the crayfish *Orconectes virilis* on aquatic macrophytes. *Freshwat. Biol.*, **24**, 81–91.

Charlesworth, B. (1980) *Evolution in Age-Structured Populations*. New York: Cambridge University Press.

Charnov, E. (1991) Evolution of life history variation among female mammals. *Proc. Natl. Acad. Sci. USA*, **88**, 1134–7.

Charnov, E. and Bull, J. (1977) When is sex environmentally determined? *Nature*, **266**, 828–30.

Charnov, E., Orians, G. and Hyatt, K. (1976) Ecological implications of resource depression. *Amer. Nat.*, **110**, 247–59.

Chase, M. and Bailey, R. (1996) Recruitment of *Dreissena polymorpha*: Does the presence and density of conspecifics determine the recruitment density and pattern in a population? *Malacologia*, **38**, 19–31.

Chatfield, J. (1986) Stream order and species diversity of Mollusca in the river Wallington, Hampshire, U.K. In *Proceedings Eighth International Malacological Congress*, ed. L. Pinter, pp. 49–52. Budapest, Hungary: Hungarian Natural History Museum.

Chaudhry, M. and Morgan, E. (1983) Circadian variation in the behavior and physiology of *Bulinus tropicus* (Gastropoda: Pulmonata). *Can. J. Zool.*, **61**, 909–14.

Chaudhry, M. and Morgan, E. (1987) Factors affecting the growth and fecundity of *Bulinus tropicus. J. Molluscan Stud.*, **53**, 52–61.

Cheatum, E. (1934) Limnological investigations on respiration, annual migratory cycle, and other related phenomena in fresh water pulmonate snails. *Trans. Am. Microsc. Soc.*, **53**, 348–407.

Cheesman, D. (1956) The snail's foot as a Langmuir trough. *Nature*, **178**, 987–8.

Chen, L., Heath, A. & Neves, R. (1999) Diurnal rhythms of gaping behaviour in seven species of freshwater mussels. *Proc. Frewat. Molluscan Conserv. Soc.*, **1**.

Cheng, T. (1986) *General Parasitol.,* Second Edition. Orlando, Florida: Academic Press.

Chernin, E. (1957) A method of securing bacteriologically sterile snails (*Australorbis glabatus*). *Proc. Soc. Exp. Biol. Med.*, **96**, 204–10.

Chernin, E. and Michelson, E. (1957a) Studies on the biological control of schistosome-bearing snails. III The effects of population density on growth and fecundity in *Australorbis glabatus. Am. J. Hyg.*, **65**, 57–70.

Chernin, E. and Michelson, E. (1957b) Studies on the biological control of schistosome-bearing snails. IV. Further observations on the effects of crowding on growth and fecundity in *Australorbis glabatus. Am. J. Hyg.*, **65**, 71–80.

Chernin, E., Michelson, E. and Augustine, D. (1956a) Studies on the biological control of schistosome-bearing snails. I: The control of *Australorbis glabatus* populations by the snail *Marisa cornuarietus* under laboratory conditions. *Am. J. Trop. Med. Hyg.*, **5**, 297–307.

Chernin, E., Michelson, E. and Augustine, D. (1956b) Studies on the biological control of schistosome-bearing snails. II: The control of *Australorbis glabatus* populations by the leech *Helobdella fusca* under laboratory conditions. *Am. J. Trop. Med. Hyg.*, **5**, 308–14.

Chernin, E. and Schork, A. (1959) Growth in axenic culture of the snail, *Australorbis glabatus. Am. J. Hyg.*, **69**, 146–60.

Chi, L., Winkler, L. and Colvin, R. (1971) Predation of *Marisa cornuarietis* on *Oncomelania formosana* eggs under laboratory conditions. *Veliger*, **14**, 184–6.

Chock, Q., Davis, C. and Chong, M. (1961) *Sepedon macropus* (Diptera: Sciomyzidae) introduced into Hawaii as a control for the liver fluke snail, *Lymnaea ollula. J. Econ. Entomol.*, **54**, 1–4.

Christensen, N., Frandsen, F. and Roushdy, M. (1980) The influence of environmental conditions and parasite-intermediate host-related factors on the transmission of *Echinostoma liei. Zeitschrift Für Parasitenkunde*, **64**, 47–63.

Chu, K., Massoud, J. and Sabraghian, H. (1966a) Host-parasite relationship of *Bulinus truncatus* and *Schistosoma haematobium* in Iran 1. Effect of the age of *B. truncatus* on the development of *S. haematobium. Bull. W.H.O.*, **34**, 113–19.

Chu, K., Massoud, J. and Sabraghian, H. (1966b) Host-parasite relationship of *Bulinus truncatus* and *Schistosoma haematobium* in Iran 3. Effect of water temperature on the ability of miracidia to infect snails. *Bull. W.H.O.*, **34**, 131–3.

Chu, K., Sabraghian, H. and Massoud, J. (1966c) Host-parasite relationship of *Bulinus truncatus* and *Schistosoma haematobium* in Iran. 2. Effect of exposure dosage of miracidia on the biology of the snail host and the development of the parasites. *Bull. W.H.O.*, **34**, 121–30.

Chu, K., Massoud, J. and Arfaa, F. (1968) Distribution and ecology of *Bulinus truncatus* in Khuzestan, Iran. *Bull. W.H.O.*, **39**, 607–37.

Chun, S. (1969) Fundamental studies on the breeding of *Lamprotula coreana*. *Bull. Pusan Fish. College*, **9**, 11–17.

Chung, B., Joo, C. and Choi, D. (1980) Seasonal variation of snail population of *Parafossarulus manchouricus* and larval trematode infection in River Kumho, Kyungpook Province, Korea. *Korean J. Parasitol.*, **18**, 54–64.

Churchill, E., Jr. and Lewis, S. (1924) Food and feeding in fresh-water mussels. *Bull. U.S. Bur. Fish. 1923–1924*, **39**, 439–71.

Clampitt, P. T. (1970) Comparative ecology of the snails *Physa gyrina* and *Physa integra*. *Malacologia*, **10**, 113–51.

Clampitt, P. (1972) Seasonal migrations and other movements in Douglas Lake pulmonate snails. *Malacol. Rev.*, **5**, 11–12.

Clampitt, P. (1973) Substratum as a factor in the distribution of pulmonate snails in Douglas Lake, Michigan. *Malacologia*, **12**, 379–99.

Clampitt, P. (1974) Seasonal migration cycle and related movements of the fresh water pulmonate snail, *Physa integra*. *Am. Midl. Nat.*, **92**, 275–300.

Clark, H. and Stein, S. (1921) Glochidia in surface towings. *Nautilus*, **35**, 16–20.

Cleland, D. (1954) A study of the habits of *Valvata piscinalis* (Müller) and the structure and function of the alimentary canal and reproductive system. *Proc. Malacol. Soc. Lond.*, **30**, 167–203.

Coffman, W. (1971) Energy flow in a woodland stream ecosystem: I. Tissue support trophic structure of the autumnal community. *Arch. Hydrobiol.*, **68**, 232–76.

Cohen, R., Dresier, P. V., Phillips, E. and Cory, R. (1984) The effect of the Asiatic clam *Corbicula fluminea*, on phytoplankton of the Potomac River, Maryland. *Limnol. Oceanogr.*, **29**, 170–80.

Coker, R., Shira, A., Clark, H. and Howard, A. (1921) Natural history and propagation of fresh-water mussels. *Bull. U.S. Bur. Fish.*, **37**, 75–181.

Combes, C. (1982) Trematodes: Antagonism between species and sterilizing effects on snails in biological control. *Parasitol.*, **84**, 151–75.

Connor, E. and McCoy, E. (1979) The statistics and biology of the species–area relationship. *Am. Nat.*, **113**, 791–833.

Connors, V. and Yoshino, T. (1990) In vitro effect of larval *Schistosoma mansoni* excretory-secretory products on phagocytosis-stimulated superoxide production in hemocytes from *Biomphalaria glabrata*. *J. Parasitol.*, **76**, 895–902.

Connors, V., Lodes, M. and Yoshino, T. (1991) Identification of *Schistosoma mansoni* sporocyst excretory-secretory antioxidant molecule and its effect on superoxide production by *Biomphalaria glabrata* hemocytes. *J. Invertebr. Pathol.*, **58**, 387–95.

Convey, L., Hanson, J. and MacKay, W. (1989) Size-selective predation on unionid clams by muskrats. *J. Wildlife Manage.*, **53**, 654–7.

Cook, P. (1949) A ciliary feeding mechanism in *Viviparus viviparus* (L.). *Proc. Malacol. Soc. Lond.*, **27**, 265–71.

Cotner, J., Gardner, W., Johnson, J., Sada, R. H., Cavaletto, J. and Heath, R. (1995) Effects of zebra mussels (*Dreissena polymorpha*) on bacterioplankton: Evidence for both size-selective consumption and growth stimulation. *J. Great Lakes Res.*, **21**, 517–28.

Cottam, C. (1939) *Food Habits of North American Diving Ducks*, Tech. Bull. 643. Washington, DC: U.S. Dept. of Agriculture. 140 pp.

Cottam, C. (1942) Supplementary notes on the food of the Limpkin. *Nautilus*, **55**, 125–8.

Counts, C. (1986) The zoogeography and history of the invasion of the United States by *Corbicula fluminea* (Bivalvia: Corbiculidae). *Am. Malacol. Bull. Special Edition*, **2**, 7–39.

Coustau, C. and Yoshino, T. (1994) Surface membrane polypeptides associated with hemocytes from *Schistosoma mansoni*-susceptible and -resistant strains of *Biomphalaria glabrata* (Gastropoda). *J. Invertebr. Pathol.*, **63**, 82–9.

Coutellec-Vreto, M.-A., Jarne, P., Guiller, A., Madec, L. & Daguzan, J. (1998) Inbreeding and fitness in the freshwater snail *Lymnaea peregra*: An evaluation over two generations of self fertilization. *Evolution*, **52**, 1635–47.

Covich, A. (1976) Recent changes in molluscan species diversity of a large tropical lake (Lago de Peten, Guatamala). *Limnol. Oceanogr.*, **21**, 51–9.

Covich, A. (1977) How do crayfish respond to plants and mollusca as alternate food resources? *Freshwat. Crayfish*, **3**, 165–79.

Covich, A. (1981) Chemical refugia from predation for thin-shelled gastropods in a sulfide-enriched stream. *Verh. Int. Ver. Theor. Angew. Limnol.*, **21**, 1632–6.

Covich, A., Dye, L. and Mattice, J. (1981) Crayfish predation on *Corbicula* under laboratory conditions. *Am. Midl. Nat.*, **105**, 181–8.

Covich, A., Crowl, T., Alexander, J. and Vaughn, C. (1994) Predator-avoidance responses in freshwater decapod-gastropod interactions mediated by chemical stimuli. *J. N. Am. Benthol. Soc.*, **13**, 283–90.

Crews, A. and Esch, G. (1986) Seasonal dynamics of *Halipegus occidualis* (Trematoda: Hemiuridae) in *Helisoma anceps* and its impact on the fecundity of the host snail. *J. Parasitol.*, **72**, 646–51.

Crews, A. and Yoshino, T. (1990) Influence of larval schistosomes in polysaccharide synthesis in albumin glands of *Biomphalaria glabrata*. *Parasitol.*, **101**, 351–9.

Crews, A. and Yoshino, T. (1991) *Schistosoma mansoni*: Influence of infection on levels of translatable mRNA and on polypeptide synthesis in the ovotestis and albumin gland of *Biomphalaria glabrata*. *Exp. Parasitol.*, **72**, 368–80.

Crowder, L. and Cooper, W. (1982) Habitat structural complexity and the interactions between bluegills and their prey. *Ecology*, **63**, 1802–13.

Crowl, T. (1990) Life-history strategies of a freshwater snail in response to stream performance and predation: Balancing conflicting demands. *Oecologia*, **84**, 238–43.

Crowl, T. and Alexander, J. (1989) Parental care and foraging ability in male waterbugs (*Belostoma flumineum*). *Can. J. Zool.*, **67**, 513–5.

Crowl, T. and Covich, A. (1990) Predator-induced life-history shifts in a freshwater snail. *Science*, **247**, 949–51.

Crowl, T. and Schnell, G. (1990) Factors determining population-density and size distribution of a fresh-water snail in streams – effects of spatial scale. *Oikos*, **59**, 359–67.

Cruz, R. and Villalobos, C. (1984) Estudios sobre la biologia de *Glabaris luteolus* (Mycetopodidae: Bivalia) IV. Biometrica y aspectos reproductivos en 28 millas, Limon, Costa Rica. *Rev. Lat. Acui. Lima – Peru*, **21**, 9–17.

Cruz, R. and Villalobos, C. (1985) Tamaño y madurez sexual de la almeja de agua dulce *Glabaris luteolus* (Lea 1852) (Mycetopodidae: Bivalvia). *Brenesia*, **24**, 371–4.

Cuker, B. (1983a) Grazing and nutrient interactions in controlling the activity and composition of the epilithic community of an arctic lake. *Limnol. Oceanogr.*, **28**, 133–41.

Cuker, B. (1983b) Competition and coexistence among the grazing snail *Lymnaea*, Chironomidae, and microcrustacea in an arctic epilithic lacustrine community. *Ecology*, **64**, 10–15.

Cummins, K. and Lauff, G. (1969) The influence of substrate particle size on the microdistribution of stream macrobenthos. *Hydrobiologia*, **34**, 145–81.

Cunjak, R. and McGladdery, S. (1991) The parasite-host relationship of glochidia (Mollusca: Margaritiferidae) on the gills of young-of-the-year Atlantic salmon (*Salmo salar*). *Can. J. Zool.*, **69**, 353–8.

Cvancara, A. (1972) Lake mussel distribution as determined with Scuba. *Ecology*, **53**, 154–7.

Cvancara, A. and Freeman, P. (1978) Diversity and distribution of mussels (Bivalvia: Unionidae) in a eutrophic reservoir, Lake Ashtabula, North Dakota. *Nautilus*, **92**, 1–8.

Daldorph, P. and Thomas, J. (1988) The chemical ecology of some British UK freshwater gastropod molluscs behavioral responses to short chain carboxylic acids and maltose. *Freshwat. Biol.*, **19**, 167–78.

Daldorph, P. and Thomas, J. (1991) The effect of nutrient enrichment on a freshwater community dominated by macrophytes and molluscs and its relevance to snail control. *J. Appl. Ecol.*, **28**, 685–702.

Darwin, C. (1882) On the dispersal of freshwater bivalves. *Nature*, **25**, 529–30.

Davis, C., Chock, Q. and Chong, M. (1961) Introduction of the liver fluke snail predator, *Sciomyza dorsata* (Sciomyzidae, Diptera), in Hawaii. *Proc. Hawaiian Entomol. Soc.*, **17**, 395–7.

Davis, G. (1969) A taxonomic study of *Semisulcospira* in Japan (Mesogastropoda: Pleuroceridae) *Malacologia*, **7**, 211–94.

Davis, G. (1971) Systematic studies of *Brotia costula episcopalis*, first intermediate host of *Paragonimus westermani* in Malaysia. *Proc. Acad. Nat. Sci. Philadelphia*, **123**, 53–86.

Davis, G., Kitikoon, V. and Temcharoen, P. (1976) Monograph on *Lithoglyphopsis aperta*, the snail host of Mekong River schistosomiasis. *Malacologia*, **15**, 241–87.

Dazo, B. (1965) The morphology and natural history of *Pleurocera acuta* and *Goniobasis livescens* (Gastropoda: Cerithiacea: Pleuroceridae). *Malacologia*, **3**, 1–80.

Dazo, B. and Moreno, R. (1962) Studies on the food and feeding habits of *Oncomelania quadrasi*, the snail intermediate host of *Schistosoma japonicum* in the Philippines. *Trans. Am. Microsc. Soc.*, **81**, 341–7.

Dazo, B., Hairston, N. and Dawood, I. (1966) The ecology of *Bulinus truncatus* and *Biomphalaria alexandrina* and its implication for the control of bilharziasis in the Egypt 49 project area. *Bull. W.H.O.*, **35**, 339–56.

Deaton, L. and Greenberg, M. (1991) The adaptation of bivalve molluscs to oligohaline and fresh waters: Phylogenetic and physiological aspects. *Malacol. Rev.*, **24**, 1–18.

de Bruin, J. and David, C. (1970) Observations on the rate of water pumping of the freshwater mussel *Anodonta cygnea zellensis* (Gme.). *Neth. J. Zool.*, **20**, 380–91.

de Coster, W. and Persoone, G. (1970) Ecological study of Gastropoda in a swamp in the neighborhood of Ghent (Belgium). *Hydrobiologia*, **36**, 65–80.

de Kock, K. and Joubert, P. (1991) Life-table experiments with *Helisoma duryi* (Wetherby) and *Biomphalaria pfeifferi* (Krauss) at constant temperatures. *S. Afri. J. Zool.*, **26**, 149–52.

de Kock, K. & van Eeden, J. (1981) Life-table studies on freshwater snails: The effect of constant temperature on the population dynamics of *Biomphalaria pfeifferi* (Krauss). Wetenskaplike Bydraes van die PU vir CHO. Reeks B: *Natuurwetenskappe*, **107**, 117.

de Kock, K. and van Eeden, J. (1985) Effect of constant temperature on population dynamics of *Bulinus tropicus* (Krauss) and *Lymnaea natalensis* (Krauss). *J. Limnol. Soc. S. Africa*, **11**, 27–31.

de Kock, K. and van Eeden, J. (1986) Effect of programmed circadian temperature fluctuations on population dynamics of *Biomphalaria pfeifferi* (Krauss). *S. Afri. J. Zool.*, **21**, 28–32.

de Kock, K., van Eeden, J. and Pretorius, S. (1986) Effect of laboratory breeding on the population dynamics of successive generations of the freshwater snail *Bulinus tropicus* (Krauss). *S. Afr. J. Sci.*, **82**, 369–72.

de Leeuw, J. and van Eerden, M. (1992) Size selection in diving tufted ducks *Aythya fuligula* explained by differential handling of small and large mussels *Dreissena polymorpha*. *Ardea*, **80**, 353–62.

Demian, E. and Ibrahim, A. (1972) Sexual dimorphism and sex ratio in the snail *Marisa cornuarietis* (L). *Bull. Zool. Soc. Egypt*, **24**, 52–63.

Demian, E. and Lufty, R. (1965a) Predatory activity of *Marisa cornuarietis* against *Bulinus truncatus*, the transmitter of urinary schistosomiasis. *Ann. Trop. Med. Parasitol.*, **59**, 331–6.

Demian, E. and Lufty, R. (1965b) Predatory activity of *Marisa cornuarietis* against *Biomphalaria alexandrina* under laboratory conditions. *Ann. Trop. Med. Parasitol.*, **59**, 337–9.

Demian, E. and Lufty, R. (1966) Factors affecting the predation of *Marisa cornuarietis* on *Bulinus truncatus*, *Biomphalaria alexandrina* and *Lymnaea cailludi*. *Oikos*, **17**, 212–30.

den Hartog, C. (1963) The distribution of the snail *Aplexa hypnorum* in Zuid-Beveland in relation to soil and salinity. *Basteria*, **27**, 8–17.

DeNicola, D. and McIntire, C. (1991) Effects of hydraulic refuge and irradiance on grazer-periphyton interactions in laboratory streams. *J. N. Am. Benthol. Soc.*, **10**, 251–62.

DeNicola, D., McIntire, C., Lamberti, G., Gregory, S. and Ashkenas, L. (1990) Temporal patterns of grazer-periphyton interactions in laboratory streams. *Freshwat. Biol.*, **23**, 475–89.

Dennis, B. and Taper, M. (1994) Density dependence in time series observations of natural populations: Estimation and testing. *Ecol. Monog.*, **64**, 205–24.

Dermott, R. and Kerec, D. (1997) Changes to the deepwater benthos of eastern Lake Erie since the invasion of *Dreissena*: 1979–1993. *Can. J. Fish. Aquat. Sci.*, **54**, 922–30.

DeVisser, J., ter Maat, A. and Zonneveld, C. (1994) Energy budgets and reproductive allocation in the simultaneous hermaphrodite pond snail, *Lymnaea stagnalis* (L.) – a trade-off between male and female function. *Am. Nat.*, **144**, 861–7.

DeWitt, R. M. (1954) The intrinsic rate of natural increase in a pond snail (*Physa gyrina* Say). *Am. Nat.*, **88**, 353–9.

DeWitt, R. (1955) The ecology and life history of the pond snail, *Physa gyrina*. *Ecology*, **36**, 40–4.

DeWitt, R. M. and Sloan, W. C. (1958) The innate capacity for increase in numbers in the pulmonate snail, *Lymnaea columella*. *Trans. Am. Microsc. Soc.*, **77**, 290–4.

DeWitt, R. M. and Sloan, W. C. (1959) Reproduction in *Physa pomilia* and *Helisoma duryi*. *Anim. Behav.*, **7**, 81–4.

DeWitt, T. J. (1991) Mating behavior of the freshwater pulmonate snail, *Physa gyrina*. *American Malacol. Bull.*, **9**, 81–4.

DeWitt, T. (1995) Gender contests in a simultaneous hermaphrodite snail: A size-advantage model of behavior. *Anim. Behav.*, **51**, 345–51.

Diamond, J. (1975) Assembly of species communities. In *Ecology and Evolution of Communities*, ed. M. Cody and J. Diamond, pp. 342–444. Cambridge: Harvard University Press.

Diehl, S. (1992) Fish predation and benthic community structure: The role of omnivory and habitat complexity. *Ecology*, **73**, 1646–61.

Dietz, T. (1985) Ionic regulation in freshwater mussels: A brief review. *Am. Malacol. Bull.*, **3**, 233–42.

Dietz, T., Wilcox, S., Byrne, R., Lynn, J. and Silverman, H. (1996) Osmotic and ionic regulation of North American zebra mussels (*Dreissena polymorpha*). *Am.. Zool.*, **36**, 364–72.

Digby, P. and Kempton, R. (1987) *Multivariate Analysis of Ecological Communities*. London/NewYork: Chapman and Hall.

Dillon, R. T., Jr. (1981) Patterns in the morphology and distribution of gastropods in Oneida Lake, NY, detected using computer-generated null hypotheses. *Am. Nat.*, **118**, 83–101.

Dillon, R. T., Jr. (1984a) Geographic distance, environmental difference, and divergence between isolated populations. *Syst. Zool.*, **33**, 69–82.

Dillon, R. T., Jr. (1984b) What shall I measure on my snails? Allozyme data and multivariate analysis used to reduce the non-genetic component of morphological variance in *Goniobasis proxima*. *Malacologia*, **25**, 503–11.

Dillon, R. T., Jr. (1986) Inheritance of isozyme phenotype at three loci in the freshwater snail, *Goniobasis proxima*: Mother-offspring analysis and an artificial introduction. *Biochem. Genet.*, **24**, 281–90.

Dillon, R. T., Jr. (1987) A new Monte Carlo method for assessing taxonomic similarity within faunal samples: Reanalysis of the gastropod community of Oneida Lake, NY. *Am. Malacol. Bull.*, **5**, 101–4.

Dillon, R. T., Jr. (1988) Evolution from transplants between genetically distinct populations of freshwater snails. *Genetica*, **76**, 111–9.

Dillon, R. T., Jr. and Benfield, E. (1982) Distribution of pulmonate snails in the New River of Virginia and North Carolina U.S.A.: Interaction between alkalinity and stream drainage area. *Freshwat. Biol.*, **12**, 179–86.

Dillon, R. T., Jr. and Davis, G. M. (1980) The *Goniobasis* of southern Virginia and northwestern North Carolina: Genetic and shell morphometric relationships. *Malacologia*, **20**, 83–98.

Dillon, R. T., Jr. and Davis, K. (1991) The diatoms ingested by freshwater snails: Temporal, spatial, and interspecific variation. *Hydrobiologia*, **210**, 233–42.

DiMaio, J. and Corkum, L. (1995) Relationship between the spatial distribution of freshwater mussels (Bivalvia: Unionidae) and the hydrological variability of rivers. *Can. J. Zool.*, **73**, 663–71.

Doums, C., Delay, B. and Jarne, P. (1994) A problem with the estimate of self-fertilization depression in the hermaphrodite freshwater snail *Bulinus truncatus*: The effect of grouping. *Evolution*, **48**, 498–504.

Doums, C., Bremond, P., Delay, B. and Jarne, P. (1996a) The genetical and environmental determination of phally polymorphism in the freshwater snail *Bulinus truncatus*. *Genetics*, **142**, 217–25.

Doums, C., Viard, F., Pernot, A., Delay, B. and Jarne, P. (1996b) Inbreeding depression, neutral polymorphism, and copulatory behavior in freshwater snails: A self-fertilization syndrome. *Evolution*, **50**, 1908–18.

Downes, B. (1986) Guild structure in water mites (*Unionicola* spp.) inhabiting freshwater mussels: Choice, competitive exclusion and sex. *Oecologia*, **70**, 457–65.

Downes, B. (1989) Host specificity, host location and dispersal. Experimental conclusions from freshwater mites *Unionicola* spp. parasitizing unionid mussels. *Parasitol.*, **98**, 189–96.

Downes, B. (1990) Host-induced morphology in mites; implications for host-parasite coevolution. *Syst. Zool.*, **39**, 162–8.

Downing, J., Amyot, J., Perusse, M. and Rochon, Y. (1989) Visceral sex, hermaphroditism, and protandry in a population of the freshwater bivalve *Elliptio complanata*. *J. N. Am. Benthol. Soc.*, **8**, 92–9.

Downing, J., Rochen, Y. and Perusse, M. H. (1993) Spatial aggregation, body size, and reproductive success in the freshwater mussel *Elliptio complanata*. *J. N. Am. Benthol. Soc.*, **12**, 148–56.

Duch, T. (1976) Aspects of the feeding habits of *Viviparus georgianus*. *Nautilus*, **90**, 7–10.

Dudgeon, D. (1980) Some aspects of the biology of *Cristaria* (*Pletholophus*) *discoidea* (Bivalvia: Unionacea) in Plover Cove Reservoir, Hong Kong. In *Proceedings of the First International Workshop on the Malacofauna of Hong Kong and Southern China*, ed. B. Morton, pp. 181–210. Hong Kong: Hong Kong University Press.

Dudgeon, D. (1983a) The effects of water level fluctuations on a gently shelving marginal zone of Plover Cove Reservoir, Hong Kong. *Arch. Hydrobiol. Suppl.*, **65**, 2–3.

Dudgeon, D. (1983b) Spatial and temporal changes in the distribution of gastropods in the Lam Tsuen River, New Territories, Hong Kong, with notes on the occurrence of the axotic snail *Biomphalaria straminea*. *Malacol. Rev.*, **16**, 91–2.

Dudgeon, D. (1986) The life cycle, population dynamics and productivity of *Melanoides tuberculata* (Müller, 1774) (Gastropoda: Prosobranchia: Thiaridae) in Hong Kong. *J. Zool. Lond.*, **208**, 37–53.

Dudgeon, D. (1989) Ecological strategies of Hong Kong Thiaridae (Gastropoda: Prosobranchia). *Malacol. Rev.*, **22**, 39–53.

Dudgeon, D. and Cheung Pui Shan, C. (1990) Selection of gastropod prey by a tropical freshwater crab. *J. Zool. Lond.*, **220**, 147–55.

Dudgeon, D. and Morton, B. (1983) The population dynamics and sexual strategy of *Anodonta woodiana* (Bivalvia: Unionacea) in Plover Cove reservoir, Hong Kong. *J. Zool. Lond.*, **201**, 161–83.

Dudgeon, D. and Morton, B. (1984) Site selection and attachment duration of *Anodonta woodiana* (Bivalvia: Unionacea) glochidia on fish hosts. *J. Zool. Lond.*, **204**, 355–62.

Dudgeon, D. and Yipp, M. (1985) The diets of Hong Kong freshwater gastropods. In *Proceedings of Second International Workshop on the Malacofauna of Hong Kong and Southern China*, ed. B. Morton and D. Dudgeon, pp. 491–509. Hong Kong: Hong Kong University Press.

Duncan, A. (1966) The oxygen consumption of *Potamopyrgus jenkinsi* (Smith) (Prosobranchiata) in different temperatures and salinities. *Verh. Int. Ver. Theor. Angew. Limnol.*, **16**, 1739–51.

Duncan, C. J. (1959) The life-cycle and ecology of the freshwater snail *Physa fontinalis* (L.). *J. Anim. Ecol.*, **28**, 97–117.

Duncan, C. J. (1960) The genital systems of the freshwater Basomatophora. *Proc. Zool. Soc. Lond.*, **135**, 339–56.

Duncan, C. J. (1975) Reproduction. In *Pulmonates*, Volume 1, ed. V. Fretter and J. Peake, pp. 309–65. New York: Academic Press.

Dundee, D. (1957) Aspects of the biology of *Pomatiopsis lapidaria* (Say). *Misc. Publ. Mus. Zool. Univ. Mich.*, **100**, 1–37.

Duobinis-Gray, L., Urban, E., Sickel, J., Owen, D. and Maddox, W. (1991) Aspidogastrid (Trematoda) parasites on unionid (Bivalvia) molluscs in Kentucky Lake. *J. Helminthol. Soc. Wash.*, **58**, 167–70.

Dussart, G. (1976) The ecology of freshwater molluscs in North West England in relation to water chemistry. *J. Molluscan Stud.*, **42**, 181–98.

Dussart, G. (1979a) Life cycles and distributions of the aquatic gastropod molluscs *Bithynia tentaculata* (L.), *Gyraulus albus* (Muller), *Planorbis planorbis* (L.), and *Lymnaea peregra* (Muller) in relation to water chemistry. *Hydrobiologia*, **67**, 223–39.

Dussart, G. (1979b) *Sphaerium corneum* (L.) and *Pisidium* spp. (Pfeiffer) – The ecology of freshwater bivalve molluscs in relation to water chemistry. *J. Molluscan Stud.*, **45**, 19–34.

Dussart, G. (1987) Effects of water flow on the detachment of some aquatic pulmonate gastropods. *Am. Malacol. Bull.*, **5**, 65–72.

Dybdahl, M. and Lively, C. (1995) Diverse, endemic and polyphyletic clones in mixed populations of a freshwater snail (*Potamopyrgus antipodarum*). *J. Evol. Biol.*, **8**, 385–98.

Dybdahl, M. and Lively, C. (1996) The geography of coevolution: Comparative population structures for a snail and its trematode parasite. *Evolution*, **50**, 2264–75.

Dybdahl, M. and Lively, C. (1998) Host-parasite coevolution: Evidence for rare advantage and time-lagged selection in a natural population. *Evolution*, **52**, 1057–66.

Eckblad, J. (1973a) Population studies of three aquatic gastropods in an intermittant backwater. *Hydrobiologia*, **41**, 199–219.

Eckblad, J. (1973b) Experimental predation studies of malacophagous larvae of *Sepedon fuscipennis* (Diptera: Sciomyzidae) and aquatic snails. *Exp. Parasitol.*, **33**, 331–42.

Eckblad, J. and Lehtinen, S. (1991) Decline in fingernail clam populations family (Sphaeriidae) from backwater lakes of the upper Mississippi River. *J. Freshwat. Ecol.*, **6**, 353–62.

Eckblad, J., Peterson, N., Ostlie, K. and Temte, A. (1977) The morphometry, benthos and sedimentation rates of a floodplain lake in Pool 9 of the Upper Mississippi River. *Am. Midl. Nat.*, **97**, 433–43.

Edgar, A. (1965) Observations on the sperm of the pelecypod *Anodontoides ferussacianus* (Lea). *Trans. Am. Microsc. Soc.*, **84**, 228–30.

Edwards, D. and Dimock, R., Jr. (1988) A comparison of the population dynamics of *Unionicola formosa* from two anodontine bivalves in a North Carolina farm pond. *J. Elisha Mitchell Sci. Soc.*, **104**, 90–8.

Edwards, D. and Dimock, R., Jr. (1995) Specificity of the host recognition behaviours of larval *Unionicola* (Acari: Unionicolidae): The effects of larval ontogeny and early larval experience. *Anim. Behav.*, **50**, 343–52.

Efford, I. and Tsumura, K. (1973) Uptake of dissolved glucose and glycine by *Pisidium*, a freshwater bivalve. *Can. J. Zool.*, **51**, 825–32.

Eichenberger, E., Schlatter, A. and Weilenmann, H. (1985) Grazing pressure as a decisive factor in the long-term succession of the benthic vegetation in artificial rivers. *Verh. Int. Ver. Theor. Angew. Limnol.*, **22**, 2332–6.

Eisenberg, R. (1966) The regulation of density in a natural population of the pond snail, *Lymnaea elodes*. *Ecology*, **47**, 889–906.

Eisenberg, R. (1970) The role of food in the regulation of the pond snail, *Lymnaea alodes*. *Ecology*, **51**, 680–4.

El-Emam, M. and Madsen, H. (1982) The effect of temperature, darkness, starvation and various food types on growth, survival and reproduction of *Helisoma duryi*, *Biomphalaria alexandrina* and *Bulinus truncatus* (Gastropoda: Planorbidae). *Hydrobiologia*, **88**, 265–75.

Eleutheriadis, N. and Lazaridoudimitriadou, M. (1995) Density and growth of freshwater prosobranch snails (*Bithynia graeca* and *Viviparus contectus*) in relation to water chemistry in Serres, northern Greece. *J. Molluscan Stud.*, **61**, 347–52.

Ellis, M. (1931a) *A Survey of Conditions Affecting Fisheries in the Upper Mississippi River*. U. S. Bur. Fisheries Circular No. 5. 18 pp.

Ellis, M. (1931b) *Some Factors Affecting the Replacement of the Commercial Fresh-Water Mussels*. U.S. Bureau Fisheries Circular No. 7. 10 pp.

Elton, C. (1946) Competition and the structure of ecological communities. *J. Anim. Ecol.*, **4**, 127–36.

Elwood, J. and Goldstein, R. (1975) Effects of temperature on food ingestion rate and absorption, retention, and equilibrium burden of phosphorus in an aquatic snail, *Goniobasis clavaeformis* Lea. *Freshwat. Biol.*, **5**, 397–406.

Elwood, J. and Nelson, D. (1972) Periphyton production and grazing rates in a stream measured with a ^{32}P material balance method. *Oikos*, **23**, 295–303.

Elwood, J., Newbold, J., Trimble, A. and Stark, R. (1981) The limiting role of phosphorus in a woodland stream ecosystem: Effects of P enrichment on leaf decomposition and primary producers. *Ecology*, **62**, 146–58.

Eriksson, F., Hornstrom, E., Mossberg, P. and Nyberg, P. (1983) Ecological effects of lime treatment of acidified lakes and rivers in Sweden. *Hydrobiologia*, **101**, 145–64.

Esch, G. and Fernandez, J. (1994) Snail-trematode interactions and parasite community dynamics in aquatic systems: A review. *Am. Midl. Nat.*, **131**, 209–37.

Estebenet, A. (1995) Food and feeding in *Pomacea canaliculata* (Gastropoda: Ampullaridae). *Veliger*, **38**, 277–83.

Estebenet, A. and Cazzaniga, N. (1992) Growth and demography of *Pomacea canaliculata* (Gastropoda: Ampullariidae) under laboratory conditions. *Malacol. Rev.*, **25**, 1–12.

Estebenet, A. and Cazzaniga, N. (1998) Sex-related differential growth in *Pomacea canaliculata* (Gastropoda: Ampullariidae). *J. Molluscan Stud.*, **64**, 119–23.

Etges, F. and Decker, C. (1963) Chemosensitivity of the miracidium of *Schistosoma mansoni* to *Australorbis glabratus* and other snails. *J. Parasitol.*, **49**, 114–16.

Etges, F. and Gresso, W. (1965) Effect of *Schistosoma mansoni* upon fecundity in *Australorbis glabratus*. *J. Parasitol.*, **51**, 757–60.

Evans, N., Whitfield, P. and Dobson, A. (1981) Parasite utilization of a host community: The distribution and occurrence of metacercarial cysts of *Echinoparyphium recurvation* (Digenea: Echinostomatidae) in seven species of mollusk at Harting Pond, Sussex. *Parasitol.*, **83**, 1–12.

Eversole, A. (1978) Life cycles, growth and population bioenergetics of the snail, *Helisoma trivolvis* (Say). *J. Molluscan Stud.*, **44**, 209–22.

Extence, C. (1981) The effect of drought on benthic invertebrate communities in a lowland river. *Hydrobiologia*, **83**, 217–24.

Fahnenstiel, G., Bridgeman, T., Lang, G., McCormich, M. and Nalepa, T. (1995a) Phytoplankton productivity in Saginaw Bay, Lake Huron: Effects of zebra mussel (*Dreissena polymorpha*) colonization. *J. Great Lakes Res.*, **21**, 464–75.

Fahnenstiel, G., Nalepa, T. and Johengen, T. (1995b) Effects of the zebra mussel (*Dreissena polymorpha*) colonization on water quality parameters in Saginaw Bay, Lake Huron. *J. Great Lakes Res.*, **21**, 435–48.

Fanslow, D., Nalepa, T. and Lang, G. (1995) Filtration rates of the zebra mussel (*Dreissena polymorpha*) on natural seston from Saginaw Bay, Lake Huron. *J. Great Lakes Res.*, **21**, 489–500.

Fauth, J., Bernardo, J., Camera, M., Resetarits, W., VanBuskirk, J. and McCollum, S. (1996) Simplifying the jargon of community ecology: A conceptual approach. *Am. Nat.*, **147**, 282–6.

Feminella, J. and Hawkins, C. (1995) Interactions between stream herbivores and periphyton: A quantitative analysis of past experiments. *J. N. Am. Benthol. Soc.*, **14**, 465–509.

Fernandez, J. and Esch, G. (1991a) Guild structures of larval trematodes in the snail *Helisoma anceps* – patterns and processes at the individual host level. *J. Parasitol.*, **77**, 528–39.

Fernandez, J. and Esch, G. (1991b) The component community structure of larval trematodes in the pulmonate snail *Helisoma anceps*. *J. Parasitol.*, **77**, 540–50.

Fisher, J. and Tevesz, M. (1976) Distribution and population density of *Elliptio complanata* in Lake Pocotopaug, Connecticut. *Veliger*, **18**, 332–8.

Foe, C. and Knight, A. (1985) The effect of phytoplankton and suspended sediment on the growth of *Corbicula fluminea* (Bivalvia). *Hydrobiologia*, **127**, 105–15.

Foe, C. and Knight, A. (1986) Growth of *Corbicula fluminea* (Bivalvia) fed artificial and algal diets. *Hydrobiologia*, **133**, 155–64.

Foote, B. (1976) Biology and larval feeding habits of three species of *Renocera* (Diptera: Sciomyzidae) that prey on fingernail clams (Mollusca: Sphaeriidae). *Ann. Entomol. Soc. Am.*, **69**, 121–33.

Forbes, G. and Crampton, H. (1942) The effect of population density upon growth and size in *Lymnaea palustris*. *Biol. Bull.*, **83**, 283–9.

Foster, G. (1973) Soil type and habitat of the aquatic snail *Lymnaea (Galba) bulinoides* Lea during the dry season. *Basteria*, **37**, 41–6.

Frandsen, F. (1979) Discussion of the relationship between *Schistosoma* and their intermediate hosts, assessment of the degree of host-parasite compatibility and evaluation of schistosome taxonomy. *Zeitschr. Parasitenkunde*, **58**, 275–96.

Frandsen, F. and Madsen, H. (1979) A review of *Helisoma duryi* in biological control. *Acta Trop.*, **36**, 67–84.

French, J. (1974) Improved methods for culturing the subspecies of *Oncomelania hupensis*, the snail hosts of *Schistosoma japonicum*, the oriental human blood fluke. *Sterkiana*, **56**, 1–20.

French, J. and Bur, M. (1993) Predation of the zebra mussel (*Dreissena polymorpha*) by freshwater drum in western Lake Erie. In *Zebra Mussels: Biology, Impacts, and Control*, ed. T. Nalepa and D. Schloesser, pp. 453–64. Boca Raton, FL: Lewis Publishers.

French, J. and Love, J. (1995) Size limitations on zebra mussels consumed by freshwater drum may preclude the effectiveness of drum as a biological controller. *J. Freshwat. Ecol.*, **10**, 379–83.

French, J. and Morgan, M. (1995) Preference of redear sunfish on zebra mussels and rams-horn snails. *J. Freshwat. Ecol.*, **10**, 49–55.

Frenzel, P. (1979) Biology and population dynamics of *Potamopyrgus jenkinsi* (Gastropoda: Prosobranchia) in the littoral of Lake Constance, West Germany. *Arch. Hydrobiol.*, **85**, 448–64.

Fretter, V. and Graham, A. (1962) *British Prosobranch Molluscs*. London: Ray Society.

Fretter, V. and Graham, A. (1964) Reproduction. In *Physiology of Mollusca*, Vol. I, ed. K. Wilson and C. Younge, pp. 127–64. New York: Academic Press.

Fryer, G. (1970) Biological aspects of parasitism of freshwater fishes by crustaceans and molluscs. *Symp. British Soc. Parasitol.*, **8**, 103–18.

Fryer, S., Oswald, R., Probert, A. and Runham, W. (1990) The effect of *Schistosoma haematobium* infection on the growth and fecundity of three sympatric species of bulinid snails. *J. Parasitol.*, **76**, 557–63.

Fuller, S. (1974) Clams and Mussels (Mollusca: Bivalvia). In *Pollution Ecology of Freshwater Invertebrates*, ed. J. Hart and S. Fuller, pp. 215–73. London: Academic Press.

Fustish, C. and Millemann, R. (1978) Glochidiosis of salmonid fishes. II. Comparison of tissue response of coho and chinook salmon to experimental infection with *Margaritifera margaritifera* (L.) (Pelecypoda: Margaritiferidae). *J. Parasitol.*, **64**, 155–7.

Gale, W. (1971) An experiment to determine substrate preference of the fingernail clam, *Sphaerium transversum* (Say). *Ecology*, **52**, 367–70.

Gale, W. (1973) Substrate preference of the fingernail clam, *Sphaerium striatinum* (Lam) (Sphaeriidae). *Southwest. Nat.*, **18**, 31–7.

Gale, W. and Lowe, R. (1971) Phytoplankton ingestion by the fingernail clam, *Sphaerium transversum* (Say), in Pool 19, Mississippi River. *Ecology*, **52**, 507–13.

Galhano, M. and Da Silva, M. (1983) The reproductive cycle of *Anodonta cygnea* L. from Mira Lagoon (Portugal). *Publ. Inst. Zool. 'Dr. Augusto Nobre' Fac. Cienc. Porto.*, **179**, 1–5.

Garcia, R. (1972) Tolerance of *Oncomelania hupensis quadrasi* to varying concentrations of dissolved oxygen and organic pollution. *Bull. W.H.O.*, **47**, 59–70.

Gardner, J., Jr., Woodall, W., Staats, A. and Napoli, J. (1976) The invasion of the Asiatic clam (*Corbicula manilensis* Phil.) in the Altamaha R., Georgia. *Nautilus*, **90**, 117–25.

Gatenby, C., Neves, R. and Parker, B. (1996) Influence of sediment and algal food on cultured juvenile freshwater mussels. *J. N. Am. Benthol. Soc.*, **15**, 597–609.

Gauch, H. (1982) *Multivariate Analysis in Community Ecology*. Cambridge, U.K.: Cambridge University Press.

Gebhardt, M. and Ribi, G. (1987) Reproductive effort and growth in the prosobranch snail, *Viviparus ater*. *Oecologia*, **74**, 209–14.

Geldiay, R. (1956) Studies on local populations of the freshwater limpet *Ancylus fluviatilis* Müller. *J. Anim. Ecol.*, **25**, 389–402.

Geraerts, W. P. M. and Joosse, J. (1984) Freshwater snails (Basommatophora). In *The Mollusca, Vol 7. Reproduction*, ed. A. S. Tompa, N. H. Verdonk and J. A. M. VandenBiggelaar, pp. 141–207. Orlando, FL: Academic Press.

Ghedotti, M., Smihula, J. and Smith, G. (1995) Zebra mussel predation by round gobies in the laboratory. *J. Great Lakes Res.*, **21**, 665–9.

Ghent, A., Singer, R. and Johnson-Singer, L. (1978) Depth distributions determined with SCUBA, and associated studies of the freshwater unionid clams, *Elliptio complanata* and *Anodonta grandis* in Lake Bernard, Ontario. *Can. J. Zool.*, **56**, 1654–63.

Ghiretti, F. and Ghiretti-Magaldi, A. (1975) Respiration. In *Pulmonates*, Vol. 1, ed. V. Fretter and J. Peake, pp. 33–52. New York: Academic Press.

Gilbert, E., Vincent, B. and Roseberry, L. (1986) Etude experimentale in situ des effets de la densité et de la quantité nourriture chez le gasteropode *Bithynia tentaculata* (Mollusca: Prosobranchia). *Can. J. Zool.*, **64**, 1696–700.

Gilbertson, D. and Jones, K. (1972) Uptake and assimilation of amino acids by *Biomphalaria glabrata* (Gastropoda: Pulmonata). *Comp. Biochem. Physiol.*, **42A**, 621–6.

Gilbertson, D., Kasim, O. and Stumpf, J. (1978) Studies on the biology of *Bulimnea megasoma* (Say) (Gastropoda: Pulmonata). *J. Molluscan Stud.*, **44**, 145–50.

Gilinsky, E. (1984) The role of fish predation and spatial heterogeneity in determining benthic community structure. *Ecology*, **65**, 455–68.

Gilliam, J., Fraser, D. and Sabat, A. (1989) Strong effects of foraging minnows on a stream benthic invertebrate community. *Ecology*, **70**, 445–52.

Gillis, P. and Mackie, G. (1994) Impact of the zebra mussel, *Dreissena polymorpha*, on populations of Unionidae (Bivalvia) in Lake St. Clair. *Can. J. Zool.*, **72**, 1260–71.

Giusti, F., Castagnolo, L., Moretti-Farina, L. and Renzoni, A. (1975) The reproductive cycle and the glochidium of *Anodonta cygnaea* L. from Lago Trasimeno (Central Italy). *Monitore Zool. Ital.*, **9**, 99–118.

Gomez, J., Vargas, M. and Malek, E. (1990) Biological control of *Biomphalaria glabrata* by *Thiara granifera* under laboratory conditions. *Trop. Med. Parasitol.*, **41**, 43–5.

Gorthner, A. (1992) Morphology, function and evolution of complex gastropod shells in long-lived lakes (German, with English abstract). *Stuttgarter Beitrage zu Naturkunde B*, **190**, 1–173.

Gracio, M. (1988) A comparative laboratory study of *Bulinus* (*Physopsis*) *globosus* uninfected and infected with *Schistosoma haematobium*. *Malacol. Rev.*, **21**, 123–7.

Graney, R., Cherry, D., Rodgers, J., Jr. and Cairns, J. (1980) The influence of thermal discharges and substrate composition on the population structure and distribution of the Asiatic clam, *Corbicula fluminea* in the New River, Va. *Nautilus*, **94**, 130–4.

Green, R. H. (1971) A multivariate statistical approach to the Hutchinsonian niche: Bivalve molluscs of central Canada. *Ecology*, **52**, 543–56.

Greenaway, P. (1970) Sodium regulation in the freshwater mollusc *Lymnaea stagnalis* (L.) (Gastropoda: Pulmonata). *J. Exp. Biol.*, **53**, 147–63.

Greenaway, P. (1971) Calcium regulation in the freshwater mollusc *Lymnaea stagnalis* (L.) (Gastropoda: Pulmonata). I. The effect of internal and external calcium concentration. *J. Exp. Biol.*, **54**, 199–214.

Gregg, W. and Rose, F. (1985) Influences of aquatic macrophytes on invertebrate community structure, guild structure, and microdistribtion in streams. *Hydrobiologia*, **128**, 45–56.

Gregory, S. (1983) Plant-herbivore interactions in stream systems. In *Stream Ecology*, ed. J. Barnes and G. Minshall, pp. 157–89. New York: Plenum.

Griffiths, R., Schloesser, D., Leach, J. and Kovalak, W. (1991) Distribution and dispersal of the zebra mussel (*Dreissena polymorpha*) in the Great Lakes region. *Can. J. Fish. Aquat. Sci.*, **48**, 1381–8.

Grime, J. (1979) *Plant Strategies and Vegetation Processes*. Chichester, England: John Wiley and Sons.

Grime, J. (1985) Towards a functional description of vegetation. In *The Population Structure of Vegetation*, ed. J. White, pp. 503–14. Dordrecht, Netherlands: W. Junk.

Grime, J. (1988) A comment on Loehle's critique of the triangular model of primary plant strategies. *Ecology*, **69**, 1618–20.

Guo, X., Hedgecock, D., Hershberger, W., Cooper, K. and Allen, S. (1998) Genetic determinants of protandric sex in the Pacific oyster, *Crassostrea gigas* Thunberg. *Evolution*, **52**, 394–402.

Guyard, A., Pointier, J.-P. and Theron, A. (1986) Le role de la competition entre *Biomphalaria straminea* et *B. glabrata* dans le declin de la schistosome intestinale en Martinique (Antilles françaises). In *Proceedings Eighth International Malacological Congress*, ed. L. Pinter, pp. 91–4. Budapest, Hungary: Hungarian Natural History Museum.

Haag, W. and Warren, M. (1997) Host fishes and reproductive biology of six freshwater mussel species from the Mobile Basin, U.S.A. *J. N. Am. Benthol. Soc.*, **16**, 576–85.

Haag, W., Berg, D., Garton, D. and Farris, J. (1993) Reduced survival and fitness in native bivalves in response to fouling by the introduced zebra mussel (*Dreissena polymorpha*) in western Lake Erie. *Can. J. Fish. Aquat. Sci.*, **50**, 13–19.

Haag, W., Butler, R. and Hartfield, P. (1995) An extraordinary reproductive strategy in freshwater bivalves: Prey mimicry to facilitate larval dispersal. *Freshwat. Biol.*, **34**, 471–6.

Ham, S. and Bass, J. (1982) The distribution of Sphaeriidae in rivers and streams of central southern England. *J. Conchol.*, **31**, 45–56.

Hambrook, J. and Sheath, R. (1987) Grazing of freshwater Rhodophyta. *J. Phycol.*, **23**, 656–62.

Hamill, S., Qadri, S. and Mackie, G. (1979) Production and turnover ratio of *Pisidium casertanum* (Pelecypoda: Sphaeriidae) in the Ottawa River near Ottawa-Hull, Canada. *Hydrobiologia*, **62**, 225–30.

Hamilton, D., Ankney, C. and Bailey, R. (1994) Predation by diving ducks on zebra mussels: An exclosure experiment. *Ecology*, **75**, 521–31.

Haniffa, M. (1982) Effects of feeding level and body size on food utilization of the freshwater snail *Pila globosa*. *Hydrobiologia*, **97**, 141–9.

Haniffa, M. and Pandian, T. (1974) Effect of body weight on feeding rate and radula size in the freshwater snail *Pila globosa*. *Veliger*, **16**, 415–18.

Hanlon, R. (1981) The influence of different species of leaf litter on the growth and food preference of the prosobranch mollusc *Potamopyrgus jenkinsi*. *Arch. Hydrobiol.*, **91**, 463–74.

Hanson, J., Chambers, P. and Prepas, E. (1990) Selective foraging by the crayfish *Orconectes virilis* and its impact on macroinvertebrates. *Freshwat. Biol.*, **24**, 69–80.

Hanson, J., Mackay, W. and Prepas, E. (1988) The effects of water depth and density on the growth of a unionid clam. *Freshwat. Biol.*, **19**, 345–56.

Hanson, J., MacKay, W. and Prepas, E. (1989) Effect of size-selective predation by muskrats (*Ondatra zebethicus*) on a population of unionid clams (*Anodonta grandis simpsoniana*). *J. Anim. Ecol.*, **58**, 15–28.

Harman, W. (1968) Replacement of pleurocerids by *Bithynia* in polluted waters of central New York. *Nautilus*, **81**, 77–83.

Harman, W. (1972) Benthic substrates: Their effect on fresh-water Mollusca. *Ecology*, **53**, 271–7.

Harman, W. (1974) Snails (Mollusca: Gastropoda). In *Pollution Ecology of Freshwater Invertebrates*, ed. C. Hart and S. Fuller, pp. 275–312. New York: Academic Press.

Harman, W. and Forney, J. (1970) Changes in the molluscan community on Oneida Lake, N.Y. between 1917 and 1967. *Limnol. Oceanogr.*, **15**, 454–60.

Harper, D., Mavuti, K. and Muchiri, S. (1990) Ecology and management of Lake Naivasha, Kenya, in relation to climatic change, alien species introductions, and agricultural development. *Environ. Conserv.*, **17**, 328–36.

Harris, R. and Charleston, W. (1977) An examination of the marsh microhabitats of *Lymnaea tormentosa* and *L. columella* (Mollusca: Gastropoda) by path analysis. *New Zealand J. Zool.*, **4**, 395–9.

Harris, R., Preston, T. and Southgate, V. (1993) Purification of an agglutinin from the haemolymph of the snail *Bulinus nasutus* and demonstration of related proteins in other *Bulinus* spp. *Parasitol.*, **106**, 127–35.

Harrison, A. and Rankin, J. (1978) Hydrobiological studies of eastern Lesser Antillean Islands. III St. Vincent: Freshwater mollusca – their distribution, population dynamics and biology. *Arch. Hydrobiol. Suppl.*, **54**, 123–88.

Harrison, A., Nduku, W. and Hooper, A. (1966) The effects of a high magnesium-to-calcium ratio on the egg-laying rate of an aquatic planorbid snail, *Biomphalaria pfeifferi*. *Ann. Trop. Med. Parasitol.*, **60**, 212–4.

Harrison, A., Williams, N. and Greig, G. (1970) Studies on the effects of calcium bicarbonate concentrations on the biology of *Biomphalaria pfeifferi* (Krauss) (Gastropoda: Pulmonata). *Hydrobiologia*, **36**, 317–27.

Hartfield, P. (1994) Status Review Summary of Seven Mobile River Basin Aquatic Snails. U.S. Fish and Wildlife Service, Jackson, MS.

Hartfield, P. and Hartfield, E. (1996) Observations on the conglutinates of *Ptychobranchus greeni* (Conrad, 1834) (Mollusca: Bivalvia: Unionoidea). *Am. Midl. Nat.*, **135**, 370–5.

Hartley, P. (1948) Food and feeding relationships in a community of fresh-water fishes. *J. Anim. Ecol.*, **17**, 1–14.

Harvey, P. and Nee, S. (1991) How to live like a mammal. *Nature*, **350**, 23–4.

Hastings, A. (1987) Can competition be detected using species co-occurrence data? *Ecology*, **68**, 117–23.

Haukioja, E. and Hakala, T. (1974) Vertical distribution of freshwater mussels (Pelecypoda, Unionidae) in southwestern Finland. *Ann. Zool. Fenn.*, **11**, 127–30.

Haukioja, E. and Hakala, T. (1978) Life-history evolution in *Anodonta piscinalis* (Mollusca, Pelecypoda). Correlations of parameters. *Oecologia*, **35**, 253–66.

Haven, D. and Morales-Alamo, R. (1970) Filtration of particles from suspension by the American oyster, *Crassostrea virginica*. *Biol. Bull.*, **139**, 248–64.

Havlik, M. and Marking, L. (1980) A quantitative analysis of naiad mollusks from the Prairie du Chein, Wisconsin dredge material site on the Mississippi River. *Bull. Am. Malacol. Union*, **1980**, 30–4.

Hawkins, C. and Furnish, J. (1987) Are snails important competitors in stream ecosystems? *Oikos*, **49**, 209–20.

Hawkins, C., Murphy, M. and Anderson, N. (1982) Effects of canopy, substrate composition, and gradient on the structure of macroinvertebrate communities in Cascade Range streams of Oregon. *Ecology*, **63**, 1840–56.

Hawkins, M. and Ultsch, G. (1979) Oxygen consumption in two species of freshwater snails (*Goniobasis*): Effects of temperature and ambient oxygen tension. *Comp. Biochem. Physiol.*, **63A**, 369–72.

Hawkins, S., Watson, D., Hill, A., Harding, S., Kyriakides, M., Hutchinson, S. and Norton, T. (1989) A comparison of feeding mechanisms in microphagous, herbivorous, intertidal, prosobranchs in relation to resource partitioning. *J. Molluscan Stud.*, **55**, 151–65.

Hayashi, K. and Otani, S. (1967) Stomach contents of a freshwater clam, *Corbicula sandai*, from Lake Biwa. *Venus*, **26**, 17–28.

Haynes, A. (1985) The ecology and local distribution of non-marine aquatic gastropods in Viti Levu, Fiji. *Veliger*, **28**, 204–10.

Haynes, A. and Taylor, B. (1984) Food finding and food preference in *Potamopyrgus jenkinsi*. *Arch. Hydrobiol.*, **100**, 479–91.

Head, G., May, R. and Pendleton, L. (1987) Environmental determination of sex in the reptiles. *Nature*, **329**, 198–9.

Heard, W. (1963) Reproductive features of *Valvata*. *Nautilus*, **77**, 64–8.

Heard, W. (1965) Comparative life histories of North American pill clams (Sphaeriidae: *Pisidium*). *Malacologia*, **2**, 381–411.

Heard, W. (1975) Sexuality and other aspects of reproduction in *Anodonta* (Pelecypoda: Unionidae). *Malacologia*, **15**, 81–104.

Heard, W. (1977) Reproduction of fingernail clams (Sphaeriidae: *Sphaerium* and *Musculium*). *Malacologia*, **16**, 421–55.

Heard, W. (1979) Hermaphroditism in *Elliptio* (Bivalvia: Unionidae). *Malacol. Rev.*, **12**, 21–8.

Heard, W. and Hendrix, S. (1964) Behavior of unionid glochidia. *Ann. Rep. Am. Malacol. Union*, **1964**, 2–3.

Heath, D. J. (1977) Simultaneous hermaphroditism; cost and benefit. *J. Theor. Biol.*, **64**, 363–73.

Hebert, P., Wilson, C., Murdoch, M. and Lazar, R. (1991) Demography and ecological impacts of the invading mollusc *Dreissena polymorpha*. *Can. J. Zool.*, **69**, 405–9.

Heeg, J. (1977) Oxygen consumption and the use of metabolic reserves during starvation and aestivation in *Bulinus* (*Physopsis*) *africanus* (Pulmonata: Planorbidae). *Malacologia*, **16**, 549–60.

Heller, J. (1979) Visual versus non-visual selection of shell colour in an Israeli freshwater snail. *Oecologia*, **98**, 98–104.

Heller, J. and Farstey, V. (1990) Sexual and parthenogenetic populations of the freshwater snail *Melanoides tuberculata* in Israel. *Israel J. Zool.*, **37**, 75–87.

Henry, R. and Simão, C. (1985) Spatial distribution of a bivalve population (*Diplodon delodontus expansus* (Kuster, 1856) in a small tropical reservoir. *Revista Brazileita de Biologia*, **45**, 407–15.

Heppleston, P. (1972) Life history and population fluctuations of *Lymnaea truncatulua* (Müller). The snail vector of fascioliasis. *J. Appl. Ecol.*, **9**, 235–48.

Hershey, A. (1990) Snail populations in arctic lakes: Competition mediated by predation? *Oecologia*, **82**, 26–32.

Hershey, A. (1992) Effects of experimental fertilization on the benthic macroinvertebrate community of an arctic lake. *J. NABS*, **11**, 204–17.

Hess, W. (1920) Notes on the biology of some common Lampryidae. *Biol. Bull.*, **38**, 39–76.

Heywood, J. and Edwards, R. (1962) Some aspects of the ecology of *Potamopyrgus jenkinsi* Smith. *J. Anim. Ecol.*, **31**, 239–50.

Hickman, C. and Morris, T. (1985) Gastropod feeding tracks as a source of data in analysis of the functional morphology of radulae. *Veliger*, **27**, 357–65.

Higashi, M., Miura, K., Tanimizu, K. and Iwasa, Y. (1981) Effect of the feeding activity of snails on the biomass and productivity of an algal community attached to a reed stem. *Verh. Int. Ver. Theor. Angew. Limnol.*, **21**, 590–5.

Hill, W. (1992) Food limitation and interspecific competition in snail-dominated streams. *Can. J. Fish. Aquat. Sci.*, **49**, 1257–67.

Hill, W., Weber, S. and Stewart, A. (1992) Food limitation of two lotic grazers: Quantity, quality and size specificity. *J. N. Am. Benthol. Soc.*, **11**, 420–32.

Hill, W., Ryon, M. and Schilling, E. (1995) Light limitation in a stream ecosystem: Responses by primary producers and consumers. *Ecology*, **76**, 1297–309.

Hillis, D. and Patton, J. (1982) Morphological and electrophoretic evidence for two species of *Corbicula* in North America. *Am. Midl. Nat.*, **108**, 74–80.

Hinch, S., Bailey, R. and Green, R. (1986) Growth of *Lampsilis radiata* (Bivalvia: Unionidae) in sand and mud: A recriprocal transplant experiment. *Can. J. Fish. Aquat. Sci.*, **43**, 548–52.

Hincks, S. and Mackie, G. (1997) Effects of pH, calcium, alkalinity, hardness, and chlorophyll on the survival, growth, and reproductive success of zebra mussel (*Dreissena polymorpha*) in Ontario lakes. *Can. J. Fish. Aquat. Sci.*, **54**, 2049–57.

Hinz, W. and Scheil, H. (1976) Substratwahlversuche an *Pisidium casertanum* und *Pisidium amnicum* (Bivalvia). *Basteria*, **40**, 89–100.

Hodasi, J. (1972) The effects of *Fasciola hepatica* on *Lymnaea truncatula*. *Parasitol.*, **65**, 359–69.

Hodgson, A. (1992) The structure and possible functions of the seminal vesicle region of the hermaphrodite duct of pulmonates. In *Proceedings Eleventh International Malacological Congress*, ed. F. Giusti and G. Manganelli, pp. 61–2. Italy: University of Siena.

Hoeh, W., Frazer, K., Naranjo-Garcia, E. and Trdan, R. (1995) A phylogenetic perspective on the evolution of simultaneous hermaphroditism in a freshwater mussel clade (Bivalvia: Unionidae: *Utterbackia*). *Malacol. Rev.*, **28**, 25–42.

Hoeh, W., Stewart, D., Sutherland, B. and Zouros, E. (1996) Multiple origins of gender-associated mitochondrial DNA lineages in bivalves (Mollusca: Bivalvia). *Evolution*, **50**, 2276–86.

Hofkin, B., Mkoji, G., Koech, D. & Loker, E. (1991) Control of schistosome-transmitting snails in Kenya by the North American crayfish *Procambarus clarkii*. *Am. J. Trop. Med. Hyg.*, **45**, 339–44.

Hofkin, B., Hofinger, D., Koech, D. and Loker, E. (1992) Predation of *Biomphalaria* and non-target molluscs by the crayfish *Procambarus clarkii*: Implications for the biological control of schistosomiasis. *Ann. Trop. Med. Parasitol.*, **86**, 663–70.

Holland-Bartels, L. (1990) Physical factors and their influence on the mussel fauna of a main channel border habitat of the upper Mississippi River. *J. N. Am. Benthol. Soc.*, **9**, 327–35.

Holmes, J. (1983) Evolutionary relationships between parasitic helminiths and their hosts. In *Coevolution*, ed. D. Futuyma and M. Slatkin, pp. 161–85. Sunderland, Massachusetts: Sinauer.

Holopainen, I. (1979) Population dynamics and production of *Pisidium* species (Bivalvia: Sphaeriidae) in the oligotrophic and mesohumic Lake Paajarvi, southern Finland. *Arch. Hydrobiol. Suppl.*, **54**, 466–508.

Holopainen, I. (1985) Feeding biology of Pisidiidae (Bivalvia) with special emphasis on functional morphology of the digestive tract. *Lammi Notes*, **12**, 5–9.

Holopainen, I. and Hanski, I. (1979) Annual energy flow in populations of two *Pisidium* species (Bivalvia: Sphaeriidae), with discussion on possible competition between them. *Arch. Hydrobiol.*, **86**, 338–54.

Holopainen, I. and Hanski, I. (1986) Life history variation in *Pisidium* (Bivalvia: Pisidiidae). *Hol. Ecol.*, **9**, 85–9.

Holopainen, I. and Jonasson, P. (1983) Long-term population dynamics and production of *Pisidium* (Bivalvia) in the profundal of Lake Esrom, Denmark. *Oikos*, **41**, 99–117.

Holopainen, I. and Jonasson, P. (1989a) Reproduction of *Pisidium* (Bivalvia, Sphaeriidae) at different depths in Lake Esrom, Denmark. *Arch. Hydrobiol.*, **116**, 85–95.

Holopainen, I. and Jonasson, P. (1989b) Bathymetric distribution and abundance of *Pisidium* (Bivalvia, Sphaeriidae) in Lake Esrom, Denmark, from 1954 to 1988. *Oikos*, **55**, 324–34.

Holopainen, I. and Lopez, G. (1989) Functional anatomy and histology of the digestive tract of fingernail clams (Sphaeriidae, Bivalvia). *Ann. Zool. Fenn.*, **26**, 61–72.

Hopf, F. and Brown, J. (1986) The bull's-eye method for testing randomness in ecological communities. *Ecology*, **67**, 1139–55.

Hornbach, D. (1992) Life history traits of a riverine population of the Asian clam *Corbicula fluminea*. *Am. Midl. Nat.*, **127**, 248–57.

Hornbach, D. and Cox, C. (1987) Environmental influences on life history traits in *Pisidium casertanum* (Bivalvia: Pisidiidae). *Am. Malacol. Bull.*, **5**, 49–64.

Hornbach, D., McLeod, M., Guttman, S. and Seilkop, S. (1980) Genetic and morphological variation in the freshwater clam *Sphaerium* (Bivalvia: Sphaeriidae). *J. Molluscan Stud.*, **46**, 158–70.

Hornbach, D., Wissing, T. and Burky, A. (1982) Life history characteristics of a stream population of the freshwater clam *Sphaerium striatinum* Lam (Bivalvia: Pisidiidae). *Can. J. Zool.*, **60**, 249–60.

Hornbach, D., Way, C., Wissing, T. and Burky, A. (1984) Effects of particle concentration and season on the filtration rates of the freshwater clam, *Sphaerium striatinum* Lam (Bivalvia: Pisidiidae). *Hydrobiologia*, **108**, 83–96.

Hornbach, D., Deneka, T. and Dado, R. (1991) Life-cycle variation of *Musculium partumeium* (Bivalvia, Sphaeriidae) from a temporary and a permanent pond in Minnesota. *Can. J. Zool.*, **69**, 2738–44.

Hornbach, D., Miller, A. and Payne, B. (1992) Species composition of the mussel assemblages in the Upper Mississippi River. *Malacol. Rev.*, **25**, 19–128.

Horst, T. and Costa, R. (1971) Distribution patterns of five selected gastropod species from McCargo Lake. *Nautilus*, **85**, 38–43.

Horvath, T., Lamberti, G., Lodge, D. and Perry, W. (1996) Zebra mussel dispersal in lake-stream systems: Source-sink dynamics? *J. N. Am. Benthol. Soc.*, **15**, 564–75.

Houp, K. (1970) Population dynamics of *Pleurocera acuta* in a central Kentucky limestone stream. *Am. Midl. Nat.*, **83**, 81–8.

Howard, A. (1915) Some exceptional cases of breeding among the Unioniidae. *Nautilus*, **29**, 4–11.

Howard, A. (1951) A river mussel parasitic on a salamander. *Nat. Hist. Miscellanea*, **70**, 1–6.

Hubendick, B. (1957) The eating function in *Lymnaea stagnalis*. *Ark. Zool.*, **10**, 511–21.

Hubendick, B. (1978) Systematics and comparative morphology of the Basommatophora. In *Pulmonates*, Vol. 2A, ed. V. Fretter and J. Peake, pp. 1–47. New York: Academic Press.

Huca, G., Brenner, R. and Niveiro, M. (1982) A study of the biology of *Diplodon delondontus* (Lam) (Bivalvia: Hyriidae). I: Ecological aspects and anatomy of the digestive tract. *Veliger*, **25**, 51–8.

Hudson, R. and Isom, B. (1984) Rearing juveniles of the freshwater mussels (Unionidae) in a laboratory setting. *Nautilus*, **98**, 129–35.

Huehner, M. (1984) Aspidogastrid trematodes from freshwater mussels in Missouri with notes on the life cycle of *Cotylaspis insignis*. *Proc. Helminthol. Soc. Wash.*, **51**, 270–4.

Huehner, M. (1987) Field and laboratory determination of substrate preferences of unionid mussels. *Ohio J. Sci.*, **87**, 29–32.

Huehner, M. and Etges, F. (1981) Encapsulation of *Aspidogater conchicola* (Trematoda: Aspidogastrea) by unionid mussels. *J. Invertebr. Pathol.*, **37**, 123–8.

Huehner, M., Hannan, K. and Garvin, M. (1989) Feeding habits and marginal organ histochemistry of *Aspidogaster conchicola* (Trematoda: Aspidogastrea). *J. Parasitol.*, **75**, 848–52.

Huffaker, C. and Messenger, P. (1964) The concept and significance of natural control. In *Biological Control of Insect Pests and Weeds*, ed. P. Debach, pp. 74–117. London: Chapman and Hall.

Hughes, R. (1986) Theories and models of species abundance. *Am. Nat.*, **128**, 879–99.

Hughes, R. and Roberts, D. (1980) Reproductive effort of winkles (*Littorina* spp) with contrasted methods of reproduction. *Oecologia*, **47**, 130–6.

Hunter, R. (1975) Growth, fecundity, and bioenergetics in three populations of *Lymnaea palustris* from upstate New York. *Ecology*, **56**, 50–63.

Hunter, R. (1980) Effects of grazing on the quantity and quality of freshwater aufwuchs. *Hydrobiologia*, **69**, 251–9.

Hunter, R. (1990) Effects of low pH and low calcium concentration on the pulmonate snail *Planorbella trivolvis*, a laboratory study. *Can. J. Zool.*, **68**, 1578–83.

Hutchinson, G. (1959) Homage to Santa Rosalia, or why are there so many kinds of animals? *Am. Nat.*, **93**, 145–59.

Huxley, J. (1942) *Evolution: The Modern Synthesis.* New York: Harper.

Hylleberg, J. (1976) Resource partitioning on basis of hydrolytic enzymes in deposit-feeding mud snails (Hydrobiidae). *Oecologia*, **23**, 116–25.

Imlay, M. and Paige, M. (1972) Laboratory growth of freshwater sponges, unionid mussels, and sphaeriid clams. *Prog. Fish-Cult.*, **34**, 210–6.

Ismail, N. and Arif, A. (1993) Population dynamics of *Melanoides tuberculata* (Thiaridae) snails in a desert spring, United Arab Emirates, and infection with larval trematodes. *Hydrobiologia*, **257**, 57–64.

Isom, B. and Hudson, R. (1982) *In vitro* culture of parasitic freshwater mussel glochidia. *Nautilus*, **96**, 147–51.

Iwanaga, Y. (1980) The growth of young snail and fecundity of *Oncomelania hupensis hupensis* in the laboratory. *Venus*, **39**, 63–8.

Jackson, D., Somers, K. and Harvey, H. (1989) Similarity coefficients: Measures of co-occurrence and association or simply measures of occurrence? *Am. Nat.*, **133**, 436–53.

Jacob, J. (1957a) Cytological studies of Melaniidae (Mollusca) with special reference to parthenogenesis and polyploidy. I. Oogenesis of the parthenogenetic species of *Melanoides. Trans. Roy. Soc. Edin.*, **63**, 341–51.

Jacob, J. (1957b) Cytological studies of Melaniidae (Mollusca) with special reference to parthenogenesis and polyploidy. II. A study of meiosis in the rare males of the polyploid race of *Melanoides tuberculata* and *Melanoides lineatus. Trans. Roy. Soc. Edin.*, **63**, 433–43.

Jacob, J. (1959a) The chromosomes of six melaniid snails. (Gastropoda: Prosobranchia). *Cytologia*, **24**, 487–97.

Jacob, J. (1959b) Sex chromosomes in melaniid snails. I. *Paludomus tanschaurica* (Gremlin) (Prosobranchia: Gastropoda). *J. Zool. Soc. India*, **11**, 17–25.

Jacoby, J. (1985) Grazing effects on periphyton by *Theodoxus fluviatilis* (Gastropoda) in a lowland stream. *J. Freshwat. Ecol.*, **3**, 265–72.

James, M. (1987) Ecology of the freshwater mussel *Hyridella menziesi* (Gray) in a small oligotrophic lake. *Arch. Hydrobiol.*, **108**, 337–48.

Janataeme, S., Uptham, E. and Kruatrachue, M. (1983) Effects of food on growth, fecundity and survival of *Indoplanorbis exustus* (Pulmonata: Planorbidae). *Malacol. Rev.*, **16**, 59–62.

Jansen, W. (1991) Seasonal prevalence, intensity of infestation, and distribution of glochidia of *Anodonta grandis simpsoniana* Lea on yellow perch, *Perca flavescens*. *Can. J. Zool.*, **69**, 964–72.

Jansen, W. and Hanson, J. (1991) Estimates of the number of glochidia produced by clams (*Anodonta grandis simpsoniana* Lea) attaching to yellow perch (*Perca flavescens*), and surviving in various ages in Narrow Lake, Alberta. *Can. J. Zool.*, **69**, 973–7.

Jarne, P. (1995) Mating system, bottlenecks and genetic polymorphism in hermaphroditic animals. *Genet. Res.*, **65**, 193–207.

Jarne, P. and Delay, B. (1990) Inbreeding depression and self-fertilization in *Lymnaea peregra* (Gastropoda: Pulmonata). *Heredity*, **64**, 169–75.

Jarne, P., Finot, L., Delay, B. and Thaler, L. (1991) Self-fertilization versus cross-fertilization in the hermaphrodite freshwater snail *Bulinus globosus*. *Evolution*, **45**, 1136–46.

Jewell, D. (1931) Observation on reproduction in the snail *Goniobasis*. *Nautilus*, **44**, 115–19.

Jeyarasasingam, U., Heyneman, D., Lim, H. and Mansour, N. (1972) Life cycle of a new echinostome from Egypt, *Echinostoma liei* sp. nov. (Trematoda: Echinostomatidae). *Parasitol.*, **65**, 203–22.

Jobin, W. and Ippen, A. (1964) Ecological design of irrigation canals for snail control. *Science*, **145**, 1324–6.

Jobin, W., Brown, R., Velez, S. and Ferguson, F. (1977) Biological control of *Biomphalaria glabrata* in major reservoirs of Puerto Rico. *Am. J. Trop. Med. Hyg.*, **26**, 1018–24.

Jobin, W., Ferguson, F. and Berrios-Duran, L. (1973) Effect of *Marisa cornuarietis* on populations of *Biomphalaria glabrata* in farm ponds of Puerto Rico. *Am. J. Trop. Med. Hyg.*, **22**, 278–84.

Johnson, B. (1952) Ciliary feeding in *Pomacea paludosa*. *Nautilus*, **66**, 3–5.

Johnson, P. and Brown, K. (1997) The role of current and light in explaining the habitat distribution of the lotic snail *Elimia semicarinata* (Say). *J. N. Am. Benthol. Soc.*, **16**, 545–61.

Johnson, S. (1992) Spontaneous and hybrid origins of parthenogenesis in *Campeloma decisum* (freshwater prosobranch snail). *Heredity*, **68**, 253–61.

Jokela, J. and Mutikainen, P. (1995) Effect of size-dependent muskrat (*Ondatra zibethica*) predation on the spatial distribution of a freshwater clam, *Andonata piscinalis* Nilsson (Unionidae, Bivalvia). *Can. J. Zool.*, **73**, 1085–94.

Jokela, J., Lively, C., Dybdahl, M. and Fox, J. (1997) Evidence for a cost of sex in the freshwater snail *Potamopyrgus antipodarum*. *Ecology*, **78**, 452–60.

Jokinen, E. (1983) *The Freshwater Snails of Connecticut*. Hartford, Connecticut: State Geology and Natural History Survey Bulletin 109. 83 pp.

Jokinen, E. (1985) Comparative life history patterns within a littoral zone snail community. *Verh. Int. Verein., Limnol.*, **22**, 3292–399.

Jokinen, E. (1987) Structure of freshwater snail communities: Species-area relationships and incidence categories. *Am. Malacol. Bull.*, **5**, 9–19.

Jokinen, E., Guerette, J. and Kortmann, R. (1982) The natural history of an ovoviviparous snail, *Viviparus georgianus* (Lea), in a soft-water eutrophic lake. *Freshwat. Invertebr. Biol.*, **1**, 1–17.

Jonasson, P. (1972) Ecology and production of the profundal benthos in relation to phytoplankton in Lake Esrom. *Oikos Suppl.*, **14**, 1–148.

Jones, H., Simpson, R. and Humphrey, C. (1986) The reproductive cycles and glochidia of fresh-water mussels (Bivalvia: Hyriidae) of the Macleay river, northern New South Wales. *Malacologia*, **27**, 185–202.

Jorgensen, C. (1966) *Biology of Suspension Feeding.* Oxford: Pergamon Press.

Jorgensen, C. (1975) Comparative physiology of suspension feeding. *Ann. Rev. Physiol.*, **37**, 57–79.

Jorgensen, C. (1983) Fluid mechancial aspects of suspension feeding. *Mar. Ecol. Prog. Ser.*, **11**, 89–103.

Joubert, P. and DeKock, K. (1990) Interaction in the laboratory between *Helisoma duryi*, a possible competitor snail, and *Biomphalaria pfeifferi*, snail host of *Schistosoma mansoni. Ann. Trop. Med. Parasitol.*, **84**, 355–9.

Joubert, P., Kruger, F., Pretorius, S. and DeKock, K. (1992) An attempt to establish *Helisoma duryi*, a possible competitor of intermediate snail hosts of schistosomiasis, in natural habitats in South Africa. *Ann. Trop. Med. Parasitol.*, **86**, 569–70.

Jude, D. (1973) Food and feeding habits of Gizzard Shad in Pool 19, Mississippi River. *Trans. Am. Fish. Soc.*, **102**, 378–83.

Kagan, I. and Geiger, S. (1965) The susceptibility of three strains of *Australorbis glabratus* to *Schistosoma mansoni* from Brazil and Puerto Rico. *J. Parasitol.*, **51**, 622–7.

Karatayev, A., Burlakova, L. and Padilla, D. (1997) The effects of *Dreissena polymorpha* (Pallas) invasion on aquatic communities in eastern Europe. *J. Shellfish Res.*, **16**, 187–203.

Karlin, A., Vail, V. and Heard, W. (1980) Parthenogenesis and biochemical variation in southeastern *Campeloma geniculum* (Gastropoda: Viviparidae). *Malacol. Rev.*, **13**, 7–15.

Karna, D. and Millemann, R. (1978) Glochidiosis of salmonid fishes. III Comparative susceptibility to natural infection with *Margaritifera margaritifera* (L.) (Pelecypoda: Margaritanidae) and associated histopathology. *J. Parasitol.*, **64**, 528–37.

Kat, P. (1982a) Effects of population density and substratum type on growth and migration of *Elliptio complanata* (Bivalvia: Unionidae). *Malacol. Rev.*, **15**, 119–27.

Kat, P. (1982b) Shell dissolution as a significant cause of mortality for *Corbicula fluminea* (Bivalvia: Corbiculidae) inhabiting acidic waters. *Malacol. Rev.*, **15**, 129–34.

Kat, P. (1983) Sexual selection and simultaneous hermaphroditism among the Unionidae (Bivalvia: Mollusca). *J. Zool.*, **201**, 395–416.

Kat, P. (1984) Parasitism and the Unionacea (Bivalvia). *Biol. Rev.*, **59**, 189–207.

Katsigianis, T. and Harman, W. (1974) Ecological notes on the trematode parasites of *Helisoma anceps* (Menke) in a eutrophic lake, including a checklist of the cercariae that occur in mollusks indigenous to Otsego County, NY. *Sterkiana*, **55**, 39–54.

Keawjam, R. (1987) The apple snails of Thailand: Aspects of comparative anatomy. *Malacol. Rev.*, **20**, 69–90.

Kehde, P. M. and Wilhm, J. L. (1972) The effects of grazing by snails on community structure of periphyton in laboratory streams. *Am. Midl. Nat.*, **87**, 8–24.

Keller, G. and Ribi, G. (1993) Fish predation and offspring survival in the prosobranch snail *Viviparus ater*. *Oecologia*, **93**, 493–500.

Kendall, S. (1953) The life history of *Limnaea truncatula* under laboratory conditions. *J. Helminth.*, **27**, 17–28.

Kendall, S. and Parfitt, J. (1965) The life history of some vectors of *Fasciola gigantica* under laboratory conditions. *Ann. Trop. Med. Parasitol.*, **59**, 10–6.

Kenkel, N. and Orloci, L. (1986) Applying metric and nonmetric multidimensional scaling to ecological studies: Some new results. *Ecology*, **67**, 919–28.

Kenmuir, D. (1981) Repetitive spawning behavior in two species of freshwater mussels (Lamellibranchiata: Unionacea) in Lake Kariba. *Trans. Zimbabwe Sci.Assoc.*, **60**, 49–56.

Kennedy, V. and Blundon, J. (1983) Shell strength in *Corbicula* sp (Bivalvia: Corbiculidae) from the Potomac River, Maryland. *Veliger*, **26**, 22–5.

Kerfoot, W. and Sih, A. (1987) *Predation: Direct and Indirect Impacts on Aquatic Communities.* Hanover, NH: University Press of New England.

Kerney, M. and Morton, B. (1970) The distribution of *Dreissena polymorpha* (Pallas) in Britain. *J. Conchol.*, **27**, 97–100.

Kershner, M. and Lodge, D. (1990) Effect of substrate architecture on aquatic gastropod-substrate associations. *J. N. Am. Benthol Soc.*, **9**, 319–26.

Kesler, D. (1981) Periphyton grazing by *Amnicola limosa*: An enclosure-exclosure experiment. *J. Freshwat. Ecol.*, **1**, 51–9.

Kesler, D. (1983a) Variation in cellulase activity in *Physa heterostropha* (Gastropoda) and other species of gastropods in a New England pond. *Am. Midl. Nat.*, **109**, 280–8.

Kesler, D. (1983b) Cellulase activity in gastropods: Should it be used in niche separation? *Freshwat. Invertebr. Biol.*, **2**, 173–9.

Kesler, D. and Bailey, R. (1993) Density and ecomorphology of a freshwater mussel (*Elliptio complanata*, Bivalvia: Unionidae) in a Rhode Island lake. *J. N.Am. Benthol. Soc.*, **12**, 259–64.

Kesler, D. and Munns, W., Jr. (1989) Predation by *Belostoma flumineum* (Hemiptera) an important cause of mortality in freshwater snails. *J. N. Am. Benthol. Soc.*, **8**, 342–50.

Kesler, D. and Tulou, C. (1980) Cellulase activity in the freshwater gastropod *Amnicola limosa*. *Nautilus*, **94**, 135–7.

Kesler, D. H., Jokinen, E. H. and Munns, W. R. (1986) Trophic preferences and feeding morphology of two pulmonate snail species from a small New England pond, U.S.A. *Can. J. Zool.*, **64**, 2570–5.

Khalil, L. (1961) On the capture and destruction of miracidia by *Chaetogaster limnaei* (Oligochaeta). *J. Helminthol.*, **35**, 269–74.

Khan, R. and Chaudhuri, S. (1984) The population and production ecology of a freshwater snail *Bellamya bengalensis* (Lamarck) (Gastropoda: Viviparidae) in an artificial lake of Calcutta, India. *Bull. Zool. Surv. India*, **5**, 59–76.

Kijviriya, V., Upatham, E., Viyanant, V. and Woodruff, D. (1991) Genetic studies of asiatic clams, *Corbicula* in Thailand: Allozymes of 21 nominal species are identical. *Am. Malacol. Bull.*, **8**, 97–106.

Kilgour, B. and Mackie, G. (1988) Factors affecting the distribution of sphaeriid bivalves in Britannia Bay, of the Ottawa River. *Nautilus*, **102**, 73–7.

Kilgour, B. and Mackie, G. (1991) Relationships between demographic features of a pill clam (*Pisidium casertanum*) and environmental variables. *J. N. Am. Benthol. Soc.*, **10**, 68–80.

King, C., Langdon, C. and Counts, C., III. (1986) Spawning and early development of *Corbicula fluminea* (Bivalvia: Corbiculidae) in laboratory culture. *Am. Malacol. Bull.*, **4**, 81–8.

Kiørbe, T. and Møhlenberg, F. (1981) Particle selection in suspension-feeding bivalves. *Mar. Ecol. Prog. Ser.*, **5**, 291–6.

Kitikoon, V. (1981a) Studies on *Tricula aperta* and related taxa, the snail intermediate hosts of *Schistosoma mekongi*. II. Methods for collecting, culturing, and maintenance. *Malacol. Rev.*, **14**, 11–35.

Kitikoon, V. (1981b) Studies on *Tricula aperta* and related taxa, the snail intermediate hosts of *Schistosoma mekongi*. III. Susceptibility studies. *Malacol. Rev.*, **14**, 37–42.

Klemm, D. (1975) Studies on the feeding relationships of leeches (Annelida: Hirudinea) as natural associates of mollusks. *Sterkiana*, **58 and 59**, 1–50, 1–20.

Knecht, A. and Walter, J. (1977) Vergleichende untersuchung der Diaten von *Lymnaea auricularia* und *L. peregra* (Gastropoda: Basommatophora) im Zurichsee. *Schweiz. Z. Hydrol.*, **39**, 299–305.

Kofron, C. and Schreiber, A. (1985) Ecology of two endangered aquatic turtles in Missouri, USA, *Kinosternon flavescens* and *Emydoidea blandingii*. *J. Herpetol.*, **19**, 27–40.

Komaru, A., Konishi, K., Nakayama, I., Kobayashi, T., Sakai, H. and Kawamura, K. (1997) Hermaphroditic freshwater clams in the genus *Corbicula* produce non-reductional spermatozoa with somatic DNA content. *Biol. Bull. (Woods Hole)*, **193**, 320–3.

Komiya, Y. (1961) The ecology of *Oncomelania nosophora*: A review. *Jap. J. Med. Sci. Biol.*, **14**, 1–9.

Kondo, A. and Tanaka, F. (1989) An experimental study of predation by the larvae of the firefly *Luciola lateralis* Mot. (Coleoptera: Lampyridae) on the apple snail *Pomacea canaliculata* Lam. (Mesogastropoda: Pilidae). *Jap. J. Appl. Ent. Zool.*, **33**, 211–6.

Kondo, T. (1987) Breeding seasons of seven species of unionid mussels (Bivalvia: Unionidae). *Venus*, **46**, 227–36.

Kondo, T. (1990) Reproductive biology of a small bivalve *Grandidieria burtoni* in Lake Tanganyika. *Venus*, **49**, 120–5.

Kraemer, L. (1970) The mantle flap in three species of *Lampsilis* (Pelecypoda: Unionidae). *Malacologia*, **10**, 225–82.

Kraemer, L. (1979) *Corbicula* (Bivalvia: Sphaeriacea) vs. indigenous mussels (Bivalvia: Unionacea) in U.S. rivers: A hard case for interspecific competition? *Am. Zool.*, **19**, 1085–96.

Kraemer, L. and Galloway, M. (1986) Larval development of *Corbicula fluminea*

(Muller) (Bivalvia: Corbiculacea): An appraisal of its heterochrony. *Am. Malacol. Bull.*, **4**, 61–79.

Kraemer, L., Swanson, C., Galloway, M. and Kraemer, R. (1986) Biological basis of behavior in *Corbicula fluminea*. II. Functional morphology of reproduction and development and review of evidence for self-fertilization. *Am. Malacol. Bull. Special Ed.*, **2**, 193–201.

Krebs, C. (1991) The experimental paradigm and long-term population studies. *Ibis*, **133**, 3–8.

Krebs, C. (1992) Population regulation revisited. *Ecology*, **73**, 714–15.

Krecker, F. (1924) Conditions under which *Goniobasis livescens* occurs in the island regon of Lake Erie. *Ohio J. Sci.*, **24**, 299–310.

Krieger, K. and Burbanck, W. (1976) Distribution and dispersal mechanisms of *Oxytrema* (= *Goniobasis*) *suturalis* Hald. (Gastropoda: Pleuroceridae) in the Yellow River, Georgia, U.S.A. *Am. Midl. Nat.*, **95**, 49–63.

Kruatrachue, M., Ratanatham, S., Jantataeme, S. and Upatham, E. (1982) Effects of various types of food and mud preparations on survivorship and egg production of *Bithynia siamensis siamensis* (Prosobranchia: Bithyniidae). *Malacol. Rev.*, **15**, 59–62.

Kryger, J. and Riisgard, H. (1988) Filtration rate capacities in six species of European freshwater bivalves. *Oecologia*, **77**, 34–8.

Kuris, A. (1980) Effect of exposure to *Echinostoma liei* miracidia on growth and survival of young *Biomphalaria glabrata* snails. *Int. J. Parasitol.*, **10**, 303–8.

Kushlan, J. (1975) Population changes of the apple snail, *Pomacea paludosa*, in the southern Everglades. *Nautilus*, **89**, 21–3.

Lack, D. (1947) *Darwin's Finches*. Cambridge: Cambridge University Press.

Lam, P. and Calow, P. (1989a) Intraspecific life-history variation in *Lymnaea peregra* (Gastropoda: Pulmonata). I. Field Study. *J. Anim. Ecol.*, **58**, 571–88.

Lam, P. and Calow, P. (1989b) Intraspecific life-history variation in *Lymnaea peregra* (Gastropoda: Pulmonata). II. Environmental or genetic variance. *J. Anim. Ecol.*, **58**, 589–602.

Laman, T., Boss, N. and Blankespoor, H. (1984) Depth distribution of seven species of gastropods in Douglas Lake, Michigan. *Nautilus*, **98**, 20–4.

Lamarche, A., Legendre, P. and Chodorowski, A. (1982) Facteurs responsables la distribution des gasteropodes dulcicoles dans le fleuve Saint-Laurent. *Hydrobiologia*, **89**, 61–76.

Lamberti, G., Ashkenas, L., Gregory, S. and Steinman, A. (1987) Effects of three herbivores on periphyton communities in laboratory streams. *J. N. Am. Benthol. Soc.*, **6**, 92–104.

Lamberti, G., Gregory, S., Askenas, L., Steinman, A. and McIntire, C. (1989) Productive capacity of periphyton as a determinant of plant-herbivore interactions in streams. *Ecology*, **70**, 1840–56.

Lassen, H. (1975) The diversity of freshwater snails in view of the equilibrium theory of island biogeography. *Oecologia*, **19**, 1–8.

Lassen, J. and Madsen, H. (1986) The role of food abundance in the interspecific competition between *Helisoma duryi* and *Bulinus truncatus* (Gastropoda: Planorbidae). In *Proceedings of the Eighth International Malacological Congress*, ed. L. Pinter, pp. 165–8. Budapest, Hungary: Hungarian Natural History Museum.

Laughlin, D. and Werner, E. (1980) Resource partitioning in two coexisting sunfish: Pumpkinseed (*Lepomis gibbosus*) and northern longear sunfish (*Lepomis megalotis peltastes*). *Can. J. Fish. Aquat. Sci.*, **37**, 1411–20.

Lauritsen, D. (1986a) Assimilation of radiolabeled algae by *Corbicula*. *Am. Malacol. Bull. Special Edition*, **2**, 219–22.

Lauritsen, D. (1986b) Filter-feeding in *Corbicula fluminea* and its effects on seston removal. *J. N. Am. Benthol. Soc.*, **5**, 165–72.

Lavrentyev, P., Gardner, W., Cavaletto, J. and Beaver, J. (1995) Effects of the zebra mussel (*Dreissena polymorpha*) on protozoa and phytoplankton from Saginaw Bay, Lake Huron. *J. Great Lakes Res.*, **21**, 545–57.

Lefevre, G. and Curtis, W. (1912) Studies on the reproduction and artificial propagation of freshwater mussels. *Bull. U.S. Bur. Fish.*, **30**, 105–201.

Leff, L., Burch, J. and McArthur, J. (1990) Spatial distribution, seston removal, and potential competitive interactions of the bivalves *Corbicula fluminea* and *Elliptio complanata*, in a coastal plain stream. *Freshwat. Biol.*, **24**, 409–16.

Leveque, C. (1972) The benthic mollusks of Lake Chad ecology study of growth and estimation of biomasses. *Cah. ORSTROM Ser. Hydrobiol.*, **6**, 3–45.

Leveque, C. and Gaborit, M. (1972) Utilization of the factorial analysis of the correspondences for the study of the population of benthic mollusks in Lake Chad. *Cah. ORSTROM Ser. Hydrobiol.*, **7**, 47–66.

Levy, M., Tunis, M. and Isserhoff, H. (1973) Population control in snails by natural inhibitors. *Nature*, **241**, 65–6.

Lewandowski, K. (1976) Unionidae as a substratum for *Dreissena polymorpha*. *Pol. Arch. Hydrobiol.*, **23**, 409–20.

Lewandowski, K. (1982) O zmiennij liczebnosci matza *Dreissena polymorpha*. *Wiad. Ekol.*, **28**, 141–54.

Lewandowski, K. (1991) Long-term changes in the fauna of family Unionidae bivalves in the Mikolajskie Lake. *Ekol. Pol.*, **39**, 265–72.

Lewandowski, K. and Stanczykowska, A. (1975) The occurrence and role of bivalves of the family Unionidae in Mikolojskie Lake. *Ekol. Pol.*, **23**, 317–34.

Lewert, R. and Para, B. (1966) The physiological incorporation of Carbon-14 in *Schistosoma mansoni* cercariae. *J. Infect. Dis.*, **116**, 171–82.

Lewis, J. and Riebel, P. (1984) The effect of substrate on burrowing in freshwater mussels (Unionidae). *Can. J. Zool.*, **62**, 2023–5.

Liang, Y. (1974) Cultivation of *Bulinus* (*Physopsis*) *globosus* and *Biomphalaria pfeifferi pfeifferi*, snail hosts of schistosomiasis. *Sterkiana*, **53/54**, 1–75.

Liang, Y. and Kitikoon, V. (1980) Cultivation of *Lithoglyphopsis aperta*, snail vector of *Schistosoma mekongi*. *Malacol. Rev. Suppl.*, **2**, 35–45.

Liang, Y. and van der Schalie, H. (1975) Cultivating *Lithoglyphopsis aperta* Temcharoen, a new snail host for *Schistosoma japonicum*, Mekong strain. *J. Parasitol.*, **61**, 915–19.

Lie, K., Joeng, K. & Heyneman, D. (1987) Molluscan host reactions to helminthic infections. In *Immune Responses in Parasitic Infections*, ed. E. Soulsby, pp. 211–70. Boca Raton, FL: CRC Press.

Liebig, J. and van der Ploeg, H. (1995) Vulnerability of *Dreissena polymorpha* larvae to predation by Great Lakes calanoid copepods: The importance of the bivalve shell. *J. Great Lakes Res.*, **21**, 353–8.

Liem, K. (1980) Adaptive significance of intra- and interspecific differences in the feeding repertoires of cichlid fishes. *Am. Zool.*, **20**, 295–314.

Lilly, M. (1953) The mode of life and the structure and functioning of the reproductive ducts of *Bithynia tentaculata* (L.). *Proc. Malacol. Soc. Lond.*, **30**, 87–110.

Lim, H. and Heyneman, D. (1972) Intramollusca inter-trematode antagonism: A review of factors influencing the host-parasite system and its possible role in biological control. *Adv. Parasitol.*, **10**, 191–268.

Little, C. (1981) Osmoregulation and excretion in prosobranch gastropods. Part I. Physiology and biochemistry. *J. Molluscan Stud.*, **47**, 221–47.

Little, C. (1985) Renal adaptations of prosobranchs to the freshwater environment. *Am. Malacol. Bull.*, **3**, 223–31.

Liu, H., Mitton, J. and Wu, S. (1996) Paternal mitochondrial DNA differentiation far exceeds maternal mitochondrial DNA and allozyme differentiation in the freshwater mussel, *Anodonta grandis grandis*. *Evolution*, **50**, 952–7.

Lively, C. (1989) Adaptation by a parasitic trematode to local populations of its snail host. *Evolution*, **43**, 1663–71.

Lively, C. (1992) Parthenogenesis in a freshwater snail: Reproductive assurance versus parasitic release. *Evolution*, **46**, 907–13.

Livshits, G. and Fishelson, L. (1983) Biology and reproduction of the freshwater snail *Melanoides tuberculata* (Gastropoda: Prosobranchia) in Israel. *Isr. J. Zool.*, **32**, 21–35.

Livshits, G., Fishelson, L. and Wise, G. (1984) Genetic similarity and diversity of parthenogenetic and bisexual populations of the freshwater snail *Melanoides tuberculata* (Gastropoda: Prosobranchia). *Biol. J. Linn. Soc.*, **23**, 41–54.

Llewellyn, J. (1965) The evolution of parasitic helminths. In *Evolution of Parasites*, ed. A. Taylor, pp. 47–8. Oxford: Blackwell Scientific.

Lodes, M., Connors, V. and Yoshino, T. (1991) Isolation and functional characterization of snail hemocyte-modulating polypeptide from primary sporocysts of *Schistosoma mansoni*. *Molec. Bioch. Parasitol.*, **49**, 1–10.

Lodge, D. (1985) Macrophyte–gastropod associations: Observations and experiments on macrophyte choice by gastropods. *Freshwat. Biol.*, **15**, 695–708.

Lodge, D. (1986) Selective grazing on periphyton: A determinant of freshwater gastropod microdistributions. *Freshwat. Biol.*, **16**, 831–41.

Lodge, D. and Kelly, P. (1985) Habitat disturbance and the stability of freshwater gastropod populations. *Oecologia*, **68**, 111–7.

Lodge, D. and Lorman, J. (1987) Reductions in submersed macrophyte biomass and species richness by the crayfish *Orconectes rusticus*. *Can. J. Fish. Aquat. Sci.*, **44**, 591–7.

Lodge, D., Brown, K., Klosiewski, S., Stein, R., Covich, A., Leathers, B. and Bronmark, C. (1987) Distribution of freshwater snails: Spatial scale and relative importance of physicochemical and biotic factors. *Am. Malacol. Bull.*, **5**, 73–84.

Lodge, D., Kershner, M., Aloi, J. and Covich, A. (1994) Effects of an omnivorous crayfish (*Orconectes rusticus*) on a freshwater littoral food web. *Ecology*, **75**, 1265–81.

Lohachit, C., Sornmani, S. and Butrcham, P. (1980) Development and maintenance of *Lithoglyphopsis aperta* in the laboratory. *Malacol. Rev. Suppl.*, **2**, 19–34.

Loker, E. (1983) A comparative study of the life histories of mammalian schistosomes. *Parasitol.*, **87**, 343–69.

Loker, E., Bayne, C. and Yul, M. (1986) *Echinostoma paraensei* hemocytes of

Biomphalaria glabrata as targets of echinostome mediated interference with host snail resistance to *Schistosoma mansoni. Exp. Parasitol.*, **62**, 149–54.

Lomolino, M. (1989) Interpretations and comparisons of constants in the species-area relationship: An additional caution. *Am. Nat.*, **133**, 277–80.

Lopez, G. and Holopainen, I. (1987) Interstitial suspension-feeding by *Pisidium* spp. (Bivalvia: Pisidiidae): A new guild in the lentic benthos? *Am. Malacol. Bull.*, **5**, 21–30.

Loreau, M. and Baluku, B. (1987a) Growth and demography of populations of *Biomphalaria pfeifferi* (Gastropoda, Pulmonata) in the laboratory. *J. Molluscan Stud.*, **53**, 171–18.

Loreau, M. and Baluku, B. (1987b) Population dynamics of the freshwater snail *Biomphalaria pfeifferi* in eastern Zaire. *J. Molluscan Stud.*, **53**, 249–65.

Love, J. and Savino, J. (1993) Crayfish (*Orconectes virilis*) predation on zebra mussels (*Dreissena polymorpha*). *J. Freshwat. Ecol.*, **8**, 253–9.

Lowe, R. L. and Hunter, R. D. (1988) Effect of grazing by *Physa integra* on periphyton community structure. *J. N. Am. Benthol. Soc.*, **7**, 29–36.

Lowe, R. & Pillsbury, R. (1995) Shifts in benthic algal community structure and function following the appearance of zebra mussels (*Dreissena polymorpha*) in Saginaw Bay, Lake Huron. *J. Great Lakes Res.*, **21**, 558–66.

Lum-Kong, A. and Kenny, J. (1989) The reproductive biology of the ampullariid snail *Pomacea urceus* (Müller). *J. Molluscan Stud.*, **55**, 53–66.

Lydeard, C., Mulvey, M. and Davis, G. (1996) Molecular systematics and evolution of reproductive traits of North American freshwater Unionacean mussels (Mollusca: Bivalvia) as inferred from 16S rRNA gene sequences. *Phil. Trans. R. Soc. Lond. B.*, **351**, 1593–603.

Mabbott, D. (1920) *Food Habits of Seven Shoal-Water Ducks*. Washington, D.C.: U.S. Dept. Agriculture Bulletin 862.

Macan, T. (1950) Ecology of freshwater mollusca in the English Lake District. *J. Anim. Ecol.*, **19**, 124–46.

Macan, T. (1961) Factors that limit the range of freshwater animals. *Biol. Rev.*, **36**, 151–98.

Macan, T. (1974) *Freshwater Ecology*, Second Ed. New York: John Wiley and Sons.

MacArthur, R. and Wilson, E. (1967) *The Theory of Island Biogeography*. New Jersey: Princeton University Press.

Machin, J. (1975) Water relationships. In *Pulmonates*, ed. V. Fretter and J. Peake, pp. 105–63. New York: Academic Press.

MacIsaac, H. (1994) Size-selective predation on zebra mussels (*Dreissena polymorpha*) by crayfish (*Orconectes propinquus*). *J. N. Am. Benthol. Soc.*, **13**, 206–16.

MacIsaac, H. (1996) Potential abiotic and biotic impacts of zebra mussels on the inland waters of North America. *Am. Zool.*, **36**, 287–99.

Mackie, G. (1976a) Trematode parasites in the Sphaeriidae clams, and the effects on three Ottawa River species. *Nautilus*, **90**, 36–41.

Mackie, G. (1976b) Sphaeriidae natality and its significance in water quality studies. *Bull. Am. Malacol. Union*, **1976**, 5–9.

Mackie, G. (1979) Growth dynamics in natural populations of Sphaeriidae clams (*Sphaerium, Musculium, Pisidium*). *Can. J. Zool.*, **57**, 441–56.

Mackie, G. (1984a) Bivalves. In *The Mollusca*, Vol. 7, *Reproduction*, ed. A. Tompa, N. Verdunk and J. vandenBiggelaar, pp. 351–418. Orlando: Academic Press.

Mackie, G. (1984b) Some effects of lake acidification on Pisidiidae in southern Ontario, Canada. *Soosiana*, **12**, 103–16.

Mackie, G. (1991) Biology of the exotic zebra mussel, *Dreissena polymorpha*, in relation to native bivalves and its potential impact in Lake St. Clair. *Hydrobiologia*, **219**, 251–68.

Mackie, G. (1993) Biology of the zebra mussel (*Dreissena polymorpha*) and observations of mussel colonization on unionid bivalves in Lake St. Clair of the Great Lakes. In *Zebra Mussels: Biology, Impacts, and Control*, ed. T. Nalepa and D. Schloesser, pp. 153–65. Boca Raton: Lewis Publishers.

Mackie, G. and Qadri, S. (1978) Effects of substratum on growth and reproduction of *Musculium securis* (Bivalvia: Sphaeriidae). *Nautilus*, **92**, 135–46.

Mackie, G., Qadri, S. and Reed, R. (1978) Significance of litter size in *Musculium securis* (Bivalvia, Sphaeriidae). *Ecology*, **59**, 1069–74.

Madenjian, C. (1995) Removal of algae by the zebra mussel (*Dreissena polymorpha*) population in western Lake Erie: A bioeneregetics approach. *Can. J. Fish. Aquat. Sci.*, **52**, 381–90.

Madsen, H. (1979a) Further laboratory studies on the interspecific competition between *Helisoma duryi* (Weth) and the intermediate hosts of *S. mansoni* : *Biomphalaria alexandrina* and *B. camerunensis*. *Hydrobiologia*, **66**, 181–92.

Madsen, H. (1979b) Preliminary observations on the role of conditioning and mechanical interference with egg masses and juveniles in the competitive relationships between *Helisoma duryi* (Wetherby) and the intermediate host of *Schistosoma mansoni* Sambon: *Biomphalaria camerunensis* (Boettger). *Hydrobiologia*, **67**, 207–14.

Madsen, H. (1982) Development of egg masses and growth of newly hatched snails of some species of intermediate hosts of schistosomiasis in water conditioned by *Helisoma duryi*. *Malacologia*, **22**, 427–34.

Madsen, H. (1983) Distribution of *Helisoma duryi* an introduced competitor of intermediate hosts of schistosomiasis in an irrigation scheme in northern Tanzania. *Acta Trop.*, **40**, 297–306.

Madsen, H. (1984) The effect of water conditioned by either *Helisoma duryi* or *Biomphalaria camerunensis* on the growth and reproduction of juvenile *Biomphalaria camerunensis* (Pulmonata: Planorbidae). *J. Appl. Ecol.*, **21**, 757–72.

Madsen, H. (1986) Studies on various aspects of food competition and interference between *Helisoma duryi* and *Bulinus truncatus* (Gastropoda: Planorbidae). In *Proceedings of the Eighth International Malacological Congress*, ed. L. Pinter, pp. 147–50. Budapest, Hungary: Hungarian Natural History Museum.

Madsen, H. (1987) Effect of calcium concentration on growth and egg laying of *Helisoma duryi*, *Biomphalaria alexandrina*, *Biomphalaria camerunensis* and *Bulinus truncatus* (Gastropoda: Planorbidae). *J. Appl. Ecol.*, **24**, 823–36.

Madsen, H. (1992) A comparative study on the food-locating ability of *Helisoma duryi*, *Biomphalaria camerunensis* and *Bulinus truncatus* (Pulmonata, Planorbidae). *J. Appl. Ecol.*, **29**, 70–8.

Madsen, H. and Frandsen, F. (1979) Studies on the interspecific competition between *Helisoma duryi* and *Biomphalaria camerunensis*. Size-weight relationships and laboratory competition experiments. *Hydrobiologia*, **66**, 17–23.

Madsen, H. and Frandsen, F. (1989) The spread of freshwater snails including those of medical and veterinary importance. *Acta Trop.*, **46**, 139–46.

Magruder, S. (1935) The anatomy of the freshwater prosobranchiate gastropod, *Pleurocera canaliculatum undulatum* (Say). *Am. Midl. Nat.*, **16**, 883–912.

Mahmoud, I. (1968) Feeding behavior in kinosternid turtles. *Herpetologica*, **24**, 300–5.

Maitland, P. (1965) The feeding relationships of salmon, trout, minnows, stone loach and three spined sticklebacks in the River Endrick, Scotland. *J. Anim. Ecol.*, **34**, 109–33.

Majumder, M. and Pal, S. (1992) Host parasite interactions between a freshwater bivalve mollusc and mites. In *Proceedings of the Tenth International Malacological Congress*, ed. C. Meier-Brook, pp. 271–5. Tübingen, Germany: Unitas Malacologia.

Malek, E. (1958) Factors conditioning the habitat of bilharziasis intermediate hosts of the family Planorbidae. *Bull. W.H.O.*, **18**, 785–818.

Malek, E. (1980) *Snail-Transmitted Parasitic Diseases*, Vol. I and II. Boca Raton, Fla.: CRC Press.

Malone, C. and Nelson, D. (1969) Feeding rates of freshwater snails *Goniobasis clavaeformis* determined with cobalt[60]. *Ecology*, **50**, 728–30.

Mancini, E. (1978) The biology of *Goniobasis semicarinata* (Say) in the Mosquito Creek drainage system, southern Indiana. Ph. D. Dissertation, p. 93. University of Louisville, KY.

Manguin, S., Vala, J. and Reidenbach, J. (1988a) Predator action of *Tetanocera ferruginea* larvae (Diptera: Sciomyzidae) in freshwater snails multi-prey systems. *Acta Oecol.*, **9**, 249–59.

Manguin, S., Vala, J. and Reidenbach, J. (1988b) Determination of food preference of *Tetanocera ferruginea* larvae (Diptera: Sciomyzidae) predator of freshwater snails. *Acta Oecol.*, **9**, 353–70.

Marker, A. (1976) The benthic algae of some streams in southern England. I. Biomass of the epilithon of some small streams. *J. Ecol.*, **64**, 343–58.

Marsden, J., Spidle, A. and May, B. (1996) Review of genetic studies of *Dreissena* sp. *Am. Zool.*, **36**, 259–70.

Marshall, J. (1973) Purification of a 1,4 Glucan hydrolase (cellulase) from the snail *Helix pomatia*. *Comp. Biochem. Physiol.*, **44B**, 981–8.

Martell, A. and Trdan, R. (1994) *Venustachoncha ellipsiformis* (Bivalvia: Unionidae) an intermediate host for *Phyllodistomum* (Trematoda: Gorgoderidae) in a Michigan stream. Paper. American Malacological Union. Houston, Tx.

Marti, H. (1986) Field observations on the population dynamics of *Bulinus globosus*, the intermediate host of *Schistosoma haematobium* in the Ifakara area, Tanzania. *J. Parasitol.*, **72**, 119–24.

Martin, T. (1981) Species-area slopes and coefficients: A caution on their interpretation. *Am. Nat.*, **118**, 823–937.

Martin, G. and Corkum, L. (1994) Predation of zebra mussels by crayfish. *Can. J. Zool.*, **72**, 1867–71.

Martin, T., Crowder, L., Dumas, C. and Burkholder, J. (1992) Indirect effects of fish on macrophytes in Bays Mountain Lake: Evidence for a littoral trophic cascade. *Oecologia*, **89**, 476–81.

Mattice, J. (1972) Production of a natural population of *Bithynia tentaculata* L. (Gastropoda, Mollusca). *Ekol. Pol.*, **20**, 525–39.

May, R. (1975) Patterns of species abundance and diversity. In *Ecology and Evolution of Communities*, ed. M. Cody and J. Diamond, pp. 81–120. Cambridge, MA: Belknap Press.

May, R. (1986) The search for patterns in the balance of nature: Advances and retreats. *Ecology*, **67**, 1115–26.

McClary, A. (1964) Surface inspiration and ciliary feeding in *Pomacea paludosa* (Prosobranchia: Mesogastropoda: Ampullariidae). *Malacologia*, **2**, 87–104.

McCollum, E., Crowder, L. and McCollum, S. (1998) Complex interactions of fish, snails, and littoral zone periphyton. *Ecology*, **79**, 1980–94.

McCormick, P. (1990) Direct and indirect effects of consumers on benthic algae in isolated pools of an ephemeral stream. *Can. J. Fish. Aquat. Sci.*, **47**, 2057–65.

McCormick, P. and Stevenson, R. (1989) Effects of snail grazing on benthic algal community structure in different nutrient environments. *J. N. Am. Benthol. Soc.*, **8**, 162–72.

McCormick, P. and Stevenson, R. (1991) Grazer control of nutrient availability in the periphyton. *Oecologia*, **86**, 287–91.

McCullough, F. (1981) Biological control of the snail intermediate hosts of human *Schistosoma* spp.: A review of its present status and future prospects. *Acta Trop.*, **38**, 5–13.

McDonald, S. (1973) Activity patterns of *Lymnaea stagnalis* (L.) in relation to temperature conditions: A preliminary study. *Malacologia*, **14**, 395–6.

McKaye, K., Stauffer, J. and Londa, S. (1986) Fish predation as a factor in the distribution of Lake Malawi gastropods. *Exp. Biol.*, **45**, 279–89.

McKillop, W. (1985) Distribution of aquatic gastropods across the Ordovician dolomite – Precambrian granite contact in southeastern Manitoba, Canada. *Can. J. Zool.*, **63**, 278–88.

McKillop, W. and Harrison, A. (1972) Distribution of aquatic gastropods across an interface between the Canadian shield and limestone formations. *Can. J. Zool.*, **50**, 1433–45.

McKillop, W. and Harrison, A. (1982) Hydrobiological studies of eastern Lesser Antillean Islands. VII. St. Lucia: Behavioral drift and other movements of freshwater marsh mollusks. *Arch. Hydrobiol.*, **94**, 53–69.

McKillop, W., Harrison, A. and Rankin, J. (1981) Hydrobiological studies of eastern Lesser Antillean Islands. VI. St. Lucia: Freshwater mollusks and the marsh environment. *Arch. Hydrobiol. Suppl.*, **58**, 357–419.

McLeod, M. (1986) Electrophoretic variation in North American *Corbicula*. *Am. Malacol. Bull. Sp. Ed.*, **2**, 125–32.

McLeod, M., Hornbach, D., Guttman, S., Way, C. and Burky, A. (1981) Environmental heterogeneity, genetic polymorphism, and reproductive strategies. *Am. Nat.*, **118**, 129–34.

McMahon, R. (1975) Growth, reproduction and bioenergetic variation in three natural populations of a freshwater limpet *Laevapex fuscus* (Adams). *Proc. Malacol. Soc. Lond.*, **41**, 331–51.

McMahon, R. (1980) Life-cycles of four species of freshwater snails from Ireland. *Am. Zool.*, **20**, 927.

McMahon, R. (1983a) Physiological ecology of freshwater pulmonates. In *The Mollusca*, Vol. 6, *Ecology*, ed. W. Russell-Hunter, pp. 359–430. New York: Academic Press.

McMahon, R. (1983b) Ecology of an invasive pest bivalve, *Corbicula*. In *The Mollusca*,Vol. 6, *Ecology*, ed. W. Russell-Hunter, pp. 505–61. New York: Academic Press.

Meenakshi, V. (1954) Studies in the physiology of digestion in *Pila virens* (Lam). *J. Anim. Morph. Physiol.*, **1**, 35–47.

Meier, M. and Meier-Brook, C. (1981) *Schistosoma mansoni*: Effect on growth, fertility, and development of distal male organs in *Biomphalaria glabrata* exposed to miracidia at different ages. *Zeitschr. Parasitol.*, **66**, 121–31.

Meier-Brook, C. (1969) Substrate relations in some *Pisidium* species (Eulamellibranchiata: Sphaeriidae). *Malacologia*, **9**, 121–5.

Meier-Brook, C. (1977) Intramarsupial suppression of fetal development in sphaeriid clams. *Malacol. Rev.*, **10**, 53–8.

Meier-Brook, C. (1978) Calcium uptake by *Marisa cornuarietis* (Gastropoda: Ampullariidae), a predator of schistosome-bearing snails. *Arch. Hydrobiol.*, **82**, 449–64.

Meier-Brook, C. and Kim, C. (1977) Notes on ciliary feeding in two Korean *Bithynia* species. *Malacologia*, **16**, 159–63.

Meier-Brook, C., Haas, D., Winter, G. and Zeller, T. (1987) Hydrochemical factors limiting the distribution of *Bulinus truncatus* (Pulmonata: Planorbidae). *Am. Malacol. Bull.*, **5**, 85–90.

Meunch, H. (1959) *Catalytic Models in Epidemiology*. Cambridge, MA: Harvard University Press.

Meyer-Lassen, J. and Madsen, H. (1989) The effect of varying relative density and varying food supply on interspecific competition between *Helisoma duryi* and *Bulinus truncatus*. *J. Molluscan Stud.*, **55**, 89–96.

Meyers, T. and Millemann, R. (1977) Glochidiosis of salmonid fishes. I. Comparative susceptibility to experimental infection with *Margaritifera margaritifera* (L.) (Pelecypoda: Margaritanidae). *J. Parasitol.*, **63**, 728–33.

Meyers, T., Millemann, R. and Fustish, C. (1980) Glochidiosis of salmonid fishes. IV. Humoral and tissue responses of coho and chinook salmon to experimental infection with *Margaritifera margaritifera* (L.) (Pelecypoda: Margaritanidae). *J. Parasitol.*, **66**, 274–81.

Michelson, E. (1957) Study on the biological control of Schistosome bearing snails. Predators and parasites of fresh-water Mollusca: A review of the literature. *Parasitol.*, **47**, 413–26.

Michelson, E. and Augustine, D. (1957) Studies on the biological control of schistosome-bearing snails. V: The control of *Biomphalaria pfeifferi* populations by the snail, *Marisa cornuarietis* under laboratory conditions (note). *J. Parasitol.*, **43**, 135.

Miller, C. (1985) Correlates of habitat favourability for benthic macroinvertebrates at five stream sites in an Appalachian mountain drainage basin, USA. *Freshwat. Biol.*, **15**, 709–33.

Miller, A. and Payne, B. (1994) Co-occurrence of native freshwater mussels (Unionidae) and the non-indigenous *Corbicula fluminea* at two stable shoals in the Ohio River, USA. *Malacol. Rev.*, **27**, 87–97.

Mills, E., Rosenberg, G., Spidle, A., Ludyanskiy, M., Pligin, Y. and May, B. (1996) A review of the biology and ecology of the quagga mussel (*Dreissana bugensis*), a second species of freshwater dreissenid introduced to North America. *Am. Zool.*, **36**, 271–86.

Minchella, D. (1985) Host life-history variation in response to parasitism. *Parasitol.*, **90**, 205–16.

Minchella, D. and Loverde, P. (1981) A cost of increased early reproductive effort in the snail *Biomphalaria glabrata*. *Am. Nat.*, **118**, 876–81.

Minshall, G., Peterson, R. and Nimz, C. (1985) Species richness in streams of different size from the same drainage basin. *Am. Nat.*, **125**, 16–38.

Mitchell, R. (1965) Population regulation of a water mite parasitic on Unionid mussels. *J. Parasitol.*, **51**, 990–6.

Mitropolskii, V. (1966a) Notes on the life-cycle and nutrition of *Sphaerium corneum* L. (Mollusca Lamellibranchia). *Tr. Inst. Biol. Vnutr. Vod., Akad. Nauk SSSR*, **12**, 125–8.

Mitropolskii, V. (1966b) One mechanisms of filtration and nutrition of sphaeriids (Mollusca, Lamellibranchia). *Tr. Inst. Biol. Vnutr. Vod., Akad. Nauk SSSR*, **12**, 129–33.

Mittelbach, G. (1984) Predation and resource partitioning in two sunfishes (Centrarchidae). *Ecology*, **65**, 499–513.

Mkoji, G. M., Koech, D., Hofkin, B., Loker, E., Kihara, J. and Kageni, F. (1992) Does the snail *Melanoides tuberculata* have a role in biological control of *Biomphalaria pfeifferi* and other medically important African pulmonates? *Ann. Trop. Med. Parasitol.*, **86**, 201–4.

Moler, P. (1994) *Siren lacertinea* diet. *Herpetol. Rev.*, **25**, 62.

Molloy, D., Karatayev, A., Burlakova, L., Kurandina, D. and Laruelle, F. (1997) Natural enemies of zebra mussels: Predators, parasites, and ecological competitors. *Rev. Fish. Sci.*, **5**, 27–97.

Monakov, A. (1972) Review of studies on feeding of aquatic invertebrates conducted at the Institute of Biology of Inland Waters, Academy of Sciences, USSR. *J. Fish. Res. Board Can.*, **29**, 363–83.

Monteiro, W., Almeida, J. and Dias, B. (1984) Sperm sharing in *Biomphalaria* snails: A new behavioral strategy in simultaneous hermaphroditism. *Nature*, **308**, 727–9.

Mooij-Vogelaar, J. and Van der Steen, W. (1973) Effects of density on feeding and growth in the pond snails *Lymnaea stagnalis* (L.). *Proc. Konink. Neder. Akad. Wetenschap., Series C*, **76**, 47–60.

Moore, I. (1964) The effects of water current on the snails *Stagnicola palustris* and *Physa propinqua*. *Ecology*, **45**, 558–64.

Morgan, E. and Last, V. (1982a) The behavior of *Bulinus africanus*: A circadian profile. *Anim. Behav.*, **30**, 557–67.

Morgan, C. and Last, V. (1982b) Observations on the feeding and digestive rhythm of the freshwater prosobranch *Melanoides tuberculata* Müller. *Zool. Anz.*, **209**, 301–93.

Morris, J. and Boag, D. (1982) On the dispersion, population structure, and life history of a basommatophoran snail, *Helisoma trivolvis*, in central Alberta. *Can. J. Zool.*, **60**, 2931–40.

Morris, T. and Corkum, L. (1996) Assemblage structure of freshwater mussels (Bivalvia: Unionidae) in rivers with grassy and forested riparian zones. *J. N. Am. Benthol. Soc.*, **15**, 576–86.

Morton, B. (1969a) Studies on the biology of *Dreissena polymorpha* Pall. 1. General Anatomy and morphology. *Proc. Malacol. Soc. London*, **38**, 301–21.

Morton, B. (1969b) Studies on the biology of *Dreissena polymorpha* Pall. 2. Correlation of the rhythms of abductor activity, feeding, digestion and excretion. *Proc. Malacol. Soc. London*, **38**, 401–14.

Morton, B. (1969c) Studies on the biology of *Dreissena polymorpha* Pall. 3. Population dynamics. *Proc. Malacol. Soc. Lond.*, **38**, 471–82.

Morton, B. (1971) Studies on the biology of *Dreissena polymorpha* Pall. V. Some aspects of filter feeding and the effect of micro-organisms upon the rate of filtration. *Proc. Malacol. Soc. Lond.*, **39**, 289–302.

Morton, B. (1982) Some aspects of the population structure and sexual strategy of *Corbicula fluminalis* (Bivalvia: Corbiculidae) from the Pearl River, P.R.C. *J. Molluscan Stud.*, **48**, 1–23.

Morton, B. (1983) Feeding and digestion in Bivalvia. In *The Mollusca*, Vol. 5, ed. A. S. M. Saleuddin and M. Wilbur, pp. 65–147. New York: Academic Press.

Morton, B. (1985) The population dynamics, reproductive strategy and life history of *Musculium lacustre* (Bivalvia: Pisidiidae) in Hong Kong. *J. Zool.*, **207**, 581–603.

Moser, W. and Willis, M. (1994) Predation on gastropods by *Placobdella* spp. (Clitellata: Rhynchobdellida). *Am. Midl. Nat.*, **132**, 399–400.

Mountford, M. (1962) An index of similarity and its application to classificatory problems. In *Progress in Soil Zoology*, ed. P. Murphy, pp. 43–50. London: Butterworth.

Mouthon, J. (1979) Structure malacologique de la riviére Aube. *Ann. Limnol.*, **15**, 299–315.

Mouthon, J. (1986) General principles for a method of assessing the overall quality of lacustrine sediments, using a simplified analysis of the mollusc community (in French). *Ann. Limnol.*, **22**, 209–17.

Møhlenberg, F. and Riisgård, H. (1978) Efficiency of particle retention in 13 species of suspension feeding bivalves. *Ophelia*, **17**, 239–46.

Muley, E. (1977) Studies on the breeding habits and development of the brood-pouch of a viviparous prosobranch: *Melania scabra*. *Hydrobiologia*, **54**, 181–6.

Mulholland, P., Newbold, J., Elwood, J. and Holm, C. (1983) The effect of grazing intensity on phosphorous spiralling in autotrophic streams. *Oecologia (Berl.)*, **58**, 358–66.

Mulholland, P., Elwood, J., Newbold, J. and Ferren, L. (1985) Effect of a leaf shredding invertebrate on organic matter dynamics and phosphorus spiralling in heterotrophic laboratory streams. *Oecologia (Berl.)*, **66**, 199–206.

Mulholland, P., Steinman, A., Palumbo, A., Elwood, J. and Kirschtel, D. (1991) Role of nutrient cycling and herbivory in regulating periphyton communities in laboratory streams. *Ecology*, **72**, 966–82.

Mulvey, M. and Vrijenhoek, R. (1981a) Genetic variation among laboratory strains of the planorbid snail *Biomphalaria glabrata*. *Biochem. Genet.*, **19**, 1169–82.

Mulvey, M. and Vrijenhoek, R. (1981b) Multiple paternity in the hermaphroditic snail, *Biomphalaria obstructa*. *J. Hered.*, **72**, 308–12.

Murdoch, W. (1994) Population regulation in theory and practice. *Ecology*, **75**, 271–87.

Murphy, G. (1942) Relationship of the freshwater mussel to trout in the Truckee River. *Calif. Fish Game*, **28**, 89–102.

Nagabhushanam, R. and Lohgaonker, A. (1978) Seasonal reproductive cycle in the mussel, *Lamellidens corrianus*. *Hydrobiologia*, **61**, 9–14.

Nagelkerke, L. and Sibbing, F. (1996) Efficiency of feeding on zebra mussels (*Dreissena polymorpha*) by common bream (*Abramis brama*),white bream (*Blicca bjoerkna*, and roach (*Rutilus rutilus*): The effects of morphology and behavior. *Can. J. Fish. Aquat. Sci.*, **53**, 2847–61.

Nalepa, T. (1994) Decline of native unionid bivalves in Lake St. Clair after infestation by the zebra mussel *Dreissena polymorpha*. *Can. J. Fish. Aquat. Sci.*, **51**, 2227–33.

Nalepa, T. and Gauvin, J. (1988) Distribution, abundance, and biomass of freshwater mussels (Bivalvia: Unionidae) in Lake St. Clair. *J. Great Lakes Res.*, **14**, 411–9.

Nalepa, T., Manny, B., Roth, J., Mozleyand, S. and Schloesser, D. (1991) Long-term decline in freshwater mussels (Bivalvia: Unionidae) of the western basin of Lake Erie. *J. Great Lakes Res.*, **17**, 214–19.

Nduku, W. and Harrison, A. (1976) Calcium as a limiting factor in the biology of *Biomphalaria pfeifferi* (Krauss) (Gastropoda: Planorbidae). *Hydrobiologia*, **49**, 143–70.

Nduku, W. and Harrison, A. (1980a) Cationic responses of organs and haemolymph of *Biomphalaria pfeifferi*, *Biomphalaria glabrata*, and *Helisoma trivolis* to cationic alterations of the medium. *Hydrobiologia*, **68**, 119–38.

Nduku, W. and Harrison, A. (1980b) Water relations and osmotic pressure in *Biomphalaria pfeifferi*, *Biomphalaria glabrata*, and *Helisoma trivolis* in response to cationic alterations of the medium. *Hydrobiologia*, **68**, 139–44.

Neff, S. (1964) Snail-killing sciomyzid flies: Application in biological control. *Verh. Int. Verein. Limnol.*, **15**, 933–9.

Negus, C. (1966) A quantitative study of growth and production of unionid mussels in the river Thames at Reading. *J. Anim. Ecol.*, **35**, 513–32.

Neumann, D. (1961) Ernährungsbiologie einer rhipidoglossen Kiemenschnecke. *Hydrobiologia*, **17**, 133–51.

Neves, R. and Odom, M. (1989) Muskrat predation on endangered freshwater mussels in Virginia. *J. Wildl. Manage.*, **53**, 934–41.

Neves, R. and Widlak, J. (1987) Habitat ecology of juvenile freshwater mussels (Bivalvia: Unionidae) in a headwater stream in Virginia. *Am. Malacol. Bull.*, **5**, 1–7.

Neves, R. and Widlak, J. (1988) Occurrence of glochidia in stream drift and on fishes of the upper North Fork Holston River, Virginia. *Am. Midl. Nat.*, **119**, 111–20.

Neves, R., Weaver, L. and Zale, A. (1985) An evaluation of host fish suitability for glochidia of *Villosa vanuxemi* and *Villosa nebulosa* (Bivalvia: Unionidae). *Am. Midl. Nat.*, **113**, 13–19.

Newell, R. and Jordan, S. (1983) Preferential ingestion of organic material by the American oyster, *Crassostrea virginica*. *Mar. Ecol. Prog. Ser.*, **13**, 47–53.

Newton, W. (1955) The establishment of a strain of *Australorbis glabratus* which combines albinism and high susceptibility to infection with *Schistosoma mansoni*. *J. Parasitol.*, **41**, 526–8.

Nichols, S. (1996) Variations in the reproductive cycle of *Dreissena polymorpha* in Europe, Russia, and North America. *Am. Zool.*, **36**, 311–25.

Nichols, S. and Wilcox, D. (1997) Burrowing saves Lake Erie clams. *Nature*, **389**, 921.

Nieder, L., Cagnin and M.Parisi, V. (1982) Burrowing and feeding behavior in the rat *Rattus norvegicus*. *Anim. Behav.*, **30**, 837–44.

Njiokou, F., Bellec, C., N'Goran, E. K., YapiYapi, G., Delay, B. and Jarne, P. (1992) Comparative fitness and reproductive isolation between two *Bulinus globosus* (Gastropoda: Planorbidae) populations. *J. Molluscan Stud.*, **58**, 367–76.

Noland, E. L. and Carriker, M. R. (1946) Observations on the biology of the snail *Lymnaea stagnalis* during twenty generations in laboratory culture. *Am. Midl. Nat.*, **36**, 467–93.

Norelius, I. (1967) Age groups and habitat of unionid mussels in a south Swedish stream. *Oikos*, **18**, 365–8.

O'Keeffe, J. (1985) Population biology of the freshwater snail *Bulinus globosus* on the Kenya coast. 1. Population fluctuations in relation to climate. *J. Appl. Ecol.*, **22**, 73–84.

Økland, J. (1964) The eutrophic Lake Borrevann (Norway) – an ecological study on shore and bottom fauna with special reference to gastropods, including a hydrographic survey. *Folia Limnol. Scandinavica*, **13**, 337 pp.

Økland, J. (1983) Factors regulating the distribution of freshwater snails (Gastropoda) in Norway. *Malacologia*, **24**, 277–88.

Økland, J. (1990) *Lakes and Snails*. Oegstgeest, Netherlands: W. Backhuys.

Oliver-Gonzalez, J., Bauman, P. and Benenson, A. (1956) Effect of the snail *Marisa cornuarietis* on *Australorbis glabratus* in natural bodies of water in Puerto Rico. *Am. J. Trop. Med. Hyg.*, **5**, 290–6.

Olsen, T., Lodge, D., Capelli, G. and Hamilton, R. (1991) Mechanisms of impact of an introduced crayfish (*Orconectes rusticus*) on littoral congeners, snails, and macrophytes. *Can. J. Fish. Aquat. Sci.*, **48**, 1853–61.

Ortmann, A. (1909) The breeding season of Unionidae in Pennsylvania. *Nautilus*, **22**, 91–95, 99–103.

Osenberg, C. (1989) Resource limitation, competition and the influence of life history in a freshwater snail community. *Oecologia*, **79**, 512–19.

Osenberg, C. and Mittelbach, G. (1989) Effects of body size on the predator-prey interaction between pumpkinseed sunfish and gastropods. *Ecol. Monog.*, **59**, 405–32.

Owen, G. (1966) Digestion. In *Physiology of the Mollusca*, Vol. II, ed. K. Wilbur and C. Younge, pp. 53–96. New York: Academic Press.

Pace, G. (1971) The hold-fast function of the preputial organ in *Helisoma*. *Malacol. Rev.*, **4**, 21–4.

Pace, G., Szuch, E. and Dapson, R. (1979) Depth distribution of three gastropods in New Mission Bay, Lake Michigan. *Nautilus*, **93**, 31–6.

Palmieri, M., Palmieri, J. and Sullivan, J. (1980) A chemical analysis of the habitat of nine commonly occurring Malaysia freshwater snails. *Malay. Nat. J.*, **34**, 39–45.

Pan, C. (1965) Studies on the host-parasite relationship between *Schistosoma mansoni* and the snail *Australorbis glabratus*. *Am. J. Trop. Med. Hyg.*, **14**, 931–76.

Panha, S. (1993) Glochidiosis and juvenile production in a freshwater pearl mussel, *Chamberlainia hainesiana*. *Invertebr. Reprod. Dev.*, **24**, 157–60.

Parashar, B. and Rao, K. (1988) Biological studies of the flesh fly *Sarcophaga misera* and its effects as a predator of the snail *Indoplanorbis exustus*. *Entomophaga*, **33**, 431–4.

Parashar, B., Kumar, A. and Rao, K. (1986) Role of food in mass cultivation of the freshwater snail *Indoplanorbis exustus*, vector of animal schistosomiasis. *J. Molluscan Stud.*, **52**, 120–4.

Pardy, R. (1980) Symbiotic algae and C incorporation in the freshwater clam, *Anodonta*. *Biol. Bull.*, **158**, 349–55.

Parisi, V. and Gandolfi, G. (1974) Further aspects of the predation by rats on various mollusc species. *Boll. Zool.*, **41**, 87–106.

Park, J.-K. and O'Foighil, D. (1998) Sphaeriids and corbiculids represent separate invasions into freshwater environments. In *World Congress of Malacology Abstracts*, ed. R. Bieler and P. Mikkelson, p. 254. Chicago: Field Museum of Natural History.

Parker, R., Hackney, C. and Vidrine, M. (1984) Ecology and reproductive strategy of a south Louisiana USA freshwater mussel *Glebula rotundata* (Unionidae: Lampsilini). *Freshwat. Invertebr. Biol.*, **3**, 53–8.

Parker, B., Patterson, M. & Neves, R. (1998) Feeding interactions between native freshwater mussels (Bivalvia: Unionidae) and zebra mussels (*Dreissena polymorpha*) in the Ohio River. *Am Malacol. Bull.*, **14**, 173–9.

Parodiz, J. and Bonetto, A. (1963) Taxonomy and zoogeographic relationships of the South American naiades (Bivalvia: Unionacea and Mutelacea). *Malacologia*, **1**, 179–213.

Paterson, C. (1983) Effect of aggregation on the respiration rate of the freshwater unionid bivalve *Elliptio complanata* (Solander). *Freshwat. Invertebr. Biol.*, **2**, 139–46.

Paterson, C. (1984) A technique for determining apparent selective filtration in the freshwater bivalve *Elliptio complanata* (Lightfoot). *Veliger*, **27**, 238–41.

Paterson, C. (1986) Particle-size selectivity in the freshwater bivalve *Elliptio complanata*. *Veliger*, **29**, 235–7.

Patrick, R. (1970) Benthic stream communities. *Am. Sci.*, **58**, 546–9.

Patterson, C. (1963) Cytological studies of pomatiopsid snails. *Am. Malacol. Union Annu. Rep.*, **30**, 13–14.

Patterson, C. (1969) The chromosomes of *Tulotoma angulata* (Streptoneura: Viviparidae). *Malacologia*, **2**, 259–65.

Patterson, C. (1973) Cytogenetics of gastropod molluscs. *Malacol. Rev.*, **6**, 141–50.

Paulinyi, H. and Paulini, E. (1972) Laboratory observations on the biological control of *Biomphalaria glabrata* by a species of *Pomacea* (Ampullariidae). *Bull. W.H.O.*, **46**, 243–7.

Payne, B. (1979) Bioenergetic budgeting of carbon and nitrogen in the life-histories of three lake populations of the prosobranch snail *Goniobasis livescens*. Ph.D. dissertation. Syracuse, NY: Syracuse University.

Payne, B. and Miller, A. (1989) Growth and survival of recent recruits to a population of *Fusconaia ebena* (Bivalvia: Unionidae) in the lower Ohio River. *Am. Midl. Nat.*, **121**, 99–104.

Payne, B., Lei, J., Miller, A. and Hubertz, E. (1995) Adaptive variations in palp and gill size of the zebra mussel (*Dreissena polymorpha*) and Asian clam (*Corbicula fluminea*). *Can. J. Fish. Aquat. Sci.*, **52**, 1130–4.

Peredo, S. and Parada, E. (1986) Reproductive cycle in the freshwater mussel *Diplodon chilensis chilensis* (Mollusca: Bivalvia). *Veliger*, **28**, 418–25.

Perera, G., Sanchez, R., Yong, M., Ferrer, J. and Amador, O. (1986) Ecology of some freshwater pulmonates from Cuba. *Malacol. Rev.*, **19**, 99–104.

Perry, W., Lodge, D. and Lamberti, G. (1997) Impact of crayfish predation on exotic zebra mussels and native invertebrates in a lake-outlet stream. *Can. J. Fish. Aquat. Sci.*, **54**, 120–5.

Pesigan, T., Hairston, N., Jauregul, J., Garcia, E., Santos, A., Santos, B. and Besa, A. (1958) Studies on *Schistosoma japonicum* infection in the Phillippines. 2. The molluscan host. *Bull. W.H.O.*, **18**, 481–578.

Peters, B. (1938) Biometrical observations on shells of *Limnaea* species. *J. Helminthol.*, **16**, 181–212.

Peters, R. (1983) *The Ecological Implications of Body Size*. Cambridge: Cambridge University Press.

Phillips, N. and Lambert, D. (1989) Genetics of *Potamopyrgus antipodarum* (Gastropoda: Prosobranchia): Evidence for reproductive modes. *New Zeal. J. Zool.*, **16**, 435–45.

Pielou, E. (1975) *Ecological Diversity*. New York: Wiley.

Piesik, Z. (1974) The role of the crayfish *Orconectes limosus* (Raf.) in extinction of *Dreissena polymorpha* (Pall.) subsisting on steelon net. *Pol. Arch. Hydrobiol.*, **21**, 401–10.

Piggott, H. and Dussart, G. (1995) Egg-laying and associated behavioural responses of *Lymnaea peregra* (Müller) and *Lymnaea stagnalis* (L.) to calcium in their environment. *Malacologia*, **37**, 13–21.

Pimentel, D. and White, P., Jr. (1959a) Physico-chemical environment of *Australorbis glabratus*, the snail intermediate host of *Schistosoma mansoni* in Puerto Rico. *Ecology*, **40**, 533–41.

Pimentel, D. and White, P., Jr. (1959b) Biological environment and habits of *Australorbis glabratus*. *Ecology*, **40**, 541–50.

Pimentel-Souza, F., Schall, V., Lautner, R., Jr., Barbosa, N., Schettino, M. and Fernandes, B. (1984) Behavior of *Biomphalaria glabrata* (Gastropoda: Pulmonata) under different lighting conditions. *Can. J. Zool.*, **62**, 2328–34.

Pinel-Alloul, B. and Magnin, E. (1971) Cycle vital et croissance de *Bithynia tentaculata* L. (Mollusca, Gastropoda, Prosobranchia) du lac St. Louis, prés de Montréal. *Can. J. Zool.*, **49**, 759–66.

Pip, E. (1978) A survey of the ecology and composition of submerged aquatic snail-plant communities. *Can. J. Zool.*, **56**, 2263–79.

Pip, E. (1985) The ecology of the freshwater gastropods on the southwestern edge of the precambrian shield. *Can. Field Nat.*, **99**, 76–85.

Pip, E. (1986a) A study of pond colonization by freshwater molluscs. *J. Molluscan Stud.*, **52**, 214–24.

Pip, E. (1986b) The ecology of freshwater gastropods in the central Canadian region. *Nautilus*, **100**, 56–66.

Pip, E. (1987a) Species richness of freshwater gastropod communities in central North America. *J. Molluscan Stud.*, **53**, 163–70.

Pip, E. (1987b) Ecological differentiation within the genus *Helisoma* (Gastropoda: Planorbidae) in central Canada. *Nautilus*, **101**, 33–44.

Pip, E. (1988) Niche congruency of freshwater gastropods in central North America with respect to six water chemistry parameters. *Nautilus*, **102**, 65–72.

Pip, E. and Stewart, J. (1976) The dynamics of two aquatic plant-snail associations. *Can. J. Zool.*, **54**, 1192–205.

Pointier, J. and McCullough, F. (1989) Biological control of the snail hosts of *Schistosoma mansoni* in the Caribbean area using *Thiara* spp. *Acta Trop.*, **46**, 147–55.

Pointier, J., Theron, A. and Imbert-Establet, D. (1988) Decline of a sylvatic focus of *Schistosoma mansoni* in Guadeloupe (French West Indies) following the competitive displacement of the snail host *Biomphalaria glabrata* by *Ampullaria glauca*. *Oecologia*, **75**, 38–43.

Pointier, J., Guyard, A. and Mosser, A. (1989) Biological control of *Biomphalaria glabrata* and *Biomphalaria straminea* by the competitor snail *Thiara tuberculata* in a transmission site of schistosomiasis in Martinique, French West Indies. *Ann. Trop. Med. Parasitol.*, **83**, 263–9.

Pointier, J., Toffart, J. and Lefévre, M. (1991) Life tables of freshwater snails of the genus *Biomphalaria* (*B. glabrata*, *B. alexandrina*, *B. straminea*) and of one of its competitors *Melanoides tuberculata* under laboratory conditions. *Malacologia*, **33**, 43–54.

Pointier, J., Theron, A. and Borel, G. (1993) Ecology of the introduced snail *Melanoides tuberculata* (Gastropoda: Thiaridae) in relation to *Biomphalaria glabrata* in the marshy forest zone of Guadeloupe, French West Indies. *J. Molluscan Stud.*, **59**, 421–8.

Pointier, J., Incani, R., Balzan, C., Chrosciechowski, P. and Prypchan, S. (1994) Invasion of the rivers of the littoral control region of Venezuela by *Thiara granifera* and *Melanoides tuberculata* (Mollusca: Prosbranchia: Thiaridae) and the absence of *Biomphalaria glabrata*, snail host of *Schistosoma mansoni*. *Nautilus*, **107**, 124–8.

Polunin, N. (1982) Effects of the freshwater gastropod *Planorbis carinatus* on reed (*Phragmites australis*) litter microbial activity in an experimental stream. *Freshwat. Biol.*, **12**, 547–52.

Ponder, W. (1988a) *Potamopyrgus antipodarium* – a molluscan coloniser of Europe and Australia. *J. Molluscan Stud.*, **54**, 271–85.

Ponder, W., ed. (1988b) Prosobranch phylogeny. *Malacol. Rev. Suppl.*, **4**.

Power, M. (1990) Effects of fish in river food webs. *Science*, **250**, 811–14.

Power, M. E., Stewart, A. J. and Matthews, W. J. (1988) Grazer control of algae in an Ozark Mountain stream: Effects of short-term exclusion. *Ecology*, **69**, 1894–8.

Prejs, A., Lewandowski, K. and Stanczykowska-Piotrowska, A. (1990) Size-selective predation by roach, *Rutilus rutilus* on zebra mussel *Dreissena polymorpha*: Field studies. *Oecologia*, **83**, 378–84.

Preston, F. (1948) The commonness, and rarity, of species. *Ecology*, **29**, 254–83.

Preston, F. (1962) The canonical distribution of commonness and rarity. *Ecology*, **43**, 185–215, 410–432.

Price, T. and Schluter, D. (1991) On the low heritability of life-history traits. *Evolution*, **45**, 853–61.

Pringle, G. and Msangi, A. (1961) Experimental study of water snails in a fishpond in Tanganyika. I. Preliminary trial of the method. *East African Med. J.*, **38**, 275–93.

Pringle, G. and Raybould, J. (1965) The experimental study of water snails in a fish pond in Tanganyika. 2. Attempts to establish reproducible conditions. *East African Med. J.*, **42**, 289–96.

Prinsloo, J. and van Eeden, J. (1969) Temperature and its bearing on the distribution and chemical control of freshwater snails. *S. Afr. Med. J.*, **43**, 1363–5.

Pulliam, H. (1988) Sources, sinks, and population regulation. *Am. Nat.*, **132**, 652–61.

Pynnönen, K. (1991) Accumulation of Ca in the freshwater unionids *Anodonta anatina* and *Unio tumidus*, as influenced by water hardness, protons, and aluminum. *J. Exp. Zool.*, **260**, 18–27.

Radke, M., Ritchie, L. and Ferguson, F. (1961) Demonstrated control of *Australorbis glabratus* by *Marisa cornuarietis* under field conditions in Puerto Rico. *Am. J. Trop. Med. Hyg.*, **10**, 370–3.

Ram, J., Fong, P. and Garton, D. (1996) Physiological aspects of zebra mussel reproduction: Maturation, spawning, and fertilization. *Am. Zool.*, **36**, 326–38.

Ram, J. and McMahon, R. (1996) Introduction: The biology, ecology, and physiology of zebra mussels. *Am. Zool.*, **36**, 239–43.

Ram, K. and Radhakrishna, Y. (1984) The distribution of freshwater mollusca in Guntur District (India) with a description of *Scaphula nagarjunai*. *Hydrobiologia*, **119**, 49–55.

Rankin, J. and Harrison, A. (1979) Hydrobiological studies of eastern Lesser Antillean Islands IV. St. Vincent: Comparison of field and laboratory populations of *Physa marmorata* Guilding (Gastropoda: Physidae). *Arch. Hydrobiol. Suppl.*, **57**, 89–116.

Raut, S. (1986) Inhibition of fish growth by the freshwater snail *Bellamya bengalensis*. *Environ. Ecol.*, **4**, 332–3.

Reavell, P. (1980) A study of the diets of some British freshwater gastropods. *J. Conchol.*, **30**, 253–71.

Reinecke, K. (1979) Feeding ecology and development of juvenile black ducks *Anas rubripes* in Maine USA. *Auk*, **96**, 737–45.

Reynoldson, T. and Piearce, B. (1979a) Feeding on gastropods by lake-dwelling *Polycelia* in the absence and presence of *Dugenia polychroa* (Turbellaria, Tricladida). *Freshwat. Biol.*, **9**, 357–67.

Reynoldson, T. and Piearce, B. (1979b) Predation on snails by three species of triclad and its bearing on the distribution of *Planaria torva* in Britain. *J. Zool.*, **189**, 459–84.

Reznick, D. (1985) Costs of reproduction: An evaluation of the empirical evidence. *Oikos*, **44**, 257–67.

Reznick, D., Bryga, H. and Endler, J. (1990) Experimentally induced life-history evolution in a natural population. *Nature*, **346**, 357–9.

Ribi, G. and Gebhardt, M. (1986) Age specific fecundity and size of offspring in the prosobranch snail, *Viviparus ater*. *Oecologia*, **71**, 18–24.

Ribi, G., Mutzner, A. and Gebhardt, M. (1986) Shell dissolution and mortality in the freshwater snail *Viviparus ater*. *Swiss J. Hydrol.*, **48**, 34–43.

Ricciardi, A., Whoriskey, F. and Rasmussen, J. (1997) The role of the zebra mussel (*Dreissena polymorpha*) in structuring macroinvertebrate communities on hard substrata. *Can. J. Fish. Aquat. Sci.*, **54**, 2596–608.

Richards, C. S. (1962) Retarded development of the male reproductive system in a Florida *Gyraulus*. *Trans. Am. Microsc. Soc.*, **81**, 347–51.

Richards, C. S. (1973) Genetics of *Biomphalaria glabrata* (Gastropoda: Planorbidae). *Malacol. Rev.*, **6**, 199–202.

Richards, C. (1975) Genetic factors in susceptibility of *Biomphalaria glabrata* for different strains of *Schistosoma mansoni*. *Parasitol.*, **70**, 231–41.

Richards, C. and Merritt, J. (1972) Genetic factors in the susceptibility of juvenile *Biomphalaria glabrata* to *Schistosoma mansoni* infection. *Am. J. Trop. Med. Hyg.*, **21**, 425–34.

Richards, C. and Merritt, J. (1975) Variation in size of *Biomphalaria glabrata* at maturity. *Veliger*, **17**, 393–5.

Richards, C. and Shade, P. (1987) The genetic variation of compatibility in *Biomphalaria glabrata* and *Schistosoma mansoni*. *J. Parasitol.*, **73**, 1146–51.

Richardson, T. and Scheiring, J. (1994) Ecological observations of two pleurocerid gastropods: *Elimia clara* (Lea) and *E. cahawbensis* (Say). *Veliger*, **37**, 284–9.

Ricklefs, R. and Travis, J. (1980) A morphological approach to the study of avian community organization. *Auk*, **97**, 321–38.

Ricklefs, R., Cochran, D. and Pianka, E. (1981) A morphological analysis of the structure of communities of lizards in desert habitats. *Ecology*, **62**, 1474–83.

Riisgard, H. (1988) Efficiency of particle retention and filtration rate in six species of northeast American bivalves. *Mar. Ecol. Prog. Ser.*, **45**, 217–23.

Ritchie, L. (1955) The biology and control of the amphibious snails that serve as intermediate hosts for *Schistosoma japonicum* in Japan. *Am. J. Trop. Med. Hyg.*, **4**, 426–41.

Robinson, J. and Wellborn, G. (1988) Ecological resistance to the invasion of a freshwater clam, *Corbicula fluminea*: Fish predation. *Oecologia*, **77**, 445–52.

Rodgers, J., Cherry, D., Clark, J., Dickson, K. and Cairns, J., Jr. (1977) The invasion of the Asiatic clam *Corbicula manilensis* in the New River, Virginia. *Nautilus*, **91**, 43–6.

Rodina, A. (1948) Bacteria as food for freshwater mollusks (in Russian). *Microbiology*, **17**, 232–9.

Rogers, J. and Korschgen, L. (1966) Foods of lesser scaups on breeding, migration, and wintering areas. *J. Wildl. Manage.*, **30**, 258–64.

Rollinson, D. (1986) Reproductive strategies of some species of *Bulinus*. In *Proceedings of the Eighth International Malacological Congress*, ed. P. Lászlo, pp. 221–6. Budapest, Hungary: Unitas Malacologia.

Rollinson, D. and Simpson, A. (1987) *The Biology of Schistosomes: From Genes to Latrines*. London: Academic Press.

Rollinson, D. and Wright, C. A. (1984) Population studies of *Bulinus cernicus* from Mauritius. *Malacologia*, **25**, 447–64.

Rollinson, D., Kane, R. and Lines, J. (1989) An analysis of fertilization in *Bulinus cernicus* (Gastropoda: Planorbidae). *J. Zool. (Lond.)*, **217**, 295–310.

Rollo, C. and Hawryluk, M. (1988) Compensatory scope and resource allocation in two species of aquatic snails. *Ecology*, **69**, 146–56.

Rooke, J. and Mackie, G. (1984a) Laboratory studies of the effects of mollusca on alkalinity of their freshwater environments. *Can. J. Zool.*, **62**, 793–7.

Rooke, J. and Mackie, G. (1984b) Mollusca of six low-alkalinity lakes in Ontario. *Can. J. Fish. Aquat. Sci.*, **41**, 777–82.

Rosemond, A., Mulholland, P. & Elwood, J. (1993) Top-down and bottom-up control of stream periphyton: Effects of nutrients and herbivores. *Ecology*, **74**, 1264–80.

Ross, M. and Ultsch, G. (1980) Temperature and substrate influences on habitat selection in 2 pleurocerid snails (*Goniobasis cahawbensis* and *Goniobasis carinifera*). *Am. Midl. Nat.*, **103**, 209–17.

Rotenberg, L., Jurberg, P. and Pieri, O. (1989) Relationship between light conditions and behavior of the freshwater snail *Biomphalaria glabrata* (Say). *Hydrobiologia*, **174**, 111–16.

Rowan, W. (1966) Autumn migration of *Helisoma trivolis* in Montana. *Nautilus*, **79**, 108–9.

Rudolph, P. H. (1979a) An analysis of copulation in *Bulinus* (*Physopsis*) *globosus* (Gastropoda: Planorbidae). *Malacologia*, **19**, 147–55.

Rudolph, P. H. (1979b) The strategy of copulation of *Stagnicola elodes* (Say) (Basommatophora: Lymnaeidae). *Malacologia*, **18**, 381–9.

Rudolph, P. H. (1983) Copulatory activity and sperm production in *Bulinus* (*Physopsis*) *globosus* (Gastropoda: Planorbidae). *J. Molluscan Stud.*, **49**, 125–32.

Rudolph, P. H. and Bailey, J. B. (1985) Copulation as females and use of allosperm in the freshwater snail genus *Bulinus* (Gastropoda: Planorbidae). *J. Molluscan Stud.*, **51**, 267–75.

Rudolph, P. and White, J. (1979) Egg-laying behaviour of *Bulinus octoploidus* Burch (Basommatophora: Planorbidae). *J. Molluscan Stud.*, **45**, 355–63.

Ruiz-Tiben, E., Palmer, J. and Ferguson, F. (1969) Biological control of *Biomphalaria glabrata* by *Marisa cornuarietis* in irrigation ponds in Puerto Rico. *Bull. W.H.O.*, **41**, 329–33.

Runham, N. (1975) Alimentary canal. In *Pulmonates*, Vol. 1, ed. V. Fritter and J. Peake, pp. 53–104. New York: Academic Press.

Russell-Hunter, W. (1961a) Annual variations in growth and density in natural populations of freshwater snails in the west of Scotland. *Proc. Zool. Soc. Lond.*, **136**, 219–53.

Russell-Hunter, W. (1961b) Life cycles of four freshwater snails in limited populations in Loch Lomond with a discussion of infraspecific variation. *Proc. Zool. Soc. Lond.*, **137**, 135–71.

Russell-Hunter, W. (1964) Physiological aspects of ecology in nonmarine molluscs. In *Physiology of Mollusca*, Vol. 1, ed. K. Wilbur and C. Yonge, pp. 83–126. New York: Academic Press.

Russell-Hunter, W. (1978) Ecology of freshwater pulmonates. In *Pulmonates*, Vol 2A, ed. V. Fretter and J. Peake, pp. 336–83. New York: Academic Press.

Russell-Hunter, W. D. and McMahon, R. F. (1976) Evidence for functional protandry in a freshwater basommatophoran limpet, *Laevapex fuscus. Trans. Am. Microsc. Soc.*, **95**, 174–82.

Russell-Hunter, W., Meadows, R., Apley, M. and Burky, A. (1968) On the use of a 'wet oxidation' method for estimates of total organic carbon in mollusc growth studies. *Proc. Malacol. Soc. Lond.*, **38**, 1–11.

Sadzikowski, M. and Wallace, D. (1976) A comparison of food habits of size classes of three sunfishes (*Lepomis macrochirus* Rafinesque, *L. gibbosus* Linnaeus, and *L. cyanellus* Rafinesque). *Am. Midl. Nat.*, **95**, 220–5.

Salánki, J. and Vero, M. (1969) Diurnal rhythm of activity in freshwater mussel (*Anodonta cygnea* L.) under natural conditions. *Ann. Inst. Biol. (Tihany) Hung. Acad. Sci.*, **36**, 95–107.

Salbenblatt, J. and Edgar, A. (1964) Valve activity in fresh-water pelecypods. *Pap. Mich. Acad. Sci., Arts and Lett.*, **49**, 177–86.

Salmon, A. and Green, R. (1983) Environmental determinants of unionid clam distribution in the middle Thames River, Ontario. *Can. J. Zool.*, **61**, 832–8.

Santos, C., Penteado, C. and Mendes, E. (1987) The respiratory responses of an amphibious snail *Pomacea lineata* (Spix, 1827), to temperature and oxygen tension variation. *Comp. Biochem. Physiol.*, **86A**, 409–15.

Schloesser, D. and Nalepa, T. (1994) Dramatic decline of unionid bivalves in offshore waters of western Lake Erie after infestation by the zebra mussel, *Dreissena polymorpha*. *Can. J. Fish. Aquat. Sci.*, **51**, 2234–42.

Schloesser, D., Nalepa, T. and Mackie, G. (1996) Zebra mussel infestation of unionid bivalves (Unionidae) in North America. *Am. Zool.*, **36**, 300–10.

Schneider, D. and Lyons, J. (1993) Dynamics of upstream migration in two species of tropical freshwater snails. *J. N. Am. Benthol. Soc.*, **12**, 3–16.

Schrag, S. J. and Read, A. F. (1992) Temperature determination of male outcrossing ability in a simultaneous hermaphrodite. *Evolution*, **46**, 1698–707.

Schrag, S. and Rollinson, D. (1994) Effects of *Schistosoma haematobium* infection on reproductive success and male outcrossing ability in the simultaneous hermaphrodite, *Bulinus truncatus* (Gastropoda: Planorbidae). *Parasitol.*, **108**, 27–34.

Schrag, S., Ndifon, G. T. and Read, A. (1994) Temperature-determined outcrossing ability in wild populations of a simultaneous hermaphrodite snail. *Ecology*, **75**, 2066–77.

Seaman, D. and Porterfield, W. (1964) Control of aquatic weeds by the snail, *Marisa cornuarietis*. *Weeds*, **12**, 87–92.

Sebestyen, O. (1938) Colonization of two new fauna-elements of Pontus origin (*Dreissena polymorpha* and *Corophium curvispinum devium*) in Lake Balaton. *Verh. Int. Verein. Limnol.*, **8 (3)**, 169–81.

Sepkoski, J. and Rex, M. (1974) Distribution of freshwater mussels: Coastal rivers as biogeographic islands. *Syst. Zool.*, **23**, 165–88.

Servos, M., Rooke, J. and Mackie, G. (1985) Reproduction of selected Mollusca in some low alkalinity lakes in south-central Ontario. *Can. J. Zool.*, **63**, 511–5.

Seshaiya, R. (1936) Notes on the comparative anatomy of some Indian Melaniidae with special reference to *Melania* (*Radina*) *crenulata* (Deshayes). *J. Annamalai Univ.*, **5**, 167.

Seshaiya, R. (1940) A free larval stage in the life history of a fluviatile gastropod. *Curr. Sci.*, **9**, 331–2.

Seshaiya, R. (1941) Tadpoles as hosts for the glochidia of the freshwater mussel. *Curr. Sci.*, **10**, 535–6.

Shaw, M. and Mackie, G. (1989) Reproductive success of *Amnicola limosa* (Gastropoda) in low alkalinity lakes in south-central Ontario. *Can. J. Fish. Aquat. Sci.*, **46**, 863–9.

Shaw, M. and Mackie, G. (1990) Effects of calcium and pH on the reproductive success of *Amnicola limosa* (Gastropoda). *Can. J. Fish. Aquat. Sci.*, **47**, 1694–9.

Sheldon, S. P. (1987) The effects of herbivorous snails on submerged communities in Minnesota lakes. *Ecology*, **68**, 1920–31.

Shiff, C. (1964a) Studies on *Bulinus (P.) globosus* in Rhodesia 1. The influence of temperature on the intrinsic rate of natural increase. *Ann. Trop. Med. Parasitol.*, **58**, 94–105.

Shiff, C. (1964b) Studies on *Bulinus (P.) globosus* in Rhodesia 2. Factors influencing the relationship between age and growth. *Ann. Trop. Med. Parasitol.*, **58**, 106–15.

Shiff, C. (1964c) Studies on *Bulinus (P.) globosus* in Rhodesia 3. Bionomics of a natural population existing in a temporary habitat. *Ann. Trop. Med. Parasitol.*, **58**, 240–55.

Shiff, C. (1966) The influence of temperature on the vertical movement of *Bulinus (P.) globosus* in the laboratory and in the field. *S. Afr. J. Sci.*, **62**, 210–14.

Shiff, C. and Garnett, B. (1967) The influence of temperature on the intrinsic rate of natural increase of the freshwater snail *Biomphalaria pfeifferi* (Krauss). *Arch. Hydrobiol.*, **62**, 429–238.

Sickel, J. (1980) Correlation of unionid mussels with bottom sediment composition in the Altamaha River, Georgia. *Bull. Am. Malacol. Union*, **1980**, 10–13.

Sickel, J. (1986) *Corbicula* population mortalities: Factors influencing population control. *Am. Malacol. Bull. Special Edition*, **2**, 89–94.

Sih, A. (1984) The behavioral response race between predator and prey. *Am. Nat.*, **123**, 143–50.

Silverman, H., Achberger, A., Lynn, J. and Dietz, T. (1995) Filtration and utilization of laboratory-cultured bacteria by *Dreissena polymorpha*, *Corbicula fluminea*, and *Carunculina texasensis*. *Biol. Bull. (Woods Hole)*, **189**, 308–19.

Silverman, H., Achberger, A., Lynn, J. and Dietz, T. (1996) Gill structure in zebra mussels: Bacterial-sized particle filtration. *Am. Zool.*, **36**, 373–84.

Simberloff, D. (1970) Taxonomic diversity of island biotas. *Evolution*, **24**, 23–47.

Simberloff, D. (1978) Using island biogeographic distributions to determine if colonization is stochastic. *Am. Nat.*, **112**, 713–26.

Skibinski, D., Gallagher, C. and Beyon, C. (1994) Sex-limited mitochondrial transmission in the marine mussel *Mytilus edulis*. *Genetics*, **138**, 801–9.

Skoog, G. (1978) The influence of natural food items on growth and egg production in brackish water populations of *Lymnaea peregra* and *Theodoxus fluviatilis* (Mollusca). *Oikos*, **31**, 340–8.

Slootweg, R. (1987) Prey selection by molluscivorous cichlids foraging on a schistosomiasis vector snail, *Biomphalaria glabrata*. *Oecologia*, **74**, 193–202.

Slootweg, R., Malek, E. and McCullough, F. (1993) The biological control of snail intermediate hosts of schistosomiasis by fish. *Rev. Fish Biol. Fish.*, **3**, 33–56.

Smit, H., bij de Vaate, A. and Fioole, A. (1992) Shell growth of the zebra mussel (*Dreissena polymorpha* Pallas) in relation to selected physiochemical parameters in the lower Rhine and some associated lakes. *Arch. Hydrobiol*, **124**, 257–80.

Smith, B. (1979) Survey of non-marine molluscs of south-eastern Australia. *Malacologia*, **18**, 103–5.

Smith, D. (1989a) Tests of feeding selectivity in *Helisoma trivolvis* (Gastropoda: Pulmonata). *Trans. Am. Microsc. Soc.*, **108**, 394–402.

Smith, D. (1989b) Radula-tooth biometry in *Helisoma trivolis* (Gastropoda, Pulmonata): Interpopulation variation and the question of adaptive significance. *Can. J. Zool.*, **67**, 1960–5.

Smith, M., Britton, J., Burke, P., Chesser, R., Smith, M. and Hagen, J. (1979) Genetic variability in *Corbicula*, an invading species. In *Proceedings of the First International Corbicula Symposium*, ed. J. Britton, pp. 243–8. Fort Worth, Tx: Texas Christian University.

Snyder, N. (1967) *An Alarm Reaction of Aquatic Gastropods to Intraspecific Extract*, Memoir 403. Ithaca, NY: Cornell University Agricultural Experiment Station. 122 pp.

Snyder, N. and Kale, H., II. (1983) Mollusk predation by snail kites in Columbia. *Auk*, **100**, 93–7.

Snyder, N. and Snyder, H. (1969) A comparative study of mollusk predation by limpkins, Everglade kites and boat-tailed grackles. *Living Bird*, **8**, 177–223.

Snyder, N. and Snyder, H. (1971) Defenses of the Florida apple snail *Pomacea paludosa*. *Behaviour*, **40**, 175–215.

Sokal, R. and Rohlf, F. (1995) *Biometry*, Third Ed. New York: W.H. Freeman and Co.

Soszka, G. (1975) The invertebrates on submerged macrophytes in three Masurian lakes. *Ekol. Pol.*, **23**, 371–91.

Sprung, M. (1989) Field and laboratory observations of *Dreissena polymorpha* larvae: Abundance, growth, mortality, and food demands. *Arch. Hydrobiol.*, **115**, 537–62.

Sprung, M. (1993) The other life: An account of present knowledge of the larval phase of *Dreissena polymorpha*. In *Zebra Mussels: Biology Impacts and Control*, ed. T. Nalepa and D. Schlosser, pp. 39–53. Boca Raton, FL: Lewis Publishers.

Sprung, M. and Rose, U. (1988) Influence of food size and food quantity on the feeding of the mussel *Dreissena polymorpha*. *Oecologia*, **77**, 526–32.

Stadler, T., Weisner, S. and Streit, B. (1995) Outcrossing rates and correlated matings in a predominantly selfing freshwater snail. *Proc. R. Soc. Lond. Series B*, **262**, 119–25.

Stanczykowska, A. (1964) On the relationship between abundance, aggregations, and 'condition' of *Dreissena polymorpha* Pall. in 36 Masurian lakes. *Ekol. Pol. Series A*, **24**, 103–12.

Stanczykowska, A. (1977) Ecology of *Dreissena polymorpha* Pall. (Bivalia) in lakes. *Pol. Arch. Hydrobiol.*, **24**, 461–530.

Stanczykowska, A. (1978) Occurrence and dynamics of *Dreissena polymorpha* (Pall.) (Bivalvia). *Verh. Int. Verein. Limnol.*, **20**, 2431–4.

Stanczykowska, A. and Lewandowski, K. (1993) Thirty years of studies of *Dreissena polymorpha* ecology in Mazurian lakes of northeastern Poland. In *Zebra Mussels: Biology, Impacts, and Control*, ed. T. Nalepa and D. Schloesser, pp. 3–37. Boca Raton, Fl: Lewis Publishers.

Stanczykowska, A., Schenker, H. and Fafaraz, Z. (1975) Comparative characteristics of populations of *Dreissena polymorpha* in 1962 and 1972 in 13 Mazurian lakes. *Bull. Pol. Sci.*, **2**, **23**, **6**, 383–90.

Stanczykowska, A., Lawaca, W., Mattice, J. and Lewandowski, L. (1976) Bivalves as a factor affecting circulation of matter in Lake Mikolajskie. *Limnologica*, **10**, 347–52.

Starobogatov, Y. (1970) *Mollusk Fauna and the Zoogeographic Partitioning of Continental Waterbodies of the Globe* (in Russian). Leningrad, USSR: Nauka Leningrad Br. Acad. Sci.

Stearns, S. (1976) Life-history tactics: A review of the ideas. *Q. Rev. Biol.*, **51**, 266–81.

Stearns, S. (1984) The effects of size and phylogeny on patterns of covariation in the life history traits of lizards and snakes. *Am. Nat.*, **123**, 56–72.

Stein, R., Kitchell, J. and Knezevic, B. (1975) Selective predation on carp (*Cyprinus carpio*) on benthic molluscs in Skadar Lake, Yugoslavia. *J. Fish Biol.*, **7**, 391–9.

Stein, R., Goodman, C. and Marshall, E. (1984) Using time and energetic measures of cost in estimating prey value for fish predators. *Ecology*, **65**, 702–15.

Sterki, V. (1895) Some notes on the genital organs of Unionidae, with reference to systematics. *Nautilus*, **9**, 91–4.

Sterki, V. (1898) Some observations on the genital organs of Unionidae with reference to classification. *Nautilus*, **12**, 18–21, 28–32.

Stern, E. (1983) Depth distribution and density of freshwater mussels (Unionidae) collected with SCUBA from the lower Wisconsin and St. Croix Rivers. *Nautilus*, **97**, 36–41.

Sterry, P., Thomas, J. and Patience, R. (1983) Behavioural responses of *Biomphalaria glabrata* (Say) to chemical factors from aquatic macrophytes including decaying *Lemna paucicostata* (Hegelm ex Englem). *Freshwat. Biol.*, **13**, 465–76.

Stiglingh, I. and van Eeden, J. (1970) Notes on the feeding behavior of *Bulinus (B.) tropicus* Basommatohora; Planorbidae. *Wetensk Bydraes Potcheftroom. Univ. B*, **22**, 1–14.

Stiven, A. and Kreiser, B. (1994) Ecological and genetic differentiation among populations of the gastropod *Goniobasis proxima* (Say) in streams separated by a reservoir in the piedmont of North Carolina. *J. Elisha Mitchell Sci. Soc.*, **110**, 53–67.

Stoddart, J. (1985) Analysis of species lineages of some Australian thiards (Thiaridae, Prosobranchia, Gastropoda) using the evolutionary species concept. *J. Malacol. Soc. Aust.*, **7**, 7–16.

Stone, N., Earll, R., Hodgson, A., Mather, J., Parker, J. and Woodward, F. (1982) The distribution of three sympatric mussel species (Bivalvia: Unionidae) in Budworth Mere, Cheshire. *J. Molluscan Stud.*, **48**, 266–74.

Storey, R. (1970) The importance of mineral particles in the diet of *Limnaea pereger* (Müller). *J. Conchol.*, **27**, 191–5.

Storey, R. (1986) Longitudinal zonation of gastropods in a chalk stream. *J. Molluscan Stud.*, **52**, 15–24.

Strayer, D. (1981) Notes on the microhabitats of unionid mussels in some Michigan streams. *Am. Midl. Nat.*, **106**, 411–15.

Strayer, D. (1983) The effects of surface geology and stream size on freshwater mussel (Bivalvia: Unionidae) distribution in southeastern Michigan, USA. *Freshwat. Biol.*, **13**, 253–64.

Strayer, D. (1993) Macrohabitats of freshwater mussels (Bivalvia: Unionacea) in streams of the Northern Atlantic Slope. *J. N. Am. Benthol. Soc.*, **12**, 236–46.

Strayer, D. and Ralley, J. (1993) Microhabitat use by an assemblage of stream-dwelling unionaceans (Bivalvia), including two rare species of *Alasmidonta. J. N. Am. Benthol. Soc.*, **12**, 247–58.

Strayer, D. and Smith, L. (1996) Relationships between zebra mussels (*Dreissena polymorpha*) and unionid clams during the early stages of the zebra mussel invasion of the Hudson River. *Freshwat. Biol.*, **36**, 771–9.

Strayer, D., Cole, J., Likens, G. and Buso, D. (1981) Biomass and annual production of the freshwater mussel *Elliptio complanata* in an oligotrophic softwater lake. *Freshwat. Biol.*, **11**, 435–40.

Streit, B. (1975) Experimentelle Untersuchungen zum Stoffhaushaltvon *Ancylus fluviatilis* (Gastropoda – Basommatophora) 1. Ingestion, assimilation, Wachstum, und Eiablage. *Arch. Hydrobiol. Suppl.*, **47**, 458–514.

Streit, B. (1981) Food searching and exploitation by a primary consumer (*Ancylus fluviatilis*) in a stochastic environment: Nonrandom movement patterns. *Rev. Suisse Zool.*, **88**, 887–95.

Strong, D., Simberloff, D., Abele, L. and Thistle, A. (1983) *Ecological Communities: Conceptual Issues and the Evidence.* Princeton, NJ: Princeton University Press.

Studier, E. and Pace, G. (1978) Oxygen consumption in the prosobranch snail *Viviparus contectoides* (Mollusca: Gastropoda). IV. Effects of dissolved oxygen level, starvation, density, symbiotic algae, substrate composition and osmotic pressure. *Comp. Biochem. Physiol. A*, **59**, 199–204.

Sturrock, B. (1966b) The influence of infection with *Schistosoma mansoni* on the growth rate and reproduction of *Biomphalaria pfeifferi. Ann. Trop. Med. Parasitol.*, **60**, 187–97.

Sturrock, B. (1967) The effect of infection with *Schistosoma haematobium* on the growth and reproduction rates of *Bulinus (Physopsis) nasutus productus. Ann. Trop. Med. Parasitol.*, **61**, 321–5.

Sturrock, R. (1966a) The influence of temperature on the biology of *Biomphalaria pfeifferi* (Krauss), an intermediate host of *Schistosoma mansoni. Ann. Trop. Med. Parasitol.*, **60**, 100–5.

Sturrock, R. (1973a) Field studies on the population dynamics of *Biomphalaria glabrata*, intermediate host of *Schistosoma mansoni* on the West Indian island of St. Lucia. *Int. J. Parasitol.*, **3**, 165–74.

Sturrock, R. (1973b) Field studies on the transmission of *Schistosoma mansoni* and on the bionomics of its intermediate host, *Biomphalaria glabrata*, on St. Lucia, West Indies. *Int. J. Parasitol.*, **3**, 175–94.

Sturrock, R. and Sturrock, B. (1972) The influence of temperature on the biology of *Biomphalaria glabrata* (Say), intermediate host of *Schistosoma mansoni* on St. Lucia, West Indies. *Ann. Trop. Med. Parasitol.*, **66**, 385–90.

Sugihara, G. (1980) Minimal community structure: An explanation of species abundance patterns. *Am. Nat.*, **116**, 770–87.

Sugihara, G. (1981) $S = CA^2$, $z = 1/4$: A reply to Connor and McCoy. *Am. Nat.*, **117**, 790–3.

Sullivan, J. and Richards, C. (1981) *Schistosoma mansoni*, NIH-Sm-PR-2 strain, in susceptible and non susceptible stocks of *Biomphalaria glabrata*: Comparative histology. *J. Parasitol.*, **67**, 702–8.

Sumner, W. and McIntire, C. (1982) Grazer-periphyton interactions in laboratory streams. *Arch. Hydrobiol.*, **93**, 135–57.

Swamikannu, X. and Hoagland, K. (1989) Effects of snail grazing on the diversity and structure of a periphyton community in a eutrophic pond. *Can. J. Fish. Aquat. Sci.*, **46**, 1698–704.

Swanson, G. and Meyer, M. (1977) Impact of fluctuating water levels on feeding ecology of breeding blue-winged teal. *J. Wildl. Manage.*, **41**, 426–33.

Tang, C.-T. (1985) A survey of *Biomphalaria straminea* (Planorbidae) for trematode infection, with a report on larval flukes from other gastropoda in Hong Kong. In *Proceedings of the Second International Workshop on the Malacofauna of Hong Kong*, ed. B. Morton and D. Dudgeon, pp. 393–408. Hong Kong: Hong Kong University Press.

Tankersley, R. (1996) Multipurpose gills: Effect of larval brooding on the feeding physiology of freshwater unionid mussels. *Invertebr. Biol.*, **115**, 242–55.

Tankersley, R. and Dimock, R., Jr. (1993) The effect of larval brooding on the filtration rate and particle-retention efficiency of *Pyganodon cataracta* (Bivalvia: Unionidae). *Can. J. Zool.*, **71**, 1934–44.

Tashiro, J. (1982) Grazing in *Bithynia tentaculata*: Age-specific bioenergetic patterns in reproductive partitioning of ingested Carbon and Nitrogen. *Am. Midl. Nat.*, **197**, 133–50.

Tashiro, J. and Colman, S. (1982) Filter feeding in the freshwater prosobranch snail *Bithynia tentaculata*: Bioenergetic partitioning of ingested Carbon and Nitrogen. *Am. Midl. Nat.*, **197**, 114–32.

Taskinen, J., Makela, T. and Tellervo-Valtonen, E. (1997) Exploitation of *Anodonta piscinalis* (Bivalvia) by trematodes: Parasite tactics and host longevity. *Ann. Zool. Fenn.*, **34**, 37–46.

Taylor, E. and Mozeley, A. (1948) A culture method for *Lymnaea truncatula*. *Nature*, **161**, 894.

Ten Winkel, E. and Davids, C. (1982) Food selection by *Dreissena polymorpha* Pallas (Mollusca: Bivalvia). *Freshwat. Biol.*, **12**, 553–8.

Thomas, G. (1959) Self-fertilization and production of young in a sphaeriid clam. *Nautilus*, **72**, 131–41.

Thomas, G. (1965) Growth in one species of sphaeriid clam. *Nautilus*, **79**, 47–54.

Thomas, J. (1986) The chemical ecology of *Biomphalaria glabrata* sugars as attractants and arrestants. *Comp. Biochem. Physiol.*, **83**, 457–60.

Thomas, J. (1987) An evaluation of the interactions between freshwater pulmonate snail hosts of human schistosomes and macrophytes. *Phil. Trans. R. Soc. Lond. B*, **315**, 75–125.

Thomas, J. (1990) Mutualistic interactions in freshwater modular systems with molluscan components. *Adv. Ecol. Res.*, **20**, 125–78.

Thomas, J. and Benjamin, M. (1974) The effects of population density on growth and reproduction of *Biomphalaria glabrata* (Say) (Gastropoda: Pulmonata). *J. Anim. Ecol.*, **42**, 31–50.

Thomas, J. and Daldorph, P. (1991) Evaluation of bioengineering approaches aimed at controlling pulmonate snails: The effects of light attenuation and mechanical removal of macrophytes. *J. Appl. Ecol.*, **28**, 532–46.

Thomas, J. and Lough, A. (1974) The effects of external calcium concentration on the rate of uptake of this ion by *Biomphalaria glabrata* (Say). *J. Anim. Ecol.*, **43**, 861–71.

Thomas, J. and Tait, A. (1984) Control of the snail hosts of schistosomiasis by environmental manipulation: A field and laboratory appraisal in the Ibadan area, Nigeria. *Phil. Trans. R. Soc. Lond. B.*, **305**, 210–53.

Thomas, J., Benjamin, M., Lough, A. and Aram, R. (1974) The effects of calcium in the external environment on the growth and natality rates of *Biomphalaria glabrata* (Say). *J. Anim. Ecol.*, **43**, 839–60.

Thomas, J., Goldsworthy, G. and Aram, R. (1975) Studies on the chemical ecology of snails: The effect of chemical conditioning by adult snails on the growth of juvenile snails. *J. Anim. Ecol.*, **44**, 1–27.

Thomas, J., Grealy, B. and Fennell, C. (1983) The effects of varying the quantity and quality of various plants on feeding and growth of *Biomphalaria glabrata* (Gastropoda). *Oikos*, **41**, 77–90.

Thomas, J., Ndifon, G. and Ukoli, F. (1985) The carboxylic-acid and amino-acid chemoreception niche of *Bulinus rohlfsi*, the snail host of *Schistosoma haematobium*. *Comp. Biochem. Physiol. C*, **82**, 91–108.

Thomas, J., Sterry, P. and Patience, R. (1984) Uptake and assimilation of short-chain carboxylic acids by *Biomphalaria glabrata* (Say), the freshwater pulmonate snail host of *Schistosoma mansoni*. *Proc. R. Soc. Lond. B*, **222**, 447–76.

Thomas, J., Sterry, P., Jones, H., Gubala, M. and Grealy, B. (1986) The chemical ecology of *Biomphalaria glabrata* sugars as phagostimulants. *Comp. Biochem. Physiol. A*, **83**, 461–76.

Thomas, J., Kowalczyk, C. and Somasundaram, B. (1990) The biochemical ecology of *Biomphalaria glabrata*, a freshwater pulmonate mollusc; the uptake and assimilation of exogenous glucose and maltose. *Comp. Biochem. Physiol. A*, **95**, 511–28.

Thomas, K. (1975) Biological control of *Salvinia* by the snail *Pila glabosa* Swainson. *Biol. J. Linn. Soc.*, **7**, 243–7.

Thompson, D. (1973) Feeding ecology of diving ducks on Keokuk Pool, Mississippi River. *J. Wildl. Manage.*, **37**, 367–81.

Thornhill, J., Jones, J. and Kusel, J. (1986) Increased oviposition and growth in immature *Biomphalaria glabrata* after exposure to *Schistosoma mansoni*. *Parasitology*, **93**, 443–50.

Thorp, J. and Bergey, E. (1981) Field experiments on responses of a freshwater, benthic macroinvertebrate community to vertebrate predators. *Ecology*, **62**, 365–75.

Threlfall, W. (1986) Seasonal occurrence of *Anodonta cataracta* (Say, 1817) glochidia on three-spined sticklebacks, *Gasterosteus aculeatus* L. *Veliger*, **29**, 231–4.

Thut, R. (1969) A study of the profundal bottom fauna of Lake Washington. *Ecol. Monog.*, **39**, 79–100.

Todd, M. (1964) Osmotic balance in *Hydrobia ulvae* and *Potamopyrgus jenkinsi* (Gastropoda: Hydrobiidae). *J. Exp. Biol.*, **41**, 665–77.

Toft, C., Aeschlimann, A. & Bolis, L. (1991) *Parasite–Host Associations, Coexistence or Conflict?* New York: Oxford University Press.

Tompa, A. (1979) Life-cycle completion of the freshwater clam *Lasmigona compressa* in an experimental host: *Lebistes reticulatus*. *Veliger*, **22**, 188–90.

Toth, M. and Baba, K. (1981) The mollusca fauna of the Tisza and its tributaries. *Tiscia (Szeged)*, **16**, 169–81.

Townsend, C. (1973) The food-finding mechanism of *Biomphalaria glabrata* (Say). *Anim. Behav.*, **21**, 544–8.

Townsend, C. (1974) The chemoreceptor sites involved in food-finding by the freshwater pulmonate snail, *Biomphalaria glabrata* (Say), with particular reference to the function of the tentacles. *Behav. Biol.*, **11**, 511–23.

Townsend, C. R. and McCarthy, T. K. (1980) On the defence strategy of *Physa fontinalis* (L.), a freshwater pulmonate snail. *Oecologia*, **46**, 75–9.

Travis, J. and Ricklefs, R. (1983) A morphological comparison of island and mainland assemblages of neotropical birds. *Oikos*, **41**, 434–41.

Trdan, R. (1981) Reproductive biology of *Lampsilis radiata siliquoidea* (Pelecypoda: Unionidae). *Am. Midl. Nat.*, **106**, 243–8.

Trdan, R. and Hoeh, W. (1982) Eurytopic host use by two congeneric species of freshwater mussel (Pelecypoda: Unionidae: Anodonta). *Am. Midl. Nat.*, **108**, 381–8.

Trigwell, J. and Dussart, G. (1998) Functional protandry in *Biomphalaria glabrata* (Gastropoda: Pulmonata), an intermediate host of *Schistosoma*. *J. Molluscan Stud.*, **64**, 253–6.

Trigwell, J., Dussart, G. and Vianey-Liaud, M. (1997) Pre-copulatory behaviour of the freshwater hermaphrodite snail *Biomphalaria glabrata* (Say, 1818) (Gastropoda: Pulmonata). *J. Molluscan Stud.*, **63**, 116–20.

Tucker, J. (1994) Colonization of unionid bivalves by the zebra mussel *Dreissena polymorpha* in Pool 26 of the Mississippi River. *J. Freshwat. Ecol.*, **9**, 129–34.

Tudorancea, C. (1969) Comparison of the populations of *Unio timidus* Philipsson from the complex of Crapina-Jijila marshes. *Ekol. Pol.*, **17**, 185–204.

Tudorancea, C. (1972) Studies on Unionidae populations from the Crapina-Jijila complex pools (Danube zone liable to inundation). *Hydrobiologia*, **39**, 527–61.

Tuersley, M. (1989) How is food arousal manifested in the pond snail *Lymnaea stagnalis*?: An overview. *J. Molluscan Stud.*, **55**, 209–16.:

Turner, H. and Corkum, K. (1979) A seasonal and ecological survey of freshwater limpet snails (Pulmonata: Ancylidae) and their digenetic trematode parasites in southeastern Louisiana USA. *Tulane Stud. Zool. Bot.*, **21**, 67–90.

Turner, R. and Roberts, T. (1978) Mollusks as prey of Ariid catfish in the Fly River. *Bull. Am. Malacol. Union*, **1978**, 33–40.

Tyrrell, M. and Hornbach, D. (1998) Selective predation by muskrats on freshwater mussels in two Minnesota rivers. *J. N. Am. Benthol. Soc.*, **17**, 301–10.

Underwood, G. and Thomas, J. (1990) Grazing interactions between pulmonate snails and epiphytic algae and bacteria. *Freshwat. Biol.*, **23**, 505–22.

Vaidya, D. (1979) Substratum as a factor in the distribution of two freshwater snails, *Viviparus bengalensis* and *Melania scabra*. *Hydrobiologia*, **65**, 17–18.

Vail, V. (1977) Comparative reproductive anatomy of 3 viviparid gastropods. *Malacologia*, **16**, 519–40.

van Cleave, H. and Altringer, D. (1937) Studies on the life cycle of *Campeloma rufum*, a freshwater snail. *Am. Nat.*, **71**, 167–84.

van den Boom-Ort, B. (1991) Uptake of sand grains into the stomach of *Biomphalaria glabrata*. In *Proc. Tenth International Malacological Congress*, ed. C. Meier-Brook, pp. 109–14. Tubingen, Germany: Unitas Malacologia.

van der Schalie, H. (1948) The commercially valuable mussels of the Grand River in Michigan. *Mich. Dept. Conservation, Misc. Publ.*, **4**, 1–42.

van der Schalie, H. (1970) Hermaphroditism among North American freshwater mussels. *Malacologia*, **10**, 93–112.

van der Schalie, H. and Berry, E. (1973) The effects of temperature on growth and reproduction of aquatic snails. *Sterkiana*, **50**, 1–92.

van der Schalie, H. and Davis, G. (1968) Culturing *Oncomelania* snails (Prosobranchia: Hydrobiidae) for studies of oriental schistosomiasis. *Malacologia*, **6**, 321–67.

van der Schalie, H. and Getz, L. (1962) Distribution and natural history of the snail *Pomatiopsis cincinnatiensis* (Lea). *Am. Midl. Nat.*, **68**, 203–31.

van der Schalie, H. and Getz, L. (1963) Comparison of temperature and moisture responses of the snail genera *Pomatiopsis* and *Oncomelania*. *Ecology*, **44**, 73–83.

van der Schalie, H. and van der Schalie, A. (1950) The mussels of the Mississippi River. *Am. Midl. Nat.*, **44**, 448–66.

van der Schalie, H. and van der Schalie, A. (1963) The distribution, ecology, and life history of the mussel *Actinonaias ellipsiformis* (Conrad), in Michigan. *Occas. Pap. Mus. Zool. Univ. Mich.*, **633**, 1–17.

van der Steen, W. (1967) The influence of environmental factors on the oviposition of *Lymnaea stagnalis* (L.) under laboratory conditions. *Arch. Neerl. Zool.*, **17**, 403–68.

van Duivenboden, Y. A. and ter Maat, A. (1988) Mating behaviour of *Lymnaea stagnalis*. *Malacologia*, **28**, 53–64.

van Duivenboden, Y. A., Pieneman, A. W. and ter Maat, A. (1985) Multiple mating suppresses fecundity in the hermaphrodite freshwater snail *Lymnaea stagnalis*: A laboratory study. *Anim. Behav.*, **33**, 1184–91.

Vannote, R., Minshall, G., Cummins, K., Sedell, J. and Cushing, C. (1980) The river continuum concept. *Can. J. Fish. Aquat. Sci.*, **37**, 130–7.

Vaught, K. (1989) *A Classification of the Living Mollusca*. Melbourne, FL: American Malacologists.

Veldhuijzen, J. (1974) Sorting and retention time of particles in the digestive system of *Lymnaea stagnalis* (L.). *Neth. J. Zool.*, **24**, 10–21.

Vermeij, G. and Covich, A. (1978) Coevolution of freshwater gastropods and their predators. *Am. Nat.*, **112**, 833–43.

Vernon, J. (1995) Low reproductive output of isolated, self-fertilizing snails: Inbreeding depression or absence of social facilitation? *Proc. R. Soc. Lond. B*, **259**, 131–6.

Vianey-Liaud, M., Dupouy, J., Lancastre, F. and Nassi, H. (1987) Genetical exchanges between one *Biomphalaria glabrata* (Gastropoda: Planorbidae) and a varying number of partners. *Mem. Inst. Oswaldo Cruz*, **82**, 457–60.

Vidrine, M. (1990) Fresh-water mussel–mite and mussel–*Ablabesmyia* associations in Village Creek, Hardin County, Texas. *Louisana J. Sci.*, **53**, 1–4.

Vieira, E. (1967) Influence of vitamin E on reproduction of *Biomphalaria glabrata* under axenic conditions. *Am. J. Trop. Med. Hyg.*, **16**, 792–6.

Vincent, B., Rioux, H. and Harvey, M. (1991) Factors affecting the structure of epiphytic gastropod communities in the St Lawrence River (Quebec, Canada). *Hydrobiologia*, **220**, 57–71.

Visser, M. (1981) Monauly versus diauly as the original condition of the reproductive system of Pulmonata and its bearing on the interpretation of the terminal ducts. *Zeitschrift Für Zoologische Systematik und Evolutionsforschung*, **19**, 59–68.

Visser, M. (1988) The significance of terminal duct structures and the role of neoteny in the evolution of the reproductive system of Pulmonata. *Zool. Scripta*, **17**, 239–52.

Vogt, R. and Guzman, S. (1988) Food partitioning in a neotropical freshwater turtle community. *Copeia*, **1988**, 37–47.

Wagner, E. and Wong, L. (1956) Some factors influencing egg laying in *Oncomelania nosophora* and *Oncomelania quadrasi*, intermediate hosts of *Schistosoma japonicum*. *Am. J. Trop. Med. Hyg.*, **5**, 544–52.

Wall, R. (1977) Seasonal movements of the pond snail, *Lymnaea catascopium*, in a northern lake. *Nautilus*, **91**, 47–51.

Wallace, C. (1979) Notes on the occurrence of males in populations of *Potamopyrgus jenkinsi*. *J. Molluscan Stud.*, **45**, 61–7.

Wallace, C. (1985) On the distribution of the sexes of *Potamopyrgus jenkinsi*. *J. Molluscan Stud.*, **51**, 290–6.

Wallace, C. (1992) Parthenogenesis, sex, and chromosomes in *Potamopyrgus*. *J. Molluscan Stud.*, **58**, 93–107.

Wallace, J., Webster, J. and Woodall, W. (1977) The role of filter feeders in flowing waters. *Arch. Hydrobiol.*, **79**, 506–32.

Walter, J. (1980) The density of the pond snails *Lymnaea auricularia* and *L. peregra* in Lake Zurich (Gastropoda: Basommatophora) (in German). *Swiss J. Hydrol.*, **42**, 65–71.

Walter, J. and Kuiper, J. (1978) Uber verbreitung und Ökologie von Sphaeriiden im Zurichsee (Mollusca: Eulamellibranchiata). *Swiss J. Hydrol.*, **40**, 60–86.

Walz, N. (1974) Rückgang der *Dreissena polymorpha* – population in Bodensee. *Gas und Wasserfach Wasser/Abwasser*, **115**, 20–4.

Walz, N. (1978a) The energy balance of the freshwater mussel *Dreissena polymorpha* Pallas in laboratory experiments and in Lake Constance. I. Pattern of activity feeding and assimilation efficiency. *Arch. Hydrobiol. Suppl.*, **55**, 83–105.

Walz, N. (1978b) The energy balance of the freshwater mussel *Dreissena polymorpha* Pallas in laboratory experiments and in Lake Constance. II. Reproduction. *Arch. Hydrobiol. Suppl.*, **55**, 106–19.

Watters, G. (1992) Unionids, fishes, and the species-area curve. *J. Biogeog.*, **19**, 481–90.

Watters, G. (1994) *An Annotated Bibliography of the Reproduction and Propagation of the Unionoidea (Primarily of North America)*. Misc. Contrib. No. 1. Columbus, Ohio: Ohio Biological Survey. 158 pp.

Watters, G. (1997) Individual-based models of mussel–fish interactions: A cautionary study. In *Conservation and Management of Freshwater Mussels. II: Initiatives for the*

Future, ed. K. Cummings, A. Buchanan, C. Mayer and T. Naimo, pp. 45–62. Rock, Island, IL: Upper Mississippi River Conservation Committee.

Watters, G. and O'Dee, S. (1996) Shedding of untransformed glochidia by fishes parasitized by *Lampsilis fasciola*: Evidence of acquired immunity in the field? *J. Freshwat. Ecol.*, **11**, 383–9.

Way, C. and Wissing, T. (1982) Environmental heterogeneity and life history variability in the freshwater clams, *Pisidium variable* (Prime) and *Pisidium compressum* (Prime) (Bivalvia: Pisidiidae). *Can. J. Zool.*, **60**, 2841–51.

Way, C., Hornbach, D., Deneka, T. and Whitehead, R. (1989) A description of the ultrastructure of the gills of freshwater bivalves, including a new structure, the frontal cirrus. *Can. J. Zool.*, **67**, 357–62.

Way, C., Hornbach, D., Miller-Way, C., Payne, B. and Miller, A. (1990) Dynamics of filter feeding in *Corbicula fluminea* (Bivalvia: Corbiculidae). *Can. J. Zool.*, **68**, 115–20.

Weaver, L., Pardue, G. and Neves, R. (1991) Reproductive biology and fish hosts of the Tennessee clubshell *Pleurobema oviforme* (Mollusca: Unionidae) in Virginia. *Am. Midl. Nat.*, **126**, 82–9.

Weber, L. and Lodge, D. (1990) Periphytic food and predatory crayfish: Relative roles in determining snail distribution. *Oecologia*, **82**, 33–9.

Webster, J. and Benfield, E. (1986) Vascular plant breakdown in freshwater ecosystems. *Ann. Rev. Ecol. Syst.*, **17**, 567–94.

Webster, J. & Woolhouse, M. (1998) Selection and strain specificity of compatibility between snail intermediate hosts and their parasitic schistosomes. *Evolution*, **52**, 1627–34.

Weiss, J. and Layzer, J. (1995) Infestations of glochidia on fishes in the Barren River, KY. *Amer. Malacol. Bull.*, **11**, 153–9.

West, K. and Cohen, A. (1996) Shell microstructure of gastropods from Lake Tanganyika, Africa: Adaption, convergent evolution, and escalation. *Evolution*, **50**, 672–81.

West, K., Cohen, A. and Baron, M. (1991) Morphology and behavior of crabs and gastropods from Lake Tanganyika, Africa: Implications for lacustrine predator-prey coevolution. *Evolution*, **45**, 589–607.

Wethington, A. R. and Dillon, R. T., Jr. (1991) Sperm storage and evidence for multiple insemination in a natural population of the freshwater snail, *Physa. Am. Malacol. Bull.*, **9**, 99–102.

Wethington, A. and Dillon, R. T., Jr. (1993) Reproductive development in the hermaphroditic freshwater snail, *Physa*, monitored with complementing albino lines. *Proc. R. Soc. Lond. B*, **252**, 109–14.

Wethington, A. and Dillon, R. T., Jr. (1996) Gender choice and gender conflict in a non-reciprocally mating simultaneous hermaphrodite, the freshwater snail, *Physa. Anim. Behav.*, **51**, 1107–18.

Wethington, A. and Dillon, R. T., Jr. (1997) Selfing, outcrossing, and mixed mating in the freshwater snail *Physa heterostropha*: Lifetime fitness and inbreeding depression. *Invertebr. Zool.*, **116**, 192–9.

Whittaker, R. (1956) Vegetation of the Great Smoky Mountains. *Ecol. Monog.*, **26**, 1–80.

Wild, S. and Lawson, A. (1937) Enemies of the land and freshwater mollusca of the British Isles. *J. Conchol.*, **20**, 351–61.

Wilken, G. and Appleton, C. (1991) Avoidance responses of some indigenous and exotic freshwater pulmonate snails to leech predation in South Africa. *S. Afr. J. Zool.*, **26**, 6–10.

Wilkinson, L. (1988) *SYSTAT: The System for Statistics*. Evanston, Illinois: SYSTAT, Inc.

Williams, C. and Gilbertson, D. (1983) Altered feeding response as a cause for the altered heartbeat rate and locomotor activity of *Schistosoma mansoni*–infected *Biomphalaria glabrata* on cercarial emergence. *J. Parasitol.*, **69**, 671–6.

Williams, G. C. (1975) *Sex and Evolution*. Princeton, New Jersey: Princeton University Press.

Williams, N. (1970a) Studies on aquatic pulmonate snails in central Africa. I. Field distribution in relation to water chemistry. *Malacologia*, **10**, 153–64.

Williams, N. (1970b) Studies on aquatic pulmonate snails in central Africa. II. Experimental investigation of field distribution patterns. *Malacologia*, **10**, 165–80.

Winkler, L. and Wagner, E. (1959) Filter paper digestion by the crystalline style in *Oncomelania*. *Trans. Am. Microsc. Soc.*, **78**, 262–8.

Witte, F., Barel, C. and Hougerhoud, J. (1990) Phenotypic plasticity of anatomical structures and its ecomorphic significance. *Neth. J. Zool.*, **40**, 278–98.

Wood, E. (1974) Some mechanisms involved in host recognition and attachment of the glochidium larvae of *Anodonta cygnea* (Mollusca: Bivalvia). *J. Zool.*, **173**, 15–30.

Woolhouse, M. (1989) The effect of *Schistosoma* infection on the mortality rates of *Bulinus globosus* and *Biomphalaria pfeifferi*. *Ann. Trop. Med. Parasitol.*, **83**, 137–41.

Woolhouse, M. (1992) Population biology of the freshwater snail *Biomphalaria pfeifferi* in the Zimbabwe highveld. *J. Appl. Ecol.*, **29**, 687–94.

Woolhouse, M. and Chandiwana, S. (1989) Spatial and temporal heterogeneity in the population dynamics of *Bulinus globosus* and *Biomphalaria pfeifferi* and in the epidemiology of their infection with schistosomes. *Parasitol.*, **98**, 21–34.

Woolhouse, M. and Chandiwana, S. (1990a) Population biology of the freshwater snail *Bulinus globosus* in the Zimbabwe highveld. *J. Appl. Ecol.*, **27**, 41–59.

Woolhouse, M. and Chandiwana, S. (1990b) Population dynamics model for *Bulinus globosus*, intermediate host for *Schistosoma haematobium*, in river habitats. *Acta Trop.*, **47**, 151–60.

Wright, C. (1960) The crowding phenomenon in laboratory colonies of freshwater snails. *Ann. Trop. Med. Parasitol.*, **54**, 224–32.

Wright, C. (1971) *Flukes and Snails*. London: George Allen and Unwin Ltd.

Wrona, M., Davies, R. and Linton, L. (1979) Analysis of the food niche of *Glossiphonia complanata* (Hirudinoidea: Glossiphoniidae). *Can. J. Zool.*, **57**, 2136–42.

Xavier, M., Fraga de Azevedo, J. and Avelino, I. (1968) Importance d'*Oscillatoria formosa* Bory dans la culture au laboratoire des mollusques vecteurs du *Schistosoma haematobium*. *Bull. Soc. Path. Exot.*, **61**, 52–66.

Yang, R. and Yoshino, T. (1990) Immunorecognition in the freshwater bivalve *Corbicula fluminea*. II. Isolation and characterization of a plasma opsonin with hemaglutinating activity. *Dev. Comp. Immunol.*, **14**, 397–404.

Yeager, B. and Neves, R. (1986) Reproductive cycle and fish hosts of the rabbit's foot mussel *Quadrula cylindrica strigillata* (Mollusca: Unionidae) in the upper Tennessee River. *Am. Midl. Nat.*, **16**, 329–40.

Yeager, M., Cherry, D. and Neves, R. (1994) Feeding and burrowing behaviors of juvenile rainbow mussels, *Villosa iris* (Bivalvia: Unionidae). *J. N. Am. Benthol. Soc.*, **13**, 217–22.

Yokley, P., Jr. (1972) Life history of *Pleuroboma cordatum* (Raf) (Bivalvia: Unionacea). *Malacologia*, **11**, 351–64.

Young, J. (1975) A laboratory study, using ^{45}Ca tracer, on the source of calcium during growth in two freshwater species of Gastropoda. *Proc. Malacol. Soc. Lond.*, **41**, 439–45.

Young, J. (1980) A serological investigation of the diet of *Helobdella stagnalis* (L.) (Hirudinea: Glossiphoniidae) in British lakes. *J. Zool.*, **192**, 467–88.

Young, J. (1981a) A serological study of the diet of British lake-dwelling *Glossiphonia complanata* (L.) (Hirudinea: Glossiphoniidae). *J. Nat. Hist.*, **15**, 475–89.

Young, J. (1981b) A comparative study of the food niches of lake-dwelling triclads and leeches. *Hydrobiologia*, **84**, 91–102.

Young, M. (1974) Seasonal variation in the occurrence of *Chaetogaster limnaei limnaei* Gr. (Oligochaeta) in two of its molluscan hosts in the Worcester-Birmingham Canal and its relationship with the digenean parasites of these mollusks. *J. Nat. Hist.*, **8**, 529–35.

Young, M., Purser, G. and Al-Mousawi, B. (1987) Infection and successful reinfection of brown trout (*Salmo trutta* (L)) with glochidia of *Margaritifera margaritifera* L. *Am. Malacol. Bull.*, **5**, 125–8.

Zale, A. and Neves, R. (1982a) Identification of a host fish for *Alasmidonta minor* (Mollusca: Unionidae). *Am. Midl. Nat.*, **107**, 386–8.

Zale, A. and Neves, R. (1982b) Reproductive biology of four freshwater mussel species (Mollusca: Unionidae) in Virginia. *Freshwat. Invertebr. Biol.*, **1**, 17–28.

Zale, A. and Neves, R. (1982c) Fish hosts of four species of lampsiline mussels (Mollusca: Unionidae) in Big Moccasin Creek, VA. *Can. J. Zool.*, **60**, 2535–42.

Zylstra, U. (1972) Uptake of particulate matter by the epidermis of the freshwater snail, *Lymnaea stagnalis*. *Neth. J. Zool.*, **22**, 229–306.

Index